Nanotechnology-based Sustainable Agriculture

Edited by

Pardeep Singh

University of Delhi
New Delhi, India

Ankit Kumar Singh

Lalit Narayan Mithila University
Bihar, India

Vipendra Kumar Singh

Indian Institute of Technology Mandi
Himachal Pradesh, India

Vijay Kumar

University of Delhi
New Delhi, India

WILEY-VCH

Editors

Prof. Dr. Pardeep Singh
University of Delhi
PGDAV College
New Delhi
India, 110065

Prof. Dr. Ankit Kumar Singh
Lalit Narayan Mithila University
Darbhanga, Bihar
India, 846004

Dr. Vipendra Kumar Singh
Indian Institute of Technology Mandi
Mandi, Himachal Pradesh
India, 175005

Dr. Vijay Kumar
University of Delhi
New Delhi
India, 110019

Cover Design: Wiley
Cover Image: © Basius77/Shutterstock

Library of Congress Cataloging-in-Publication Data:
applied for

British Library Cataloguing-in-Publication Data
A catalogue record for this book is available from the British Library.

Bibliographic information published by the Deutsche Nationalbibliothek
The Deutsche Nationalbibliothek lists this publication in the Deutsche Nationalbibliografie; detailed bibliographic data are available on the Internet at <http://dnb.d-nb.de/>.

Print ISBN: 9783527354559
ePDF ISBN: 9783527849932
ePub ISBN: 9783527849925
oBook ISBN: 9783527849949

Set in STIXTwoText 9.5/12.5pts by Lumina Datamatics

Printed and bound by
CPI Group (UK) Ltd, Croydon, CR0 4YY

C9783527354559_250425

The manufacturer's authorized representative according to the EU General Product Safety Regulation is Wiley-VCH GmbH, Boschstr. 12, 69469 Weinheim, Germany, e-mail: Product_Safety@wiley.com.

Contents

List of Contributors

Arjun Kumar Mehara
Department of Botany
Lalit Narayan Mithila University
Darbhanga
Bihar
India

Anuradha Kumari
Department of Botany
Lalit Narayan Mithila University
Darbhanga
Bihar
India

Neeraj K. Verma
CSIR-Indian Institute of Toxicology Research
Lucknow
Uttar Pradesh
India

Abhishek Rai
University Department of Chemistry
Lalit Narayan Mithila University
Darbhanga
Bihar
India

Prachi Marwaha
Department of Home Science
Lalit Narayan Mithila University
Darbhanga
Bihar
India

Mayank Kumar Singh
The National Dendrimer and Nanotechnology Center
Mount Pleasant, MI
USA

Central Michigan University-College of Medicine
Mount Pleasant, MI
USA

Ankit Kumar Singh
University Department of Botany
Lalit Narayan Mithila University
Darbhanga
Bihar
India

Katina Chachei
Analytical Environmental and Chemistry Laboratory (AECL)
Institute of Environment and Sustainable Development
Banaras Hindu University
Varanasi
Uttar Pradesh
India

Sonali Ranjan
Chemical Engineering and Technology
Indian Institute of Technology
Banaras Hindu University
Varanasi
Uttar Pradesh
India

Kirpa Ram
Analytical Environmental and Chemistry Laboratory (AECL)
Institute of Environment and Sustainable Development
Banaras Hindu University
Varanasi
Uttar Pradesh
India

Ram Sharan Singh
Chemical Engineering and Technology
Indian Institute of Technology
Banaras Hindu University
Varanasi
Uttar Pradesh
India

Aaradhya Pandey
School of Biotechnology
Gautam Buddha University
Greater Noida
Uttar Pradesh
India

Pragya Tiwari
School of Biotechnology
Gautam Buddha University
Greater Noida
Uttar Pradesh
India

Eti Sharma
School of Biotechnology
Gautam Buddha University
Greater Noida
Uttar Pradesh
India

Mohd Anas
Interdisciplinary Nanotechnology Centre
Aligarh Muslim University
Aligarh
India

Mohammad Umar
Institute of Plant Genetics and Biotechnology
Plant Science and Biodiversity Centre
Slovak Academy of Sciences
Nitra
Slovakia

Abdul Razzak
Institute of Forest Ecology
Slovak Academy of Sciences
Technical University in Zvolen
Zvolen
Slovakia

Nisha Kumari
Department of Soil Science and Agricultural Chemistry
Bihar Agricultural University Sabour
Bhagalpur
Bihar
India

Abhishek Tiwari
Department of Soil Science and Agricultural Chemistry
Bihar Agricultural University Sabour
Bhagalpur
Bihar
India

Ingle Sagar Nandulal
Department of Soil Science and Agricultural Chemistry
Bihar Agricultural University Sabour
Bhagalpur
Bihar
India

Sai Parasar Das
Department of Soil Science and Agricultural Chemistry
Bihar Agricultural University Sabour
Bhagalpur
Bihar
India

Bhabani Prasad Mondal
Department of Soil Science and Agricultural Chemistry
Bihar Agricultural University Sabour
Bhagalpur
Bihar
India

Bipin Bihari
Department of Soil Science and Agricultural Chemistry
Bihar Agricultural University Sabour
Bhagalpur
Bihar
India

Pritam Ganguly
Department of Soil Science and Agricultural Chemistry
Bihar Agricultural University Sabour
Bhagalpur
Bihar
India

Chandini
Department of Agronomy
Bihar Agricultural University Sabour
Bhagalpur
Bihar
India

Randeep Kumar
Department of Soil Science and Agricultural Chemistry
Bihar Agricultural University Sabour
Bhagalpur
Bihar
India

Anuradha Kumari
School of Environmental Sciences
Jawaharlal Nehru University
New Delhi
India

Anumanchi Sree Manogna
SASTRA Deemed University
Thanjavur
Tamil Nadu
India

Prabhat Kumar
School of Environmental Sciences
Jawaharlal Nehru University
New Delhi
India

Ilora Ghosh
School of Environmental Sciences
Jawaharlal Nehru University
New Delhi
India

Akanksha Rout
Department of Biological Sciences
BITS Pilani
K. K. Birla Goa Campus
Goa
India

Komal Jalan
Department of Botany
University of Delhi
Delhi
India

Pradipta Banerjee
Department of Surgery
University of Pittsburgh
Pittsburgh, PA
USA

Abirami Geetha Natarajan
Department of Chemical Engineering
Sri Venkateswara College of Engineering
Sriperumbudur
India

Department of Chemical and Petroleum Engineering

Schulich School of Engineering
University of Calgary
Calgary
Canada

Kripa V
Department of Chemical Engineering
Sri Venkateswara College of Engineering
Sriperumbudur
India

Jothi Ganesan M
Department of Chemical Engineering
Sri Venkateswara College of Engineering
Sriperumbudur
India

Philip Bernstein Saynik
Department of Chemical Engineering
Sri Venkateswara College of Engineering
Sriperumbudur
India

Swagata Lakshmi Dhali
Department of Genetics and Microbiology
Faculty of Science
Charles University
Prague
Czech Republic

Moumita Pal
Department of Biosciences and Biomedical Engineering
Indian Institute of Technology Indore
Indore
India

Vikky Kumar Mahto
University Department of Botany
Lalit Narayan Mithila University
Darbhanga
Bihar
India

Ved Prakash
Ecotoxicology Laboratory
Regulatory Toxicology Group
CSIR-Indian Institute of Toxicology Research
Lucknow
Uttar Pradesh
India

Academy of Scientific and Innovative Research (AcSIR)
Ghaziabad
India

Vipendra Kumar Singh
School of Biosciences and Bioengineering
Indian Institute of Technology Mandi
VPO Kamand
Mandi
Himachal Pradesh
India

Department of Biochemistry and Molecular Medicine
The George Washington University
Washington
DC
USA

Vikas Kumar Singh
Department of Science and Humanities
Supaul College of Engineering
Supaul
Bihar
India

Poorna Sneha M
Kusuma School of Biological Sciences
Indian Institute of Technology Delhi
New Delhi
India

Mohit Biju
School of Biological and Marine Sciences
University of Plymouth
Plymouth
UK

Aishwarya Thomas
Department of Biotechnology
CMS College Kottayam (Autonomous)
Kottayam
India

Parvathi Balachandran
Department of Biotechnology
CMS College Kottayam (Autonomous)
Kottayam
India

Puja Kumari
Department of Plant Pathology
Bihar Agricultural University
Sabour
Bhagalpur
Bihar
India

ICAR-National Research Centre on Litchi
Mushahari
Muzaffarpur
Bihar
India

Sawant Shraddha Bhaskar
Department of Plant Pathology
Bihar Agricultural University
Sabour
Bhagalpur
Bihar
India

Jeetu Narware
Department of Mycology and Plant Pathology
IAS
Banaras Hindu University
Varanasi
Uttar Pradesh
India

Abhijeet Ghatak
Department of Plant Pathology
Bihar Agricultural University
Sabour
Bhagalpur
Bihar
India

Umesh Kumar
Laboratory of Ecotoxicology
Department of Botany
Institute of Science
Banaras Hindu University
Varanasi
India

Prince Kumar Singh
Laboratory of Ecotoxicology
Department of Botany
Institute of Science
Banaras Hindu University
Varanasi
India

Parvati Madheshiya
Laboratory of Ecotoxicology
Department of Botany
Institute of Science
Banaras Hindu University
Varanasi
India

Indrajeet Kumar
Department of Botany
Government Degree College
Pawanikala
Sonbhadra
India

Ranjani Ravikumar
Department of Chemical Engineering
Sri Venkateswara College of Engineering
Sriperumbudur
India

Jayavardhini M
Department of Chemical Engineering
Sri Venkateswara College of Engineering
Sriperumbudur
India

Sai Sidharth A
Department of Chemical Engineering
Sri Venkateswara College of Engineering
Sriperumbudur
India

Vikky Rajulapati
School of Science and Technology
Örebro University
Örebro
Sweden

Preface

Due to poor agricultural production, deterioration of natural resources, significant post-harvest wastes, lack of or low-value inclusion, and population growth, establishing food security in poor or developing countries is challenging. Researchers have endeavored to integrate modern technology in the last few years to heighten supply and reduce the gap between demand and supply. Nanotechnology is one of the most reliable technologies that could improve agricultural productivity by augmenting nanofertilizers, utilizing more effective pesticides and herbicides, modulating soil properties, and spotting diseases. Nanotechnology can benefit sustainable agricultural practices by minimizing the negative impacts of agricultural practices on human health and the environment. It also increases agricultural productivity and food security and nurtures economic and social equity. Taking this idea as a fundamental part of our work, we would like to introduce nanotechnology-based sustainable agriculture on a large scale by displaying their general scientific basis and considerable interconnections. The subsequent chapters have inspired ideas and supplied the reader with a better understanding of the subject. This is an escorted tour of the discovery of nanotechnology and sustainable agriculture, which we hope will trigger the reader's interest so that they can be more thorough with the subject.

In the last few years, we have spotted a tidal wave for fabricating nanoparticles for sustainable agriculture directly associated with nanotechnology. Therefore, the basic ideas have been addressed, and a brief blueprint is illustrated in Chapter 1. Nanocomposites have emerged as favorable materials for increasing crop plants' sustainable productivity and growth. These materials, which integrate nanoparticles with other structures or compounds, can provide enhanced plant defense mechanisms, sustained delivery of nutrients, and decreased negative environmental impacts, contributing to sustainable agricultural practices. General snapshots of these ideas have been illustrated in Chapter 2.

Chapter 3 describes nano-biofertilizers' formulations, characterization, and fabrications and their role in global food security. We have entered the nano era, which is highly recognized as an essential tool for stress management in crop plants. By increasing plants' ability to cope with different environmental stressors such as salinity, heavy metals, toxicity, and drought, nanotechnology supports more adaptable crop production and sustainable agricultural practices covered in Chapter 4.

In Chapter 5, the authors have introduced the interaction of nanomaterials with microbes and plants to improve crop productivity for sustainable agriculture. Specific nanomaterials, such as nanofertilizers, can enhance nutrient accessibility to plants and laterally soil microbes, such as oxides of iron and zinc oxide nanoparticles can provide micronutrients and promote plant growth, which in turn helps a more vital and different microbial community in the rhizosphere. Silver nanoparticles have recently gathered attention in agriculture for their ability to increase crop growth and yield. The latest applications of silver nanoparticles in sustainable agriculture are given in Chapter 6. It provides a snapshot of emerging roles and distinct mechanisms by which silver nanoparticles enhance plant productivity. In Chapter 7, the rapid increase in the concentration of inorganic and organic contaminants due to growing human interference revealed a threat to the ecosystems. This chapter deals with the impacts of nanomaterials on the physiology of various crop plants, molecular and biochemical responses, enzymatic activity, and stress responses. It also deals with gene expression, signaling pathways, and case studies. Chapter 8 provides a detailed look at nano-bioremediation and its significance as a sustainable soil remediation and detoxification solution. Nanoparticles of Chitosan have become one of the most efficient nanosorbents for detoxifying different pollutants from the soil, offering an auspicious, eco-friendly solution for soil improvement. Nanoparticles of chitosan are highly efficient for pollutant adsorbent and removal due to high reactivity, surface area, and pollutant-binding potential. Chapter 9 deals with applying nanotechnology as an efficient and effective solution for the remediation of numerous heavy metals. Because plant-based nanomaterials display a creative and eco-friendly route for remediating heavy metal-polluted soils derived from biowastes, plant extracts, and other plant-based biomolecules, these nanomaterials can successfully immobilize, adsorb, and decrease the bioavailability of harmful heavy metals such as chromium (Cr), cadmium (Cd), lead (Pb), arsenic (As), and mercury (Hg). Plant-derived nanomaterials offer many benefits due to their minimal environmental toxicity, excellent biocompatibility, and renewability compared to conventional materials.

Chapter 10 describes the urbanization and industrialization that resulted in a heavy load of organic contaminants in fertile agricultural land. In the last few years, carbon quantum dots have sought attention due to their excellent potential for the degradation of various organic contaminants, such as insecticides, pesticides, dyes, and other industrial chemicals.

Chapter 11 explains that carbon quantum dots are one of the most reliable and sustainable tools for sensing several toxic metals in agricultural soils. Their ease of functionalization, excellent sensitivity, and biocompatibility make them a better candidate for various environmental monitoring applications. With constant research into enhancing their stability and performance, carbon quantum dots can be crucial in managing and detecting soil pollutants, thus providing more constructive monitoring and environmental remediation. Chapter 12 looks at the impact of nanopathology on breaking barriers for conventional disease protection. Chapter 13 has described the use of different types of nanoparticles in plant disease management and other emerging strategies for mitigating different plant diseases.

Nanomaterials-based chemicals are gradually being utilized in sustainable agriculture to increase crop production, condition soil health, and decrease the environmental effects of synthetic and other conventional chemicals in agricultural practices. However, using nanomaterials in sustainable agriculture also introduces risks and challenges, requiring careful inspection and risk assessment to verify their effective and safe use. Therefore, Chapter 14 has discussed the challenges and possible risks associated with chemicals containing nanomaterials to improve crop productivity for sustainable agriculture.

The advancement in nanotechnology and the emergence of a new class of nanomaterials open opportunities for potential applications in sustainable agriculture. Nanotechnology has attracted scientific communities and gained momentum in agriculture during the last few decades, but there is still a significant knowledge gap among scientific communities. Therefore, nanotechnologies, such as phyto-nanotechnology, nanobiotechnology, nanoremediation, nanofertilizer, agro-nanobiotechnology, pest control, and crop productivity, are described in different chapters. Undoubtedly, this book provides holistic coverage of recent advancements in nanotechnology-based sustainable agriculture. It is an updated and ideal book for undergraduate or postgraduate students and researchers or scientists involved in improved crop production, preservation, nanotechnology, remediation technologies, and human and environmental health without hampering sustainable development goals. Nanotechnology is a crucial turning point in our understanding of sustainable agriculture. Such technology has its pros and cons, or, in other words, of a notable imaginary character: every technology has its dark side. However, our future goals of sustainable agriculture depend on how we use the novel nanomaterials and what risks they bring to sustainable agricultural productivity and our environment.

The primary aim of this book is to provide a thorough knowledge of plant-derived nanomaterials to achieve the goals of sustainable agriculture. Illustrating a different cluster of studies, this book provides precious assets for researchers, professors, policymakers, stakeholders, and others devoted to constructing a more sustainable agriculture. The insights and strategies illustrated in this book provide a comprehensive roadmap for harnessing the power of nanomaterials, providing actionable solutions to the most pressing challenges of conventional approaches. By embracing plant-derived nanomaterial approaches, we can work towards creating a sustainable and prosperous future for all, where environmental stewardship and social equity go hand in hand.

1

Fabrication of Nanomaterials and Their Potential Advantage for Sustainable Agriculture

Arjun Kumar Mehara[1], Anuradha Kumari[1], Neeraj K. Verma[2], Prachi Marwaha[3], Abhishek Rai[4], Mayank Kumar Singh[5, 6], and Ankit Kumar Singh[1]

[1] *University Department of Botany, Lalit Narayan Mithila University, Darbhanga, Bihar, India*
[2] *CSIR-Indian Institute of Toxicology Research, Lucknow, Uttar Pradesh, India*
[3] *Department of Home Science, Lalit Narayan Mithila University, Darbhanga, Bihar, India*
[4] *Department of Chemistry, Lalit Narayan Mithila University, Darbhanga, Bihar, India*
[5] *The National Dendrimer and Nanotechnology Center, Mount Pleasant, MI, USA*
[6] *Central Michigan University-College of Medicine, Mount Pleasant, MI, USA*

1.1 Introduction

The current global population is anticipated to be 8.12 billion and is increasing day by day. The developed countries with highly advanced economies and highly developed technological infrastructure grow food in surplus, while developing countries, because of unawareness of farming technology and harsh environmental impacts, lead to food crises. To overcome this food shortage or to feed such a huge population, there's no other way than intensifying the growth of agricultural products and improving food processing and distribution. Earlier, traditional or chemical fertilizers were used in agriculture for optimizing crop growth and productivity. Fertilizers, once indispensable for boosting food, fodder, and fuel production, have become a double-edged sword in modern agriculture, contributing to environmental challenges (Bhardwaj et al. 2022). The excessive and prolonged use of conventional fertilizers has led to severe environmental hazards, such as emission of greenhouse gases, groundwater contamination, eutrophication of water bodies, soil degradation, and harmful for the beneficial organisms (Meena et al. 2017; Chaitra et al. 2021). The inefficient utilization of fertilizer nutrients by plants hampers the pursuit of sustainable agriculture practices. The rapid release of nutrients from chemical fertilizers often exceeds plant uptake, leading to nutrient losses and reduced bioavailability. To address the environmental consequences of traditional fertilizers and improve nutrient use efficiency, there is a critical need for innovative, eco-friendly fertilizer solutions with minimal leaching potential (Alkhader 2023; Van Eerd et al. 2018). This can be achieved by introducing new technologies in the field of agriculture. Among several modern technologies (biotechnology, industrial, information, nanotechnology [NT], etc.),

Nanotechnology-based Sustainable Agriculture, First Edition. Edited by Pardeep Singh et al.

NT, although an evolving one, is the one with immense potential to transform the system of food production. With having some unique properties, such as small size and high surface area to volume ratio, enhanced physical strength, improved electrical conductance, distinctive optical characteristics, stability, heightened reactivity, and notable magnetic properties, nanoparticles (NPs) are currently offering interesting innovative solutions for sustainable agricultural production, such as nanopesticides, nano-fungicides, nanofertilizers (NFs), and nanosensors, to enhance crop growth, quality, and protection (Periakaruppan et al. 2023). The development of NPs through green synthesis methods using plants and microorganisms has emerged as a more sustainable and eco-friendly approach, as fabricated NPs through physical and chemical methods have had an adverse impact on the ecosystem and enhanced input efficiency in agriculture (Gupta et al. 2023). These NPs, including nanosensors and nanobarcodes, have significantly contributed to modern agricultural practices by elevating crop yield, nutrient utilization, and disease resistance, along with minimizing waste production (Bhandari et al. 2023; Yasmine et al. 2023). By utilizing NT in agriculture, we can deal with the challenges caused by population explosion, environmental damage, and food security concerns, paving the way for a more efficient and sustainable agricultural system (León-Silva et al. 2018). Hence, NT offers a promising approach to enhance horticulture crop production by utilizing NFs, nanopesticides, and nano-biofertilizers to support overall plant growth via higher uptake and bioavailability and lower rain-fastness. These innovative solutions address the growing food demand while promoting sustainability through reduced resource consumption and environmental impact (Feregrino-Perez et al. 2018).

1.1.1 Shortcomings of Conventional Agriculture

The soil microbiome supports multiple ecosystem processes. Plant growth-promoting bacteria (or PGPB), such as *Azospirillum, Bacillus*, and *Rhizobium*, as well as soil fungal communities, including mycorrhizal fungi and super-parasites, can be found in the rhizosphere, on the root surface, or associated with it. They stabilize the soil microbiome and aid in biogeochemical cycles and bioremediation. They are also capable of enhancing the growth of plants by processes such as biological nitrogen fixation, phosphate solubilization, and stress alleviation, including countering salt stress in crops through the modulation of 1-aminocyclopropane-1-carboxylate (ACC) deaminase expression and production of phytohormones and siderophores, while also protecting them from diseases and abiotic stresses (Frąc et al. 2022; Del Carmen Orozco-Mosqueda, Glick, and Santoyo 2020). In walnut trees, it has been observed by Bai et al. (2020) that the use of chemical fertilizers for long duration results in excess ammonium–nitrogen and available phosphorus. The excess of NH_4^+ causes soil acidification and altered bacterial communities, while available phosphorus diminishes fungal diversity. In contrast, nonfertilized soils have higher organic matter, nitrate-nitrogen, total nitrogen, pH, and total phosphorus levels than fertilized soils, and naturally grown walnut trees foster beneficial bacteria like *Burkholderia, Nitrospira*, and *Pseudomonas*, along with fungi such as *Trichoderma, Phomopsis*, and *Chaetomium*, which enhance nutrient mobilization and plant growth (Bai et al. 2020). Lin et al. (2024) investigated the presence and diversity of antibiotic resistance genes (ARGs) in the soil and rhizosphere of maize and found that chemical fertilization led to lower but more diverse ARGs compared to straw-return while promoting mobile genetic elements (MGEs) in the rhizosphere. Metagenomic analysis identified *Pseudomonas, Bacillus*, and *Streptomyces* as key biomarkers for ARG accumulation. A total of 509 isolates belonging to these three genera from the rhizosphere showed high multiresistance, especially in *Pseudomonas*. The co-occurrence of specific ARGs and class I integrons (LR134330, LS998783, CP065848, and LT883143) in *Pseudomonas* sp. contigs suggests a complex link between chemical fertilization and antibiotic resistance. Overall, the chemical fertilizers may influence the resistance of the maize rhizosphere, potentially increasing the risk of multidrug-resistant bacteria that could impact animal and human health.

Pesticides play a critical and indispensable role in agriculture to enhance crop production and controlling weeds and pests. They are classified as fungicides, insecticides, herbicides, and rodenticides. However, pesticide use has created environmental contamination in the long term as their residues can persist in the environment and agricultural crops through processes like leaching, adsorption, and runoff. These pesticide residues can disrupt the stability of ecosystems and produce potential health risks to humans and animals. For example, dichlorodiphenyltrichloroethane (DDT), an organochlorine compound, known for their high toxicity, bioaccumulation, and slow degradation, build up in the tissues and also harm the different ecosystems. The contaminants of soil and water resources, chlorpyrifos and its metabolites such as TCP (3,5,6-trichloropyridinol), pose potential health risks to human health, especially as endocrine disruptors (Tudi et al. 2021). Pesticides also impact agrobiodiversity as less than 1% of them reach their target pests, while most target nontarget organisms like honeybees, earthworms, parasitoids, and predators, crucial to the functioning of agricultural ecosystems (Elhamalawy, Bakr, and Eissa 2024). The use of chemical pesticides and fertilizers on a large scale for better yield and productivity in agriculture has put tremendous pressure on the environment, questioning its sustainability in the near future. The synthetic fertilizers and pesticides can enhance crop yields in the short term, but in the long term, they may degrade soil and water, disrupt local ecosystems, impact human health, and also contribute to the development of pesticide-resistant pests, creating a cycle of dependency on them.

1.2 Fabrication Techniques for Nanomaterials

NPs can be fabricated by different methods, such as biological, chemical and physical techniques (Figure 1.1). These methods come under two prominent approaches (Top-down and Bottom-up) utilized for the synthesis of nanomaterials. The top-down approach downscales bulk material into nano-sized material mechanically. The physical methods are termed as top-down approaches. The bottom-up approach self-assembles themselves to create nano-sized material. Biological and chemical methods are some methods of bottom-up approach (Mabrouk et al. 2021).

1.2.1 Top-down Approaches

The top-down/physical/destructive methods reduce the larger bulk molecule into the smaller molecule. These smaller molecules are further fragmented to form NPs. Some of the techniques of top-down/destructive methods are mechanical milling or ball milling, laser ablation, sputtering, thermal decomposition methods, lithography, and arc-discharge methods (Mekuye and Abera 2023).

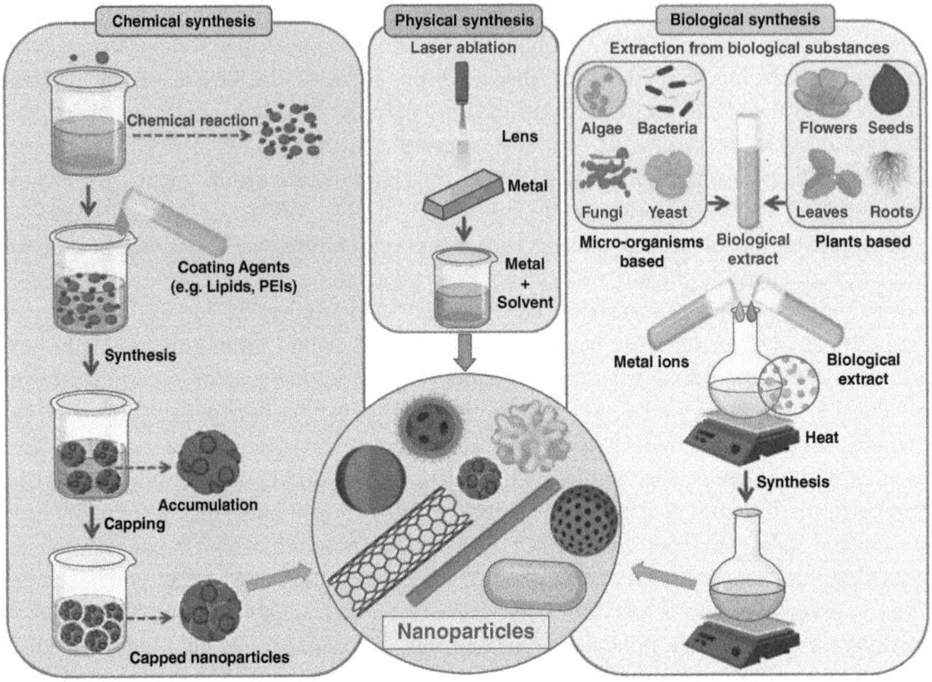

Figure 1.1 Different approaches (chemical, physical, and biological) for synthesis of nanomaterials.

1.2.1.1 Mechanical Milling or Ball Milling

Mechanical milling is the most inexpensive and simplest mechanical method for producing nano-metric scale particles from bulk materials. The purpose of milling is to break particles into smaller sizes and allow their blending in next phases. The efficiency of the milling and alloying process hinges on the energy transferred from the ball to powder particles. The energy transferred is influenced by various parameters, such as type of mill, milling speed, dry or wet milling, the powder used, temperature, and duration of milling (Baig, Kammakakam, and Falath 2021).

The predominant factors like cold-welding, plastic deformation, and fracture play a key role in high-energy ball milling process. In the deformation stage, the particle's shape is changed. The cold-welding initiates an increase in size, while fracture causes decrease in the particle's size. The three-staged process results in the making of alloying particles, which are finely dispersed in a grain-refined matrix. The mechanical milling method is leveraged in the synthesis of different types of oxide, aluminum/copper/magnesium/nickel-based nanoscale alloys, aluminum alloys strengthened by oxide and carbide particles, wear-resistant spray coatings, and numerous types of nano-composite materials (Namakka et al. 2023; Ratso et al. 2021; Liu et al. 2023). Carbon nanomaterials formed through the ball milling technique are utilized in environmental remediation, energy storage, and conversion (Kovalev, Kochetov, and Chuev 2021). Nagesha et al. ball milled powder of iron, nickel, and cobalt with an impact of 48% speed of ball milling (Nagesha et al. 2023). The mechanical/ball milling is a simple, eco-friendly, cost-effective, and high yield method (Wang et al. 2023).

1.2.1.2 Nanolithography

Nanolithography is another fabrication method utilized for creating nano-metric scale structure. Lithography on the basis of use of mask/template is categorized into two types. One is mask lithography, which utilizes mask/templates/molds to transfer nano-patterns over exposed large surface areas of wafers, resulting in the synthesis of high-throughput devices. Some of the examples of mask lithography are nanoimprint lithography and photolithography (Sharma et al. 2022; Salem et al. 2022). Another is maskless lithography, which creates nano-patterns without using masks. Such methods generate patterns in a serial manner. It causes ultrahigh-resolution patterning of different shapes having feature size smaller than a few nanometers. Nevertheless, the amount of material passing through this system is limited due to its leisurely serial nature. In maskless lithography, a small area of wafers is exposed in one step resulting in low throughput. And this low throughput removes any possibility of mass production. Some examples of maskless lithography are scanning probe, focussed ion beam, and electron beam lithography (EBL; Sharma et al. 2022; Paras et al. 2022). Panikar et al. (2023) assembled 30 nm Au NPs using simple stamping method upon EBL substrate. McMullen et al. developed a 5–40 nm wide electrode pair using a single step of EBL (McMullen, Mishra, and Slinker 2022). Nanosphere lithography is one of the recent advancements in the world of interfacial technology. Recently, using nanosphere lithography, polystyrene nanoscale meshes were created (Sharma et al. 2023). The meshes were shaped in honeycombed structures. Maskless lithography is inexpensive and easy to use (Brady et al. 2019).

1.2.1.3 Laser Ablation Method

Laser ablation technique of top-down approach synthesizes nanomaterial by utilizing laser beams as an energy source. In laser ablation, a high-intensity laser beam is focused on target material. This high-intensity laser beam increases the temperature of the irradiated spot, and the target material or precursor gets vaporized, resulting in the formation of plasma (Altammar 2023). This laser-induced plasma is produced because of haphazard collisions among nearby molecules, clusters, and atoms (Kim et al. 2017). The laser ablation technique does not use any toxic or hazardous chemicals or stabilizing agents (Ijaz et al. 2020). This method is used to produce different kinds of NPs, such as metallic NPs, metallic oxide NPs, semiconductor quantum dots, nanowires, and nanocarbons (Su and Chang 2017; Baig et al. 2021). Laser ablation is a simple, laboratory safe, and environmentally friendly method and has the ability to synthesize nanomaterials with complex stoichiometry, high purity (~90%), and uniform size distribution (Meidanchi and Jafari 2019).

1.2.1.4 Thermal Decomposition

Thermal decomposition method is an endothermic process in which the chemical bond of the molecule is decomposed by heat (Zeng, Xuan, and Li 2023). The temperature at which the molecule gets fragmented or decomposed is termed as decomposition temperature (Ealia and Saravanakumar 2017). The rate at which chemical bonds of the molecule get decomposed for NP production is measured by thermogravimetric analysis (Odularu 2018). Ahab et al. (2016) used thermal decomposition technique and synthesized NPs of gadolinium oxide. Zinc dehydrate acetate, along with ethylene glycol, was thermally decomposed at a temperature of 500 °C for 240 minutes to fabricate 35.4 nm-sized NPs of

ZnO_2, as reported by Rathore and Kaurav (2022), and its fabrication was confirmed by conducting X-ray diffraction (XRD), energy dispersive X-ray spectroscopy, and field emission scanning electron microscopy analyses. Tomar and Jeevanandam (2022) used thermal decomposition method and synthesized NPs of zinc ferrite ($ZnFe_2O_4$). Che et al. (2022) fabricated NPs of 99.92% pure CuO. The thermal decomposition technique is favorable for the fabrication of NPs used in cancer treatment (Odularu 2018), the utilization and storage of solar energy (Zeng et al. 2023), and the formation of nanoscale electrodes and catalysts (Yu et al. 2023), among others.

1.2.1.5 Sputtering Method

Sputtering is a nanofabrication method, in which tiny clusters of atoms are expelled off from the target surface by bombarding with high-energy gas or plasma particles (Inoue et al. 2023; Wirecka et al. 2022). The deposition of nanoscale particles is followed by the process of annealing. Substrate, temperature, thickness of layer, and annealing time help in determining the shape and size of nanoscale particles (Ijaz et al. 2020). Pleskunov et al. (2021) employed single-step plasma-based method for the fabrication of NPs. For the production of tantalum oxynitride NPs, direct current (DC) reactive magnetron sputtering techniques were used. The structural characteristics of platinum layer were modified using conventional magnetron sputtering technique at varying argon pressure (0.3–3.2 Pa; Sandhya et al. 2021). Sputtering technique is fascinating because it is cheaper than EBL and also because the sputtered nanomaterials' make-up is akin to the target material and has less impurity (Baig et al. 2021).

1.2.1.6 Arc-discharge Method

Arc-discharge is the classical method of fabrication of NPs. This is one of the most fascinating methods to develop carbon-based NPs, such as few-layer graphene, carbon nano-horns, fullerenes, and carbon nanotubes (CNTs; Kafle 2019; Zhang et al. 2019). In this method, carbon rod vaporization is produced by an electric arc between two graphite rods. The graphite rods are enclosed in a closed chamber filled with helium. The pressure of helium in the enclosed chamber is maintained as availability of oxygen prevents fullerene formation (Baig et al. 2021). The yield of nanoscale material fabricated is greatly influenced by the factors such as the distance between the graphite rods, its diameter, and the medium in which the rods are immersed. Other than these factors, pressure, temperature, voltage, and current also affect the yield of the synthesized nanomaterial (El-Khatib et al. 2018). As the growth mechanism of carbon-based nanomaterials differs from each other, their position of collection is different in the arc-discharge method. Karpov et al. (2019) studied the effectiveness on residual stress by the technological parameter during the fabrication of NPs and also established its correlation with voltage of arc plasma generator and NP magnetization on gaseous mixture pressure. They developed a single-step process in low-pressure arc-discharge to synthesize copper oxide NPs. Utilizing the arc-discharge methods, 19.60 and 32.97-nm-sized nanoscale particles of copper and argon were formed and analyzed using XRD, transmission electron microscopy (TEM), and energy-dispersive X-ray spectroscopy (EDX) methods (El-Khatib et al. 2018). Tseng et al. (2018) developed Ag^+ and AgO nanoscale particles by submerged arc-discharge method.

1.2.2 Bottom-up Approach

The Bottom-up approach, also termed as constructive method, progressively builds atoms into clusters and subsequently to NPs. This method includes both chemical and biological/green synthesis methods. Some of the chemical and biological methods are sol–gel method, spinning method, hydrothermal method, chemical vapor deposition (CVD), and biological synthesis by plant.

1.2.2.1 CVD Method

CVD method involves congealment of a thin coating of gaseous reactants on the surface of substrate in a reaction chamber. At an ambient temperature, chemical reaction ensues when combining gas and heated substrate collide with each other (Ijaz et al. 2020). The thin coating of the product accumulated on the substrate as a result of the reaction, was recovered, and then used. The substrate temperature greatly affects the CVD process. Lumen et al. (2021) utilized CVD method for the synthesis of porous silica NPs. This method circumvents the requirement of a highly pure monocrystalline silicon substrate as an initial material. In case of CVD-based synthesis of graphene, multilayer graphene is produced using Co and Ni catalysts, whereas monolayer graphene is produced by Cu catalyst (Baig et al. 2021). Ideal CVD precursors are volatile, pure, stable, inexpensive, nonhazardous, and long-lasting. CVD is a fascinating approach as it fabricates high-quality pure nanoscale particles that are homogeneous and stiff (Machac et al. 2020).

1.2.2.2 Sol–Gel Method

Sol–gel method is a simple, eco-friendly, and most preferred wet-chemical method for fabrication of NPs (Won et al. 2023; Solunke et al. 2023). In the word "sol–gel," sol is a colloidal dispersion consisting of solid particles suspended within a continuous liquid phase and gel is a dispersion of liquid within a solid matrix. In the sol–gel method, metal alkoxide and chloride used as precursors results in a stable solution that is thermally decomposed, condensed, and hydrolyzed (Altammar 2023). The precursor is distributed uniformly in host liquid by sonication, shaking, or stirring. The resultant solution is carried out and phase separation is performed by utilizing different techniques like filtration, sedimentation, and centrifugation (Ijaz et al. 2020). The morphology and size of nanoscale particles are controlled by optimizing the process parameters like solution pH, gelation/calcination temperature, and precursor concentration (Alabada et al. 2023). The flexibility in adjusting morphological characteristics makes sol–gel synthesis a highly effective technique for synthesizing photocatalysts (Habte et al. 2019). This method, processed at low temperature, provides uniform quality NPs and easily produces complex nanomaterials (Araoyinbo et al. 2018).

1.2.2.3 Spinning Method

The fabrication of nanoscale particles, using the spinning method, is processed by utilizing a spinning disc reactor (SDR). A rotating disc, consisting of stainless steel or copper with smooth and polished surface, is housed within a chamber/reactor where temperature, rotation speed, and other physical parameters can be regulated (Mekuye and Abera 2023). SDR is capable of continuous production of commercial amounts of the solid product. The morphological characteristics of nanoscale particles fabricated via SDR are influenced by several parameters, including disc surface properties, liquid-to-precursor ratio, disc rotation speed, feed location, and flow rate of liquid (Ijaz et al. 2020). The key factors that make SDR technology successful and more efficient are optimal heat and mass transfer, exceptional liquid–liquid mixing facilitating efficient micro-mixing, the ability to operate continuously, and suitability for gas–liquid reactions (Chianese, Picano, and Stoller 2021).

1.2.2.4 Hydrothermal Method

Fabrication of NPs by hydrothermal technique is the most widely used solution-reaction-based method. The synthesis of NPs by hydrothermal process is carried out at a wide range of temperature, i.e. from room temperature to significantly elevated temperature and pressure (Gan et al. 2020). The size and morphology of the NP synthesized by the hydrothermal process are maintained by controlling variable parameters such as pH, pressure, reactant concentration, temperature, and additives (Li et al. 2016). Chang et al. (2020) synthesized versatile nanostructures of ZnO like one-dimensional nanorods, two-dimensional nanoplatelets, and three-dimensional multi-branched flower-like particles utilizing hydrothermal processes and maintaining the pH of the precursor. Fabricated ZnO NPs were analyzed using selected area electron diffraction (SAED), XRD, and TEM. Zhongguan et al. (2023) fabricated graphene oxide by one-pot hydrothermal process. Okada, Kuno, and Yamada (2023) using microwave-assisted hydrothermal technique fabricated monoclinic vanadium dioxide (VO_2). Hydrothermal techniques serve as a versatile tool for enhancing other fabrication techniques (Yaghoobi, Asjadi, and Sanikhani 2023) and producing magnetite (Silva et al. 2023). There are several advantages of this process that makes it more attractive: less hazardous, environment friendly, low cost, use of simple equipment, etc.

1.3 Green Synthesis of Nanomaterials

In recent years, there has been a surge in research to develop eco-friendly methods for the synthesis of well-characterized NPs. Green/biological synthesis is one such method. The NP fabrication technique "green synthesis" employs biological entities, such as prokaryotes, complex eukaryotes, organic acid, primary or secondary metabolite, and enzymes, to reduce metallic ion to their elemental form (Kumari, Dhand, and Padma 2021; Parveen, Banse, and Ledwani 2016). This method is favored over conventional physical and chemical NP fabrication methods owing to its economic viability, reduced environmental impact, and enhanced safety for humans and the environment. It's also inexpensive, eco-friendly (Jiang et al. 2022), and prevents the use of toxic and hazardous chemicals. Besides, it has several other advantages: it's readily adaptable to mass production, produces highly stable products, and reduces processing time (Malhotra and Alghuthaymi 2022).

1.3.1 Nanomaterial Synthesis Using Microorganism

Biological synthesis of green NPs using microorganisms, like bacteria, fungi, and yeast, have emerged as prominent platforms. Due to certain properties like rapid growth rate, ease in cultivation, and ability to adapt in ambient conditions made these microorganisms a potential source of green synthesis (Ali et al. 2020). The wide variety of microorganisms can act as potential biofactories for the synthesis of several metallic NPs, such as copper, zinc, silver, gold, and iron. NPs prepared by Magnetotactic bacteria (found on the ocean floor) are specialized for synthesis of Magnetic NPs (10 to 20 nm). However, synthesis of Au NPs by photosynthetic bacteria, and Ag NPs by Fusarium oxysporum are some of the examples of green nanomaterial fabricated using microbes (Mekuye and Abera 2023).

1.3.2 Nanomaterial Synthesis Using Bacteria

In recent times, the primary focus of research has been on utilizing prokaryotes for metallic NPs production (Omran 2020). The synthesis of nontoxic, biologically derived NPs by utilizing bacteria is gaining attention due to their unique and superior properties, such as ubiquity, adaptation to harsh environments, ease of growth, ease of cultivation, and inexpensiveness, which make them ideal for biomedical applications. Nanomaterial synthesized by bacteria can utilize extracellular or intracellular methods (Marooufpour et al. 2019).

The extracellular mechanism involves the nitrate reductase synthesis method. This nitrate reductase enzyme is either secreted by the cell or is present in the cell wall and is responsible for the reduction of metal ions (Yusof et al. 2019). Nicotinamide adenine dinucleotide (NADH)-dependent reductase catalyzes the bioreduction of silver ions into Ag NPs. Electrons donated by NADH are oxidized to NAD^{+} during the process. The enzyme undergoes simultaneous oxidation as silver ions are reduced to nanosilver (Prathna et al. 2010). Some of the bacterial species, such as *Bacillus subtilis, Bacillus licheniformis, Ochrobactrum anthropi, Pseudomonas stutzeri*, and actinobacter, are utilized for the production of Ag NPs (Dikshit et al. 2021). Rauf et al. (2017) in their study stated that *Staphylococcus aureus* via extracellular biosynthesis mechanism can produce ZnO NPs. The bacterial electrokinetic potential and alkalinity of the solution play a key role in the reduction of metal ions. For the reduction of metal ions, several extracellular enzymes act as electron shuttles. *Mycobacterium paratuberculosis, Geobacter fermentans*, and *Shewanella oneidensis* are some bacteria that release such enzymes, which act as electron shuttles and reduce Fe^{3+} ions (Priya et al. 2021).

The intracellular mechanism for the synthesis of nanomaterial is different from the extracellular mechanism of microorganism. The production of NPs through an intracellular mechanism involves three steps: trapping, bioreduction, and capping. The intracellular synthesis of nanomaterial involves the targeted transport of specific ions through negatively charged cellular environments. This process is influenced by biomolecules, enzymes, and coenzymes, which facilitate the reduction of metal ions into NPs (Slavin et al. 2017). Król et al. (2018) produced ZnO nanocomposite via intracellular biosynthesis mechanism utilizing *Lacticaseibacillus paracasei* bacterial strain and precursor Zn nitrate. Magnetite NPs were produced intracellularly by sulfate-reducing bacterial strain *Desulfovibrio magneticus*, which grows and respires in the presence of fumarate. Gold NPs were fabricated via an intracellular mechanism using *Lactobacillus kimchicus* DCY51^{T} (Markus et al. 2016).

1.3.3 Nanomaterial Synthesis Using Actinomycetes

Actinomycetes, while understudied, have emerged as a promising group of organisms for metal NP synthesis (Golinska et al. 2014). It has the ability to produce varieties of secondary metabolites during their saprophytic existence. Actinomycetes synthesize NPs characterized by good polydispersity, excellent stability, and broad-spectrum antimicrobial properties (Mabrouk, Elkhooly, and Amer 2021). It can serve as bio-nanofactories for the fabrication of various NPs (Aswani, Reshmi, and Suchithra 2019). Species of *Streptomyces* like *Streptomyces zaomyceticus* and *Streptomyces pseudogriseolus* by both extracellular and intracellular mechanisms produces CuO NPs of size 78 and 80 nm, respectively (Hassan et al. 2019). Silver NPs of size 30 and 60 nm were synthesized using strains of *Streptomyces rimosus* and *Streptomyces chrestomyceticus* for the control of phytopathogens (Zwar et al. 2022). Some of the actinomycetes, such as *Nocardia farcinica, Streptomyces viridogens, Streptomyces hygroscopicus, Rhodococcus* sp., and *Thermoactinomycetes* sp., were identified for producing silver NPs.

1.3.4 Green Synthesis of NP Using Fungi

Mycotechnology is another biological approach, which is used in the synthesis of various NPs. Fungi are considered as an ideal organism for synthesizing metal and sulfide-based NPs due to their abundant intracellular enzymes, mycelial structure, potential for large-scale cultivation, ease of handling, superior tolerance of heavy metal, and economic viability benefits (Ahmed et al. 2022). It was observed that Ag^+ ions are trapped on the mycelial surface via electrostatic interaction with anionic carboxyl groups, facilitated by enzymes present in the mycelial cell wall (Altammar 2023). These enzymes then further reduced the trapped Ag^+ ion to Ag^+ bionanomaterial. Fungus-based nanomaterials can be synthesized both via extracellular and intracellular mechanisms. It was observed that the fabrication of NPs by fungi was influenced by several factors. The age of the culture affects the count of NPs, change in pH results in change in the shape of nanomaterial, and increase in temperature results in higher growth rate and accumulation of NPs of gold (Ma et al. 2017). Silver NPs of size 20–55 nm were produced extracellularly by different strains of *Penicillium aculeatum*. This fungal strain can be used as a potential bioresource for eco-friendly and cost-effective production of silver NPs. Chatterjee et al. (2020) fabricated 20 to 40 nm-sized superamagnetic Fe_3O_4 NPs by utilizing *Aspergillus niger* isolated from mangroves. Vijayanandan and Balakrishnan (2018) used the strain of *Aspergillus nidulans* and developed NPs of cobalt oxide. Another species of *Aspergillus, A. niger* with excellent antimicrobial properties were used in the synthesis of zinc oxide NPs. Researchers have observed varieties of filamentous fungal strains for the production of metallic NPs, such as AuO, AgO, and FeO of required shape, size, and ionic charge. Some of these observed fungal species are *Hormoconis resinae, Humicola* sp., *A. niger, F. oxysporum, Phoma* sp. *Phanerochaete chrysosporium, Pestalotiopsis* sp., *Trichoderma* sp., and *Penicillium* sp. (Rai et al. 2021).

1.3.5 Nanomaterial Synthesis Using Plant Extract

NPs can also be synthesized using different parts of plants, like roots, leaves, barks, flowers, shoots, fruits, and their extracts (Aboyewa et al. 2021; Mekuye and Abera 2023). Plants are the most eco-friendly resource and inexpensive to grow as well as available in abundance. The synthesis involves combining plant extracts and metal salt solution at room temperature to create NPs. The plant extract contains several phytochemicals, such as organic acids, quinones, and flavones, which work as natural reducing agents in the synthesis of NPs. NPs of palladium and platinum were fabricated using plant extracts of various plants, such as *Ocimum sanctum, Curcuma longa, Pulicaria glutinosa, Anogeissus latifolia, Glycine max*, and *Diospyros kaki* (Siddiqi and Husen 2016). Utilizing the extracts of *Pelargonium graveolens* and *Medicago sativa*, different shapes of gold NPs were fabricated (Mekuye and Abera 2023). Armendariz et al. (2004) were the first to report the biosynthesis of rod-shaped NPs using biomaterials. They characterized the Au NPs synthesized from wheat biomass using a 0.3 mM potassium tetrachloroaurate solution at pH levels ranging from 2 to 6 and at room temperature. Ag NPs of approximately 58 nm were fabricated from leaves of *Polyalthia longifolia* (Kumar et al. 2016). Gardea-Torresdey et al. (2002, 2003) first documented the intracellular biosynthesis of Au and Ag NPs within living plants. The extracts of sunflower (*Helianthus annuus*) and *Brassica juncea* were used to fabricate Co, Zn, Cu, Ni, and Ag NPs. Rate of synthesis of NPs from plants is influenced by several factors, such as plant extract type, its concentration, metal salt concentration, temperature, pH, and reaction duration (Ali et al. 2020).

1.4 Nanomaterials as Controlled Delivery System for Actives and Sustainable Agriculture

NT has revolutionized the agriculture sector by offering sustainable solutions for precision agriculture and providing us eco-friendly and resource-efficient technologies. Nanomaterials have immense potential to deal with crop productivity, the requirement of nutrients and fertilizers, soil health, plant disease, detection of pathogens and pests responsible for the disease, pesticide delivery, etc. by synthesizing the essential nanomaterials. Nanomaterials fabricated in the range of 1–100 nm are utilized for sustainable agriculture in the form of NFs, nanocides (nanopesticides, nano-bacteriocides, etc.), nanosensors to improve land and crop productivity, to detect and cure pathogenic disease, to site-specific and controlled release of fertilizers and pesticides, to improve soil health, etc. This section deals with NPs used in the form of NFs, nanocides, and nanosensors for improved, efficient, and productive agriculture.

1.4.1 Carrier-based Nanomaterials

The current methods of biomolecule delivery systems in agriculture and plants face various limitations, like low delivery efficiency, complex procedures, and dependence on specific plant species. In order to address challenges related to environmental and global food security issues, NT has revolutionized genetic engineering in plants and agricultural sectors (delivery of active biomolecules [herbicides/insecticide], including proteins, RNA, and DNA into plants) with improved delivery efficiencies, biocompatibility, and plant regeneration (Zhang, Ying, and Ping 2021).

A man-made, spheroidal nanomaterial known as "Dendrimer" has gained attention due to its multifunctional, three-dimensional structure, typically ranging from 1 to 10 nm in size. These dendrimers can be mathematically designed and synthesized (Figure 1.2). Structurally, they consist of a central core, an interior cavity, and surface functional groups.

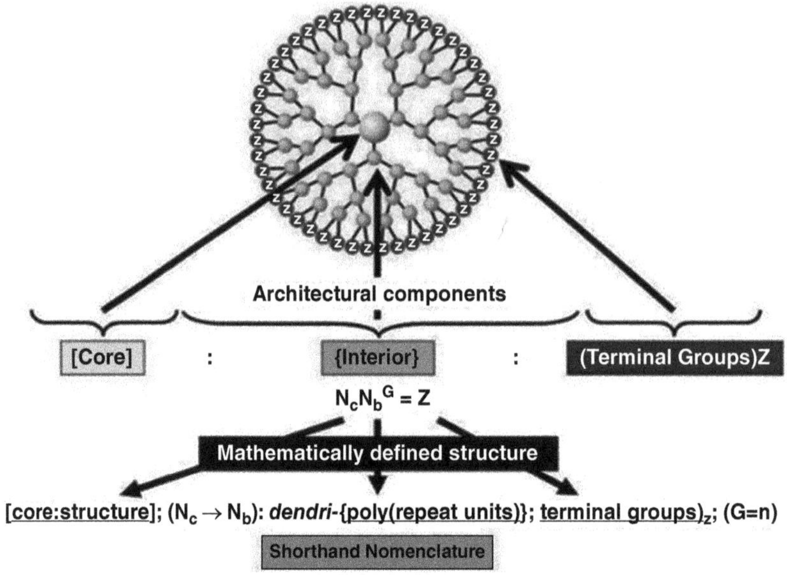

Figure 1.2 Architectural three-dimensional components of PAMAM dendrimer. *Source:* Graphic provided by NanoSynthons (i.e. Dendrimer Nanotechnology Company located in Mount Pleasant, MI, USA).

The interior cavity is capable of entrapping guest molecules, while the surface groups offer customizable sites for various physical or chemical interactions allowing for tailored applications across fields, such as medicine, biotechnology, materials science, and agriculture. Historically, Tomalia-type poly(amidoamine) (PAMAM) dendrimers are recognized as the first dendrimer family to be synthesized and become commercially available (https://www.sigmaaldrich.com).

However, a second commercial family of dendrimers was developed by Tomalia (2005), while managing Starpharma's Dendritic Nanotechnologies Inc. (DNT) at Mount Pleasant, Michigan, USA. This family is called Priostar® dendrimers; Starpharma further enhanced the water solubility of various crop protection actives (herbicides) like trifluralin and 2,4-dichlorophenoxyacetic acid that is used to kill weeds and regulate plant growth. The formulation of Trifluralin with dendrimer enhanced its aqueous solubility by 25,000 folds from 0.22 mg/L to 5.43 g/L. Meanwhile, the dendrimer also provided another benefit related to improving the shelf-life of products and greater stability in the field, leading to improved control of pests. However, with 2,4-dichlorophenoxyacetic acid, these dendrimers provided numerous benefits over commercially existing products, such as improved weed control, increased seedling density, and reduced environmental impact. Again, glyphosate, a well-known broad-spectrum systemic herbicide and crop desiccant, was formulated with these dendrimers resulting in a 150–250% improvement in rain-fastness in glasshouse trials (i.e. making the active more resistant to being washed away in poor weather when compared to a non-dendrimer control glyphosate formulation). Other glasshouse studies with these dendrimers have successfully witnessed the role of dendrimers in improving the efficacy of an active and systemic control of a fungicide. Amazingly, this dendrimer has also been successful in confirming the enhanced soil permeability of a well-known insecticide (i.e. imidacloprid) and making it more effective at targeting pests and diseases that current formulations cannot control due to poor soil permeation. These dendrimers-based imidacloprid study confirms the potential of dendrimers to elevate the soil permeation up to 7.5 cm, as compared to only 2.5 cm without dendrimers. However, native imidacloprid could reach only 10% at the bottom, whereas dendrimer-based imidacloprid showed significant accumulation, up to 40%, in the depth of the soil. Thus, dendrimer technology can be explored as nano excipients in the area of agriculture-related crop protection (https://starpharma.com).

Furthermore, other NT-based carriers (Figure 1.3) demonstrate a unique opportunity to manipulate some beneficial plant traits and allow natural resistance to pests and disease, as well as enhanced production of crop yields with longer shelf lives. In the past decade, mesoporous silica nanoparticles (MSNs) arranged in a honeycomb-like structure have attracted more and more attention for their potential use for specific delivery within plant organelles. Due to porous nature, MSNs provide a high surface area that can be easily modified with functional groups, thanks to the presence of exposed silanol groups. Silica particles with their high biocompatibility and inertness, as well as chemically and thermally stable mesoporous structures with variable pore sizes (2–10 nm in diameter), make MSNs suitable for hosting a diverse range of guest molecules, including those necessary for gene and chemical delivery (Miyamoto and Numata 2023; Law, Miyamoto, and Numata 2023).

CNTs, cylindrical or tube-like structures of carbon allotropes, possess distinctive physicochemical qualities. CNTs as microtubes inside the plant body can improve seed germination by aiding in water uptake and also act as a plant growth stimulator, increasing dry biomass and lengths of root/shoot. As a growth promoter and slow-release fertilizer, CNT is emerging as a new nano-carbon fertilizer in agriculture (Vithanage et al. 2017).

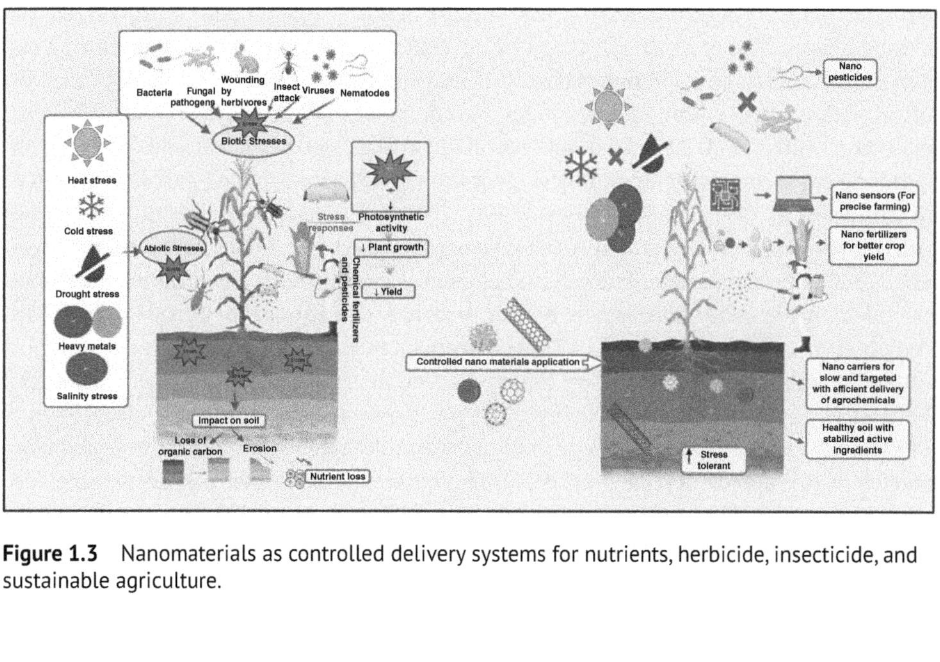

Figure 1.3 Nanomaterials as controlled delivery systems for nutrients, herbicide, insecticide, and sustainable agriculture.

An ornamental plant, Lily (*Lilium* spp.), has inefficient traditional genetic transformation methods. The highest germination percentage (88.32%) of *Lilium* regale pollen was observed while using nanomagnetic beads. Pollen magnetofection utilizes magnetic MPs as gene carriers to directly produce transgenic seeds without the need for tissue culture. The green fluorescent protein (GFP) activity reached an average of 69.66% in transformed pollen. A significant β-glucuronidase (GUS) activity, with a transformation efficiency of 56.34%, was also observed in transgenic seedlings, demonstrating that pollen magnetofection is effective in plant breeding transformation (Zhang et al. 2023).

1.4.1.1 Nanopesticides

The development of pesticide nanoformulations incorporating polymers, copolymers, silica, nanocarriers, and metal NPs is a rapidly expanding field within the pesticide industry. The main objective of nanopesticides development is to enhance pesticide efficacy, prolong its action, and minimize environmental impact, revolutionizing pests and disease management in crop production (Pérez-de-Luque and Rubiales 2009). Nanopesticides are categorized as follows: polymer-coated organic molecules and carrier-less inorganic molecules, on the basis of formulation of nanopesticides (Zhang et al. 2019). Polymer-coated pesticides are often used in the field of agriculture because of its features like targeted delivery, enhanced stability and solubility, etc. (Nair et al. 2010; Zhang et al. 2019). Huang et al. (2018) fabricated light-responsive nanopesticides by encapsulating pesticides within a PEO-PMAA copolymer matrix. Bhan, Mohan, and Srivastava (2014) demonstrated enhanced larvicidal efficacy of polyethylene glycol (PEG)-encapsulated temephos and imidacloprid nanopesticides compared to conventional formulation. Against a wide variety of phytopathogens, such as *Fusarium culmorum, Botrytis cinerea, Magnaporthe grisea, Bipolaris sorokiniana, Pythium ultimum*, and *Trichoderma* sp., silver NP encapsulated nanopesticides exhibit antimicrobial activity (Park et al. 2006; Sharon, Choudhary, and Kumar 2010). Porous silica nanomaterials offer a promising platform for controlled pesticides' delivery, as exemplified by the successful encapsulation and controlled release of validamycin (Liu et al. 2006). Copper NPs exhibit potent broad-spectrum antimicrobial activity against both Gram-positive and Gram-negative bacteria, as well as a range of fungal pathogens, including *Fusarium* sp., *Xanthomonas oryzae, Phytophthora infestans*, and others (Cioffi et al. 2005; Giannousi et al. 2014). Polymeric nanocapsules coated with both fungicides and herbicides mitigate environmental impact, applied and demonstrated by Campos et al. (2015) and De Oliveira et al. (2015), respectively. Both studies reported superior performance of nanoformulations compared to conventional products. A diverse range of nanopesticide formulation have been developed, including nanosphere, mesoporous metal oxide NPs, nanocapsule, nanosuspension, nanoemulsion, and solid lipid NPs (Ayoub et al. 2018; Zhao et al. 2017). Abamectin nanoemulsion demonstrated superior wettability and rapid penetration on cabbage leaves, characterized by a low contact angle (Feng et al. 2020). Nanocapsule delivery system demonstrated superior efficacy against *Rhizoctonia solani* compared to conventional formulation, highlighting the synergistic benefits of this approach (Cui et al. 2020). Thorough environmental risk assessment and regulatory compliance are essential prerequisites for the commercialization of nanomaterials. To mitigate potential environmental risks, comprehensive nanopesticides regulations should be implemented at the federal level.

1.4.1.2 Nanofertilizers

NFs are formulated by coating nutrients within active materials, enabling controlled and gradual release into the soil. NFs can mitigate nutrient loss through leaching and volatilization, enhancing nutrient use efficiency and soil fertility. This ultimately leads to improved crop productivity and long-term soil health (Rautela et al. 2021). NPs act like an adjuvant by conjugating easily with the plant because of having large surface area. NFs have the potential to significantly enhance crop growth by enhancing seed emergence, metabolism of nitrogen, photosynthesis, and production of carbohydrates and protein, while also improving stress tolerance (Rop et al. 2018). Leveraging their unique properties, NFs demonstrate superior efficacy compared to traditional polymer-coated slow-release fertilizers (Duhan et al. 2017). While NFs offer several benefits, there are also some challenges associated with their use. These NFs containing microelement when administered in high dose may result in phytotoxic effects. According to several literature, silver-containing NFs have detrimental effects on plant growth, primarily through root epidermal damage. This disruption inhibits root development, leading to overall plant weakness and reduced yield (Goswami, Yadav, and Mathur 2019; Izadiyan et al. 2020). NFs are typically produced by encapsulating fertilizer components within a NP matrix. This can be encapsulated through three techniques: by encapsulating nutrients within nonporous structure, by coating with a thin layer of polymer film, and by delivering it in the form of nanoscale particles or emulsion (Rai, Acharya, and Dey 2012). These NFs are further classified into two types based on nutrient availability which is further described as: (i) micronutrient NFs and (ii) macronutrient NFs. Some of the NFs fabricated and used in the agriculture sector are listed in Table 1.1.

1.4.1.3 Nanosensors

Nanosensors are accurate, efficient, and cost-effective devices that are revolutionizing farming by identifying pathogens, pesticides, pollutants, heavy metal ions, and more, along with monitoring humidity and temperature. They are helpful in monitoring precise crop progression, optimizing planting and harvesting schedules, and also detecting crop diseases efficiently. Biofilms, which are sessile and complex bacterial communities, can be detected by nanosensors optically with high sensitivity and spatial resolution, aiding in disease prevention (Pu et al. 2020). Sheikh et al. (2024) reported nanosensors' role in managing biotic and abiotic stresses in plants, diagnosing diseases, and providing accurate detection methods. Yin et al. (2021) developed a nanosensor platform for monitoring soil moisture levels using gold NPs, aiding in precision irrigation and water conservation. To promote healthier crops, integration of nano-functionalized materials, like hydrogels, will help in enhancing water retention and nutrient delivery (Das et al. 2024).

1.4.1.4 Stimuli-responsive Nanocarriers

Designing controlled stimuli-responsive nanomaterials (NFs, nanopesticides, etc.) in response to biotic and abiotic stimuli (e.g. enzymes, light, pH, redox conditions, and temperature) has been researched and developed in the past decade. These nanocarriers achieve controlled release through structural changes triggered by external stimuli or by breaking chemical or hydrogen bonds between the nanocarrier and the pesticide or nutrient-active ingredient.

Table 1.1 List of nanoformulations (nanofertilizers, nanocides, and nanosensors) fabricated with plant species and their critical impact on crops.

Nanoformulations with particle size (nm)	Plant species	Impact on crops	References
Nanofertilizers			
CaB (80 nm)	*Lactuca sativa* and *Cucurbita pepo*	Boosts productivity of plant and enhances root and shoot dry weight	Meier et al. (2020)
CaO (~69.9 nm)	*Arachis hypogaea*	Boosts root growth and augments Ca deposition in plant	Deepa et al. (2014)
NPK chitosan (26.2–30.6 nm)	*Triticum aestivum*	Reduces life cycle of crop and enhances yield and vegetative growth	Abdel-Aziz et al. (2016)
N (biosynthesis; 1–2 nm)	*Pennisetum americanum*	Boosts biomass production and augments soil microbes responsible for P and N metabolism	Thomas, Rathore, and Tarafdar (2016)
MgO	*Solanum lycopersicum*	Prevents wilt outbreak	Imada et al. (2015)
Mg	*Vigna unguiculata*	Enhances growth and yield and promotes photosynthesis	Delfani et al. (2014)
S (20 ± 4 nm)	*Cucurbita pepo*	Promotes stem and root growth	Salem et al. (2015)
Urea HAP (~18 nm)	*Oryza sativa*	Maintains sustained release of nitrogen; enhances N and K content and increases yield	Kottegoda et al. (2017)
P (15.8 ± 7.4 nm)	*Glycine max*	Enhances plant growth, biomass, and yield	Liu and Lal (2014)
Mo (100–250 nm)	*Cicer arietinum*	Minimizes K leaching and enhances biomass yield	Taran et al. (2014)
ZnO (35 nm)	*S. lycopersicum*	Enhances root and shoot growth	Faizan et al. (2018)
ZnO (90 nm)	*Lactuca sativa*	Enhances biomass production and photosynthesis efficiency	Xu et al. (2017)
ZnO (23 nm)	*Vigna radiata*	Improves photosynthetic efficiency and plant height	Raliya, Tarafdar, and Biswas (2016)
ZnO (16–35 nm)	*Abelmoschus esculentus* L. Moench	Enhances photosynthetic pigments and reduces soluble sugar and proline concentration	Alabdallah and Alzahrani (2020)
SiO_2 (12 nm)	*S. lycopersicum*	Promotes seed germination	Siddiqui and Al-Whaibi (2014)

Nanoformulations with particle size (nm)	Plant species	Impact on crops	References
SiO_2 (12 nm)	*Cucurbita pepo* L.	Produces enhanced proline, chlorophyll content, and transpiration rate	Siddiqui et al. (2014)
SiO_2 (10–20 nm)	*Fragaria* × *ananassa*	Enhances absorption of K, Ca, Mg, Fe, Mn, and Si. Limits biochemical and anatomical behavior due to salt stress	Avestan et al. (2019)
SiO_2 (20–30 nm)	*Hordeum vulgare*	Enhances carotenoid and chlorophyll II content	Ghorbanpour, Mohammadi, and Kariman (2020)
SiO_2 (20–40 nm)	*Musa acuminata*	Promotes shoot development and photosynthesis; balances Na^+ and K^+	Mahmoud et al. (2020)
TiO_2 (10–25 nm)	*Linum usitatissimum*	Enhances chlorophyll and carotenoid concentration and reduces H_2O_2 level	Aghdam et al. (2015)
TiO_2 (30–50 nm)	*Triticum vulgare* L.	Enhances Zn, P, N, and Cu content	Dağhan et al. (2020)
TiO_2	*Spinacia oleracea*	Promotes N_2 fixation and enhances dry weight of plant	Zheng et al. (2005)
TiO_2 (30–50 nm)	*S. lycopersicum*	Enhances chlorophyll II content and photosynthetic efficiency	Tiwari et al. (2017)
TiO_2 (10–25 nm)	*Triticum aestivum*	Increases seed dry weight, chlorophyll, and carotenoid content and enhances transpiration rate and antioxidative enzyme activity	Faraji and Sepehri (2020)
Fe_2O_3 (20 nm)	*Arachis hypogaea*	Enhances chlorophyll II content, biomass, and plant growth	Rui et al. (2016)
Fe_2O_3 (50 nm)	*Spinacia oleracea*	Enhances plant biomass	Jeyasubramanian et al. (2016)
Fe_2O_3 (27.3–34.62 nm)	*Mentha piperita* L.	Deduces level of malondialdehyde (MDA) and proline	Askary, Amirjani, and Saberi (2016)
Fe_2O_3 (6 nm)	*O. sativa*	Improves root growth	Alidoust and Isoda (2013)

(Continued)

Table 1.1 (Continued)

Nanoformulations with particle size (nm)	Plant species	Impact on crops	References
Fe_3O_4 (13 nm)	*Hordeum vulgare* L.	Enhances plant growth and improves chlorophyll II and protein content	Tombuloglu et al. (2019)
Fe_3O_4 (40–53 nm)	*Fragaria* × *ananassa* Duch.	Improves growth parameters of plantlets	Mozafari, Havas, and Ghaderi (2017)
CuO (20–40 nm)	*S. lycopersicum*	Enhances sugar and chlorophyll concentration and root length	Singh et al. (2017)
CuO (10 nm)	*Zea mays*	Improves biomass and seed growth	Saharan et al. (2016)
CuO (<50 nm)	*Zea mays*	Enhances plant growth	Adhikari et al. (2015)
AgNPs (20 nm)	*Trigonella foenum-graecum*	Enhances seed dry weight and germination rate	Hojjat and Kamyab (2017)
AgNPs (47 ± 7 nm)	*Arabidopsis thaliana*	Promotes root growth and mitotic cycle metabolism	Syu et al. (2014)
AgNPs (25 nm)	*Populus deltoides*	Promotes root growth	Wang et al. (2013)
AgNPs (20 nm)	*S. lycopersicum*	Enhances root growth and promotes germination of seed and its dry weight	Almutairi (2016)
AgNPs (21 nm)	*Trigonella foenum-graecum*	Enhances plant growth	Jasim et al. (2017)
AgNPs (15–29 nm)	*Triticum aestivum*	Promotes leaf growth and enhances proline and soluble sugar concentration	Mohamed et al. (2017); Iqbal et al. (2017)
Au (10 nm)	*Cucumis sativus*	Promotes root growth	Lin and Xing (2007)
Au, Ag, Cu, etc. (14–35 nm)	*Pennisetum glaucum*	Promotes seed germination and growth	Parveen at al. (2016)
Selenium and nano-selenium (N-Se; 8–15 nm)	*Lycopersicum esculentum*	Enhances shoot growth and dry weight and enhances root diameter, volume, and dry weight	Haghighi, Abolghasemi, and Da Silva (2014)
Se-NP (10–40 nm)	*Sorghum bicolor* L.	Augments membrane integrity and antioxidant system	Djanaguiraman et al. (2018)
Nanocides (nanopesticides, nano-bacteriocides, etc.)			
Ag-based NPs coated with chitosan (50 nm)	*Prunus persica*	Reduces bacterial canker disease caused by *Pseudomonas syringae*	Shahryari, Rabiei, and Sadighian (2020)
Ag-based NPs	*Punica granatum*	Controls blight disease caused by *Xanthomonas axonopodis*	As et al. (2018)

Nanoformulations with particle size (nm)	Plant species	Impact on crops	References
AgNPs (5–100 nm)	*Nicotiana tabacum*	Prevents infection by *Phytophthora parasitica* and *Phytophthora capsici* and enhances antibacterial efficiency	Ali et al. (2015)
Si-based NPs (<50 nm)	*Zea mays*	Protects plant from severe damage by *Sitophilus oryzae, Rhyzopertha dominica*, and *Tribolium castaneum*	El-Naggar et al. (2020)
Si NPs with or without chitosan coating (36 nm)	*Citrullus lanatus*	Suppresses disease severity caused by *Fusarium* wilt	Buchman et al. (2019)
$TiO_2/Cu_2(OH)_2CO_3$ NPs (20–50 nm)	Various crops	Prevents crop diseases caused by *Escherichia coli* and *Fusarium graminearum*	Liu et al. (2020)
Cu NPs/Thiophanate methyl (25 nm)	Greek orchards	Suppresses growth of benzimidazole-resistant *Monilinia fructicola*	Malandrakis, Kavroulakis, and Chrysikopoulos (2021)
Star polymer (SPc; 17.4–144.6 nm)	*Solanum tuberosum*	Activates endocytosis pathway and effective against potato late blight disease	Wang et al. (2021)
Polylactic acid/ dinotefuran/avermectin co-delivery nanoparticles (245.7 nm)	*Pyrus communis* L.	Increases bioactivity at lower concentration and is more effective against *Grapholita molesta* Busck	An et al. (2022)
Phyto-fabricated ZnO (20–30 nm)	*Brassica*	Is effective against Alternaria blight disease caused by *Alternaria brassicae*	Dhiman et al. (2021)
Metarhizium anisopliae–chitosan NPs (75.83 nm)	*Brassica oleracea*	Decreases the pathogenicity of *Plutella xylostella*	Wu et al. (2021)
Neem oil-zein NPs (198 nm)	*Phaseolus vulgaris* L.	Is highly effective against the disease caused by *Acanthoscelides obtectus*	Pascoli et al. (2020)
ZnO NPs (76.2–183.8 nm)	*S. officinarum*	Shows larvicidal efficacy against *Holotrichia* sp. (white grubs)	Shukla, Gaurav, and Singh (2020)
TiO_2/Fe_2O_3 NPs (20/40–100 nm)	*Nicotiana benthamiana*	Suppresses viral infection caused by *Turnip mosaic* virus	Hao et al. (2018)

(Continued)

Table 1.1 (Continued)

Nanoformulations with particle size (nm)	Plant species	Impact on crops	References
Cymbopogon martinii essential oil-chitosan (100 nm)	*Zea mays*	Shows antifungal and anti-mycotoxin activities and reduces the growth of *F. graminearum*	Kalagatur et al. (2018)
Azadirachtin-zinc oxide and chitosan (33.1 and 78.8 nm)	*Arachis hypogaea* L.	Growth of *Caryedon serratus* was successfully controlled	Jenne et al. (2018)
Mg $(OH)_2$ NPs (100 nm)	*S. lycopersicum*	Prevents growth of *E. coli*, *P. syringae*, and *X. alfaalfae*	Huang et al. (2018)
Nanosensors			
Au NPs-based (20 nm)	*S. tuberosum*	Detects bacterial wilt *R. solanacearum*	Khaledian et al. (2017)
TiO_2-Acetylcholinesterase (29 nm)	*Zea mays, Anacardium occidentale, O. sativa*	Detects pesticide dichlorvos causing toxic effects on humans and the environment	Cui et al. (2018)
Cd and Au NPs-quantum dots (27 and 4 nm)	Citrus plant	Detects the infection caused by *Citrus tristeza*	Safarnejad et al. (2017)
ZnO NPs-multiwalled carbon nanotubes (23 and <10 nm)	–	Detects *E. coli* and Gram-positive and Gram-negative bacteria	Meraat et al. (2020)
Primer-mediated asymmetric PCR-Au NPs (15 nm)	*S. tuberosum*	Detects *Phytophthora infestans* which cause potato late blight disease	Zhan et al. (2018)

In agriculture, soil pH influences the nutrient and pesticide fate and efficacy. A pH of 7.0 and above can reduce nitrogen loss from applied urea fertilizers to agricultural soil due to volatilization. Also, optimization and release of pesticides at a specific pH level that matches pest or infected crop conditions can improve its efficacy. pH-responsive nanofertilizers and pesticides often use natural and synthetic polymers with carboxyl or amino groups that react to pH changes. Carboxyl groups, at a pH below their pKa, remain protonated (–COOH) and form hydrogen among these groups and with other functional groups, such as hydroxyls, hindering the nutrient or pesticide-active ingredients release from the polymer matrix. When the pH exceeds the pKa, carboxyl groups become predominantly deprotonated (–COO–) causing swelling, which provides pore channels for accelerated release due to electrostatic repulsion.

1.4.2 Carrier-free Nanomaterials

The rapid development and expansion of carrier-free nanomaterials composed of pure active material can be directly used in crops or soil without the need for a secondary carrier substance, as discussed above (i.e. Dendrimers, CNTs, MSNs, etc.). In contrast to conventional carrier-based nanomaterials, the active materials are often entrapped, encapsulated, or chemically attached with carriers to improve stability and delivery. However, a carrier-free nanomaterials bypass this extra component, making the active material itself responsible for the intended function.

1.4.2.1 Micronutrient NFs

Micronutrient NFs represent advanced agricultural inputs that enhance nutrient delivery efficiency compared to conventional methods. Micronutrient NFs offer numerous advantages for plant health, including enhanced nutrient absorption, increased productivity, and improved resilience to biotic and abiotic stresses. These are required in trace amounts, but their deficiency can significantly disrupt plant physiological processes. Some of the micronutrients required by the plants are boron, zinc, copper, nickel, iron, manganese, and titanium (Yadav, Yadav, and Abd-Elsalam 2023).

Zinc is a vital micronutrient for the growth and development of all living organisms. Zinc functions as a crucial cofactor for numerous enzymes in plants, playing a pivotal role in protein synthesis, carbohydrate production, auxin biosynthesis, and overall plant defense mechanisms (Noreen et al. 2018). Awan et al. (2021) developed ZnO-NFs (zinc oxide NFs) using a green synthesis approach involving zinc sulfate solution and black seed (*Nigella sativa* L.) extract. On introducing ZnO NFs into broccoli, they observed significant enhancement in seed germination, seedling weight, root and shoot growth, leaf number, plant height, etc. Mattiello et al. (2015) studied enhancement in shoot growth and yield of cucumber when cultivated in a nutrient solution containing rubbery nanosuspension of zinc. In a recent study, *Citrus aurantium* fruit treated with 20 mg/L of zinc oxide NPs showed an increase in biochemical composition and growth parameters (Punitha et al. 2023).

Boron, an essential micronutrient, plays a vital role in plant growth and development by contributing to cell wall formation, pollen development, and nutrient translocation (Davarpanah et al. 2016). Boron-based NFs are produced by encapsulating borates within a NP matrix composed of material like humic acid. These formulations can be applied in both liquid and solid forms for both soil and foliar applications. Their unique properties enable the penetration of plant cells for direct nutrient delivery (Gehlout et al. 2022). Foliar application of zinc and boron, applied at optimal concentration (i.e. high concentration of Zn and low concentration of B), significantly increased pomegranate fruit yield (*Punica granatum* cv. Ardestani; Shang et al. 2019). While studying the effect of boron NPs on mung beans, Ibrahim and Farttoosi (2019) observed an enhancement in plant length, productivity of seed, and the number of pods.

Iron is a critical micronutrient that plays a pivotal role in plant metabolism, supporting processes, such as RNA synthesis, enzyme function, and photosynthesis (Mushtaq et al. 2020). It exhibits low soil availability due to limited solubility. Iron in reduced or excess quantities negatively impacts plant growth and development by disrupting essential metabolic processes (Palmqvist et al. 2017). Iron oxide and sulfide NPs represent the simplest form of iron fertilizers. Their small size enables superior soil penetration and plant uptake compared to conventional iron fertilizers. To enhance stability and longevity, iron NPs can be encapsulated within a protective biopolymer or lipid coating. Iron NFs offer a sustainable approach to enhance agricultural productivity by boosting crop yields, restoring soil health, reducing contamination, mitigating the effect of acid rain, and safeguarding water quality (Pitambara, Archana, and Shukla 2019; Cui et al. 2017). Dola et al. (2022) observed that foliar application of Fe_3O_4 NPs on *G. max* effectively enhanced the drought stress tolerance, deducing the loss caused by it. The application of Fe_3O_4 also promotes the productivity, quality, and growth of the plants.

Copper NPs outperform bulk copper in nutrient delivery, enhancing plant growth. As a vital micronutrient for protein and enzyme function, copper is efficiently delivered to plant cells via these NPs, ensuring rapid and effective nutrient uptake. Copper NFs offer a safe and efficient method for improving plant health and productivity (Yadav, Yadav, and Abd-Elsalam 2023; Khatri and Bhateria 2023). Van Nguyen et al. (2021), in their study, highlighted the efficacy of copper NP priming in mitigating drought stress in maize. The study demonstrated a reduction in reactive oxygen species (ROS) accumulation, suggesting an upregulation of ROS scavenging enzymes. Kasana et al. (2017) demonstrated the antimicrobial efficacy of biosynthesized copper NPs, but emphasized the need for further research to assess their yield-enhancing potential in field conditions. These copper NPs exhibit antimicrobial and antifungal properties, enhancing plant resistance to pests and disease.

Silicon, the second most abundant element on Earth, plays a vital role in plant health and development. Silica NPs mitigate salt stress, which is one of the biggest hurdles in enhancing crop production in agriculture, by enhancing K uptake and reducing Na uptake (Rizwan et al. 2015). Chitosan-Si NFs exhibited sustained Si release, promoting maize growth and yield. These NFs, when applied through foliar spray, activate antioxidant defense mechanisms by balancing oxidative stress through improved oxygen and hydrogen peroxide levels (Kumaraswamy et al. 2021). Tereshchenko et al. (2017) observed enhanced plant resistance to various stresses, while studying the impact of silicon dioxide NPs on hulless oat (monocotyledons) and lucerne (dicotyledons) under field conditions. Fatemi, Pour, and Rizwan (2020) studied the effect of silicon NPs on *Coriandrum sativum* L. They applied Si NPs through foliar spray on coriander plants and observed reduction in lead, malondialdehyde, flavonoid, and vitamin C concentration.

Titanium dioxide, known for its photocatalytic properties, is commercially used as pigment in cosmetics, paints, and other industries. Titanium dioxide NFs protect plants against several pests and diseases because of its antimicrobial properties. When treated with 0.01% $nTiO_2$, significant rise in plant growth parameters, including length of root and shoot, leaf area, and biomass accumulation were observed, compared to alkaline condition (Pitambara et al. 2019). TiO_2-based NFs successfully applied to enhance the growth of various crops like lettuce, tomato, spinach, watermelon, millet, lemna minor, wheat, and bean. These studies consistently reported a rise in germination rate, photosynthetic efficiency, chlorophyll content, biomass, and nutrient uptake (Zahra et al. 2015). TiO_2 NPs synthesized using *Citrus medica* L. were applied through foliar spray on *Capsicum annuum* L. As a result of utilization of TiO_2 NPs, the plant becomes resistant to disease and shows enhancement in growth parameters such as yield, biomass, and productivity (Prakashraj et al. 2021).

1.4.2.2 Macronutrient NFs

The essential micronutrients like calcium, magnesium, nitrogen, potassium, phosphorus, hydrogen, and sulfur have been encapsulated in nanomaterials to create NFs and capable of precise nutrient delivery to crops, resulting in the enhancement of plant growth (Chhipa 2016;

Ditta and Arshad 2016). These macronutrient-based NFs mitigate eutrophication, offer superior nutrient absorption and leaching resistance, and enhance crop yield and quality (Mejias et al. 2021; Sharma et al. 2022; Sarkar et al. 2022).

Nitrogen is a critical macronutrient essential for plant growth and development. As a key component of chlorophyll, it plays a pivotal role in photosynthesis. Nitrogen deficiency leads to chlorosis, characterized by yellowing of plant leaves. The nitrogen molecules encapsulated with various NPs like metal oxide, graphene, and nanotubes enhance the nitrogen availability in the soil and make the plant more accessible to nutrients (Erisman et al. 2008). The controlled release of nitrogen from these fertilizers minimizes nutrient leaching and runoff, reducing environmental risk in aqueous ecosystems. The nitrogen-encapsulated NFs sustain nutrient availability for approximately 1200 hours compared to the 300–350 hours duration of traditional fertilizer. Manikandan and Subramanian (2016) suggested that zeolite-encapsulated NFs enhance the efficiency of nitrogen. A novel nitrogen NF was developed using urea-coated hydroxyapatite nanorods for targeted and sustained nutrient release (Kottegoda et al. 2017).

Like nitrogen, phosphorus is also critical macronutrients essential for growth and development of plants. Phosphorus NFs offer a sustainable and efficient approach to nutrient management, surpassing traditional fertilizers in delivering essential nutrients to crops. Employing slow-release phosphorus NFs can optimize phosphorus utilization throughout the crop lifecycle, thereby conserving this essential resource (Saraiva et al. 2022). Nano-rock phosphate demonstrated comparable phosphorus utilization efficiency to superphosphate while offering cost-benefits in maize cultivation (Adhikari et al. 2014). For the synthesis of fungal-mediated P-NPs, tricalcium phosphate was utilized as a precursor by Tarafdar, Raliya, and Rathore (2012). Phosphorus NFs derived using natural phosphorite, produced using ultrasonic dispersion, demonstrate effective seed treatment benefits (Sharonova et al. 2015). Hydroxyapatite NPs coated with spherical carboxymethyl cellulose significantly enhanced soybean growth and yield by improving phosphorus availability compared to conventional soluble phosphate fertilizers (Liu and Lal 2014). Research on Ca-phosphate-based phosphorus fertilizers has highlighted the influence of ionic strength, pH, calcium-to-phosphorus ratio, and NP size on hydroxyapatite solubility (Tang and Fei 2021) and nearly doubles the nutrient use efficiency.

Magnesium also plays a crucial role in photosynthetic processes, activation of enzymes, and in synthesizing protein. Chlorophyll, the green pigment essential for photosynthesis, incorporates magnesium as a key component (Liao et al. 2021). Some of the Mg-based NFs, like Mg-oxide and Mg-sulfate, effectively enhance the growth and productivity of crops, nutritional values of vegetable and fruits, and bolster natural defenses against disease and pests. Anand et al. (2020) investigated the potential of magnesium oxide NPs to enhance mung bean seed germination. MgO NFs release free radicals. The antioxidant enzymes neutralize the free radicals and protect the chloroplast membrane (Cai et al. 2018). Delfani et al. (2014) conducted a factorial experiment involving three replications to evaluate the interactive effects of varying iron and magnesium concentrations. They observed the significant enhancement in yield, Fe content in leaves, Mg content in stem, chlorophyll content, and membrane integrity. Varamin et al. (2018) studied the impact of chitosan and Mg-based NFs under stress condition on sesame plants and observed significant enhancement in chlorophyll, proline, protein, soluble sugar, and carotenoid level, while also boosting enzyme activity.

Sulfur is also a vital macronutrient essential for numerous plant processes, including protein synthesis, activation of enzymes, vitamin production, and hormone regulation. Like Mg, sulfur also enhances plant defense against pests and disease (Ingenbleek and Kimura 2013). There are several sulfur-based NFs. Some of these are produced by encapsulation of elemental sulfur or sulfur compounds like urea and potassium sulfate within a nanomaterial matrix (Li et al. 2019). Sulfate-based NFs enhance the uptake and utilization of sulfur in groundnut, leading to improved root, shoot, kernel, and shell development as compared to conventional sulfur sources. Thirunavukkarasu et al. (2018) observed that when applied sulfur-encapsulated NFs, uptake of sulfur by root, kernel, and shoot was more compared to that of traditional fertilizer.

The microscopic size potassium nanofertilizers, often called nano potassium, allows for deeper soil penetration, facilitating direct nutrient delivery to plant roots. This results in enhanced absorption rate, leading to faster and more efficient nutrient delivery to plants compared to traditional fertilizers. Other than these, potassium-encapsulated NFs exhibit superior leaching resistance and water solubility compared to traditional potassium fertilizers, ensuring efficient nutrient retention within the soil (Sheoran et al. 2021). Foliar application of monopotassium significantly enhanced chlorophyll biosynthesis in peanuts.

1.4.2.3 Nano-biofertilizers

Biofertilizers are microbial formulations that enhance soil fertility by fixing atmospheric nitrogen, producing growth hormones, and solubilizing phosphorus. Hence, the integration of biofertilizers with NT has led to the development of nano-biofertilizers (Kumari and Singh 2019; Simarmata et al. 2016). Biofertilizers offer a sustainable and eco-friendly alternative to traditional chemical fertilizers by providing essential plant nutrients through microbial activity. The synergistic interaction between microbes and NPs, along with their controlled release and transport properties, are three crucial determinants of nano-biofertilizer efficacy. Biofertilizers encompass diverse microbial groups, including mycorrhizal fungi, phosphorus-solubilizing bacteria, nitrogen-fixing bacteria, and plant growth–promoting rhizobacteria (PGPR), but PGPR, like rhizospheric and endophytic bacteria, have been primarily explored in this domain (Ali et al. 2021) as it enhances plant growth by improving nutrient uptake, bolstering stress tolerance, and activating plant defense mechanism against pathogens (Castro et al. 2018; Redman et al. 2002; Batista et al. 2018; Waller et al. 2005). Nano-biofertilizers are fabricated by encapsulating nutrients or growth-promoting bacteria within a polymer matrix through nanoencapsulation methods to optimize nutrient uptake and minimize losses. Chitosan and Zeolite are commonly used encapsulating agents (Iqbal 2019). Gahoi et al. (2021) engineered a nanocomposite biofertilizer by encapsulating bacterial endospores and iron–carbon nanofibers coated with homoserine lactone within an activated carbon matrix and applied on legumes and nonlegumes plants. After 30 days of the experiment, significant enhancement in biomass, length of root, chlorophyll content, and protein accumulation was observed. In an investigation, *Pseudomonas putida*, when applied to gold NPs coated with PGPRs, showed no observable changes on the plant because of their harmful effect (Duhan et al. 2017) on microorganism. However, enhancement of growth parameters in *B. subtilis, Paenibacillus elgii*, and *Pseudomonas fluorescens* was noticed (Shukla et al. 2015). The emulsified hydro-repellent silica nanoscale particles have demonstrated enhanced biofertilizer delivery and extended cell longevity by preventing moisture loss (Kaushik and Djiwanti 2017). Due to the large size of microbe compared to NPs, macroscopic filters constructed from multiwalled CNTs are used for adsorption and capture of *Escherichia coli* (Srivastava et al. 2004). Gatahi et al. (2016) observed that *Ralstonia solanacearum*, a wilt pathogen, infected tomato plant when treated with nano-biofertilizers, and the infected plants exhibited enhanced wilt resistance. The fungal and nematode infectious rabi crops can be recovered by applying nanoclay-coated biofertilizer encapsulated with *Pseudomonas* sp. and *Trichoderma* sp. They also enhance the plant tolerance to abiotic stress. Nano-biofertilizer formulations leverage the unique surface chemistry of NPs, characterized by a net negative charge with both positive and hydrophobic domains (Kurdish 2019). These properties influence NP aggregation and binding to bacterial cell membranes, primarily through hydrophobic interactions (Hayden et al. 2012). Bacterial-NP interactions involve not only electrostatic force but also surface chemical reactions, including interactions with the phospholipid membrane (Palmqvist et al. 2015). Smart seeds, encapsulating specific bacterial strains, offer enhanced crop performance by reducing seeding rates. These seeds can be evenly distributed across fields and programmed to germinate under optimal environmental conditions (El-Ramady et al. 2018; Chinnamuthu and Boopathi, 2009). Nano-biofertilizer offers distinct advantages over conventional biofertilizers: (i) nanoformulation significantly enhances the longevity of biofertilizers; (ii) it also enhances biofertilizer resilience to desiccation, heat stress, and ultraviolet (UV) radiation; (iii) nano-biofertilizers offers precise control over nutrient delivery and release; (iv) nanoformulation enhances rhizosphere bacterial activity, leading to improved soil condition; (v) improves plant resistance against abiotic and biotic stresses; (vi) retains soil hydration; and (vii) enhances siderophore production (Shukla et al. 2015; Malusá, Sas-Paszt, and Ciesielska 2012).

1.5 Challenges and Future Outlook

Despite the benefits of NT, there are some demerits too, influenced by different factors, such as shape, size, chemical composition, and doses. Nanomaterials interact with organic matter and particles in soil, like altering its structure and nutrient cycling processes (Sun et al. 2021). Some studies have shown that while these nanomaterials improve nutrient delivery and promote plant growth, certain NPs, like silver and copper NPs, can have adverse effects on beneficial soil earthworms, aquatic organisms, and microbes, potentially disrupting ecological balances and negatively impacting biodiversity in agricultural ecosystems. Also, there are concerns regarding the potential accumulation of NPs in the environment and food chain, posing risks to nontarget organisms, ecosystems, and human health (Francis et al. 2024; Parada et al. 2018). In lettuce leaves, the accumulation of copper NPs negatively impacts the plant growth and photosynthesis, affects antioxidant system activity, and increases ROS accumulation (Xiong et al. 2021). Similarly, Yusefi-Tanha et al. (2024) observed size-dependent toxicity of soil-applied nCuO (copper oxide NPs) in soybean (*G. max* L.). The smaller nCuO exhibited higher inhibitory effects on photosynthetic pigments, seed yield, and nutrient accumulation, with increased root and seed Cu bioaccumulation leading to oxidative stress. Structural alterations in seed oil bodies and protein storage vacuoles were also observed during TEM analysis. In zebrafish embryos, developmental abnormalities have been observed due to oxidative stress when administering gold NPs at concentrations of 1 mg/L (Strungaru et al. 2018). Similarly, silver NPs have been found to reduce the reproduction rates and survival of *Lumbriculus variegatus* at concentrations of 1098 mg/kg (Francis et al. 2024). Another NP, CNT, can enhance seed germination and plant growth, but they may also form ROS that can potentially harm plant tissues, consequently leading to cell death. Overtime, CNT accumulation in soil also impacts the microbial diversity, population, and composition in soil (Vithanage et al. 2017). The judicious release of nanomaterials into the ecosystems and understanding their sustainability and biodegradation in environmental fate will definitely benefit all its stakeholders in the long term and will mitigate the ecological imbalances with robust risk management strategies (Wahab et al. 2024).

1.6 Conclusion

NT has emerged and exhibits great potential towards each and every aspect of modern agriculture practices and food industries. To ensure food security, various NT modalities have been developed as sustainable tools for agriculture production. This chapter discusses the drawbacks of conventional agriculture practices and advantages of NT for sustainable agriculture. Further in the preceding major sections, this chapter provides a plethora of nanomaterial fabrication techniques and subsequent utilization of these advanced nanoscale materials like NFs, nanocides, and nanosensors to ensure slow and sustained delivery of active agents, endorsing soil permeability, low rain-fastness, and higher nutrient uptake. Apart from the beneficial aspect, there are some issues also associated with NT in agriculture. Difficulty in scalability during nanomaterial synthesis is one of them. Owing to their higher reactivity, nanomaterials may also pose phytotoxicity and bioaccumulation in target crops or nearby other flora and fauna. However, as compared to controlled studies in a laboratory set-up, the physicochemical behavior of these nanomaterials in an open ecosystem are poorly investigated. Although green nanotechnologies described in this chapter are advantageous in agriculture, in few cases, synthesized nanomaterials using green methods are not green at all. As far as ecosystem health is concerned, nanotoxicity linkage with populations including humans must also be considered. All these aforementioned possible facts must be considered while designing and developing nanoscale materials for agricultural application. Additionally, cost-effective NT is the need of the hour for sustainable development in the field of agriculture, and these developing NT shall be thoroughly tested for the benefits and risks associated with their sustainability and biodegradability in the environment.

Acknowledgments

The authors are grateful to Ms. Linda S. Nixon, Vice President of Operations at NanoSynthons (Mount Pleasant, MI, USA), for providing the dendrimer graphics and editing this manuscript.

References

Abdel-Aziz, H.M.M., Hasaneen, M.N.A., and Omer, A.M. (2016). Nano chitosan-NPK fertilizer enhances the growth and productivity of wheat plants grown in sandy soil. *Spanish Journal of Agricultural Research* 14 (1): e0902. doi: 10.5424/sjar/2016141-8205.

Aboyewa, J.A., Sibuyi, N.R.S., Meyer, M. et al. (2021). Green synthesis of metallic nanoparticles using some selected medicinal plants from Southern Africa and their biological applications. *Plants* 10 (9): 1929. doi: 10.3390/plants10091929.

Adhikari, T., Kundu, S., Meena, V. et al. (2014). Utilization of nano rock phosphate by maize (*Zea mays* L.) crop in a vertisol of central India. *Journal of Agricultural Science and Technology A* 4 (5): 384–394. https://krishi.icar.gov.in/jspui/handle/123456789/36064

Adhikari, T., Sarkar, D., Mashayekhi, H. et al. (2015). Growth and enzymatic activity of maize (*Zea mays* L.) plant: solution culture test for copper dioxide nano particles. *Journal of Plant Nutrition* 39 (1): 99–115. doi: 10.1080/01904167.2015.1044012.

Aghdam, M.T.B., Mohammadi, H., and Ghorbanpour, M. (2015). Effects of nanoparticulate anatase titanium dioxide on physiological and biochemical performance of *Linum usitatissimum* (Linaceae) under well-watered and drought stress conditions. *Revista Brasileira De Botânica* 39 (1): 139–146. doi: 10.1007/s40415-015-0227-x.

Ahab, A., Rohman, F., Iskandar, F. et al. (2016). A simple straightforward thermal decomposition synthesis of PEG-covered Gd_2O_3 (Gd_2O_3@PEG) nanoparticles. *Advanced Powder Technology* 27 (4): 1800–1805. doi: 10.1016/j.apt.2016.06.012.

Ahmed, S.F., Mofijur, M., Rafa, N. et al. (2022). Green approaches in synthesizing nanomaterials for environmental nanobioremediation: technological advancements, applications, benefits and challenges. *Environmental Research* 204: 111967. doi: 10.1016/j.envres.2021.111967.

Alabada, R., Kadhim, M.M., Abbas, Z.S. et al. (2023). Investigation of effective parameters in the production of alumina gel through the sol-gel method. *Case Studies in Chemical and Environmental Engineering* 8: 100405. doi: 10.1016/j.cscee.2023.100405.

Alabdallah, N.M. and Alzahrani, H.S. (2020). The potential mitigation effect of ZnO nanoparticles on [*Abelmoschus esculentus* L. Moench] metabolism under salt stress conditions. *Saudi Journal of Biological Sciences* 27 (11): 3132–3137. doi: 10.1016/j.sjbs.2020.08.005.

Ali, M., Kim, B., Belfield, K.D. et al. (2015). Inhibition of phytophthora parasitica and *P. capsici* by silver nanoparticles synthesized using aqueous extract of *Artemisia absinthium*. *Phytopathology* 105 (9): 1183–1190. doi: 10.1094/phyto-01-15-0006-r.

Ali, M.A., Ahmed, T., Wu, W. et al. (2020). Advancements in plant and microbe-based synthesis of metallic nanoparticles and their antimicrobial activity against plant pathogens. *Nanomaterials* 10 (6): 1146. doi: 10.3390/nano10061146.

Ali, S. S., Darwesh, O. M., Kornaros, M. et al. (2021). Nano-biofertilizers: Synthesis, advantages, and applications. In: *Biofertilizers*, 359–370. Woodhead Publishing. doi: 10.1016/b978-0-12-821667-5.00007-5.

Alidoust, D. and Isoda, A. (2013). Phytotoxicity assessment of γ-Fe_2O_3 nanoparticles on root elongation and growth of rice plant. *Environmental Earth Sciences* 71 (12): 5173–5182. doi: 10.1007/s12665-013-2920-z.

Alkhader, A. (2023). Nanofertilizers as an alternative to inorganic fertilizers: a review. *African Journal of Food Agriculture Nutrition and Development* 23 (122): 23956–23974. doi: 10.18697/ajfand.122.23145.

Almutairi, Z.M. (2016). Influence of silver nano-particles on the salt resistance of tomato (*Solanum lycopersicum*) during germination. *International Journal of Agriculture and Biology* 18 (02): 449–457. doi: 10.17957/ijab/15.0114.

Yasmine, R., Ahmad, J., Qamar, S. et al. (2023). Engineered nanomaterials for sustainable agricultural production, soil improvement, and stress management: an overview. In: *Engineered Nanomaterials for Sustainable Agricultural Production, Soil Improvement and Stress Management*, 1–23. doi: 10.1016/b978-0-323-91933-3.00022-2.

Altammar, K.A. (2023). A review on nanoparticles: characteristics, synthesis, applications, and challenges. *Frontiers in Microbiology* 14: 1155622. doi: 10.3389/fmicb.2023.1155622.

An, C., Sun, C., Li, N. et al. (2022). Nanomaterials and nanotechnology for the delivery of agrochemicals: strategies towards sustainable agriculture. *Journal of Nanobiotechnology* 20 (1): 11. doi: 10.1186/s12951-021-01214-7.

Anand, K.V., Anugraga, A., Kannan, M. et al. (2020). Bio-engineered magnesium oxide nanoparticles as nano-priming agent for enhancing seed germination and seedling vigour of green gram (*Vigna radiata* L.). *Materials Letters* 271: 127792. doi: 10.1016/j.matlet.2020.127792.

Araoyinbo, A.O., Abdullah, M.M.A.B., Salleh, M.A.A.M. et al. (2018). Phase study of titanium dioxide nanoparticles prepared via sol-gel process. *IOP Conference Series Materials Science and Engineering* 343: 012011. doi: 10.1088/1757-899x/343/1/012011.

Armendariz, V., Herrera, I., Peralta-Videa, J.R. et al. (2004). Size controlled gold nanoparticle formation by *Avena sativa* biomass: use of plants in nanobiotechnology. *Journal of Nanoparticle Research* 6 (4): 377–382. doi: 10.1007/s11051-004-0741-4.

As, S., Hh, S., Ps, M. et al. (2018). Control of bacterial blight disease of pomegranate using silver nanoparticles. *Journal of Nanomedicine & Nanotechnology* 09 (03): 500. doi: 10.4172/2157-7439.1000500.

Askary, M., Amirjani, M.R., and Saberi, T. (2016). Comparison of the effects of nano-iron fertilizer with iron-chelate on growth parameters and some biochemical properties of *Catharanthus roseus*. *Journal of Plant Nutrition* 40 (7): 974–982. doi: 10.1080/01904167.2016.1262399.

Aswani, T., Reshmi, S., and Suchithra, T.V. (2019). Actinomycetes: its realm in nanotechnology. In: *Microbial Nanobionics*: Volume 1, *State-of-the-Art*, 127–140. Cham: Springer. doi: 10.1007/978-3-030-16383-9_5.

Avestan, S., Ghasemnezhad, M., Esfahani, M. et al. (2019). Application of nano-silicon dioxide improves salt stress tolerance in strawberry plants. *Agronomy* 9 (5): 246. doi: 10.3390/agronomy9050246.

Awan, S., Shahzadi, K., Javad, S. et al. (2021). A preliminary study of influence of zinc oxide nanoparticles on growth parameters of *Brassica oleracea* var italic. *Journal of the Saudi Society of Agricultural Sciences* 20 (1): 18–24. doi: 10.1016/j.jssas.2020.10.003.

Ayoub, H.A., Khairy, M., Elsaid, S. et al. (2018). Pesticidal activity of nanostructured metal oxides for generation of alternative pesticide formulations. *Journal of Agricultural and Food Chemistry* 66 (22): 5491–5498. doi: 10.1021/acs.jafc.8b01600.

Bai, Y., Chang, Y., Hussain, M. et al. (2020). Soil chemical and microbiological properties are changed by long-term chemical fertilizers that limit ecosystem functioning. *Microorganisms* 8 (5): 694. doi: 10.3390/microorganisms8050694.

Baig, N., Kammakakam, I., and Falath, W. (2021). Nanomaterials: a review of synthesis methods, properties, recent progress, and challenges. *Materials Advances* 2 (6): 1821–1871. doi: 10.1039/d0ma00807a.

Batista, B.D., Lacava, P.T., Ferrari, A. et al. (2018). Screening of tropically derived, multi-trait plant growth-promoting rhizobacteria and evaluation of corn and soybean colonization ability. *Microbiological Research* 206: 33–42. doi: 10.1016/j.micres.2017.09.007.

Bhan, S., Mohan, L., and Srivastava, C. (2014). Relative larvicidal potentiality of nano-encapsulated temephos and imidacloprid against culex quinquefasciatus. *Journal of Asia-Pacific Entomology* 17 (4): 787–791. doi: 10.1016/j.aspen.2014.07.006.

Bhandari, G., Dhasmana, A., Chaudhary, P. et al. (2023). A perspective review on green nanotechnology in agro-ecosystems: opportunities for sustainable agricultural practices & environmental remediation. *Agriculture* 13 (3): 668. doi: 10.3390/agriculture13030668.

Bhardwaj, A.K., Arya, G., Kumar, R. et al. (2022). Switching to nanonutrients for sustaining agroecosystems and environment: the challenges and benefits in moving up from ionic to particle feeding. *Journal of Nanobiotechnology* 20 (1): 19. doi: 10.1186/s12951-021-01177-9.

Brady, B., Wang, P. H., Steenhoff, V., & Brolo, A. G. (2019). Nanostructuring solar cells using metallic nanoparticles. In: *Metal Nanostructures for Photonics*, 197–221. Elsevier. doi: 10.1016/b978-0-08-102378-5.00009-x.

Buchman, J.T., Elmer, W.H., Ma, C. et al. (2019). Chitosan-coated mesoporous silica nanoparticle treatment of *Citrullus lanatus* (Watermelon): enhanced fungal disease suppression and modulated expression of stress-related genes. *ACS Sustainable Chemistry & Engineering* 7 (24): 19649–19659. doi: 10.1021/acssuschemeng.9b04800.

Cai, L., Chen, J., Liu, Z. et al. (2018). Magnesium oxide nanoparticles: effective agricultural antibacterial agent against *Ralstonia solanacearum*. *Frontiers in Microbiology* 9: 790. doi: 10.3389/fmicb.2018.00790.

Campos, E.V.R., De Oliveira, J.L., Da Silva, C.M.G. et al. (2015). Polymeric and solid lipid nanoparticles for sustained release of carbendazim and tebuconazole in agricultural applications. *Scientific Reports* 5 (1): 13809. doi: 10.1038/srep13809.

Castro, R.A., Dourado, M.N., De Almeida, J.R. et al. (2018). Mangrove endophyte promotes reforestation tree (*Acacia polyphylla*) growth. *Brazilian Journal of Microbiology* 49 (1): 59–66. doi: 10.1016/j.bjm.2017.04.002.

Chaitra, P., Kalia, A., Ahuja, R. et al. (2021). Importance of nano fertilizers in sustainable agriculture. *Environmental Sciences and Ecology: Current Research* 2 (5): 1–2. doi: 10.54026/esecr/1029.

Chang, T., Lu, Y., Yang, M. et al. (2020). Multibranched flower-like ZnO particles from eco-friendly hydrothermal synthesis as green antimicrobials in agriculture. *Journal of Cleaner Production* 262: 121342. doi: 10.1016/j.jclepro.2020.121342.

Chatterjee, S., Mahanty, S., Das, P. et al. (2020). Biofabrication of iron oxide nanoparticles using manglicolous fungus *Aspergillus niger* BSC-1 and removal of Cr(VI) from aqueous solution. *Chemical Engineering Journal* 385: 123790. doi: 10.1016/j.cej.2019.123790.

Che, J., Zhang, W., Ma, B. et al. (2022). An efficient process for recovering copper as CuO nanoparticles from acidic waste etchant via chemical precipitation and thermal decomposition: turning waste into value-added product. *Journal of Cleaner Production* 369: 133404. doi: 10.1016/j.jclepro.2022.133404.

Chhipa, H. (2016). Nanofertilizers and nanopesticides for agriculture. *Environmental Chemistry Letters* 15 (1): 15–22. doi: 10.1007/s10311-016-0600-4.

Chianese, A., Picano, A., and Stoller, M. (2021). Spinning disk reactor to produce nanoparticles: applications and best operating variables. *Chemical Engineering Transactions* 84: 121–126. doi: 10.3303/cet2184021.

Chinnamuthu, C.R. and Boopathi, P.M. (2009). Nanotechnology and agroecosystem. *Madras Agricultural Journal* 96: 17–31. https://www.cabdirect.org/cabdirect/abstract/20093323550

Cioffi, N., Torsi, L., Ditaranto, N. et al. (2005). Copper nanoparticle/polymer composites with antifungal and bacteriostatic properties. *Chemistry of Materials* 17 (21): 5255–5262. doi: 10.1021/cm0505244.

Cui, H., Yi, Q., Yang, X. et al. (2017). Effects of hydroxyapatite on leaching of cadmium and phosphorus and their availability under simulated acid rain. *Journal of Environmental Chemical Engineering* 5 (4): 3773–3779. doi: 10.1016/j.jece.2017.07.001.

Cui, S., Ling, P., Zhu, H. et al. (2018). Plant pest detection using an artificial nose system: a review. *Sensors* 18 (2): 378. doi: 10.3390/s18020378.

Cui, J., Sun, C., Wang, A. et al. (2020). Dual-functionalized pesticide nanocapsule delivery system with improved spreading behavior and enhanced bioactivity. *Nanomaterials* 10 (2): 220. doi: 10.3390/nano10020220.

Dağhan, H., Gülmezoglu, N., Köleli, N. et al. (2020). Impact of titanium dioxide nanoparticles (TiO2-NPs) on growth and mineral nutrient uptake of wheat (*Triticum vulgare* L.). *Biotech Studies* 29 (2): 69–76. doi: 10.38042/biost.2020.29.02.03.

Das, H., Singh, S.K., Priyadarshini, E. et al. (2024). Nanotechnology facilitated real time soil monitoring for optimized crop production. *Asian Journal of Soil Science and Plant Nutrition* 10 (2): 582–607. doi: 10.9734/ajsspn/2024/v10i2315.

Davarpanah, S., Tehranifar, A., Davarynejad, G. et al. (2016). Effects of foliar applications of zinc and boron nano-fertilizers on pomegranate (*Punica granatum* cv. Ardestani) fruit yield and quality. *Scientia Horticulturae* 210: 57–64. doi: 10.1016/j.scienta.2016.07.003.

Deepa, M., Sudhakar, P., Nagamadhuri, K.V. et al. (2014). First evidence on phloem transport of nanoscale calcium oxide in groundnut using solution culture technique. *Applied Nanoscience* 5 (5): 545–551. doi: 10.1007/s13204-014-0348-8.

De Oliveira, J.L., Campos, E.V.R., Da Silva, C.M.G. et al. (2015). Solid lipid nanoparticles co-loaded with simazine and atrazine: preparation, characterization, and evaluation of herbicidal activity. *Journal of Agricultural and Food Chemistry* 63 (2): 422–432. doi: 10.1021/jf5059045.

Del Carmen Orozco-Mosqueda, M., Glick, B.R., and Santoyo, G. (2020). ACC deaminase in plant growth-promoting bacteria (PGPB): an efficient mechanism to counter salt stress in crops. *Microbiological Research* 235: 126439. doi: 10.1016/j.micres.2020.126439.

Delfani, M., Firouzabadi, M.B., Farrokhi, N. et al. (2014). Some physiological responses of black-eyed pea to iron and magnesium nanofertilizers. *Communications in Soil Science and Plant Analysis* 45 (4): 530–540. doi: 10.1080/00103624.2013.863911.

Dhiman, S., Singh, S., Varma, A. et al. (2021). Phytofabricated zinc oxide nanoparticles as a nanofungicide for management of Alternaria blight of Brassica. *BioMetals* 34 (6): 1275–1293. doi: 10.1007/s10534-021-00342-9.

Dikshit, P.K., Kumar, J., Das, A.K. et al. (2021). Green synthesis of metallic nanoparticles: applications and limitations. *Catalysts* 11 (8): 902. doi: 10.3390/catal11080902.

Ditta, A. and Arshad, M. (2016). Applications and perspectives of using nanomaterials for sustainable plant nutrition. *Nanotechnology Reviews* 5 (2): 209–229. doi: 10.1515/ntrev-2015-0060.

Djanaguiraman, M., Belliraj, N., Bossmann, S.H. et al. (2018). High-temperature stress alleviation by selenium nanoparticle treatment in grain sorghum. *ACS Omega* 3 (3): 2479–2491. doi: 10.1021/acsomega.7b01934.

Dola, D.B., Mannan, M.A., Sarker, U. et al. (2022). Nano-iron oxide accelerates growth, yield, and quality of *Glycine max* seed in water deficits. *Frontiers in Plant Science* 13: 992535. doi: 10.3389/fpls.2022.992535.

Duhan, J.S., Kumar, R., Kumar, N. et al. (2017). Nanotechnology: the new perspective in precision agriculture. *Biotechnology Reports* 15: 11–23. doi: 10.1016/j.btre.2017.03.002.

Ealia, S.A.M. and Saravanakumar, M.P. (2017). A review on the classification, characterisation, synthesis of nanoparticles and their application. *IOP Conference Series Materials Science and Engineering* 263: 032019. doi: 10.1088/1757-899x/263/3/032019.

Elhamalawy, O., Bakr, A., and Eissa, F. (2024). Impact of pesticides on non-target invertebrates in agricultural ecosystems. *Pesticide Biochemistry and Physiology* 202: 105974. doi: 10.1016/j.pestbp.2024.105974.

El-Khatib, A.M., Badawi, M.S., Ghatass, Z.F. et al. (2018). Synthesize of silver nanoparticles by arc discharge method using two different rotational electrode shapes. *Journal of Cluster Science* 29 (6): 1169–1175. doi: 10.1007/s10876-018-1430-2.

El-Naggar, M.E., Abdelsalam, N.R., Fouda, M.M. et al. (2020). Soil application of nano silica on maize yield and its insecticidal activity against some stored insects after the post-harvest. *Nanomaterials* 10 (4): 739. doi: 10.3390/nano10040739.

El-Ramady, H., El-Ghamry, A., Mosa, A. et al. (2018). Nanofertilizers vs. biofertilizers: new insights. *Environment, Biodiversity and Soil Security* 2 (1): 40–50. doi: 10.21608/jenvbs.2018.3880.1029.

Erisman, J.W., Sutton, M.A., Galloway, J. et al. (2008). How a century of ammonia synthesis changed the world. *Nature Geoscience* 1 (10): 636–639. doi: 10.1038/ngeo325.

Faizan, M., Faraz, A., Yusuf, M. et al. (2018). Zinc oxide nanoparticle-mediated changes in photosynthetic efficiency and antioxidant system of tomato plants. *Photosynthetica* 56 (2): 678–686. doi: 10.1007/s11099-017-0717-0.

Faraji, J. and Sepehri, A. (2020). Exogenous nitric oxide improves the protective effects of TiO_2 nanoparticles on growth, antioxidant system, and photosynthetic performance of wheat seedlings under drought stress. *Journal of Soil Science and Plant Nutrition* 20 (2): 703–714. doi: 10.1007/s42729-019-00158-0.

Fatemi, H., Pour, B.E., and Rizwan, M. (2020). Foliar application of silicon nanoparticles affected the growth, vitamin C, flavonoid, and antioxidant enzyme activities of coriander (*Coriandrum sativum* L.) plants grown in lead (Pb)-spiked soil. *Environmental Science and Pollution Research* 28 (2): 1417–1425. doi: 10.1007/s11356-020-10549-x.

Feng, J., Chen, W., Liu, Q. et al. (2020). Development of abamectin-loaded nanoemulsion and its insecticidal activity and cytotoxicity. *Pest Management Science* 76 (12): 4192–4201. doi: 10.1002/ps.5976.

Feregrino-Perez, A.A., Magaña-López, E., Guzmán, C. et al. (2018). A general overview of the benefits and possible negative effects of the nanotechnology in horticulture. *Scientia Horticulturae* 238: 126–137. doi: 10.1016/j.scienta.2018.03.060.

Frąc, M., Hannula, E.S., Bełka, M. et al. (2022). Soil mycobiome in sustainable agriculture. *Frontiers in Microbiology* 13: 1033824. doi: 10.3389/fmicb.2022.1033824.

Francis, D.V., Abdalla, A.K., Mahakham, W. et al. (2024). Interaction of plants and metal nanoparticles: exploring its molecular mechanisms for sustainable agriculture and crop improvement. *Environment International* 190: 108859. doi: 10.1016/j.envint.2024.108859.

Gahoi, P., Omar, R.A., Verma, N. et al. (2021). Rhizobacteria and acylated homoserine lactone-based nanobiofertilizer to improve growth and pathogen defense in cicer arietinum and triticum aestivum plants. *ACS Agricultural Science & Technology* 1 (3): 240–252. doi: 10.1021/acsagscitech.1c00039.

Gan, Y.X., Jayatissa, A.H., Yu, Z. et al. (2020). Hydrothermal synthesis of nanomaterials. *Journal of Nanomaterials* 2020: 1–3. doi: 10.1155/2020/8917013.

Gardea-Torresdey, J.L., Gomez, E., Peralta-Videa, J.R. et al. (2003). Alfalfa sprouts: a natural source for the synthesis of silver nanoparticles. *Langmuir* 19 (4): 1357–1361. doi: 10.1021/la020835i.

Gardea-Torresdey, J.L., Parsons, J.G., Gomez, E. et al. (2002). Formation and growth of Au nanoparticles inside live Alfalfa plants. *Nano Letters* 2 (4): 397–401. doi: 10.1021/nl015673.

Gatahi, D.M., Wanyika, H., Kihurani, A.W. et al. (2016). Use of bio-nanocomposites in enhancing bacterial wilt plant resistance, tomato production and water conservation in greenhouse farming. In: *Scientific Conference Proceedings* 1. http://journals.jkuat.ac.ke/index.php/jscp/article/view/1234.

Gehlout, S., Priyam, A., Afonso, L., Schultze, G. A., and Singh, P. P. (2022). Application of metallic nanoparticles as agri inputs: modulation in nanoparticle design and application dosage needed. Deakin University #. In: *Nanotechnology in Agriculture and Environmental Science*, 16–54. Boca Raton, FL: CRC Press. doi: 10.1201/9781003323945-4. https://hdl.handle.net/10779/DRO/DU:21343005.v1

Ghorbanpour, M., Mohammadi, H., and Kariman, K. (2020). Nanosilicon-based recovery of barley (*Hordeum vulgare*) plants subjected to drought stress. *Environmental Science Nano* 7 (2): 443–461. doi: 10.1039/c9en00973f.

Giannousi, K., Sarafidis, G., Mourdikoudis, S. et al. (2014). Selective synthesis of Cu_2O and Cu/Cu_2O NPs: antifungal activity to yeast *Saccharomyces cerevisiae* and DNA interaction. *Inorganic Chemistry* 53 (18): 9657–9666. doi: 10.1021/ic501143z.

Golinska, P., Wypij, M., Ingle, A.P. et al. (2014). Biogenic synthesis of metal nanoparticles from actinomycetes: biomedical applications and cytotoxicity. *Applied Microbiology and Biotechnology* 98 (19): 8083–8097. doi: 10.1007/s00253-014-5953-7.

Goswami, P., Yadav, S., and Mathur, J. (2019). Positive and negative effects of nanoparticles on plants and their applications in agriculture. *Plant Science Today* 6 (2): 232–242. doi: 10.14719/pst.2019.6.2.502.

Gupta, A., Rayeen, F., Mishra, R. et al. (2023). Nanotechnology applications in sustainable agriculture: an emerging eco-friendly approach. *Plant Nano Biology* 4: 100033. doi: 10.1016/j.plana.2023.100033.

Habte, L., Shiferaw, N., Mulatu, D. et al. (2019). Synthesis of nano-calcium oxide from waste eggshell by sol-gel method. *Sustainability* 11 (11): 3196. doi: 10.3390/su11113196.

Haghighi, M., Abolghasemi, R., and Da Silva, J.A.T. (2014). Low and high temperature stress affect the growth characteristics of tomato in hydroponic culture with Se and nano-Se amendment. *Scientia Horticulturae* 178: 231–240. doi: 10.1016/j.scienta.2014.09.006.

Hao, Y., Yuan, W., Ma, C. et al. (2018). Engineered nanomaterials suppress Turnip mosaic virus infection in tobacco (*Nicotiana benthamiana*). *Environmental Science Nano* 5 (7): 1685–1693. doi: 10.1039/c8en00014j.

Hassan, S.E., Fouda, A., Radwan, A.A. et al. (2019). Endophytic actinomycetes *Streptomyces* spp. mediated biosynthesis of copper oxide nanoparticles as a promising tool for biotechnological applications. *JBIC Journal of Biological Inorganic Chemistry* 24 (3), 377–393. doi: 10.1007/s00775-019-01654-5.

Hayden, S.C., Zhao, G., Saha, K. et al. (2012). Aggregation and interaction of cationic nanoparticles on bacterial surfaces. *Journal of the American Chemical Society* 134 (16): 6920–6923. doi: 10.1021/ja301167y.

Hojjat, S.S. and Kamyab, M. (2017). The effect of silver nanoparticle on Fenugreek seed germination under salinity levels. *Russian Agricultural Sciences* 43 (1): 61–65. doi: 10.3103/s1068367417010189.

https://starpharma.com/news/view/view/318/starpharma-sells-agrochemicals-business-to-agrium-for-35m

https://www.sigmaaldrich.com/US/en/technical-documents/technical-article/materials-science-and-engineering/drug-delivery/hybridized-multi-dendrimer-platforms

Huang, B., Chen, F., Shen, Y. et al. (2018). Advances in targeted pesticides with environmentally responsive controlled release by nanotechnology. *Nanomaterials* 8 (2): 102. doi: 10.3390/nano8020102.

Huang, Z., Rajasekaran, P., Ozcan, A. et al. (2018). Antimicrobial magnesium hydroxide nanoparticles as an alternative to CU biocide for crop protection. *Journal of Agricultural and Food Chemistry* 66 (33): 8679–8686. doi: 10.1021/acs.jafc.8b01727.

Ibrahim, N. K., PhD, Al Farttoosi, H.A.K., and Field Crops Department, College of Agriculture, University of Karbala, Iraq. (2019). Response of mung bean to boron nanoparticles and spraying stages (*Vigna radiata* L.). *Plant Archives* 9 (Suppl 2): 712–715. https://www.plantarchives.org/SPL%20ISSUE%20SUPP%202,2019/126%20(712-715).pdf

Ijaz, I., Gilani, E., Nazir, A. et al. (2020). Detail review on chemical, physical and green synthesis, classification, characterizations and applications of nanoparticles. *Green Chemistry Letters and Reviews* 13 (3): 223–245. doi:10.1080/17518253.2020.1802517.

Imada, K., Sakai, S., Kajihara, H. et al. (2015). Magnesium oxide nanoparticles induce systemic resistance in tomato against bacterial wilt disease. *Plant Pathology* 65 (4): 551–560. doi: 10.1111/ppa.12443.

Ingenbleek, Y. and Kimura, H. (2013). Nutritional essentiality of sulfur in health and disease. *Nutrition Reviews* 71 (7): 413–432. doi: 10.1111/nure.12050.

Inoue, M., Ren, C.P., Sagami, H. et al. (2023). Application of TiO_2 nanotubes as bulky supports of highly active CO_2 methanation catalysts prepared via the polygonal barrel-sputtering method. *Materials Letters* 349: 134787. doi: 10.1016/j.matlet.2023.134787.

Iqbal, M. A. (2019). Nano-fertilizers for sustainable crop production under changing climate: a global perspective. *Sustainable Crop Production* 8: 1–13.

Iqbal, M., Raja, N.I., Mashwani, Z. et al. (2017). Effect of silver nanoparticles on growth of wheat under heat stress. *Iranian Journal of Science and Technology Transactions a Science* 43 (2): 387–395. doi: 10.1007/s40995-017-0417-4.

Izadiyan, Z., Shameli, K., Miyake, M. et al. (2020). Cytotoxicity assay of plant-mediated synthesized iron oxide nanoparticles using Juglans regia green husk extract. *Arabian Journal of Chemistry* 13 (1): 2011–2023. doi: 10.1016/j.arabjc.2018.02.019.

Jasim, B., Thomas, R., Mathew, J. et al. (2017). Plant growth and diosgenin enhancement effect of silver nanoparticles in Fenugreek (*Trigonella foenum-graecum* L.). *Saudi Pharmaceutical Journal* 25 (3): 443–447. doi: 10.1016/j.jsps.2016.09.012.

Jenne, M., Kambham, M., Tollamadugu, N. et al. (2018). The use of slow releasing nanoparticle encapsulated Azadirachtin formulations for the management of *Caryedon serratus* O. (groundnut bruchid). *IET Nanobiotechnology* 12 (7): 963–967. doi: 10.1049/iet-nbt.2017.0276.

Jeyasubramanian, K., Thoppey, U.U.G., Hikku, G.S. et al. (2016). Enhancement in growth rate and productivity of spinach grown in hydroponics with iron oxide nanoparticles. *RSC Advances* 6 (19): 15451–15459. doi: 10.1039/c5ra23425e.

Jiang, Y., Zhou, P., Zhang, P. et al. (2022). Green synthesis of metal-based nanoparticles for sustainable agriculture. *Environmental Pollution* 309: 119755. doi: 10.1016/j.envpol.2022.119755.

Kafle, B.P. (2019). Introduction to nanomaterials and application of UV–visible spectroscopy for their characterization. In: *Chemical Analysis and Material Characterization by Spectrophotometry*, 147–198. Elsevier eBooks. doi: 10.1016/b978-0-12-814866-2.00006-3.

Kalagatur, N.K., Ghosh, O.S.N., Sundararaj, N. et al. (2018). Antifungal activity of chitosan nanoparticles encapsulated with *Cymbopogon martinii* essential oil on plant pathogenic fungi *Fusarium graminearum*. *Frontiers in Pharmacology* 9: 610. doi: 10.3389/fphar.2018.00610.

Karpov, I., Ushakov, A., Demin, V. et al. (2019). Investigation of the residual stresses effect on the magnetic properties of CuO nanoparticles synthesized in a low-pressure arc discharge plasma. *Journal of Magnetism and Magnetic Materials* 490: 165492. doi: 10.1016/j.jmmm.2019.165492.

Kasana, R.C., Panwar, N.R., Kaul, R.K. et al. (2017). Biosynthesis and effects of copper nanoparticles on plants. *Environmental Chemistry Letters* 15 (2): 233–240. doi: 10.1007/s10311-017-0615-5.

Kaushik, S. and Djiwanti, S.R. (2017). Nanotechnology for enhancing crop productivity. In *Nanotechnology* (ed. R. Prasad, M. Kumar, and V. Kumar), 249–262. Springer eBooks. doi: 10.1007/978-981-10-4573-8_11.

Khaledian, S., Nikkhah, M., Shams-Bakhsh, M. et al. (2017). A sensitive biosensor based on gold nanoparticles to detect *Ralstonia solanacearum* in soil. *Journal of General Plant Pathology* 83 (4): 231–239. doi: 10.1007/s10327-017-0721-z.

Khatri, A. and Bhateria, R. (2023). Efficacy of nanofertilizers over chemical fertilizers in boosting agronomic production. *Nature Environment and Pollution Technology* 22 (2): 767–776. doi: 10.46488/nept.2023.v22i02.019.

Kim, M., Osone, S., Kim, T. et al. (2017). Synthesis of nanoparticles by laser ablation: a review. *KONA Powder and Particle Journal* 34: 80–90. doi: 10.14356/kona.2017009.

Kottegoda, N., Sandaruwan, C., Priyadarshana, G. et al. (2017). Urea-hydroxyapatite nanohybrids for slow release of nitrogen. *ACS Nano* 11 (2): 1214–1221. doi: 10.1021/acsnano.6b07781.

Kovalev, D.Y., Kochetov, N., and Chuev, I. (2021). Fabrication of high-entropy carbide (TiZrHfTaNb)C by high-energy ball milling. *Ceramics International* 47 (23): 32626–32633. doi: 10.1016/j.ceramint.2021.08.158.

Król, A., Railean-Plugaru, V., Pomastowski, P. et al. (2018). Mechanism study of intracellular zinc oxide nanocomposites formation. *Colloids and Surfaces a Physicochemical and Engineering Aspects* 553: 349–358. doi: 10.1016/j.colsurfa.2018.05.069.

Kumar, V., Bano, D., Mohan, S. et al. (2016). Sunlight-induced green synthesis of silver nanoparticles using aqueous leaf extract of *Polyalthia longifolia* and its antioxidant activity. *Materials Letters* 181: 371–377. doi: 10.1016/j.matlet.2016.05.097.

Kumaraswamy, R., Saharan, V., Kumari, S. et al. (2021). Chitosan-silicon nanofertilizer to enhance plant growth and yield in maize (*Zea mays* L.). *Plant Physiology and Biochemistry* 159: 53–66. doi: 10.1016/j.plaphy.2020.11.054.

Kumari, R. and Singh, D.P. (2019). Nano-biofertilizer: an emerging eco-friendly approach for sustainable agriculture. *Proceedings of the National Academy of Sciences India Section B Biological Sciences* 90 (4): 733–741. doi: 10.1007/s40011-019-01133-6.

Kumari, S.C., Dhand, V., and Padma, P.N. (2021). Green synthesis of metallic nanoparticles: a review. In: *Nanomaterials*, 259–281. Elsevier eBooks. doi: 10.1016/b978-0-12-822401-4.00022-2.

Kurdish, I.K. (2019). Interaction of microorganisms with nanomaterials as a basis for creation of high-efficiency biotechnological preparations. In: *Nanotechnology in the Life Sciences*, 259–287. doi: 10.1007/978-3-030-17061-5_11.

Law, S.S.Y., Miyamoto, T., and Numata, K. (2023). Organelle-targeted gene delivery in plants by nanomaterials. *Chemical Communications* 59 (47): 7166–7181. doi: 10.1039/d3cc00962a.

León-Silva, S., Arrieta-Cortes, R., Fernández-Luqueño, F. et al. (2018). Design and production of nanofertilizers. In *Agricultural Nanobiotechnology: Modern Agriculture for a Sustainable Future*, 17–31. Springer eBooks. doi: 10.1007/978-3-319-96719-6_2.

Li, H., Du, J., Liu, J. et al. (2016). Smart superstructures with ultrahigh pH-sensitivity for targeting acidic tumor microenvironment: instantaneous size switching and improved tumor penetration. *ACS Nano* 10 (7): 6753–6761. doi: 10.1021/acsnano.6b02326.

Li, T., Gao, B., Tong, Z. et al. (2019). Chitosan and graphene oxide nanocomposites as coatings for controlled-release fertilizer. *Water Air & Soil Pollution* 230 (7): 1–9. doi: 10.1007/s11270-0 19-4173-2.

Liao, Y., Huang, Y., Carvalho, R. et al. (2021). Magnesium oxide nanomaterial, an alternative for commercial copper bactericides: field-scale tomato bacterial spot disease management and total and bioavailable metal accumulation in soil. *Environmental Science & Technology* 55 (20): 13561–13570. doi: 10.1021/acs.est.1c00804.

Lin, D. and Xing, B. (2007). Phytotoxicity of nanoparticles: inhibition of seed germination and root growth. *Environmental Pollution* 150 (2): 243–250. doi: 10.1016/j.envpol.2007.01.016.

Lin, X., Xu, G., Li, Y. et al. (2024). Chemical fertilizers promote dissemination of ARGs in maize rhizosphere: an overlooked risk revealed after 37-year traditional agriculture practice. *The Science of the Total Environment* 941: 173737. doi: 10.1016/j.scitotenv.2024.173737.

Liu, B., Mu, L., Zhang, J. et al. (2020). $TiO_2/Cu_2(OH)_2CO_3$ nanocomposite as efficient antimicrobials for inactivation of crop pathogens in agriculture. *Materials Science and Engineering C* 107: 110344. doi: 10.1016/j.msec.2019.110344.

Liu, M., Chen, J., Li, B. et al. (2023). Preparation of microcrystalline graphite/zinc ferrite composites with enhanced and tunable electromagnetic wave absorption using a high-temperature ball milling method. *Materials Research Bulletin* 161: 112170. doi: 10.1016/j.materresbull.2023.112170.

Liu, R. and Lal, R. (2014). Synthetic apatite nanoparticles as a phosphorus fertilizer for soybean (*Glycine max*). *Scientific Reports* 4 (1): 5686. doi: 10.1038/srep05686.

Liu, X., Feng, Z., Zhang, F. et al. (2006). Preparation and testing of cementing and coating nano-subnanocomposites of slow/controlled-release fertilizer. *Agricultural Sciences in China* 5 (9): 700–706. doi: 10.1016/s1671-2927(06)60113-2.

Lumen, D., Wang, S., Mäkilä, E. et al. (2021). Investigation of silicon nanoparticles produced by centrifuge chemical vapor deposition for applications in therapy and diagnostics. *European Journal of Pharmaceutics and Biopharmaceutics* 158: 254–265. doi: 10.1016/j.ejpb.2020.11.022.

Ma, L., Su, W., Liu, J. et al. (2017). Optimization for extracellular biosynthesis of silver nanoparticles by *Penicillium aculeatum* Su1 and their antimicrobial activity and cytotoxic effect compared with silver ions. *Materials Science and Engineering C* 77: 963–971. doi: 10.1016/j.msec.2017.03.294.

Mabrouk, M., Das, D.B., Salem, Z.A. et al. (2021). Nanomaterials for biomedical applications: production, characterisations, recent trends and difficulties. *Molecules* 26 (4): 1077. doi: 10.3390/molecules26041077.

Mabrouk, M., Elkhooly, T.A., and Amer, S.K. (2021). Actinomycete strain type determines the monodispersity and antibacterial properties of biogenically synthesized silver nanoparticles. *Journal of Genetic Engineering and Biotechnology* 19 (1): 57. doi: 10.1186/s43141-021-00153-y.

Machac, P., Cichon, S., Lapcak, L. et al. (2020). Graphene prepared by chemical vapour deposition process. *Graphene Technology* 5 (1–2): 9–17. doi: 10.1007/s41127-019-00029-6.

Mahmoud, L.M., Dutt, M., Shalan, A.M. et al. (2020). Silicon nanoparticles mitigate oxidative stress of in vitro-derived banana (*Musa acuminata* 'Grand Nain') under simulated water deficit or salinity stress. *South African Journal of Botany* 132: 155–163. doi: 10.1016/j.sajb.2020.04.027.

Malandrakis, A.A., Kavroulakis, N., and Chrysikopoulos, C.V. (2021). Copper nanoparticles against benzimidazole-resistant *Monilinia fructicola* field isolates. *Pesticide Biochemistry and Physiology* 173: 104796. doi: 10.1016/j.pestbp.2021.104796.

Malhotra, S.P.K. and Alghuthaymi, M.A. (2022). Biomolecule-assisted biogenic synthesis of metallic nanoparticles. In: *Agri-Waste and Microbes for Production of Sustainable Nanomaterials*, 139–163. Elsevier eBooks. doi: 10.1016/b978-0-12-823575-1.00011-1.

Malusá, E., Sas-Paszt, L., and Ciesielska, J. (2012). Technologies for beneficial microorganisms inocula used as biofertilizers. *The Scientific World Journal* 2012: 1–12. doi: 10.1100/2012/491206.

Manikandan, A. and Subramanian, K. (2016). Evaluation of zeolite based nitrogen nano-fertilizers on maize growth, yield and quality on inceptisols and alfisols. *International Journal of Plant & Soil Science* 9 (4): 1–9. doi: 10.9734/ijpss/2016/22103.

Markus, J., Mathiyalagan, R., Kim, Y. et al. (2016). Intracellular synthesis of gold nanoparticles with antioxidant activity by probiotic *Lactobacillus kimchicus* DCY51 T isolated from Korean kimchi. *Enzyme and Microbial Technology* 95: 85–93. doi: 10.1016/j.enzmictec.2016.08.018.

Marooufpour, N., Alizadeh, M., Hatami, M. et al. (2019). Biological synthesis of nanoparticles by different groups of bacteria. In: *Nanotechnology in the Life Sciences*, 63–85. Springer. doi: 10.1007/978-3-030-16383-9_3.

Mattiello, E.M., Ruiz, H.A., Neves, J.C. et al. (2015). Zinc deficiency affects physiological and anatomical characteristics in maize leaves. *Journal of Plant Physiology* 183: 138–143. doi: 10.1016/j.jplph.2015.05.014.

McMullen, R., Mishra, A., and Slinker, J.D. (2022). Straightforward fabrication of sub-10 nm nanogap electrode pairs by electron beam lithography. *Precision Engineering* 77: 275–280. doi: 10.1016/j.precisioneng.2022.06.004.

Meena, D.S., Gautam, C., Patidar, O.P. et al. (2017). Nano fertilizers is a new way to increase nutrients use efficiency in crop production. https://bioinfopublication.org/pages/article.php?id=BIA0003483

Meidanchi, A. and Jafari, A. (2019). Synthesis and characterization of high purity Ta_2O_5 nanoparticles by laser ablation and its antibacterial properties. *Optics & Laser Technology* 111: 89–94. doi: 10.1016/j.optlastec.2018.09.039.

Meier, S., Moore, F., Morales, A. et al. (2020). Synthesis of calcium borate nanoparticles and its use as a potential foliar fertilizer in lettuce (*Lactuca sativa*) and zucchini (*Cucurbita pepo*). *Plant Physiology and Biochemistry* 151: 673–680. doi: 10.1016/j.plaphy.2020.04.025.

Mejias, J.H., Salazar, F., Amaro, L.P. et al. (2021). Nanofertilizers: a cutting-edge approach to increase nitrogen use efficiency in grasslands. *Frontiers in Environmental Science* 9: 635114. doi: 10.3389/fenvs.2021.635114.

Mekuye, B. and Abera, B. (2023). Nanomaterials: an overview of synthesis, classification, characterization, and applications. *Nano Select* 4 (8): 486–501. doi: 10.1002/nano.202300038.

Meraat, R., Issazadeh, K., Ziabari, A.A. et al. (2020). Rapid detection of *Escherichia coli* by ß-galactosidase biosensor based on ZnO NPs and MWCNTs: a comparative study. *Current Microbiology* 77 (10): 2633–2641. doi: 10.1007/s00284-020-02040-0.

Miyamoto, T. and Numata, K. (2023). Advancing biomolecule delivery in plants: harnessing synthetic nanocarriers to overcome multiscale barriers for cutting-edge plant bioengineering. *Bulletin of the Chemical Society of Japan* 96 (9): 1026–1044. doi: 10.1246/bcsj.20230147.

Mohamed, A.K.S.H., Qayyum, M.F., Abdel-Hadi, A.M. et al. (2017). Interactive effect of salinity and silver nanoparticles on photosynthetic and biochemical parameters of wheat. *Archives of Agronomy and Soil Science* 63 (12): 1736–1747. doi: 10.1080/03650340.2017.1300256.

Mozafari, A.A., Havas, F., and Ghaderi, N. (2017). Application of iron nanoparticles and salicylic acid in in vitro culture of strawberries (Fragaria × ananassa Duch.) to cope with drought stress. *Plant Cell Tissue and Organ Culture* (PCTOC) 132 (3): 511–523. doi: 10.1007/s11240-017-1347-8.

Mushtaq, T., Shah, A.A., Akram, W. et al. (2020). Synergistic ameliorative effect of iron oxide nanoparticles and *Bacillus subtilis* S4 against arsenic toxicity in *Cucurbita moschata*: polyamines, antioxidants, and physiochemical studies. *International Journal of Phytoremediation* 22 (13): 1408–1419. doi: 10.1080/15226514.2020.1781052.

Nagesha, K., Arunkumar, D., Kumar, G.M. et al. (2023). Parametric study on four station ball mill for synthesis of ultrafine powders. *Materials Today Proceedings*. doi: 10.1016/j.matpr.2023.04.360.

Nair, R., Varghese, S.H., Nair, B.G. et al. (2010). Nanoparticulate material delivery to plants. *Plant Science* 179 (3): 154–163. doi: 10.1016/j.plantsci.2010.04.012.

Namakka, M., Rahman, M.R., Said, K.A.M.B. et al. (2023). A review of nanoparticle synthesis methods, classifications, applications, and characterization. *Environmental Nanotechnology Monitoring & Management* 20: 100900. doi: 10.1016/j.enmm.2023.100900.

Noreen, S., Fatima, Z., Ahmad, S. et al. (2018). Foliar application of micronutrients in mitigating abiotic stress in crop plants. In: *Plant Nutrients and Abiotic Stress Tolerance*, 95–117. Springer eBooks. doi: 10.1007/978-981-10-9044-8_3.

Odularu, A.T. (2018). Metal nanoparticles: thermal decomposition, biomedicinal applications to cancer treatment, and future perspectives. *Bioinorganic Chemistry and Applications* 2018: 1–6. doi: 10.1155/2018/9354708.

Okada, M., Kuno, M., and Yamada, Y. (2023). Fast synthesis of monoclinic VO_2 nanoparticles by microwave-assisted hydrothermal method and application to high-performance thermochromic film. *Solar Energy Materials and Solar Cells* 255: 112311. doi: 10.1016/j.solmat.2023.112311.

Omran, B.A. (2020). Prokaryotic Microbial Synthesis of Nanomaterials (The World of Unseen). In: *Nanobiotechnology: A Multidisciplinary Field of Science. Nanotechnology in the Life Sciences*, 37–79. Cham: Springer. doi: 10.1007/978-3-030-46071-6_2.

Palmqvist, N.G.M., Bejai, S., Meijer, J. et al. (2015). Nano titania aided clustering and adhesion of beneficial bacteria to plant roots to enhance crop growth and stress management. *Scientific Reports* 5 (1): 10146. doi: 10.1038/srep10146.

Palmqvist, N.M., Seisenbaeva, G.A., Svedlindh, P. et al. (2017). Maghemite nanoparticles acts as nanozymes, improving growth and abiotic stress tolerance in *Brassica napus*. *Nanoscale Research Letters* 12 (1): 1–9. doi: 10.1186/s11671-017-2404-2.

Panikar, S.S., Cialla-May, D., Haaß, T. et al. (2023). Large-scale assembly and pattern transfer of SERS-active nanoparticles for application in drug monitoring of methotrexate in blood serum. *Vibrational Spectroscopy* 124: 103470. doi: 10.1016/j.vibspec.2022.103470.

Parada, J., Rubilar, O., Fernández-Baldo, M.A. et al. (2018). The nanotechnology among US: are metal and metal oxides nanoparticles a nano or mega risk for soil microbial communities? *Critical Reviews in Biotechnology* 39 (2): 157–172. doi: 10.1080/07388551.2018.1523865.

Paras, Yadav, K., Kumar, P. et al. (2022). A review on low-dimensional nanomaterials: nanofabrication, characterization and applications. *Nanomaterials* 13 (1): 160. doi: 10.3390/nano13010160.

Park, H., Kim, S., Kim, H. et al. (2006). A new composition of nanosized silica-silver for control of various plant diseases. *The Plant Pathology Journal* 22 (3): 295–302. doi: 10.5423/ppj.2006.22.3.295.

Parveen, K., Banse, V., and Ledwani, L. (2016). Green synthesis of nanoparticles: their advantages and disadvantages. *AIP Conference Proceedings*. doi: 10.1063/1.4945168.

Pascoli, M., De Albuquerque, F.P., Calzavara, A.K. et al. (2020). The potential of nanobiopesticide based on zein nanoparticles and neem oil for enhanced control of agricultural pests. *Journal of Pest Science* 93 (2): 793–806. doi: 10.1007/s10340-020-01194-x.

Pérez-de-Luque, A. and Rubiales, D. (2009). Nanotechnology for parasitic plant control. *Pest Management Science* 65 (5): 540–545. doi: 10.1002/ps.1732.

Periakaruppan, R., Romanovski, V., Thirumalaisamy, S.K. et al. (2023). Innovations in modern nanotechnology for the sustainable production of agriculture. *ChemEngineering* 7 (4): 61. doi: 10.3390/chemengineering7040061.

Pitambara, N., Archana, N., and Shukla, Y.M. (2019). Nanofertilizers: a recent approach in crop production. In: *Nanotechnology for Agriculture: Crop Production & Protection*, 25–58. Springer eBooks. doi: 10.1007/978-981-32-9374-8_2.

Pleskunov, P., Košutová, T., Vaidulych, M. et al. (2021). The sputter-based synthesis of tantalum oxynitride nanoparticles with architecture and bandgap controlled by design. *Applied Surface Science* 559: 149974. doi: 10.1016/j.apsusc.2021.149974.

Prakashraj, R., Vijayakumar, S., Punitha, V.N. et al. (2021). Fabricated TiO_2 nanofertilizers for foliar assimilation to enhance yield and disease resistance in *Capsicum annuum* L. *Journal of Plant Growth Regulation* 41 (8): 3387–3394. doi: 10.1007/s00344-021-10520-1.

Prathna, T.C., Lazar Mathew, N.C., Ashok, M. et al. (2010). Biomimetic synthesis of nanoparticles: science, technology & applicability. In: *Biomimetics Learning from Nature*. InTech eBooks. doi: 10.5772/8776.

Priya, N., Naveen, N., Kaur, K. et al. (2021). Green synthesis: an eco-friendly route for the synthesis of iron oxide nanoparticles. *Frontiers in Nanotechnology* 3: 655062. doi: 10.3389/fnano.2021.655062.

Pu, H., Xu, Y., Sun, D. et al. (2020). Optical nanosensors for biofilm detection in the food industry: principles, applications and challenges. *Critical Reviews in Food Science and Nutrition* 61 (13): 2107–2124. doi: 10.1080/10408398.2020.1808877.

Punitha, V.N., Vijayakumar, S., Alsalhi, M.S. et al. (2023). Biofabricated ZnO nanoparticles as vital components for agriculture revolutionization–a green approach. *Applied Nanoscience* 13 (9): 5821–5827. doi: 10.1007/s13204-023-02765-x.

Rai, M., Bonde, S., Golinska, P. et al. (2021). Fusarium as a novel fungus for the synthesis of nanoparticles: mechanism and applications. *Journal of Fungi* 7 (2): 139. doi: 10.3390/jof7020139.

Rai, V., Acharya, S., and Dey, N. (2012). Implications of nanobiosensors in agriculture. *Journal of Biomaterials and Nanobiotechnology* 03 (02): 315–324. doi: 10.4236/jbnb.2012.322039.

Raliya, R., Tarafdar, J.C., and Biswas, P. (2016). Enhancing the mobilization of native phosphorus in the mung bean rhizosphere using ZNO nanoparticles synthesized by soil fungi. *Journal of Agricultural and Food Chemistry* 64 (16): 3111–3118. doi: 10.1021/acs.jafc.5b05224.

Rathore, R. and Kaurav, N. (2022). The structural and optical properties of ZnO nanoparticles synthesized via thermal decomposition. *Materials Today Proceedings* 54: 624–627. doi: 10.1016/j.matpr.2021.10.207.

Ratso, S., Zitolo, A., Käärik, M. et al. (2021). Non-precious metal cathodes for anion exchange membrane fuel cells from ball-milled iron and nitrogen doped carbide-derived carbons. *Renewable Energy* 167: 800–810. doi: 10.1016/j.renene.2020.11.154.

Rauf, M.A., Owais, M., Rajpoot, R. et al. (2017). Biomimetically synthesized ZnO nanoparticles attain potent antibacterial activity against less susceptible *S. aureus* skin infection in experimental animals. *RSC Advances* 7 (58): 36361–36373. doi: 10.1039/c7ra05040b.

Rautela, I., Dheer, P., Thapliyal, P. et al. (2021). Current scenario and future perspectives of nanotechnology in sustainable agriculture and food production. *Plant Cell Biotechnology and Molecular Biology* 22: 99–121.

Redman, R.S., Sheehan, K.B., Stout, R.G. et al. (2002). Thermotolerance generated by plant/fungal symbiosis. *Science* 298 (5598): 1581. doi: 10.1126/science.1078055.

Rizwan, M., Ali, S., Ibrahim, M. et al. (2015). Mechanisms of silicon-mediated alleviation of drought and salt stress in plants: a review. *Environmental Science and Pollution Research* 22 (20): 15416–15431. doi: 10.1007/s11356-015-5305-x.

Rop, K., Karuku, G.N., Mbui, D. et al. (2018). Formulation of slow release NPK fertilizer (cellulose-graft-poly(acrylamide)/nano-hydroxyapatite/soluble fertilizer) composite and evaluating its N mineralization potential. *Annals of Agricultural Sciences* 63 (2): 163–172. doi: 10.1016/j.aoas.2018.11.001.

Rui, M., Ma, C., Hao, Y. et al. (2016). Iron oxide nanoparticles as a potential iron fertilizer for peanut (*Arachis hypogaea*). *Frontiers in Plant Science* 7. doi: 10.3389/fpls.2016.00815.

Safarnejad, M.R., Samiee, F., Tabatabie, M. et al. (2017). Development of quantum dot-based nanobiosensors against citrus tristeza virus (CTV). *DOAJ (DOAJ: Directory of Open Access Journals)*. https://doaj.org/article/580aece42ca14905ad2bf2638a0d0d09

Saharan, V., Kumaraswamy, R.V., Choudhary, R.C. et al. (2016). CU-Chitosan nanoparticle mediated sustainable approach to enhance seedling growth in maize by mobilizing reserved food. *Journal of Agricultural and Food Chemistry* 64 (31): 6148–6155. doi: 10.1021/acs.jafc.6b02239.

Salem, N.M., Albanna, L.S., Awwad, A.M. et al. (2015). Green synthesis of nano-sized sulfur and its effect on plant growth. *Journal of Agricultural Science* 8 (1): 188. doi: 10.5539/jas.v8n1p188.

Salem, S., Hammad, E., Mohamed, A. et al. (2022). A comprehensive review of nanomaterials: types, synthesis, characterization, and applications. *Biointerface Research in Applied Chemistry* 13 (1): 41. doi: 10.33263/briac131.041.

Sandhya, M., Ramasamy, D., Sudhakar, K. et al. (2021). Ultrasonication an intensifying tool for preparation of stable nanofluids and study the time influence on distinct properties of graphene nanofluids – a systematic overview. *Ultrasonics Sonochemistry* 73: 105479. doi: 10.1016/j.ultsonch.2021.105479.

Saraiva, R., Ferreira, Q., Rodrigues, G.C. et al. (2022). Phosphorous nanofertilizers for precise application in rice cultivation as an adaptation to climate change. *Climate* 10 (11): 183. doi: 10.3390/cli10110183.

Sarkar, M.R., Rashid, M.H., Rahman, A. et al. (2022). Recent advances in nanomaterials based sustainable agriculture: an overview. *Environmental Nanotechnology Monitoring & Management* 18: 100687. doi: 10.1016/j.enmm.2022.100687.

Shahryari, F., Rabiei, Z., and Sadighian, S. (2020). Antibacterial activity of synthesized silver nanoparticles by sumac aqueous extract and silver-chitosan nanocomposite against *Pseudomonas syringae* pv. *syringae*. *Journal of Plant Pathology* 102 (2): 469–475. doi: 10.1007/s42161-019-00478-1.

Shang, Y., Hasan, M.K., Ahammed, G.J. et al. (2019). Applications of nanotechnology in plant growth and crop protection: a review. *Molecules* 24 (14): 2558. doi: 10.3390/molecules24142558

Sharma, B., Lakra, U., Sharma, R. et al. (2022). A comprehensive review on nanopesticides and nanofertilizers—a boon for agriculture. In: *Nano-enabled Agrochemicals in Agriculture*, 273–290. Elsevier eBooks. doi: 10.1016/b978-0-323-91009-5.00026-4.

Sharma, E., Rathi, R., Misharwal, J. et al. (2022). Evolution in lithography techniques: microlithography to nanolithography. *Nanomaterials* 12 (16): 2754. doi: 10.3390/nano12162754.

Sharma, P., Ponte, F., Lima, M. et al. (2023). Plasma etching of polycarbonate surfaces for improved adhesion of Cr coatings. *Applied Surface Science* 637: 157903. doi: 10.1016/j.apsusc.2023.157903.

Sharon, M., Choudhary, A.K., and Kumar, R. (2010). Nanotechnology in agricultural diseases and food safety. *The Journal of Phytology* 2 (4): 78–82. https://www.cabdirect.org/abstracts/20113034565.html

Sharonova, N.L., Yapparov, A.K., Khisamutdinov, N.S. et al. (2015). Nanostructured water-phosphorite suspension is a new promising fertilizer. *Nanotechnologies in Russia* 10 (7–8): 651–661. doi: 10.1134/s1995078015040187.

Sheikh, Z.N., Raina, S., Bhat, U.H. et al. (2024). Deciphering the potential role of nano-based biosensors in growth, development and stress management of crop plants. *South African Journal of Botany* 167: 542–553. doi: 10.1016/j.sajb.2024.02.035.

Sheoran, P., Goel, S., Boora, R. et al. (2021). Biogenic synthesis of potassium nanoparticles and their evaluation as a growth promoter in wheat. *Plant Gene* 27: 100310. doi: 10.1016/j.plgene.2021.100310.

Shukla, S.K., Kumar, R., Mishra, R.K. et al. (2015). Prediction and validation of gold nanoparticles (GNPs) on plant growth promoting rhizobacteria (PGPR): a step toward development of nano-biofertilizers. *Nanotechnology Reviews* 4 (5): 439–448. doi: 10.1515/ntrev-2015-0036.

Shukla, G., Gaurav, S.S., and Singh, A. (2020). Synthesis of mycogenic zinc oxide nanoparticles and preliminary determination of its efficacy as a larvicide against white grubs (*Holotrichia* sp.). *International Nano Letters* 10 (2): 131–139. doi: 10.1007/s40089-020-00302-0.

Siddiqi, K.S. and Husen, A. (2016). Green synthesis, characterization and uses of palladium/platinum nanoparticles. *Nanoscale Research Letters* 11 (1): 1–13. doi: 10.1186/s11671-016-1695-z.

Siddiqui, M.H. and Al-Whaibi, M.H. (2014). Role of nano-SiO_2 in germination of tomato (*Lycopersicum esculentum* seeds Mill.). *Saudi Journal of Biological Sciences* 21 (1): 13–17. doi: 10.1016/j.sjbs.2013.04.005.

Siddiqui, M.H., Al-Whaibi, M.H., Faisal, M. et al. (2014). Nano-silicon dioxide mitigates the adverse effects of salt stress on *Cucurbita pepo* L. *Environmental Toxicology and Chemistry* 33 (11): 2429–2437. doi: 10.1002/etc.2697.

Silva, J.M.M., Feuser, P.E., Cercená, R. et al. (2023). Obtention of magnetite nanoparticles via the hydrothermal method and effect of synthesis parameters. *Journal of Magnetism and Magnetic Materials* 580: 170925. doi: 10.1016/j.jmmm.2023.170925.

Simarmata, T., Hersanti, N., Turmuktini, T. et al. (2016). Application of bioameliorant and biofertilizers to increase the soil health and rice productivity. *HAYATI Journal of Biosciences* 23 (4): 181–184. doi: 10.1016/j.hjb.2017.01.001.

Singh, A., Singh, N., Hussain, I. et al. (2017). Effect of biologically synthesized copper oxide nanoparticles on metabolism and antioxidant activity to the crop plants *Solanum lycopersicum* and *Brassica oleracea* var. *botrytis*. *Journal of Biotechnology* 262: 11–27. doi: 10.1016/j.jbiotec.2017.09.016.

Slavin, Y.N., Asnis, J., Häfeli, U.O. et al. (2017). Metal nanoparticles: understanding the mechanisms behind antibacterial activity. *Journal of Nanobiotechnology* 15 (1): 1–20. doi: 10.1186/s12951-017-0308-z.

Solunke, A., Barote, V.K., Sonawane, B. et al. (2023). Sol-gel synthesis of Fe-rich cobalt ferrite nanoparticles and influence of pH concentration. *Materials Today Proceedings* 92: 1225–1230. doi: 10.1016/j.matpr.2023.05.327.

Srivastava, A., Srivastava, O.N., Talapatra, S. et al. (2004). Carbon nanotube filters. *Nature Materials* 3 (9): 610–614. doi: 10.1038/nmat1192.

Strungaru, S., Plavan, G., Ciobica, A. et al. (2018). Acute exposure to gold induces fast changes in social behavior and oxidative stress of zebrafish (*Danio rerio*). *Journal of Trace Elements in Medicine and Biology* 50: 249–256. doi: 10.1016/j.jtemb.2018.07.013.

Su, S. S. and Chang, I. (2017). Review of production routes of nanomaterials. In: *Commercialization of Nanotechnologies–A Case Study Approach*, 15–29. Springer eBooks. doi: 10.1007/978-3-319-56979-6_2.

Sun, H., Lei, C., Xu, J. et al. (2021). Foliar uptake and leaf-to-root translocation of nanoplastics with different coating charge in maize plants. *Journal of Hazardous Materials* 416: 125854. doi: 10.1016/j.jhazmat.2021.125854.

Syu, Y., Hung, J., Chen, J. et al. (2014). Impacts of size and shape of silver nanoparticles on *Arabidopsis* plant growth and gene expression. *Plant Physiology and Biochemistry* 83: 57–64. doi: 10.1016/j.plaphy.2014.07.010.

Tang, S. and Fei, X. (2021). Refractory calcium phosphate-derived phosphorus fertilizer based on hydroxyapatite nanoparticles for nutrient delivery. *ACS Applied Nano Materials* 4 (2): 1364–1376. doi: 10.1021/acsanm.0c02921.

Taran, N.Y., Gonchar, O.M., Lopatko, K.G. et al. (2014). The effect of colloidal solution of molybdenum nanoparticles on the microbial composition in rhizosphere of *Cicer arietinum* L. *Nanoscale Research Letters* 9 (1): 1–8. doi: 10.1186/1556-276x-9-289.

Tarafdar, J.C., Raliya, R., and Rathore, I. (2012). Microbial synthesis of phosphorous nanoparticle from tri-calcium phosphate using *Aspergillus tubingensis* TFR-5. *Journal of Bionanoscience* 6 (2): 84–89. doi: 10.1166/jbns.2012.1077.

Tereshchenko, N., Zmeeva, O., Makarov, B. et al. (2017). The influence of silicon oxide nanoparticles on morphometric parameters of monocotyledons and dicotyledons in soil and climatic conditions of Western Siberia, as well as on microbiological soil properties. *BioNanoScience* 7 (4): 703–711. doi: 10.1007/s12668-017-0419-x.

Thirunavukkarasu, M., Subramanian, K., Kannan, P. et al. (2018). Response of nano-sulphur to the groundnut. *International Journal of Chemical Studies* 6 (3): 2067–2072. https://www.chemijournal.com/archives/2018/vol6issue3/PartAD/6-3-386-750.pdf

Thomas, E., Rathore, I., and Tarafdar, J.C. (2016). Bio-inspired synthesis of nitrogen nanoparticles and its application on Pearl Millet (*Pennisetum americanum* L) cv. HHB 67. *Journal of Bionanoscience* 10 (4): 300–306. doi: 10.1166/jbns.2016.1374.

Tiwari, M., Sharma, N.C., Fleischmann, P. et al. (2017). Nanotitania exposure causes alterations in physiological, nutritional and stress responses in tomato (*Solanum lycopersicum*). *Frontiers in Plant Science* 8: 633. doi: 10.3389/fpls.2017.00633.

Tomalia, D.A. (2005). Birth of a new macromolecular architecture: dendrimers as quantized building blocks for nanoscale synthetic polymer chemistry. *Progress in Polymer Science* 30 (3–4): 294–324. doi: 10.1016/j.progpolymsci.2005.01.007.

Tomar, D. and Jeevanandam, P. (2022). Synthesis of $ZnFe_2O_4$ nanoparticles with different morphologies via thermal decomposition approach and studies on their magnetic properties. *Journal of Magnetism and Magnetic Materials* 564: 170033. doi: 10.1016/j.jmmm.2022.170033.

Tombuloglu, H., Slimani, Y., Tombuloglu, G. et al. (2019). Uptake and translocation of magnetite (Fe_3O_4) nanoparticles and its impact on photosynthetic genes in barley (*Hordeum vulgare* L.). *Chemosphere* 226: 110–122. doi: 10.1016/j.chemosphere.2019.03.075.

Tseng, K., Chou, C., Liu, T. et al. (2018). Interactive relationship between silver ions and silver nanoparticles with PVA prepared by the submerged arc discharge method. *Advances in Materials Science and Engineering* 2018: 1–9. doi: 10.1155/2018/3240959.

Tudi, M., Ruan, H.D., Wang, L. et al. (2021). Agriculture development, pesticide application and its impact on the environment. *International Journal of Environmental Research and Public Health* 18 (3): 1112. doi: 10.3390/ijerph18031112.

Van Eerd, L., Turnbull, J., Bakker, C. et al. (2018). Comparing soluble to controlled-release nitrogen fertilizers: storage cabbage yield, profit margins, and N use efficiency. *Canadian Journal of Plant Science* 98 (4): 815–829. doi: 10.1139/cjps-2017-0277.

Van Nguyen, D., Nguyen, H.M., Le, N.T. et al. (2021). Copper nanoparticle application enhances plant growth and grain yield in maize under drought stress conditions. *Journal of Plant Growth Regulation* 41 (1): 364–375. doi: 10.1007/s00344-021-10301-w.

Varamin, J.K., Fanoodi, F., Sinaki, J.M. et al. (2018). Physiological response of sesame (*Sesamum indicum* L.) to application of chitosan and magnesium-nano fertilizers under irrigation cut-off in a sustainable agriculture system. *Plant Physiology* 9 (1): 2629–2639. doi: 10.22034/ijpp.2018.545665.

Vijayanandan, A.S. and Balakrishnan, R.M. (2018). Biosynthesis of cobalt oxide nanoparticles using endophytic fungus *Aspergillus nidulans*. *Journal of Environmental Management* 218: 442–450. doi: 10.1016/j.jenvman.2018.04.032.

Vithanage, M., Seneviratne, M., Ahmad, M. et al. (2017). Contrasting effects of engineered carbon nanotubes on plants: a review. *Environmental Geochemistry and Health* 39 (6): 1421–1439. doi: 10.1007/s10653-017-9957-y.

Wahab, A., Muhammad, M., Ullah, S. et al. (2024). Agriculture and environmental management through nanotechnology: eco-friendly nanomaterial synthesis for soil-plant systems, food safety, and sustainability. *The Science of the Total Environment* 926: 171862. doi: 10.1016/j.scitotenv.2024.171862.

Waller, F., Achatz, B., Baltruschat, H. et al. (2005). The endophytic fungus Piriformospora indica reprograms barley to salt-stress tolerance, disease resistance, and higher yield. *Proceedings of the National Academy of Sciences of the United States of America* 102 (38): 13386–13391. doi: 10.1073/pnas.0504423102.

Wang, J., Koo, Y., Alexander, A. et al. (2013). Phytostimulation of poplars and arabidopsis exposed to silver nanoparticles and Ag^+ at sublethal concentrations. *Environmental Science & Technology*, 47 (10): 5442–5449. doi: 10.1021/es4004334.

Wang, P., Liu, T., Zhao, S. et al. (2023). One-step ball-milling synthesis of cesium tungsten bronze nanoparticles and near-infrared shielding performance. *Ceramics International* 49 (13): 21393–21401. doi: 10.1016/j.ceramint.2023.03.268.

Wang, X., Zheng, K., Cheng, W. et al. (2021). Field application of star polymer-delivered chitosan to amplify plant defense against potato late blight. *Chemical Engineering Journal* 417: 129327. doi: 10.1016/j.cej.2021.129327.

Wirecka, R., Lachowicz, D., Berent, K. et al. (2022). Ion distribution in iron oxide, zinc and manganese ferrite nanoparticles studied by XPS combined with argon gas cluster ion beam sputtering. *Surfaces and Interfaces* 30: 101865. doi: 10.1016/j.surfin.2022.101865.

Won, J.H., Kim, M.K., Oh, H. et al. (2023). Scalable production of visible light photocatalysts with extended nanojunctions of WO3/g-C3N4 using zeta potential and phase control in sol-gel process. *Applied Surface Science* 612: 155838. doi: 10.1016/j.apsusc.2022.155838.

Wu, J., Du, C., Zhang, J. et al. (2021). Synthesis of metarhizium anisopliae–chitosan nanoparticles and their pathogenicity against *Plutella xylostella* (Linnaeus). *Microorganisms* 10 (1): 1. doi: 10.3390/microorganisms10010001.

Xiong, T., Zhang, S., Kang, Z. et al. (2021). Dose-dependent physiological and transcriptomic responses of lettuce (*Lactuca sativa* L.) to copper oxide nanoparticles—insights into the phytotoxicity mechanisms. *International Journal of Molecular Sciences* 22 (7): 3688. doi: 10.3390/ijms22073688.

Xu, J., Luo, X., Wang, Y. et al. (2017). Evaluation of zinc oxide nanoparticles on lettuce (*Lactuca sativa* L.) growth and soil bacterial community. *Environmental Science and Pollution Research* 25 (6): 6026–6035. doi: 10.1007/s11356-017-0953-7.

Yadav, A., Yadav, K., and Abd-Elsalam, K.A. (2023). Nanofertilizers: types, delivery and advantages in agricultural sustainability. *Agrochemicals* 2 (2): 296–336. doi: 10.3390/agrochemicals2020019.

Yaghoobi, M., Asjadi, F., and Sanikhani, M. (2023). A facile one-step green hydrothermal synthesis of paramagnetic Fe_3O_4 nanoparticles with highly efficient dye removal. *Journal of the Taiwan Institute of Chemical Engineers* 144: 104774. doi: 10.1016/j.jtice.2023.104774.

Yin, H., Cao, Y., Marelli, B. et al. (2021). Soil sensors and plant wearables for smart and precision agriculture. *Advanced Materials* 33 (20): 2007764. doi: 10.1002/adma.202007764.

Yu, T., Xing, A., Wang, S. et al. (2023). Growth of FeOCl nanoparticles via thermal decomposition under reduced pressure and rotation. *Journal of Crystal Growth* 609: 127147. doi: 10.1016/j.jcrysgro.2023.127147.

Yusof, H.M., Mohamad, R., Zaidan, U.H. et al. (2019). Microbial synthesis of zinc oxide nanoparticles and their potential application as an antimicrobial agent and a feed supplement in animal industry: a review. *Journal of Animal Science and Biotechnology/ Journal of Animal Science and Biotechnology* 10 (1): 1–22. doi: 10.1186/s40104-019-0368-z.

Yusefi-Tanha, E., Fallah, S., Pokhrel, L.R. et al. (2024). Role of particle size-dependent copper bioaccumulation-mediated oxidative stress on *Glycine max* (L.) yield parameters with soil-applied copper oxide nanoparticles. *Environmental Science and Pollution Research* 31 (20): 28905–28921. doi: 10.1007/s11356-024-33070-x.

Zahra, Z., Arshad, M., Rafique, R. et al. (2015). Metallic nanoparticle (TiO_2 and Fe_3O_4) application modifies rhizosphere phosphorus availability and uptake by *Lactuca sativa*. *Journal of Agricultural and Food Chemistry* 63 (31): 6876–6882. doi: 10.1021/acs.jafc.5b01611.

Zeng, J., Xuan, Y., and Li, Q. (2023). Direct solar-thermal scalable-decomposition of methanol flowing through a nanoparticle-packed bed reactor under outdoor environment. *Energy* 280: 128187. doi: 10.1016/j.energy.2023.128187.

Zhan, F., Wang, T., Iradukunda, L. et al. (2018). A gold nanoparticle-based lateral flow biosensor for sensitive visual detection of the potato late blight pathogen, *Phytophthora infestans*. *Analytica Chimica Acta* 1036: 153–161. doi: 10.1016/j.aca.2018.06.083.

Zhang, D., Ye, K., Yao, Y. et al. (2019). Controllable synthesis of carbon nanomaterials by direct current arc discharge from the inner wall of the chamber. *Carbon* 142: 278–284. doi: 10.1016/j.carbon.2018.10.062.

Zhang, M., Ma, X., Jin, G. et al. (2023). A modified method for transient transformation via pollen magnetofection in lilium germplasm. *International Journal of Molecular Sciences* 24 (20): 15304. doi: 10.3390/ijms242015304.

Zhang, Q., Ying, Y., and Ping, J. (2021). Recent advances in plant nanoscience. *Advanced Science* 9 (2): 2103414. doi: 10.1002/advs.202103414.

Zhang, X., Xu, Z., Wu, M. et al. (2019). Potential environmental risks of nanopesticides: application of $Cu(OH)_2$ nanopesticides to soil mitigates the degradation of neonicotinoid thiacloprid. *Environment International* 129: 42–50. doi: 10.1016/j.envint.2019.05.022.

Zhang, Y., Chen, W., Jing, M. et al. (2019). Self-assembled mixed micelle loaded with natural pyrethrins as an intelligent nano-insecticide with a novel temperature-responsive release mode. *Chemical Engineering Journal* 361: 1381–1391. doi: 10.1016/j.cej.2018.10.132.

Zhao, X., Cui, H., Wang, Y. et al. (2017). Development strategies and prospects of nano-based smart pesticide formulation. *Journal of Agricultural and Food Chemistry* 66 (26): 6504–6512. doi: 10.1021/acs.jafc.7b02004.

Zheng, L., Hong, F., Lu, S. et al. (2005). Effect of nano-TiO_2 on strength of naturally aged seeds and growth of Spinach. *Biological Trace Element Research* 104 (1): 83–92. doi: 10.1385/bter:104:1:083.

Zhongguan, H., Qiang, Z., Sen, L. et al. (2023). Cost-effective one-spot hydrothermal synthesis of graphene oxide nanoparticles for wastewater remediation: AI-enhanced approach for transition metal oxides. *Chemosphere* 337: 139064. doi: 10.1016/j.chemosph.

Zwar, I.P., Trotta, C.D.V., Ziotti, A.B.S. et al. (2022). Biosynthesis of silver nanoparticles using actinomycetes, phytotoxicity on rice seeds, and potential application in the biocontrol of phytopathogens. *Journal of Basic Microbiology* 63 (1): 64–74. doi: 10.1002/jobm.202200439.

2

Effect of Nanocomposites on Sustainable Growth of Crop Plants and Productivity

Katina Chachei[1], *Sonali Ranjan*[2], *Kirpa Ram*[1], *and Ram Sharan Singh*[2]

[1] *Analytical Environmental and Chemistry Laboratory (AECL), Institute of Environment and Sustainable Development, Banaras Hindu University, Varanasi, Uttar Pradesh, India*
[2] *Chemical Engineering and Technology, Indian Institute of Technology, Banaras Hindu University, Varanasi, Uttar Pradesh, India*

2.1 Introduction

The surge in the world population with an estimate of ~10 billion by 2050 has posed a challenge to global food security and the sustainable use of resources for agricultural production (Ahmed et al. 2023). Over the years, we have witnessed crop yield reduction due to biotic and abiotic factors leading to potential health risks and food security. Abiotic stress contributes to the substantial loss of crop production with an estimate of at least 50% crop loss, consisting of heat stress, drought, salinity, and other abiotic pressures (Figure 2.1; Kajla et al. 2015). The total number of spikes/plant and grain/spike count and weight of thousand grains are reduced during drought stress (Das et al. 2023). In addition, abiotic stresses trigger various responses in plants, such as growth rate, productivity, cell metabolism, and alter gene expression (Shinde, Kawar, and Dalvi 2024). The yield of crops is also largely affected by biotic factors, such as weeds, pests, and pathogens, accounting for up to 20–40% of crop loss globally (Popp, Pető, and Nagy 2013; Zhang and Goss 2022). Consequently, crop productivity is diminishing everyday due to pests, diseases, loss of fertile land, unprecedented weather conditions, etc. (Mirbolook 2024). To address these challenges, conventional agrochemicals are widely used, but their effectiveness is reduced as a substantial amount is lost through leaching, surface runoff, and volatilization (Figure 2.2). Despite numerous efforts, the nutrient use efficiency (NUE) of traditional fertilizer remains low, around 30–35% for nitrogen (N), 18–20% for phosphorus (P), and 35–40% for potassium (K; Shang et al. 2019). Approximately 95% of additionally supplied micronutrients, such as zinc (Zn), iron (Fe), copper (Cu), and molybdenum (Mo), are also lost to the environment (Nagaraju et al. 2022). Further studies have shown that more than 99.9% of pesticides fail to be delivered at specific sites, which leads to adverse effects on the health of the soil, water, and air and increased resistance in pathogenic species (Pimentel 1995).

Nanotechnology-based Sustainable Agriculture, First Edition. Edited by Pardeep Singh et al.

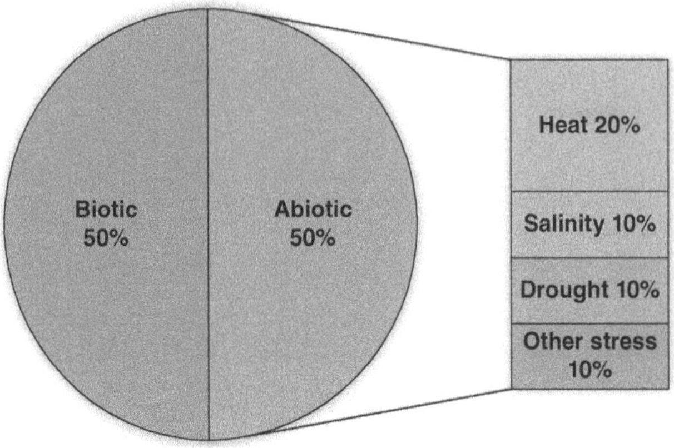

Figure 2.1 Contribution of biotic and abiotic factors to crop losses (%) and the share by various abiotic stresses.

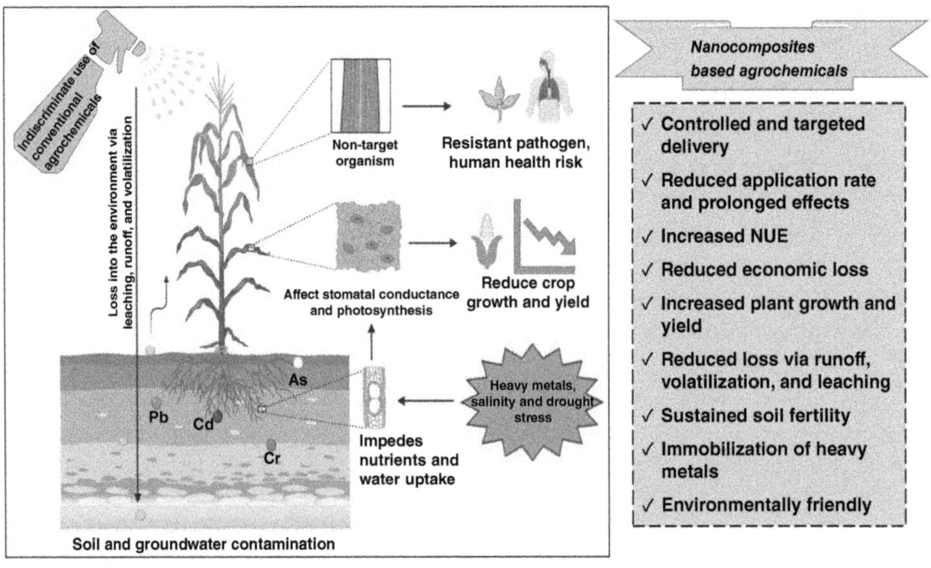

Figure 2.2 The indiscriminate application of conventional agrochemicals results in numerous environmental consequences compared to nanocomposite-based agrochemicals.

The combined effect of the green revolution in the 1960s resulted in a doubling of food over the last 50 years (Popp et al. 2013). Similarly, nanotechnology has immensely benefited the agricultural sector today and is considered the sixth most revolutionary technology in modern times (Knell 2011). The use of nanoagrochemicals is becoming progressively more popular in the agro-industrial sector. It has gained attention as a potential application in agricultural science, particularly in tackling climate change, enhancing NUE, and abiotic stress management (Manzoor and Ali 2022). There are different classes of nanotechnology based on their types of materials, shapes, and sizes, such as nanoparticles (NPs), nanomaterials, nanofibers, and nanocomposites (NCs). Out of these, NCs have emerged as the most popular among researchers and industrialists. NCs are hybrid materials produced by combining two or more different materials with unique physicochemical properties, where at least one phase has dimensions on a nanometer scale (<100 nm). In NC, each constituent matrix influences the bulk properties of the aggregate material. They emerged as the most promising class due to their biodegradability, hydrophilic nature, and capacity to achieve sustainable development goals (SDGs) like SDG12 (sustainable production and consumption patterns) and SDG14 (sustainable use of water resources); (Chakraborty et al. 2023).

NC plays an important role in the controlled-release (CR) system, an emerging nano-based technology that has attracted worldwide attention, particularly in agriculture, food, pharmaceuticals, and cosmetics (Guha et al. 2020). It is reported that agrochemicals account for about 77% of the CR nanoformulations used (Camara et al., 2019). In agriculture, a CR system is the regulated transfer of active ingredients to a specific site, to provide precise agrochemical inputs for a longer duration (Guha et al. 2020). In contrast to traditional agrochemicals, this system reduces the application rate of agrochemicals, resulting in lesser phytotoxicity, higher NUE, and a mitigated impact on the environment. The nutrients in the CR system are released through diffusion, burst release, degradation, etc. Usually, natural biopolymers, particularly polysaccharides like alginate, starch, cellulose, pectin, and chitosan, are used as vehicles to encapsulate or absorb active ingredients to elaborate CR systems for agrochemicals. They play a key role in crop productivity with efficient delivery of agrochemicals and genetic material (Mallakpour and Hussain 2021). However, compared to the development of nanotechnology in various fields such as medicine and cosmetics, the progress of nanotechnology, particularly NCs, in agriculture is still inadequate, although the momentum is changing at the moment. There has been an increase in studies on the utility of NC as a potent source of agrochemicals showing enhanced properties over traditional agrochemicals. This chapter discusses the recent development and research conducted in the application and effects of NCs as fertilizer, pesticides, herbicides, sensors, etc. Along with its advantages, we have also highlighted the safety concerns that may arise in the environment due to unregulated applications of NPs. Finally, we highlight the challenges impeding its widespread use and adoption and its future prospects.

2.2 Types of NCs and Its Uptake Through Roots and Leaves in Plants

NCs are composed of hybrid materials from the mixtures of polymers and inorganic solids (clays to oxides) with superior properties compared to bulk materials. They are broadly classified as polymer-based NC and non-polymer NC (Figure 2.3). The non-polymer NC can be further categories into three classes (Sen 2020):

1. Metal-based NC: they are formed either by alloy or core–shell structure characterized by low melting point and enhanced magnetic and electrical properties.
2. Ceramic-based NC: they have more than one solid phase where one of it is in nanometric scale (<50–100 nm) characterized by magnetic, chemical, optical, and mechanical properties.
3. Ceramic-ceramic-based NC: they are multifunctional materials made of nonmetallic components. They are employed in catalysis, photodegradation of dyes, and biomedical applications.

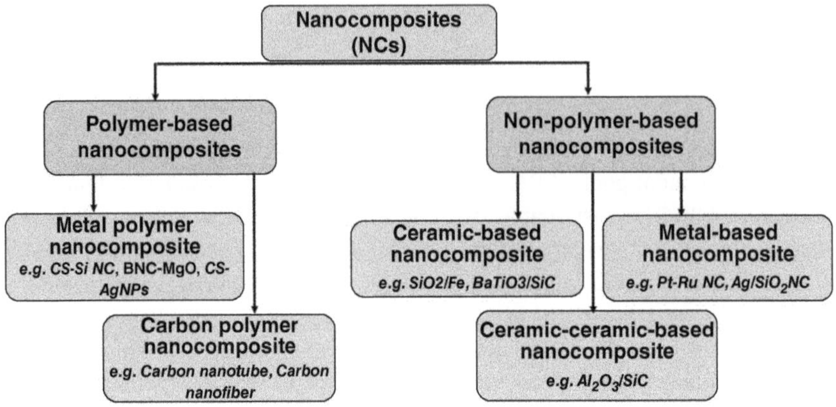

Figure 2.3 Types of nanocomposites. *Source:* Sen, 2020 / IntechOpen / CC BY 3.0.

The polymer NCs contain NPs or nanofillers dispersed in the polymer matrix. Among other NC, polymer-based nanomaterials have gained wide attention in current research and development. They are broadly grouped into two classes: metal-based polymer and carbon-based polymer.

2.2.1 Metal-based Polymer Composites

Various metallic nanomaterials, such as Cu, Fe, titanium dioxide (TiO_2), silicon dioxide (SiO_2), and zinc oxide (ZnO), are widely applied for crop growth and protection. However, they tend to build up in the roots and decrease translocation to the aerial parts, in addition to their difficulty in direct application to land due to instability and tendency to agglomerate (Ashfaq et al. 2019). Thus, hybrid nanomaterials with nanomaterials as a filler and a polymeric matrix have been proven to have better biocompatibility, CR, stability, and reduced toxicity. Metal-based polymer composites are synthesized by using metal/metal oxide encapsulated with polymers to produce the desired material. Improving material properties through filler–polymer interactions at the interface and ensuring the even distribution of nanomaterials throughout the polymer matrix.

2.2.2 Carbon-based Polymer Composites

Significant advancement has been achieved in the production of carbon-based nanomaterials, such as graphene, fullerenes, carbon nanotubes (CNTs), carbon nanofibers (CNFs), and activated carbon (Ashfaq et al. 2019). Nanotubes and CNFs are known to have the potential to penetrate seed coat and translocate substances from root to shoot and leaves. The CNFs–metal NC helps slow the release of metal NPs, such as Cu, Zn, and Fe, which can act as micronutrients for the plants and enhance water uptake and germination rates.

Most agrochemical-based NCs are supplied to the plants by foliar or soil application. Upon application of NCs via foliar, it reaches the shoot and leaves of the plants via stomata, hydathodes, or any injury, while NCs in soil can get into the plants via root hairs, lenticels, etc. In the root cells, the NCs entry may take place via apoplastic, i.e., through cell walls and symplastic pathways and through plasmodesmata connections into the

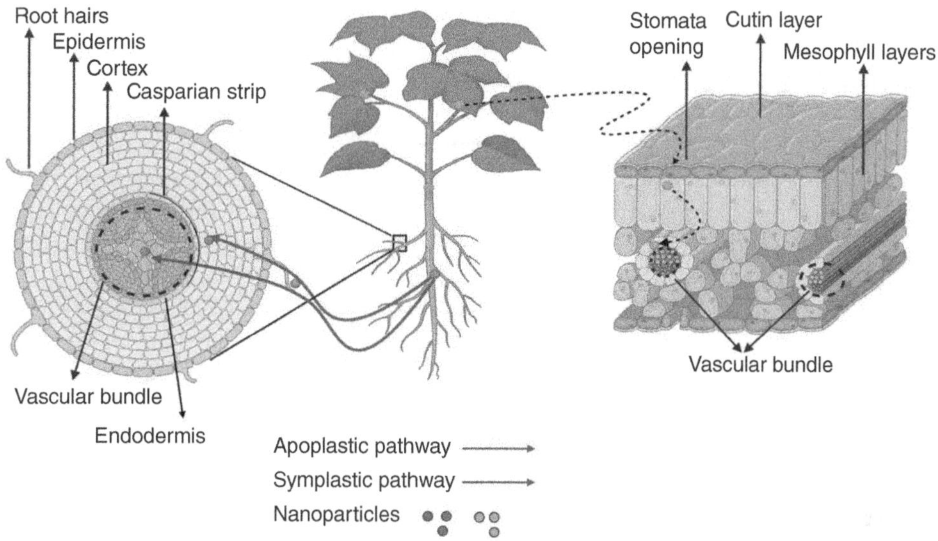

Figure 2.4 Uptake of nanoparticles through the roots and leaves of plants.

cytoplasm of the cell (Figure 2.4; Chakraborty et al. 2023). Nanomaterials may also accumulate at the Casparian strip, which blocks further entry by the apoplastic pathway unless translocated via a symplastic pathway toward the shoot and root. However, once it reaches the root vascular bundle, the particles get translocated to the aerial parts of the plants. The interaction between plants and NCs, such as accumulation, uptake, and translocation, also rely on a variety of factors, such as dimensions, size and shape, surface charge, stability, chemical nature, and functional group (Ashfaq et al. 2019). Among these, surface charge and size tend to influence the absorption and movement of micronutrients and NPs. The negatively charged nanomaterial is favored for translocation within the plants as the plant cell walls are also negatively charged. However, positively charged nanomaterials tend to attract the cell wall and accumulate within the region where it is applied, particularly in the root, leading to higher concentrations and phytotoxicity. The size of the nanomaterials in the range of 20–200 nm favors uptake and translocation inside the plants. The movement from plant root to shoot and leaves takes place through the cell wall network and plasmodesmata. Their movement may be facilitated by one of the driving forces, such as capillary action and osmotic forces. The alteration of nanomaterials by the addition of functional groups and coating may enhance the adsorption and accumulation potential inside the plants.

2.3 Application and Effects of NCs in Plant Development and Productivity

NCs are widely utilized for monitoring and enhancing the crop life cycle due to their intrinsic unique properties, such as thermal stability and electrical, chemical, and mechanical strength (Kumar et al. 2021). They offer solutions that minimize nutrient losses, lower agrochemical dosages, extend effectiveness, provide low-cost production, boost crop yields by improving nutrient uptake, and ensure health protection (Menossi, Casalongué, and Alvarez 2022). Therefore, they are widely used in fertilizers for controlled release of nutrients, pesticides against plant diseases, NC sensors for monitoring and managing crops, and chitosan- and biochar-based NCs (BNCs) for encapsulating NPs for seed priming and improving soil physicochemical properties (Figure 2.5). Various studies have investigated the positive effects (Table 2.1) and negative aspects (Table 2.2) of the application of NCs in the plant growth and development, including its yield. In the following sections, we have presented each one of the applications in detail.

2.3.1 Positive Effects on the Application of NCs in Plant

Various positive effects of NCs include enhanced rates of germination, increased vegetative biomass of seedlings in plants, improved root and shoot lengths and ratios in crops like maize and wheat, and increased photosynthetic activity and nitrogen metabolism in a few crops (Figure 2.6; Singh Sekhon 2014). The nanomaterials component of NC can originate from both inorganic and organic sources. Among inorganic materials, Zn and metal oxide NPs, such as ZnO, silver oxide (Ag_2O), and TiO_2, are extensively used and described. Zinc serves as a regulatory cofactor in protein synthesis and is necessary for the synthesis of tryptophan, enzymes, cell division, maintaining the integrity of membrane structures, and photosynthesis (Rahman et al. 2022). Zn treatment can boost rice's chlorophyll content and N absorption, resulting in increased biomass production (Parashar et al. 2023). Fe is vital in plants for chlorophyll synthesis, respiration, and N fixation. Due to its phyto-unavailability form, crops frequently suffer from Fe deficiency. Similarly, Cu is also a micronutrient needed in trace amounts but crucial for overall plant growth. Nickel (Ni) is necessary at the ideal

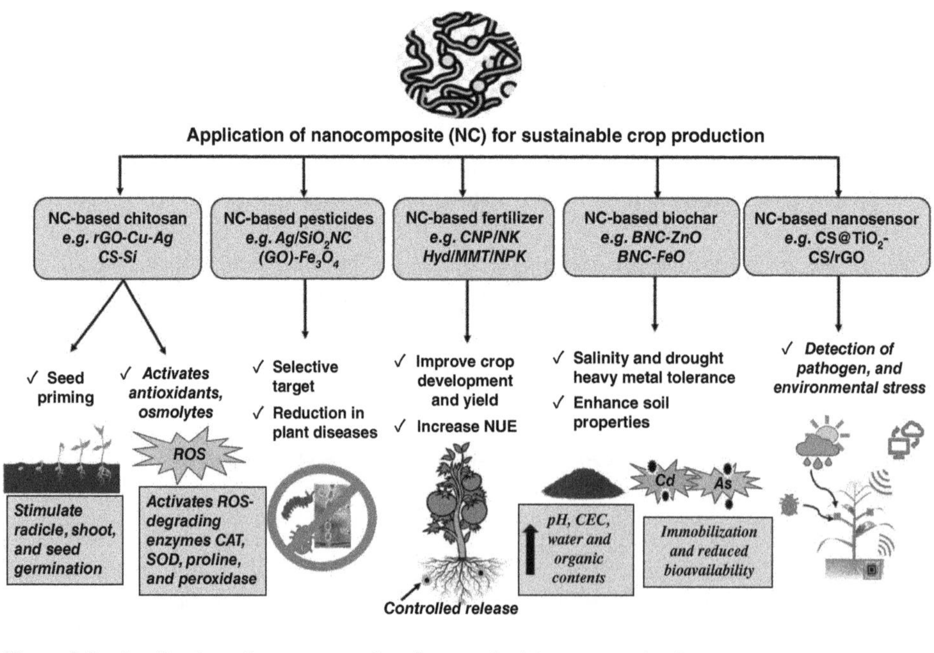

Figure 2.5 Application of nanocomposites for sustainable crop production.

Table 2.1 Positive effects of nanocomposites on the growth and productivity of plants.

Nanocomposites	Crop	Positive effects	References
Urea was coated individually and also in combination with different nanocomposites like calcium phosphate (CP), gypsum powder (GP), calcium nitrate (CN), and gelatin (G) for CR of N	Sorghum	Maximum increase in plant height, chlorophyll, dry matter yield, N uptake, and apparent N recovery was observed in calcium nitrate-gelatin (CNG).	Khan et al. (2021)
Zeolite-based nanocomposite for N, P, and K delivery	*Ipomoea aquatica*	Higher N, P, and K uptake by plants compared to conventional urea.	Guha et al. (2020)
Plants with Mg and Fe deficiencies were supplied with liposomes (100 nm) loaded with Mg or Fe	Tomato (*Solanum lycopersicum*)	Display normal turgor pressure in the leaves. Recover from chlorosis and epinasty, nutrient deficiency, and activate plant growth mechanisms.	Karny et al. (2018)
pH-dependent ferrous fertilizer with attapulgite coating	Maize (*Zea mays*)	Enhanced plant growth and chlorophyll pigment contents.	Wang et al. (2016)
Zinc-oxide-multiwalled carbon nanotubes (ZnO)/MWCNT nanocomposite compared with MWCNT, ZnNPs in arid region	Onion (*Allium cepa*)	15 μg/mL (ZnO)/MWCNT nanocomposite showed highest germination compared to 2, 5, 10, 20, and 40 μg/mL. Enhanced growth of root, shoot lengths, biomass, and germination percentage in less water in ZnO/MWCNT medium.	Kumar et al. (2018)
Chitosan (Ch) and three chitosan–metal NCs (Ch-CuO, Ch-ZnO, Cu-Ag) and fungicide copper-oxy-chloride (CuOCl) as nanoagrochemicals against *Fusarium oxysporum* f. sp. *ciceri*	Chickpea (*Cicer arietinum* L.)	Ch-CuO and Ch-ZnO were the most effective at different levels, i.e. 50, 100, and 200 μg/mL. At 100 μg/mL, the disease reduction was 46.67% in Ch-CuO, 40% in Ch-ZnO, and 33.33% in Ch-Ag and Ch.	Kaur, Duhan, and Thakur (2018)
Biochar-based magnesium oxide and manganese oxide NCs under salt stress	Safflower (*Carthamus tinctorius* L.)	Decreased the sodium absorption rate than non-biochar treatment, by 28% and 27% under 6 and 12 dS/m saline state respectively. Pronounced growth of lateral root compared to the main root accelerated the flowering time and enhanced plant biomass, seed yield, oil yield, etc..	Ghassemi-Golezani and Farhangi-Abriz (2021)

(Continued)

Table 2.1 (Continued)

Nanocomposites	Crop	Positive effects	References
Graphene oxide–nanometal composite (Ag and Cu) against *Xanthomonas euvesicatoria*	Tomato and pepper	50 and 500 μg/mL of the nanocomposite resulted in lower infection against bacterial spot by 15.6 and 16.7%, respectively, compared to the commercial copper-based bactericide Kocide® 2000.	Byteŝníková et al. (2022)
Silver/silicon dioxide nanocomposite (Ag/SiO_2 NC) against *Botrytis cinerea*	French beans (*Vicia faba* L.)	Act as a growth promoter and effective antifungal by increasing the total phenolic content and defense enzymes (peroxidase and polyphenol oxidase activity).	Baka and El Zahed (2022)
Biochar-based nanocomposite with manganese dioxide (BC-MnO_2) and zinc oxide (BC-ZnO) on the immobilization and bioavailability of HMs (Pb, Cd, Zn, and Ni)	Ryegrass (*Lolium perenne* L.)	Increased shoots and roots dry weight compared to control. Higher reduction in HMs' uptake compared to biochar alone.	Ghandali et al. (2024)
Chitosan–silicon nanofertilizer (CS-Si NF) compared to SiO_2	Maize	0.04–0.12% (w/v) of the nanocomposite showed ~3.7 times higher SVI than SiO_2. 0.08% of the NC resulted in 43.4% higher yield/plot. 0.04% of the nanocomposite gave 45% higher test weight.	Kumaraswamy et al. (2021)
Nickel chitosan nanoconjugate (NiCNC) against *Fusarium* rot of wheat, compared with chitosan nanoparticles (CNPs) and commercial fungicide Mancozeb.	Wheat	0.04 mg/mL NiCNC showed total elimination of the pathogen. Damages and disrupts the fungus membrane surface. 83.33% reduction in rot disease incidence in wheat seedlings.	Chouhan et al. (2022)
Chitosan MMT NC hydrogel as slow-released fertilizer	—	Total P and K release increased from 22.0% to 94.9% and 9.6% to 31.4% over pure chitosan gel.	Dou et al. (2022)
Gold NP-based sensors and carbon nanotube		Detect and monitor concentration of insecticides like dichlorodiphenyltrichloroethane (DDT) and methyl parathion.	Nagaraju et al. (2022)

amount for wheat seedlings and increase in chlorophyll content (Chouhan et al. 2022). It is also an essential component of various metalloenzymes, such as urease, superoxide dismutase activity (SOD), and acetyl coenzyme A synthase. Therefore, its deficiency affects

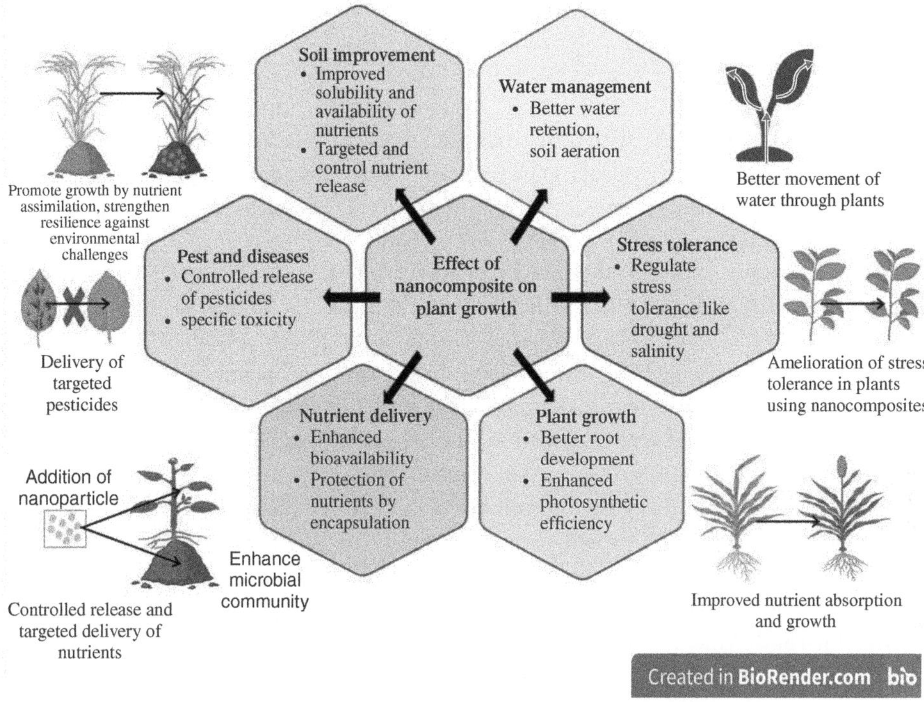

Figure 2.6 Positive effects of nanocomposites on plant growth and development.

their activity, including N metabolism and Fe uptake. Furthermore, AgNPs have a wide range of antifungal activity against diverse fungal species, like *Candida, Aspergillus*, and *Fusarium*. Its use as an antifungal is particularly promising as it is less toxic to humans and specific target to fungal cells without harming other cells.

2.3.1.1 NC-based Fertilizer

The application of fertilizers is inevitable for enhancing agricultural productivity. It is estimated that the need for macronutrient fertilizers will cross 263 Mt by 2050 (Jakhar et al. 2022). However, the indiscriminate use of fertilizers leads to various issues, including soil acidification, emission of greenhouse gases, water eutrophication, groundwater pollution, health risks, and economic losses. Besides that, CFs are reported to have poor NUE with only ~40% of applied get utilized by plants and the rest is lost to the surrounding region resulting in an imbalance of soil nutrients (Kumaraswamy et al. 2021). The NUE also depends on various factors, such as precipitation, temperature, irradiance, pathogens, and agronomic practices like tillage, drainage, crop rotation, soil type, and crop genotypes (Chakraborty et al. 2023). Therefore, it is time to develop fertilizers that are target-specific, less toxic, cost-effective, and have higher NUE for sustainable crop productivity and yield. Nanocomposite-based plant fertilizers (NCPFs) have emerged as a viable option and sustainable solution with its unique characteristic that offers many benefits to the crop, farmers, and the environment (Shang et al. 2019). The slow- and controlled-release nanofertilizers (SRNF and CRNF) are more effective compared to traditional nanofertilizers as they control the delivery of the active ingredient as per the requirement of the crop. NCPFs can improve the overall well-being of crop's morphological, biochemical, and physiological indices. They enhance photosynthetic activity, nutrient uptake efficiency, phytohormone regulation, and plant defense mechanism. Under stress conditions, nanofertilizers may lead to a reduction in oxidative stress, which improves the stress tolerance of the crop; activate the production of various biochemicals, such as proline content and chlorophyll; regulate salt levels; and reduce malondialdehyde (MDA) and hydrogen peroxide (H_2O_2) accumulation (Shalaby et al. 2022).

A study developed NCs composed of magnetic carbon nanofibers (Fe-CNFs) and acylated homoserine lactones (AHLs), which acted as fertilizer as well as plant defense regulator in chickpea (*Cicer arietinum*) plants (Gupta, Kumar, and Verma 2019). Treatment of seeds with 1.71 mg/L NC for three days resulted in a significant ($p < 0.01$) increase in the germination index, vigor index compared to the control, and AHLs alone treated seeds. Furthermore, when seeds exposed to NC were grown under oxidative (5 mM H_2O_2) and salinity (200 mM NaCl) stresses, normal germination rates were observed, while other treatment types showed decreased germination rates. In addition, Fe uptake, plant biomass, total chlorophyll, and protein were reported to be higher than control. A similar study conducted recently also showed a pH-responsive dual pesticide–fertilizer NC (Wang et al. 2023). The NC was fabricated using copper ions (Cu^{2+}), ZnO, and fungicides prochloraz (Pro) and hymexazol (Hym) into mesoporous silica nanocarriers (Cu-MSNs). This NC has a great response to acid levels and slowly releases the components in three days ensuring proper absorption by crop, effective against *Rhizoctonia solani*, increased chlorophyll and cellulose content, as well as low environmental toxicity.

Another study utilized agricultural waste biochar (BC) as carbon NPs with macronutrient N and K to obtain CNP/NK NC for common beans (*Phaseolus vulgaris* L.; Salama et al. 2021). The NC was used as a foliar fertilizer for the plants 25 days after sowing at 0, 10, 20, 30, and 40 mg/L and their development and yield were monitored for two subsequent seasons. At 20 mg/L concentration of NC, significant growth in the height of the shoot, root, number of leaves and flowers, and yield, including the total phenol and flavonoid contents, was observed.

A mixed nanocomposite (MNC) fertilizer was studied for its influence on rice's nutritional and antioxidant properties as well as endophytes (Rahman et al. 2022). Several endophytes like *Penicillium* spp., *Aspergillus* spp., *Rhizopus* spp., and *Fusarium* spp. known to improve plant nutrient uptakes were identified in MNC-treated fertilizer. As a result, there was a significant ($p < 0.05$) increase in growth parameters like shoot length, height of plant, and biochemical parameters like proteins, chlorophyll pigments, carbohydrates, vitamin A, and fibers compared to CF. Another study showed that NC composed of extruded thermoplastic starch/urea (TPSUr) blends as a matrix in hydroxyapatite particles (Hap) was able to control the release of urea for a longer duration and also increase P availability for plants (Giroto et al. 2017).

For a sustainable agroecosystem, the role of microorganisms for plant growth and productivity is inevitable. Recently, there has been an emphasis on biofertilizers, which have mutual benefit to microbes, crops, and soil. In an interesting study, a multipotent micronutrient solubilizing bacterium (MSB) was formulated with soya chunks (carrier) and cerium oxide (CeO_2) as a micronutrient, to observe the NC effect on fenugreek (Sonali et al. 2022).

Increase in the shoot and root length was observed at just 75 ppm of CeO_2 in NC, which also further minimized the build-up of Ce in soil. This study thus shows the ability of a sustainable NC with MSB bioinoculants for application in micronutrient-deficient areas for enhancing yield. Thus, the application of NCPFs not only boosts crop growth and yield but also stimulates soil properties, including microbial community. Use of nano-chitosan-urea composite (NCUC) resulted in a significant increase in soil enzymes such as dehydrogenase, organic carbon content, and the availability of K compared to conventional urea (CU). Furthermore, the microbial group, namely, ammonia oxidizing bacteria, which influences the nitrification process, was lower in the NCUC than in CU indicating a slow release process of NH_4^+ and NO_3^- in the soil (Kondal et al. 2021).

A novel slow-release NPK fertilizer was formulated consisting of montmorillonite (MMT) within the superabsorbent matrix alginate-g-poly(acrylic acid-co-acrylamide) (Hyd/MMT/NPK NC); (Rashidzadeh and Olad 2014). The rate of NC releasing fertilizer was 14.66% on day one, 28.54% in a week, and 57.66% in a month indicating that the NC displays excellent slow-release properties of NPK. In addition, the water retention ratio of soil with the NC was also higher with 90.44% and 78.09%, while the soil without the NC was 71.64% and 53.27% on the 15th and 30th day, respectively. It was also observed that administration of chitosan and MMT nanoclay NC fertilizer led to a daily release of around 3%, with over 50% of the total loaded fertilizer released in 15 days (Dou et al. 2022). In contrast, almost 90% of the CF was washed out during the first irrigation, which resulted in significant economic loss and posed detrimental impacts on the environment. Therefore, the development and widespread adoption of a cheap, eco-friendly and CR fertilizer is necessary for long-term crop productivity and sustainable agriculture.

2.3.1.2 NC-based Pesticide: Fungicide, Bactericide, and Herbicide

Plants are prone to biotic stresses, such as fungi, bacteria, moths, and viruses, which hamper their development, growth, and yield, hence necessitating the use of pesticides for effective and large-scale production. However, the application of conventional pesticides leads to a variety of problems, including contamination of surroundings, harm to nonspecific species, the emergence of resistant pathogens, bioaccumulation via the food chain, and adverse effects on humans. The application of NC-based pesticides for pest management has proven to be quite successful not only in warding off pest but also in protecting the nontarget species and the environment. Nanomaterials used in pesticide formulation show essential characteristics for sustainable agro-environmental systems, such as increased stiffness, permeability, heat stability, solubility, crystallinity, and biodegradability (Shang et al. 2019).

Plant cell walls are the primary defense and their degradation causes susceptibility to pathogens. Fungus encodes pectinases or pectin-degrading enzymes that act as a carbon source for pathogens to further penetrate and colonize. Fungi infections are reported to cause substantial loss to crops. For example, chocolate spot disease in *Vicia faba* caused by *Botrytis cinerea* is considered to be one of the most serious fungal infection and can result in a loss of 60–80% among vulnerable species and up to 34% among resistant species in Africa (Omar 2021). A study reported that this disease caused a reduction in root biomass, length, density, and root/shoot ratio relative to the control plant. However, the application of Ag/SiO_2 NC was found to increase the root biomass, length, and root/shoot ratio, including enhanced crop yield (Baka and El Zahed 2022). Furthermore, at 40 ppm of Ag/SiO_2 NC minimal inhibition concentration (MIC) values and above, it was found to be effective against *B. cinerea* compared to the lower ones. The Ag/SiO_2 NC inhibitory rate values were observed to be close to Dithane M-45 MIC values, which are recommended as strong fungicides against this disease. Ag/SiO_2 NC-treated plants also showed a considerable increase in proline and phenol concentration compared to untreated infected plants and Dithane M-45–treated plants. This indicates the potential of Ag/SiO_2 NC for application as an antifungal to control chocolate spot disease and minimize yield losses of faba beans across the world.

The metal NPs used in the formulation of NCs have the potential to generate reactive oxygen species (ROS), which is effective against microbial activities. Copper NPs (CuNPs) are reported to produce free radicals that result in "multiple hit damage" in cells and disrupt the pathogen metabolic pathways (Chowdhury et al. 2024). A higher antibacterial activity was observed with an inhibition zone of 13 mm and antifungal activities of 77% in the NC of CuNPs and silver NPs (AgNPs) against *Xanthomonas oryzae* pv. *oryzae* and *R. solani* compared to the individual NP. *X. oryzae* pv. *oryzae* and *R. solani* cause leaf blight disease in rice and to date they remain a major threat to rice crops. Similarly, the application of biogenic AgNPs showed a reduction of lesions of bacterial leaf blight with a maximum inhibition rate observed in those treated with AgNP suspension at 100 mg/L (Ahmed et al. 2020).

Graphene oxide (GO)-Fe_3O_4 NC was found to effectively suppress fungal spore germination and prevent the growth of downy mildew in grapevine plants with no significant toxic effects even at higher concentration (~1000 μg/mL; Wang et al. 2017). This study suggested that at 50 μg/mL GO-Fe_3O_4, the NC was efficient for protection against the pathogen. Further at 250 μg/mL GO-Fe_3O_4 as foliar application, the severity of the disease was also reduced. Another study also reported that reduced graphene oxide (rGO) with Cu and Ag NPs (rGO-Cu-Ag) NC demonstrated antibacterial activity against *Xanthomonas euvesicatoria* even at a relatively low concentration, i.e. 50 μg/mL. The rGO helped to stabilize Cu and Ag NPs and prevent their aggregation, which helps NC adhere to plant leaves (Bytešníková et al. 2022). Results from scanning electron microscope (SEM) and DNA content analyses showed that the NCs exert their antibacterial activity by inducing stress on the bacterial membrane, disrupting cell wall integrity, and causing leakage and release of cellular and genetic material, which also inhibited the replication of bacteria. Interestingly, no visible negative impacts or inhibitory effects were observed on the growth and development of plants even at a concentration of 500 μg/mL. Few other studies on graphene have also reported a positive influence on tomato plants with enhanced seed germination and increased root fresh weight and length (Zhang et al. 2015).

One study fabricated humic acid (HA) and boron (B) with Fe_2O_3 NC to test antifungal efficacy against *Fusarium oxysporum*, the causal agent for wilt disease in cucumber (El-Batal et al. 2023). It was observed that the application of Fe_2O_3 NPs-B (0.25 mM) and Fe_2O_3 NPs-HA (0.125 mM) drastically reduced the severity of the disease by 83.33%. The disease indexes were reduced by 20.83% and 25% and the protection against the disease by 75.0% and 69.99% when Fe_2O_3 NPs-B and Fe_2O_3 NPs-HA were applied, respectively. In addition, there was an overall improvement in the morphological and physiological traits in the cucumber plant with an increase in the antioxidant enzymes, osmolytes, photosynthetic pigments, soluble protein contents and peroxidase (POD), and polyphenol oxidase (PPO) enzymes activities.

Similar to other agrochemicals, the spraying of conventional herbicides has various adverse consequences due to their nondegradable, insolubility, bioavailability, and soil sorption characteristics, which can find way into the surrounding environment. However, when paraquat was formulated with alginate/chitosan NPs, it resulted in controlled release with a prolonged effect of herbicides. These NCs exhibited slow release, target-specific, and minimized the adverse effect of the herbicide (Silva et al. 2011).

2.3.1.3 Biochar-based NCs

BC is a solid carbon-rich material obtained in the absence (pyrolysis) or limited supply (gasification) of oxygen from agricultural waste or biomass feedstock having properties like high surface area, porosity, and cation exchange capacity (CEC; Ghassemi-Golezani and Rahimzadeh 2022). They elevate crop productivity by improving soil characteristics, such as enhanced water retention and high charge density, which tend to retain large amounts of nutrients (Chausali, Saxena, and Prasad 2021). BC from biomass contains a considerable number of macronutrients, such as carbon (C), N, phosphorous (P), sulfur (S), K, Mg, Mn, and Ca, and micronutrients, such as Zn, Cu, and Fe. Nanoformulations of BC can give rise to a new complex material with the properties of both materials.

Biochar-based metal oxide NCs combine the benefits of BC with nanomaterials, forming an idle compound/material for soil amendment in abiotic stress, such as salinity and heavy metals (HMs). They are carbon-rich materials that supply required nutrients under salt stress and improve the movement of salt ions out of the root zone. Under salt stress conditions, even moderate salt-tolerant plants are impeded in their productivity. However, agronomic practices like amendment of soil properties using BNCs have proven to improve performance and salt tolerance as well. The soil CEC is enhanced when BNCs are applied by increasing soil pH, and activating nutrient exchange in the rhizosphere. Furthermore, the oxygen-containing functional groups on BC surfaces are known to immobilize metal ions by forming precipitates and complexes (Ghandali et al. 2024).

Experiments were conducted using BNC with Mg and Mn (12.5 g BNC-MgO + 12.5 g BNC-MnO per kg soil) in saline soil in safflower plants (*Carthamus tinctorius* L.); (Farhangi-Abriz and Ghassemi-Golezani 2021). The positive effects observed in the plants and soil include increased pH and CEC of the soil, water content, K, Ca Mg, Mn, chlorophyll, water use efficiency, and plant development. Furthermore, there was a reduction in the sodium adsorption ratio (SAR), exchangeable sodium percentage (ESP), sodium absorption rate of plants, and osmolyte production under 6 and 12 dS/m salinities. Another similar study reported that the application of BNCs combined with Fe and Zn (15 g/kg BNC-FeO + 15 g/kg BNC-ZnO) resulted in increased oil in dill seeds (*Anethum graveolens* L.) under high salt stress (Ghassemi-Golezani and Rahimzadeh 2022). In addition, maximum antioxidant activity was also observed along with a reduction in Na^+ accumulation in leaf tissues of the plants. Therefore, the application of BNC coupled with metal oxide is strongly recommended for enhancing plant growth under saline and nonsaline conditions (Ghassemi-Golezani and Farhangi-Abriz 2021).

Another important abiotic stress is the accumulation of HMs in soil. High concentration of HMs in plants can affect their physiology by lowering their biomass. The addition of modified BC to the soil stabilizes HMs, regulates their bioavailability, and minimizes the translocation from root to shoot, leading to enhanced development of plant root and shoot and subsequently increasing its dry weight and grain yield. Ghandali et al. (2024) compared BC and modified BC (BC-MnO_2 and BC-ZnO) and showed that BC-MnO_2 treatment yielded higher dry biomass of ryegrass roots and shoots. This is attributed to BC-MnO_2 having a higher specific surface area, amorphous structure, and more exchangeable cations reducing the bioavailability of HMs in the soil and its impact on ryegrass plant. Compared to unmodified BC, those modified BC with NPs have a higher specific surface area. Additionally, they were also found to be more effective in reducing the transfer of HMs preventing their translocation to the shoot of the plant. It was also found that amendment of soil with $MnFe_2O_4$-BC caused enhanced Sb and Cd immobilization in *Lolium multiflorum* Lam. and reduces its bioavailability and accessible forms in comparison to the application of BC alone (Wang et al. 2019). Thus, BNCs are cost-effective amendment techniques to mitigate HM-contaminated soils and enhance plant growth and development.

2.3.1.4 NC-based Materials as Sensors

Nanosensors or nanobiosensors are transducers composed of surfaces with nanostructures, nanomaterials with physical confinement at the nanoscale, or NPs manufactured via nanofabrication. They use nanotechnology, which provides the field of study with possible alternatives to traditional biosensors by lowering the effort required for standard laboratory procedures and techniques (Beegum and Das 2022). Utilizing nanomaterials, including CNTs, GO, quantum dots, metallic gold, and iron oxide NPs, the transducer functions as a sensor. The nanoconfinement, flexible shape, enhanced optical, mechanical, electrical, and thermal properties, and high specific surface area of the nanomaterials all contribute to a considerably increased transducing capability (Srivastava, Dev, and Karmakar 2018). Plant-system-interrogating nanosensors have the potential to advance the basic knowledge of plant biology (Medintz 2006).

Nanosensors are employed in agriculture for real-time crop condition monitoring utilizing piezoelectric nanosensors to measure temperature, pH, nutrient requirements, and moisture content in the soil (Dimri, Pathak, and Sharma 2020). According to a study by Sharma et al. (2021), the plant sap monitoring was done using an electrochemical biosensor nutrient sensor (rGO aerogel/ion specific electrode (ISE) for the detection of nitrate and calcium ion concentration. Nanosensors have great sensitivity and low detection limit, making ideal for smart agriculture, particularly for the identification of crop pests and pesticide residues. They are able to recognize particular pathogens, like bacteria, viruses, and fungus. Diseases like bacterial spot disease in Solanaceae plants, *Xanthomonas axonopodis* pv. *vesicatoria*, can be identified using sensors that combine fluorescent silica NPs and antibody molecules. Various NCs were developed and utilized for the identification of pathogens, which include gold NPs for identifying pathogenic bacteria *Pantoea stewartii*, gold nanorods to identify the pathogenic viruses Odontoglossum ring-spot virus and Cymbidium mosaic virus, and ZnO chitosan NC membrane to identify *Trichoderma harzianum* (Kashyap et al. 2019). In addition, a lateral flow immunoassay nanobiosensor platform was created by utilizing graphene NPs (GNPs) to detect the soil-borne pathogen *Ralstonia solanacearum* in potato plants (Zain et al. 2023). Moreover, *Phytophthora* inoculum that causes the blight disease of tomato and potato plants has been found using a GNP-based "lateral flow biosensor" (LFBS; Zain et al. 2023).

Besides, numerous nanosensors are also deployed to identify a range of plant hormones, including auxin, ethylene, and strigolactones (Shaw and Honeychurch 2022). Plant hormones, sometimes referred to as phytohormones, are signal molecules that are produced by plants and play a role in both biotic and abiotic stress tolerance as well as a plant's growth and development from embryogenesis to reproductive development. Plant hormones and signaling pathways can be studied in vivo using nanosensors. Ang et al. (2024) clarify the relationship between ROS and salicylic acid (SA) pathways in response to various environmental stimuli and stresses by developing optical nanosensors based on single-walled carbon nanotubes (SWCNT) for the purpose of real-time SA detection in plants. A study by John et al. (2022) suggests that through the use of remote glucose detection in algae and entire plants, as well as epifluorescence microscopy of H_2O_2 and nitric oxide (NO) in leaf sections, optical nanosensors have been shown to report plant signaling molecules that communicate and initiate plant stress responses.

It has also been observed that NPs can improve plants' tolerance to stress. Son et al. (2023) discovered a surface-enhanced Raman scattering (SERS) nanosensor that may be used to detect several stress-related compounds in growing plants in real time. Alova et al. (2020) developed oxygen (O_2) nanosensors that monitor the reduced oxygen content in *Chara corallina* plants by employing quartz micropipettes with carbon- and platinum-coated ends. In addition, nanosensors can help prevent soil and water contamination by identifying dangerous chemicals and contaminants in the surrounding environment (Chandna et al. 2022). They have the ability to sense various environmental pollutants, such as HMs, pesticide residues, and antibiotics (Shakeel et al. 2022). The residues of pesticides, such as carbaryl and parathion, cause water contamination and pose toxic effects on plants, animals, including human beings. Studies have reported the use of NC sensor based on zirconium oxide-cerium oxide-polyacrylic acid ($ZrO_2/CeO_2/$PAA) for the detection of methyl parathion (MP; Wu et al. 2021). The sensors designed for pesticide residue detection provide exceptional sensitivity, low detection thresholds, high specificity, rapid response time, and compact dimensions (Khot et al. 2012). According to the study by Zain et al. (2023) for target monitoring, a variety of NP types, including magnetic and metal oxides (TiO_2, MgO, and ZnO), are used in conjunction with nanobiosensors. TiO_2 porous sol–gel film has attracted significant interest for use as a protein immobilization matrix on account of its high specific surface area, inexpensive cost, nontoxicity, strong chemical and thermal stability, semi-conductivity, and exceptional biocompatibility (Cui et al. 2013). Cui et al. (2018) reported that by simply adsorbing acetylcholinesterase (AChE) on chitosan, TiO_2 sol–gel, and rGO-based multilayered immobilization matrix (denoted as CS@TiO_2-CS/rGO), a very stable electrochemical biosensor for the detection of organophosphorus pesticides (OPs) is developed. Malathion was also detected at a level of 0.01 nM by the electrochemical nanosensor grounded using CuONPs decorated with three-dimensional (3D) graphene NC, but at a startling 0.001 ng/mL by the electrochemical aptasensor made using chitosan–iron oxide NC (Sharma et al. 2021).

2.3.1.5 Biopolymer-based NCs

Despite the alluring benefits of NCs, the controlled manipulation of their properties with remarkable properties may sometimes be detrimental to the environment and threaten both biotic and abiotic components. Therefore, scientists have now turned to biodegradable polymers. Biopolymers are sustainable products as they are nontoxic and biodegradable in nature. Polymers originating from organic compounds like starch have been often used as raw materials, because it is readily available and comparatively easier to synthesize biodegradable nanomaterials. However, various natural polysaccharides, including alginate, hyaluronic acid, cellulose, hemicelluloses, and cationic chitin/chitosan, have been used (Yu et al. 2021). These natural biopolymers, such as starch, alginate, cellulose, pectin, and chitosan, are utilized to deliver encapsulated active ingredients for efficient targeting of the agrochemicals (Guha et al. 2020). They also have good absorption of water (biopolymer-based hydrogels, super absorbent) in soil (Behera and Mahanwar 2020). They are considered to be the best choice for the delivery of agrochemicals with maximum efficiency in minimum usage.

Although bio-based polymer is a good alternative to synthetic-based polymers, some of their characteristics should be refined to achieve the desired product and for wider application. This requires the formulation of multiphase materials, such as blends or composites. Blending them can result in an excellent bio-based composite with beneficial properties for the desired application. Thus, bio-NCs are materials derived from the integration of inorganic or organic reinforcement with one or more dimensions in the nanoscale range (10^{-9} m) into a bio-based polymer. The integration of bio-based polymers with nanofillers gives bio-NCs improved mechanical, thermal, and barrier properties (Mallakpour and Hussain 2021). They also cause the spectra–radiometric characteristics in the UV and visible ranges to decrease. This characteristic is particularly advantageous for farming since it prevents the growth of weeds. Among the several biopolymers, we have discussed below chitosan NC which is of great interest among the researchers and industrialists (Yu et al. 2021).

2.3.1.6 Chitosan-based NCs

N-acetyl-D-glucosamine and D-glucosamine units with an amino (NH_2) group and two hydroxyl (OH) groups in each repeating glycosidic unit make up chitosan, a polycationic polymer (Yu et al. 2021). In short, it is the deacetylated form of chitin. It is biodegradable and biocompatible with low solubility and thermal stability making it an ideal choice for NCs as film or agrochemical delivery (Mallakpour and Hussain 2021). Its structural characteristic makes it an ideal material for bio-NCs given that there are functional groups available along its backbone for reacting with other polymers and metal ions (Deshpande et al. 2017).

Chitosan-based NC's main applications in agriculture are as an elicitor (You and Chan 2015), as a carrier for fertilizers (Zungu et al. 2023), and plant growth regulators and stress inhibitors (Das et al. 2023; Gomes et al. 2021). They also possess excellent antifungal and antibacterial properties (Al-Zahrani and Al-Garni 2023; Chouhan et al. 2022). *Helichrysum odoratissimum* aqueous extract was used to nucleate Zn ionic species to develop ZnO NPs (Zungu et al. 2023).

These NCs were further incorporated into tripolyphosphate (TPP)-cross-linked chitosan (CS/TPP) nanostructures and used as priming agents for maize (*Zea mays* L.) seeds. Seed priming quickens the germination rate. The NCs exhibited a slow release of Zn (II) ions in water for about two weeks. Also, priming maize seeds with this solution enhanced the growth of the radicle, root, and shoot. Additionally, in salt-stressed seeds, low concentration (0.01%) CS/TPP NPs and Zn (II)-loaded CS/TPP NPs were reported to improve the early development of the maize plant. A similar study of maize seed priming with chitosan encapsulated Cu^{2+} promoted the initial development, increased shoot growth, antioxidant enzymes, and photosynthetic electron transport in comparison to the control, particularly under high temperatures and relative humidity (Gomes et al. 2021).

Chitosan has also emerged as a potential biostimulator and elicitor in agriculture to counteract the adverse effects of drought stress. Water-deficit stress impedes plant metabolism and generates ROS, including superoxide anion (O_2^-), H_2O_2, and hydroxyl radical ($^{\bullet}OH$), causing cell death by disrupting the cell membrane lipids and proteins of plant organelles, particularly pigment molecules like chlorophyll (You and Chan 2015). Water stress condition led to the reduction of 45% and 50% crop yield in resistant and susceptible varieties of wheat, respectively (Das et al. 2023). However, the foliar application of zinc-chitosan-SA (ZCS) NPs at 100 mg/L significantly ($p < 0.05$) improves relative water content (RWC) and enhances ROS scavenging enzymes, such as SOD and glutathione reductase, and metabolites, such as ascorbate, proline, and MDA, in wheat under drought condition. As a result, there was a considerable increase in the yield up to 63% and 41% in two tolerant cultivars and 50% and 46% in two susceptible cultivars. This was attributed to adjustment in the amount of chlorophyll and carotenoid, antioxidants, and osmolytes.

Another study also showed that the encapsulation of NO donor, namely, S-nitrosoglutathione (GSNO), with chitosan and free GSNO attenuated the effects of water deficit on sugarcane plants (Silveira et al. 2019). However, the entrapment or encapsulation of GSNO with chitosan protects NO from rapid degradation for a slow and CR, which prolongs its effects and also causes higher photosynthetic rate and increased root/shoot ratio as compared to free GSNO. The slow release of NO also facilitated carboxylation activity during water stress, as rubisco protein is considered a possible target of nitrosylation. In addition, the increase in root/shoot ratio was due to NO influence on auxin content, which is responsible for the growth and development of lateral and adventitious roots that subsequently improved the water uptake in plants to cope during drought stress.

Chitosan is also found to ameliorate plants during salinity stress. The substantial accumulation of Na^+ and Cl^- in saline soil leads to the cell ionic imbalance causing reduction in the leaf photosynthetic area. Increased concentration of Na^+ and Cl^- and ROS in plants degrades chlorophyll and impairs the photosynthetic pigments. The uptake of nutrients like sodium (N), K, Calcium (Ca), magnesium (Mg), and manganese (Mn) are also limited causing nutrient imbalance. Foliar application of chitosan is known to enhance N uptake and translocation, thereby increasing chlorophyll pigments, enhance stomatal conductance, and transpiration rate. The application of chitosan-based SA NCs in grape (*Vitis vinifera* cv. "Sultana") under salinity stress increased the proline, MDA, H_2O_2, and activated antioxidant enzymes, such as SOD and catalase (CAT; Aazami et al. 2023).

The incorporation of metals with chitosan provides nutritional benefits and helps crops to combat stresses. CS-Si NF demonstrated a controlled release of silicon (Si) and stimulated growth and productivity in maize plants (Kumaraswamy et al. 2021). The study experimented seed priming and foliar application with different concentrations of CS-Si NF. The seed primed with 0.04–0.12% (w/v) of the NC exhibited about a fourfold increase in seedling vigor index (SVI) than SiO_2. While foliar application of CS-Si NF activated antioxidant enzymes for scavenging ROS species in the leaf that help expedite photosynthesis compared to the application of SiO_2 alone.

A study reported the biosynthesis of AgNPs with the help of endophytic fungus, namely, *Curvularia kusanoi*, and encapsulation with chitosan (CS-AgNPs); (Al-Zahrani and

Al-Garni 2023). The antifungal activity of the biosynthesized NC was tested against fungal pathogens, namely, *Curvularia lunata, Aspergillus fumigatus*, and *Cladosporium* sp. at different concentrations (50, 25, 12.5, and 6.25 mg/L) using the mycelial growth inhibition (MGI) method. The highest inhibition of these fungal species was observed at 50 mg/L of CS-AgNPs followed by 25, 12.5, and 6.25 mg/L. The mycelial growth at 50 mg/L CS-AgNPs was reduced by 97% for *C. lunata*, 89% for *Cladosporium* sp., and 83% for *A. fumigatus*. Whereas, at lower concentration of 6.25 mg/L, minimum inhibition rate was observed with 60%, 62%, and 44% reduced mycelial growth for *C. lunata, Cladosporium* sp., and *A. fumigatus*, respectively. This study also indicates the prospective of fungal proteins for the synthesis of AgNPs that are cost-effective, have greater control, and less toxic than physical and chemical methods.

2.4 Adverse Effects of NCs on Crop Productivity and Sustainability

Scientific innovations in agriculture provide numerous benefits in various ways, yet they are also associated with risks to crop and human health (Nagaraju et al. 2022). Similar to the conventional agrochemicals, nanoagrochemicals are expected to pose serious risks if not regulated leading to water, soil, and air pollution; transfer of chemical residues to food products; disruption in plant functions; and equilibrium shift in microbial communities. Some of the adverse effects reported include changes in photosynthetic parameters, transcriptome changes, DNA degradation, chromosomal aberrations, and destruction of chloroplasts and vascular bundles (Table 2.2; Jampílek and Kráľová 2019). Therefore, they can lead to phytotoxicity and genotoxicity, like chromosome fragmentation, genetic mutation, ROS production, biomass reduction, and retarded growth (Behl et al. 2022). The phytotoxicity mechanisms of NPs are still largely unknown. NC moves within the plants via symplastic (i.e. within cells through plasmodesmata, or sieve) or apoplastic pathways (i.e. between cell walls, extracellular spaces, or xylem vessels) or a mix of both. Once NCs are absorbed, they tend to accumulate in various tissues and organs of the plant, including edible (grains and fruits) and nonedible parts (leaves and flowers); (Servin and White 2016). After which, they incite biochemical, morphological, molecular, and physiological changes in the crops (Khan and Rizvi 2014). They also move through the vascular system to other tissues and cause oxidative stress, exert influence on antioxidant enzyme activity, and ultimately impact root elongation and plant seed germination (Scherer et al. 2019).

Numerous research works have looked into how NPs' widespread use and their effects on the environment could result in "nanopollution or invisible pollution" (Jan et al. 2022). It has been noted in numerous research works that the effects of nanomaterials depend on the size and concentrations. For example, higher toxicity is observed in 5–10 nm AgNPs size as they tend to easily translocate and accumulate in various parts of the plant (Nagaraju et al. 2022). The build-up of NPs has detrimental effects on nutrient intake and plant development, which can lower the crop productivity (Gardea-Torresdey, Rico, and White 2014). Rico et al. (2013) observed that CeO_2 NPs changed amino acids, fatty acids, phenolics, and nonreducing sugars, which caused toxicity in rice plants. Following the administration of CeO_2 NPs, similar harmful effects

Table 2.2 Adverse effects of nanocomposites on plant growth and development.

Nanocomposites	Crop	Adverse effects	References
Carbon nanotubes (CNTs)	Tomato	Reduction in growth rate. Inhibition of seed germination.	Khodakovskaya et al. (2009)
Titanium dioxide NPs (TiO_2 NPs)	Wheat	Transmission electron microscopy (TEM) pictures revealed that shoots treated with TiO_2 possessed black particles present in cortical cells and were able to pass through the cell wall. This blockage can impair to absorb essential elements from soil.	Du et al. (2011)
Multiwalled CNT, aluminum, alumina, zinc, and zinc oxide, NPs (ZnO NPs)	Rapeseed (*Brassica napus*), radish (*Raphanus sativus*), ryegrass, and cucumber (*Cucumis sativus*)	Suspensions of 2000 mg/L nano-ZnO completely inhibited root development from elongation. For radish, the estimated half-inhibitory concentrations (IC50) of nano-Zn and nano-ZnO were close to 50 mg/L, whereas for rape and ryegrass, they were approximately 20 mg/L.	Lin and Xing (2007), Hasan et al. (2021)
Silver NP (AgNPs)	*Arabidopsis thaliana*	Stunted growth, chlorosis, and smaller roots.	Navarro et al. (2008)
Silver NPs (AgNPs)	Lettuce (*Lactuca sativa*)	Prevented root development and impaired seed germination.	Rico et al. (2011)
Zinc oxide NP (ZnO NPs)	Mung bean (*Vigna radiata*) and chickpea	Treatment of 20 and 1 ppm of ZnO shows retarded growth and development patterns.	Mahajan, Dhoke, and Khanna (2011)
Cerium oxide NP (CeO_2 NPs)	Corn, tomato, and cucumber (*Cucumis sativus*)	Corn: At 2000 mg/L, there was a ~30% reduction in both root elongation and germination rate. Tomato: At 2000 mg/L, germination decreased by 30%. Cucumber: 20% less germination at 2000 mg/L.	Moreno et al. (2010)
Cerium-oxide-engineered NP (CeO_2 ENPs)	Kidney bean	Proteome analysis on seeds shows alterations in nutritional makeup and quality of seeds. Overexpression of proteins linked to stress and downregulation of genes related to storage of nutrients.	Majumdar et al. (2015)
Cerium-oxide-engineered NP (CeO_2 ENPs),	Tomato	Reduced development and poor seedling growth.	Wang et al. (2013)
Zinc oxide NP (ZnO NP)	Cabbage (*Brassica oleracea*)	A dosage of 500–1000 μg/mL inhibits root growth and germination by 40.6%.	Pokhrel and Dubey (2013)

were seen in transgenic cotton (Bt 29317) along with a decrease in the number of micronutrients, namely, Mg, Zn, Fe, and P (Li et al. 2014).

Nanomaterials may also induce the production of ROS that cause oxidative damage to cells and chemicals, diminish photosynthesis, and change the way genes are transcriptionally expressed. Growing *Arabidopsis thaliana* seedlings in a medium with a water-soluble fullerene malonic acid derivative (FMAD) has been shown to limit root development (Liu et al. 2010). TiO_2 is another NP whose application has expanded significantly in recent years, making it one of the most widely utilized NP in agriculture (Azim et al. 2023). However, TiO_2 NPs also exhibited some detrimental effects on transpiration and water transport in maize, whereas Ag and CeO_2 NPs were genotoxic to onions and soybeans, respectively (Asli and Neumann 2009; Kumari, Mukherjee, and Chandrasekaran 2009; Soran et al. 2021). The ideal concentration of TiO_2 NPs for a notable boost in parsley (*Petroselinum crispum*) germination, root and shoot length, fresh biomass, and vigor index is 30 mg/L. The lengths of the shoots and seedlings were reduced by the increase in nano-sized TiO_2 concentrations beyond 10 ppm. At 100 ppm nano-sized TiO_2, the lowest shoot length was attained. Thus, it is understood that there is a relatively low ideal concentration of NPs, and as concentrations rise, adverse consequences become apparent. Furthermore, various studies have confirmed the relationships between TiO_2 NPs and reduced shoot biomass in wheat, delayed germination and root elongation in maize and narbon beans (*Vicia narbonensis* L.), and DNA damage in tobacco (*Nicotiana tabacum*) and onion (Du et al. 2011; Ruffini Castiglione et al. 2014; Ghosh, Bandyopadhyay, and Mukherjee 2010).

Other studies have also reported that exposure of NPs can induce negative effects in plant growth and yield (Table 2.2). For instance, treatment of SiO_2 NP resulted in growth suppression in *Bacillus thuringiensis* (Bt)-transgenic cotton (Lee et al. 2008). Rice seedlings showed a considerable drop in total chlorophyll and carotenoid contents following a week of exposure to AgNPs (Nair and Chung 2014). AgNPs can build up in mustard (*Brassica* sp.) seedlings and severely impede photosynthesis (Vishwakarma et al. 2017). According to Yanga, Cao, and Rui (2017), applying AgNPs to tomatoes reduced their biomass and length of roots. In CuONPs, we generally observed two main effects on plants. First, they have the potential to release Cu, which is a necessary element but can be hazardous in high concentrations. Secondly, they have the potential to enter the plant and induce particle stress (Ruttkay-Nedecky et al. 2017; Velicogna et al. 2020). They also altered the structure of the roots and hindered the growth of wheat plants cultivated in a sand matrix (Dimkpa et al. 2012).

According to Tortella et al. (2024), the coexistence of atrazine (at a field dose of 3 mg/kg) and Cu-based NPs (at 0.05 and 0.15% w/w) significantly affected the abundance of nitrifying bacteria and archaea communities in a soil–plant system, which had a detrimental effect on the system's capacity for nitrification. Moreover, NPs with antimicrobial properties may harm nontargeted or even beneficial microorganisms, which threaten the health of agroecosystems. Furthermore, microbes vital for the idle functioning and structural integrity of agriculture ecosystem that regulate nutrient flow, decomposition of organic matter, and formation of symbiosis with plant may be impacted. Besides that, the metal-based NCs are also a concern as they can accumulate as HMs contaminating soil and water, thereby posing environmental risks (Saberi Riseh et al. 2024).

2.5 Challenges and Future Prospects in Application of NCs on Crop Plants

There has been a tremendous advancement in application of nanotechnology over the last two decades. However, challenges still exist, and the consumption of foods that are treated with nanoagrochemicals may lead to potential health risk in vital organs of our body, such as liver, heart, brain, and blood (Wypij et al. 2024). The concern is also on the migration of the NPs into water, soil, and air. Hence, its application warrants further research for food safety and environmental security. Furthermore, in adoption of any novel technology, public perception plays a critical role. Users should be informed of the advantages, along with its limitations, so that they use it judiciously to realize its full potential. Some of the prominent challenges and future prospective in application of NCs are highlighted in Table 2.3.

Table 2.3 Challenges and future prospective in the application of nanocomposites.

Parameters	Challenges/limitations	Future prospective
Toxicity	• Their small size is prone to health risks as they can easily disperse and get into the body via inhalation • Potential toxicity and their fate in the environment remain an open question	• Research on ecotoxicity, phytotoxicity should be prioritized • Fate, distribution, and transport of engineered NPs in the environment should be properly addressed
Cost	• Synthesis of NCs is expensive	• Utilize low cost and naturally available material
Bioaccumulation	• Transfer and bioaccumulation of NPs through the food chain • Accumulation (frequency and quantity), transformation, and NP speciation are still not known	• Understand the interactions between NCs and target species as well as nontarget species
Upscale and market	• Most studies are conducted in lab and have low production/demand for the market	• Move from lab to field-based trials and ultimately to farmers • Collaboration between various disciplines
Synthesis	• Complex methodology in the synthesis remains a challenge for large-scale production	• Develop inexpensive, biodegradable, and simpler methodology
Standards and regulation	• No specific set of standards and policy regulation in some countries	• Implement stricter regulation to avoid nanotoxicity and nanopollution in the future
Environmental safety	• Long-term impact of NCs on the soil, water, air, flora, and fauna is unknown	• Ensure controlled release of active ingredients and stability of formulations under varying environmental conditions
Public awareness and acceptance	• Public skepticism on accepting new technologies	• Build capacity and disseminate the pros and cons of the technologies, creating awareness and acceptance

2.6 Conclusion

NCs, compared to conventional agrochemicals, are the materials of twenty-first century because they have higher NUE and augment crop yield with lesser inputs and lower detrimental effects on the environment. They present best approaches to enhance crop productivity and soil quality and restore fertility for a sustainable agriculture and food security. In addition, they help to reduce biotic stresses like pest and pathogens and abiotic stresses like drought, salinity, and HMs and mitigate food wastage. They can also be used as nanosensors for crop management and precise detection of pest and diseases. Thus, their uses are at a staggering growth rate of 25% per year and have totally revolutionize the agriculture sector with its immense potential by offering alternative solutions and cutting the abysmal usage of conventional agrochemicals. Although NCs have numerous benefits, few studies have reported their negative effects, especially at high concentrations. Beyond certain level, they can exhibit phytotoxicity; inhibit root and shoot growth and seed germination; generate ROS and DNA degradation; and reduce photosynthesis. The unregulated application may perhaps lead to hazardous consequences on crops as stable NPs from the NCs migrate into soil, air, and water bodies. Other challenges impeding their widespread adoption include high cost of some NCs, complex methodology, limited field studies, and public concern on biosafety and ecotoxicity. Nevertheless, significant improvement has been made over the last two decades positioning NCs as a promising avenue in attainment of agricultural and environmental sustainability. Therefore, it is crucial to develop cost-effective materials with an emphasis on biopolymers, shifting from laboratory settings to field applications, and extending awareness to the public on the effects of NCs.

Acknowledgement

K.C. and S.R. are thankful to Union Grants Commission (UGC), Government of India, for funding this work in the form of UGC-Junior Research Fellowship (UGC-JRF).

Author Contribution

K.C.: conceptualization, writing manuscript, editing, and visualization. S.R.: writing manuscript and visualization. K.R.: supervision, final editing, and review. R.S.S.: review.

References

Aazami, M.A., Maleki, M., Rasouli, F. et al. (2023). Protective effects of chitosan based salicylic acid nanocomposite (CS-SA NCs) in grape (*Vitis vinifera* cv. 'Sultana') under salinity stress. *Scientific Reports* 13 (1): 883. doi: 10.1038/s41598-023-27618-z.

Ahmed, A., He, P., He, P. et al. (2023). Environmental effect of agriculture-related manufactured nano-objects on soil microbial communities. *Environment International* 173: 107819. doi: 10.1016/j.envint.2023.107819.

Ahmed, T., Shahid, M., Noman, M. et al. (2020). Silver nanoparticles synthesized by using *Bacillus cereus* SZT1 ameliorated the damage of bacterial leaf blight pathogen in rice. *Pathogens* 9 (3): 160. doi: 10.3390/pathogens9030160.

Al-Zahrani, S.S. and Al-Garni, S.M. (2023). Fabrication of antifungal AgNPs capped with chitosan using endophytic fungus *Curvularia kusanoi*. *Biocatalysis and Agricultural Biotechnology* 51: 102769. doi: 10.1016/j.bcab.2023.102769.

Alova, A., Erofeev, A., Gorelkin, P. et al. (2020). Prolonged oxygen depletion in microwounded cells of *Chara corallina* detected with novel oxygen nanosensors. *Journal of Experimental Botany* 71 (1): 386–398.

Ang, M.C.Y., Saju, J.M., Porter, T.K. et al. (2024). Decoding early stress signaling waves in living plants using nanosensor multiplexing. *Nature Communications* 15 (1): 2943.

Ashfaq, M., Talreja, N., Chuahan, D. et al. (2019). Polymeric nanocomposite-based agriculture delivery system: emerging technology for agriculture (Chapter 5). In: *Genetic Engineering – A Glimpse of Techniques and Applications* (ed. F. Jamal). IntechOpen. doi: 10.5772/intechopen.89702.

Asli, S. and Neumann, P.M. (2009). Colloidal suspensions of clay or titanium dioxide nanoparticles can inhibit leaf growth and transpiration via physical effects on root water transport. *Plant, Cell and Environment* 32 (5): 577–584. doi: 10.1111/j.1365-3040.2009.01952.x.

Azim, Z., Singh, N.B., Singh, A. et al. (2023). A review summarizing uptake, translocation and accumulation of nanoparticles within the plants: current status and future prospectus. *Journal of Plant Biochemistry and Biotechnology* 32 (2): 211–224.

Baka, Z. A. and El Zahed, M.M. (2022). Antifungal activity of silver/silicon dioxide nanocomposite on the response of faba bean plants (*Vicia faba* L.) infected by *Botrytis cinerea*. *Bioresources and Bioprocessing* 9 (1): 102. doi: 10.1186/s40643-022-00591-7.

Beegum, S. and Das, S. (2022). Nanosensors in agriculture. In: *Agricultural Nanobiotechnology* (ed. S. Ghosh, S. Thongmee, and A. Kumar), 465–478. Amsterdam: Elsevier.

Behera, S. and Mahanwar, P.A. (2020). Superabsorbent polymers in agriculture and other applications: a review. *Polymer-Plastics Technology and Materials* 59 (4): 341–356. doi: 10.1080/25740881.2019.1647239.

Behl, T., Kaur, I., Sehgal, A. et al. (2022). The dichotomy of nanotechnology as the cutting edge of agriculture: nano-farming as an asset versus nanotoxicity. *Chemosphere* 288: 132533. doi: 10.1016/j.chemosphere.2021.132533.

Bytešníková, Z., Pečenka, J., Tekielska, D. et al. (2022). Reduced graphene oxide-based nanometal-composite containing copper and silver nanoparticles protect tomato and pepper against *Xanthomonas euvesicatoria* infection. *Chemical and Biological Technologies in Agriculture* 9 (1): 84. doi: 10.1186/s40538-022-00347-7.

Camara, M.C., Campos, E.V.R., Monteiro, R.A. et al. (2019). Development of stimuli-responsive nano-based pesticides: emerging opportunities for agriculture. *Journal of Nanobiotechnology* 17 (1): 1–19. doi: 10.1186/s12951-019-0533-8.

Chakraborty, R., Mukhopadhyay, A., Paul, S. et al. (2023). Science of the total environment nanocomposite-based smart fertilizers: a boon to agricultural and environmental sustainability. *Science of the Total Environment* 863: 160859. doi: 10.1016/j.scitotenv.2022.160859.

Chandna, S., Gogde, K., Kaur, R. et al. (2022). Nano-biosensors for plant biomass: concept and applications. In: *The Role of Nanoparticles in Plant Nutrition Under Soil Pollution: Nanoscience in Nutrient Use Efficiency* (ed. V.D. Rajput, K.K. Verma, N. Sharma et al.), 199–221. Cham: Springer.

Chausali, N., Saxena, J., and Prasad, R. (2021). Nanobiochar and biochar based nanocomposites: advances and applications. *Journal of Agriculture and Food Research* 5: 100191. doi: 10.1016/j.jafr.2021.100191.

Chouhan, D., Dutta, A., Kumar, A. et al. (2022). Application of nickel chitosan nanoconjugate as an antifungal agent for combating *Fusarium* rot of wheat. *Scientific Reports* 12 (1): 14518. doi: 10.1038/s41598-022-18670-2.

Chowdhury, A.R., Kumar, R., Mahanty, A. et al. (2024). Inhibitory role of copper and silver nanocomposite on important bacterial and fungal pathogens in rice (*Oryza sativa*). *Scientific Reports* 14 (1): 1779. doi: 10.1038/s41598-023-49918-0.

Cui, H., Wu, W., Li, M. et al. (2018). Biosensors and bioelectronics graphene nanocomposites for detection of organophosphate pesticides. *Biosensors and Bioelectronics* 99: 223–229. doi: 10.1016/j.bios.2017.07.068.

Cui, H., Zhang, K., Zhang, Y. et al. (2013). Biosensors and bioelectronics immobilization of glucose oxidase into a nanoporous TiO_2 film layered on metallophthalocyanine modified vertically-aligned carbon nanotubes for efficient direct electron transfer. *Biosensors and Bioelectronics* 46: 113–118. doi: 10.1016/j.bios.2013.02.029.

Das, D., Bisht, K., Chauhan, A. et al. (2023). Morpho-physiological and biochemical responses in wheat foliar sprayed with zinc-chitosan-salicylic acid nanoparticles during drought stress. *Plant Nano Biology* 4: 100034. doi: 10.1016/j.plana.2023.100034.

Deshpande, P., Dapkekar, A., Oak, M.D. et al. (2017). Zinc complexed chitosan/TPP nanoparticles: a promising micronutrient nanocarrier suited for foliar application. *Carbohydrate Polymers* 165: 394–401.

Dimkpa, C. O., McLean, J. E., Latta, D. E. et al. (2012). CuO and ZnO nanoparticles: Phytotoxicity, metal speciation, and induction of oxidative stress in sand-grown wheat. *Journal of Nanoparticle Research* 14 (9). doi: 10.1007/s11051-012-1125-9.

Dimri, A., Pathak, N., and Sharma, S. (2020). Nanosensors for root zone parameters influencing plant growth. In: *Nanomaterials for Agriculture and Forestry Applications* (ed. A. Husen and M. Jawaid), 387–406. Amsterdam: Elsevier.

Dou, Z., Bini Farias, M.V., Chen, W. et al. (2022). Highly degradable chitosan-montmorillonite (MMT) nano-composite hydrogel for controlled fertilizer release. *Frontiers of Environmental Science & Engineering* 17 (5): 53. doi: 10.1007/s11783-023-1653-9.

Du, W., Sun, Y., Ji, R. et al. (2011). TiO_2 and ZnO nanoparticles negatively affect wheat growth and soil enzyme activities in agricultural soil. *Journal of Environmental Monitoring* 13 (4): 822–828. doi: 10.1039/c0em00611d.

El-Batal, A.I., El-Sayyad, G.S., Al-shammari, B.M. et al. (2023). Protective role of iron oxide nanocomposites on disease index, and biochemical resistance indicators against *Fusarium oxysporum* induced-cucumber wilt disease: in vitro, and in vivo studies. *Microbial Pathogenesis* 180: 106131. doi: 10.1016/j.micpath.2023.106131.

Farhangi-Abriz, S. and Ghassemi-Golezani, K. (2021). Changes in soil properties and salt tolerance of safflower in response to biochar-based metal oxide nanocomposites of magnesium and manganese. *Ecotoxicology and Environmental Safety* 211: 111904. doi: 10.1016/j.ecoenv.2021.111904.

Gardea-Torresdey, J.L., Rico, C.M., and White, J.C. (2014). Trophic transfer, transformation, and impact of engineered nanomaterials in terrestrial environments. *Environmental Science & Technology* 48 (5): 2526–2540.

Ghandali, M.V., Safarzadeh, S., Ghasemi-Fasaei, R. et al. (2024). Heavy metals immobilization and bioavailability in multi-metal contaminated soil under ryegrass cultivation as affected by ZnO and MnO_2 nanoparticle-modified biochar. *Scientific Reports* 14 (1): 10684. doi: 10.1038/s41598-024-61270-5.

Ghassemi-Golezani, K. and Farhangi-Abriz, S. (2021). Biochar-based metal oxide nanocomposites of magnesium and manganese improved root development and productivity of safflower (*Carthamus tinctorius* L.) under salt stress. *Rhizosphere* 19: 100416. doi: 10.1016/j.rhisph.2021.100416.

Ghassemi-Golezani, K. and Rahimzadeh, S. (2022). The biochar-based nanocomposites influence the quantity, quality and antioxidant activity of essential oil in dill seeds under salt stress. *Scientific Reports* 12 (1): 21903. doi: 10.1038/s41598-022-26578-0.

Ghosh, M., Bandyopadhyay, M., and Mukherjee, A. (2010). Genotoxicity of titanium dioxide (TiO_2) nanoparticles at two trophic levels: plant and human lymphocytes. *Chemosphere* 81 (10): 1253–1262.

Giroto, A.S., Guimarães, G.G.F., Foschini, M. et al. (2017). Role of slow-release nanocomposite fertilizers on nitrogen and phosphate availability in soil. *Scientific Reports* 7 (1): 46032. doi: 10.1038/srep46032.

Gomes, D.G., Pelegrino, M.T., Ferreira, A.S. et al. (2021). Seed priming with copper-loaded chitosan nanoparticles promotes early growth and enzymatic antioxidant defense of maize (*Zea mays* L.) seedlings. *Journal of Chemical Technology & Biotechnology* 96 (8): 2176–2184.

Guha, T., Gopal, G., Kundu, R. et al. (2020). Nanocomposites for delivering agrochemicals: a comprehensive review. *Journal of Agricultural and Food Chemistry* 68 (12): 3691–3702. doi: 10.1021/acs.jafc.9b06982.

Gupta, G.S., Kumar, A., and Verma, N. (2019). Bacterial homoserine lactones as a nanocomposite fertilizer and defense regulator for chickpeas. *Environmental Science: Nano* 6 (4): 1246–1258. doi: 10.1039/C9EN00199A.

Hasan, M., Mehmood, K., Mustafa, G. et al. (2021). Phytotoxic evaluation of phytosynthesized silver nanoparticles on lettuce. *Coatings* 11 (2): 225.

Jakhar, A.M., Aziz, I., Kaleri, A.R. et al. (2022). Nano-fertilizers: a sustainable technology for improving crop nutrition and food security. *NanoImpact* 27: 100411. doi: 10.1016/j.impact.2022.100411.

Jampílek, J. and Kráľová, K. (2019). Impact of nanoparticles on photosynthesizing organisms and their use in hybrid structures with some components of photosynthetic apparatus. In: *Plant Nanobionics: Volume 1, Advances in the Understanding of Nanomaterials Research and Applications* (ed. R. Prasad), 255–332. Cham: Springer.

Jan, N., Majeed, N., Ahmad, M. et al. (2022). Nano-pollution: why it should worry us. *Chemosphere* 302: 134746.

John, S.A., Chattree, A., Ramteke, P.W. et al. (2022). 20 – Nanosensors for plant health monitoring. In: *Micro and Nano Technologies* (ed. A. Denizli, T.A. Nguyen, S. Rajendran et al.), 449–461. Amsterdam: Elsevier. doi: 10.1016/B978-0-12-824554-5.00012-4.

Kajla, M., Yadav, V.K., Khokhar, J. et al. (2015). Increase in wheat production through management of abiotic stresses: a review. *Journal of Applied and Natural Science* 7 (2): 1070–1080.

Karny, A., Zinger, A., Kajal, A. et al. (2018). Therapeutic nanoparticles penetrate leaves and deliver nutrients to agricultural crops. *Scientific Reports* 8 (1): 7589. doi: 10.1038/s41598-018-25197-y.

Kashyap, P.L., Kumar, S., Jasrotia, P. et al. (2019). Nanosensors for plant disease diagnosis: current understanding and future perspectives. In: *Nanoscience for Sustainable Agriculture* (ed. R. Pudake, N. Chauhan, and C. Kole), 189–205. Cham: Springer.

Kaur, P., Duhan, J.S., and Thakur, R. (2018). Comparative pot studies of chitosan and chitosan-metal nanocomposites as nano-agrochemicals against *Fusarium* wilt of chickpea (*Cicer arietinum* L.). *Biocatalysis and Agricultural Biotechnology* 14: 466–471. doi: 10.1016/j.bcab.2018.04.014.

Khan, M.R. and Rizvi, T.F. (2014). Nanotechnology: scope and application in plant disease management. *The Plant Pathology Journal* 13 (3): 214–231.

Khan, O., Bilal Khan Niazi, M., Abbas Shah, G. et al. (2021). Green synthesis and evaluation of calcium-based nanocomposites fertilizers: a way forward to sustainable agricultural. *Journal of the Saudi Society of Agricultural Sciences* 20 (8): 519–529. doi: 10.1016/j.jssas.2021.06.005.

Khodakovskaya, M., Dervishi, E., Mahmood, M. et al. (2009). Carbon nanotubes are able to penetrate plant seed coat and dramatically affect seed germination and plant growth. *ACS Nano* 3 (10): 3221–3227.

Khot, L.R., Sankaran, S., Maja, J.M. et al. (2012). Applications of nanomaterials in agricultural production and crop protection: a review. *Crop Protection* 35: 64–70. doi: 10.1016/j.cropro.2012.01.007.

Knell, M. (2011). Nanotechnology and the sixth technological revolution. In: *Nanotechnology and the Challenges of Equity, Equality and Development* (ed. S. Cozzens, and J. Wetmore), 127–143. Dordrecht: Springer.

Kondal, R., Kalia, A., Krejcar, O. et al. (2021). Chitosan-urea nanocomposite for improved fertilizer applications: the effect on the soil enzymatic activities and microflora dynamics in N cycle of potatoes (*Solanum tuberosum* L.). *Polymers* 13 (17): 2887. doi: 10.3390/polym13172887.

Kumar, A., Choudhary, A., Kaur, H. et al. (2021). Smart nanomaterial and nanocomposite with advanced agrochemical activities. *Nanoscale Research Letters* 16 (1): 1–26. doi: 10.1186/s11671-021-03612-0.

Kumar, V., Sachdev, D., Pasricha, R. et al. (2018). Zinc-supported multiwalled carbon nanotube nanocomposite: a synergism to micronutrient release and a smart distributor to promote the growth of onion seeds in arid conditions. *ACS Applied Materials & Interfaces* 10 (43): 36733–36745. doi: 10.1021/acsami.8b13464.

Kumaraswamy, R.V, Saharan, V., Kumari, S. et al. (2021). Chitosan-silicon nanofertilizer to enhance plant growth and yield in maize (*Zea mays* L.). *Plant Physiology and Biochemistry* 159: 53–66. doi: 10.1016/j.plaphy.2020.11.054.

Kumari, M., Mukherjee, A., and Chandrasekaran, N. (2009). Genotoxicity of silver nanoparticles in *Allium cepa*. *Science of the Total Environment* 407 (19): 5243–5246.

Lee, W., An, Y., Yoon, H. et al. (2008). Toxicity and bioavailability of copper nanoparticles to the terrestrial plants mung bean (*Phaseolus radiatus*) and wheat (*Triticum aestivum*): plant agar test for water-insoluble nanoparticles. *Environmental Toxicology and Chemistry: An International Journal* 27 (9): 1915–1921.

Li, X., Gui, X., Rui, Y. et al. (2014). Bt-transgenic cotton is more sensitive to CeO_2 nanoparticles than its parental non-transgenic cotton. *Journal of Hazardous Materials* 274: 173–180.

Lin, D. and Xing, B. (2007). Phytotoxicity of nanoparticles: inhibition of seed germination and root growth. *Environmental Pollution* 150 (2): 243–250. doi: 10.1016/j.envpol.2007.01.016.

Liu, Q., Zhao, Y., Wan, Y. et al. (2010). Study of the inhibitory effect of water-soluble fullerenes on plant growth at the cellular level. *ACS Nano* 4 (10): 5743–5748.

López-Moreno, M. L., de la Rosa, G., Hernández-Viezcas, J. A. et al. (2010). X-ray absorption spectroscopy (XAS) corroboration of the uptake and storage of CeO_2 nanoparticles and assessment of their differential toxicity in four edible plant species. *Journal of Agricultural and Food Chemistry* 58 (6): 3689–3693.

Mahajan, P., Dhoke, S.K., and Khanna, A.S. (2011). Effect of nano-ZnO particle suspension on growth of mung (*Vigna radiata*) and gram (*Cicer arietinum*) seedlings using plant agar method. *Journal of Nanotechnology* 2011 (1): 696535.

Majumdar, S., Almeida, I.C., Arigi, E.A. et al. (2015). Environmental effects of nanoceria on seed production of common bean (*Phaseolus vulgaris*): a proteomic analysis. *Environmental Science & Technology* 49 (22): 13283–13293.

Mallakpour, S. and Hussain, C.M. (2021). *Handbook of Consumer Nanoproducts*. Singapore: Springer. doi: 10.1007/978-981-15-6453-6.

Manzoor, N. and Ali, L. (2022). Recent advancements and development in nano-enabled agriculture for improving abiotic stress tolerance in plants. *Frontiers in Plant Science* 13 (July): 1–12. doi: 10.3389/fpls.2022.951752.

Medintz, I.L. (2006). Recent progress in developing FRET-based intracellular sensors for the detection of small molecule nutrients and ligands. *Trends in Biotechnology* 24 (12): 539–542.

Menossi, M., Casalongué, C., and Alvarez, V.A. (2022). Bio-nanocomposites for modern agricultural applications. In: *Handbook of Consumer Nanoproducts*, 1201–1237. Singapore: Springer.

Mirbolook, A. (2024). Perspective chapter: novel slow-release nanocomposite fertilizers (Chapter 23). In: *Nanocomposites – Properties, Preparations and Applications* (ed. D.V. Parvulescu and D.E.M.M. Anghel). London: IntechOpen. doi: 10.5772/intechopen.114830.

Nagaraju, G., Saritha, G., Anju, T. et al. (2022). Nanotechnology – big impact: how nanotechnology is changing the future of agriculture? Bureau of Indian Standards Drug Controller General of India. *Journal of Agriculture and Food Research* 10 (November): 100457. doi: 10.1016/j.jafr.2022.100457.

Nair, P.M.G. and Chung, I.M. (2014). Physiological and molecular level effects of silver nanoparticles exposure in rice (*Oryza sativa* L.) seedlings. *Chemosphere* 112: 105–113.

Navarro, E., Baun, A., Behra, R. et al. (2008). Environmental behavior and ecotoxicity of engineered nanoparticles to algae, plants, and fungi. *Ecotoxicology* 17 (5): 372–386. doi: 10.1007/s10646-008-0214-0.

Omar, S.A.M. (2021). The importance of faba bean (*Vicia faba* L.) diseases in Egypt. In: *Mitigating Environmental Stresses for Agricultural Sustainability in Egypt* (ed. H. Awaad, M. Abu-hashim, and A. Negm), 371–388. Cham: Springer International Publishing. doi: 10.1007/978-3-030-64323-2_13.

Parashar, R., Afzal, S., Mishra, M. et al. (2023). Improving biofortification success rates and productivity through zinc nanocomposites in rice (*Oryza sativa* L.). *Environmental Science and Pollution Research* 30 (15): 44223–44233. doi: 10.1007/s11356-023-25293-1.

Pimentel, D. (1995). Amounts of pesticides reaching target pests: environmental impacts and ethics. *Journal of Agricultural and Environmental Ethics* 8 (1): 17–29. doi: 10.1007/BF02286399.

Pokhrel, L.R. and Dubey, B. (2013). Evaluation of developmental responses of two crop plants exposed to silver and zinc oxide nanoparticles. *Science of the Total Environment* 452: 321–332.

Popp, J., Pető, K., and Nagy, J. (2013). Pesticide productivity and food security. A review. *Agronomy for Sustainable Development* 33 (1): 243–255. doi: 10.1007/s13593-012-0105-x.

Rahman, M.H., Hasan, M.N., Amin, R. et al. (2022). Mixed nanocomposite fertilizers influencing endophytic symbiosis and nutritional and antioxidant properties of *Oryza sativa* as a sustainable alternative for commercial fertilizers. *ACS Omega* 7 (8): 6787–6794. doi: 10.1021/acsomega.1c06395.

Rashidzadeh, A. and Olad, A. (2014). Slow-released NPK fertilizer encapsulated by NaAlg-g-poly(AA-co-AAm)/MMT superabsorbent nanocomposite. *Carbohydrate Polymers* 114: 269–278. doi: 10.1016/j.carbpol.2014.08.010.

Rico, C.M., Hong, J., Morales, M.I. et al. (2013). Effect of cerium oxide nanoparticles on rice: a study involving the antioxidant defense system and in vivo fluorescence imaging. *Environmental Science & Technology* 47 (11): 5635–5642.

Rico, C.M., Majumdar, S., Duarte-Gardea, M. et al. (2011). Interaction of nanoparticles with edible plants and their possible implications in the food chain. *Journal of Agricultural and Food Chemistry* 59 (8): 3485–3498.

Riseh, R. S., Vatankhah, M., Hassanisaadi, M. et al. (2024). Nanocomposites in agriculture as pesticides for plant protection: a review. *Advances in Natural Sciences: Nanoscience and Nanotechnology* 15 (2): 23003.

Ruffini Castiglione, M., Giorgetti, L., Cremonini, R. et al. (2014). Impact of TiO_2 nanoparticles on *Vicia narbonensis* L.: potential toxicity effects. *Protoplasma* 251: 1471–1479.

Ruttkay-Nedecky, B., Krystofova, O., Nejdl, L. et al. (2017). Nanoparticles based on essential metals and their phytotoxicity. *Journal of Nanobiotechnology* 15: 1–19.

Saberi Riseh, R., Vatankhah, M., Hassanisaadi, M. et al. (2024). Nanocomposites in agriculture as pesticides for plant protection: a review. *Advances in Natural Sciences: Nanoscience and Nanotechnology* 15 (2): 023003. doi: 10.1088/2043-6262/ad4bad.

Salama, D.M., Abd El-Aziz, M.E., El-Naggar, M.E. et al. (2021). Synthesis of an eco-friendly nanocomposite fertilizer for common bean based on carbon nanoparticles from agricultural waste biochar. *Pedosphere* 31 (6): 923–933. doi: 10.1016/S1002-0160(21)60024-3.

Scherer, M.D., Sposito, J.C.V., Falco, W.F. et al. (2019). Cytotoxic and genotoxic effects of silver nanoparticles on meristematic cells of *Allium cepa* roots: a close analysis of particle size dependence. *Science of the Total Environment* 660: 459–467.

Sen, M. (2020). Nanocomposite materials (Chapter 6). In: *Nanotechnology and the Environment* (ed. M. Sen). London: IntechOpen. doi: 10.5772/intechopen.93047.

Servin, A.D. and White, J.C. (2016). Nanotechnology in agriculture: next steps for understanding engineered nanoparticle exposure and risk. *NanoImpact* 1: 9–12. doi: 10.1016/j.impact.2015.12.002.

Shakeel, A., Rizwan, K., Farooq, U. et al. (2022). Chemosphere polymer based nanocomposites: a strategic tool for detection of toxic pollutants in environmental matrices. *Chemosphere* 303 (P1): 134923. doi: 10.1016/j.chemosphere.2022.134923.

Shalaby, T.A., Bayoumi, Y., Eid, Y. et al. (2022). Can nanofertilizers mitigate multiple environmental stresses for higher crop productivity? *Sustainability* 14 (6). doi: 10.3390/su14 063480.

Shang, Y., Hasan, K., Ahammed, G.J. et al. (2019). Applications of nanotechnology in plant growth and crop protection: a review. *Molecules* 24 (2558): 1–24. https://www.ncbi.nlm.nih.gov/pmc/articles/PMC6680665/pdf/molecules-24-02558.pdf

Sharma, P., Pandey, V., Sharma, M.M.M. et al. (2021). A review on biosensors and nanosensors application in agroecosystems. *Nanoscale Research Letters* 16 (1): 1–24. doi: 10.1186/s1167 1-021-03593-0.

Shaw, D.S. and Honeychurch, K.C. (2022). Nanosensor applications in plant science. *Biosensors* 12 (9): 675. doi: 10.3390/bios12090675.

Shinde, N.A., Kawar, P.G., and Dalvi, S.G. (2024). Chitosan-based nanoconjugates: a promising solution for enhancing crops drought-stress resilience and sustainable yield in the face of climate change. *Plant Nano Biology* 7: 100059. doi: 10.1016/j.plana.2024.100059.

Silva, M. dos S., Cocenza, D.S. et al. (2011). Paraquat-loaded alginate/chitosan nanoparticles: preparation, characterization and soil sorption studies. *Journal of Hazardous Materials* 190 (1): 366–374. doi: 10.1016/j.jhazmat.2011.03.057.

Silveira, N.M., Seabra, A.B., Marcos, F.C.C. et al. (2019). Encapsulation of S-nitrosoglutathione into chitosan nanoparticles improves drought tolerance of sugarcane plants. *Nitric Oxide* 84: 38–44. doi: 10.1016/j.niox.2019.01.004.

Singh Sekhon, B. (2014). Nanotechnology in agri-food production: an overview. *Nanotechnology, Science and Applications* 7 (2): 31–53. doi: 10.2147/NSA.S39406.

Son, W.K., Choi, Y.S., Han, Y.W. et al. (2023). In vivo surface-enhanced Raman scattering nanosensor for the real-time monitoring of multiple stress signalling molecules in plants. *Nature Nanotechnology* 18 (2): 205–216.

Sonali, J.M.I., Kavitha, R., Kumar, P.S. et al. (2022). Application of a novel nanocomposite containing micro-nutrient solubilizing bacterial strains and CeO_2 nanocomposite as bio-fertilizer. *Chemosphere* 286: 131800.

Soran, M.-L., Lung, I., Opriş, O. et al. (2021). The effect of TiO_2 nanoparticles on the composition and ultrastructure of wheat. *Nanomaterials* 11 (12): 3413.

Srivastava, A.K., Dev, A., and Karmakar, S. (2018). Nanosensors and nanobiosensors in food and agriculture. *Environmental Chemistry Letters* 16 (1): 161–182. doi: 10.1007/s10311-017-0 674-7.

Tortella, G., Rubilar, O., Fincheira, P. et al. (2024). Copper nanoparticles as a potential emerging pollutant: divergent effects in the agriculture, risk-benefit balance and integrated strategies for its use. *Emerging Contaminants* 10 (4): 100352.

Velicogna, J.R., Schwertfeger, D.M., Beer, C. et al. (2020). Phytotoxicity of copper oxide nanoparticles in soil with and without biosolid amendment. *NanoImpact* 17: 100196.

Vishwakarma, K., Shweta, Upadhyay, N. et al. (2017). Differential phytotoxic impact of plant mediated silver nanoparticles (AgNPs) and silver nitrate ($AgNO_3$) on *Brassica* sp. *Frontiers in Plant Science* 8: 1501.

Wang, M., Hu, Z., Yang, T. et al. (2023). A dual pesticide–fertilizer silicon-based nanocomposite to synergistically control fungal disease and provide nutrition. *Environmental Science: Nano* 10 (12): 3462–3475. doi: 10.1039/D3EN00500C.

Wang, M., Zhang, G., Zhou, L. et al. (2016). Fabrication of pH-controlled-release ferrous foliar fertilizer with high adhesion capacity based on nanobiomaterial. *ACS Sustainable Chemistry & Engineering* 4 (12): 6800–6808. doi: 10.1021/acssuschemeng.6b01761.

Wang, Q., Ebbs, S.D., Chen, Y. et al. (2013). Trans-generational impact of cerium oxide nanoparticles on tomato plants. *Metallomics* 5 (6): 753–759.

Wang, X., Cai, A., Wen, X. et al. (2017). Graphene oxide-Fe_3O_4 nanocomposites as high-performance antifungal agents against *Plasmopara viticola*. *Science China Materials* 60 (3): 258–268.

Wang, Y.-Y., Ji, H.-Y., Lyu, H.-H. et al. (2019). Simultaneous alleviation of Sb and Cd availability in contaminated soil and accumulation in *Lolium multiflorum* Lam. After amendment with Fe–Mn-modified biochar. *Journal of Cleaner Production* 231: 556–564. doi: 10.1016/j.jclepro.2019.04.407.

Wu, X., Wei, J., Wu, C. et al. (2021). ZrO_2/CeO_2/polyacrylic acid nanocomposites with alkaline phosphatase-like activity for sensing. *Spectrochimica Acta Part A: Molecular and Biomolecular Spectroscopy* 263: 120165. doi: 10.1016/j.saa.2021.120165.

Wypij, M., Trzcińska-Wencel, J., Golińska, P. et al. (2024). The strategic applications of natural polymer nanocomposites in food packaging and agriculture: chances, challenges, and consumers' perception. *Frontiers in Chemistry* 2023: 1–18. doi: 10.3389/fchem.2022.1106230.

Yanga, J., Cao, W., and Rui, Y. (2017). Interactions between nanoparticles and plants: phytotoxicity and defense mechanisms. *Journal of Plant Interactions* 12 (1): 158–169. doi: 10.1080/17429145.2017.1310944.

You, J. and Chan, Z. (2015). ROS regulation during abiotic stress responses in crop plants. *Frontiers in Plant Science* 6: 1092. doi: 10.3389/fpls.2015.01092.

Yu, J., Wang, D., Geetha, N. et al. (2021). Current trends and challenges in the synthesis and applications of chitosan-based nanocomposites for plants: a review. *Carbohydrate Polymers* 261: 117904. doi: 10.1016/j.carbpol.2021.117904.

Zain, M., Ma, H., Nuruzzaman, M. et al. (2023). Nanotechnology based precision agriculture for alleviating biotic and abiotic stress in plants. *Plant Stress* 10 (August): 100239. doi: 10.1016/j.stress.2023.100239.

Zhang, M., Gao, B., Chen, J. et al. (2015). Effects of graphene on seed germination and seedling growth. *Journal of Nanoparticle Research* 17: 1–8.

Zhang, Y. and Goss, G. G. (2022). Nanotechnology in agriculture: comparison of the toxicity between conventional and nano-based agrochemicals on non-target aquatic species. *Journal of Hazardous Materials* 439: 129559. doi: 10.1016/j.jhazmat.2022.129559.

Zungu, B., Kamdem Paumo, H., Gaorongwe, J.L. et al. (2023). Zn nutrients-loaded chitosan nanocomposites and their efficacy as nanopriming agents for maize (*Zea mays*) seeds. *Frontiers in Chemistry* 11: 1243884.

3

Role of Nanofertilizers in Sustainable Growth of Crop Plants and Production

Aaradhya Pandey, Pragya Tiwari, and Eti Sharma

School of Biotechnology, Gautam Buddha University, Greater Noida, Uttar Pradesh, India

3.1 Introduction

Ecological catastrophes along with concerns related to food and nutrients are world-level problems that require to be solved through rapid discoveries at both scientific and registration levels. Modernization at the industrial level and increasing pollution can lead enhancement of harmful pollutants in the environment. The research performed by Adam et al. stated that the world's pollution in 2021 will be approximately 11.0 billion by 2100 (Adam 2021). Thus, the present food manufacturing has to be enhanced by 60–70%, which can help to achieve the needs of continuously expanding pollution such as nutrient-enriched food along with the high level of yield and growth of a crop (Babu et al. 2022). The evolution in the field of agriculture is synchronized with the evolution of humans. The agriculture process in a traditional way requires a regular utilization of fertilizers. Thus, conventional fertilizers have played a vital role in advanced agricultural procedures at the time of the green revolution. Their application in conventional agriculture procedures can lead to enhancement in crop yield and growth along with high nutrient value (Rajput et al. 2021). However, there are many advantages of chemical fertilizers, but still, there are several disadvantages, such as their production can create a formation of certain harmful chemical products or by-products that can lead to air pollution and unutilized nutrients of chemical fertilizers have adverse effects on the soil health and its fertility and also have an impact on groundwater (Chandini et al. 2019). For example, when unused urea is released into the ecological system, it has the potential to act as an air, soil, and water pollutant. The volatilized form of urea, "nitrous oxide," works as a greenhouse gas, which releases a kind of ammonia gas that takes part in air pollution and global warming and nitrate form that leaches through the soil to water causing an impact on the quality of drinking water (Kashyap et al. 2023). Due to this, the agricultural sector faces a major climate change, which affects agriculture and also endangers food security (Saraiva et al. 2023). Therefore, to ensure sustainable agriculture, developing and implementing smart agricultural techniques using new technologies is crucial. This involves the necessity for the creation of current and creative

Nanotechnology-based Sustainable Agriculture, First Edition. Edited by Pardeep Singh et al.

fertilizers that have great effectiveness and few adverse effects such as nanofertilizer (NF). NFs use modern nanotechnology to fertilize crops efficiently and sustainably, making them a possible solution to address these difficulties, providing increased efficiency and lower environmental impact (Yadav, Yadav, and Abd-Elsalam 2023b).

NFs are made up of nanostructured particles, usually <100 nm in size. NFs improve nutrient uptake, controlled release kinetics, and nutrient use efficiency by taking advantage of nanoparticles' (NPs) unique qualities, such as their high surface area-to-volume ratio and tailored surface functions (Avila-Quezada et al. 2022). These traits can promote crop development along with their growth and also reduce the loss of nutrients that occur via volatilization and leaching, lowering the environmental impact of fertilizer application. Moreover, nanostructured materials are used in agriculture to discharge pest-control chemicals, monitor soil health, and provide plant nourishment (Nongbet et al. 2022a).

In addition to the aforementioned benefits, they must be applied to the soil in smaller quantities, improving application convenience and lowering transportation costs. However, like any other fertilizer, the usage of NFs has several restrictions and drawbacks (Shukla et al. 2019). Currently, 383 articles can be retrieved from PubMed by searching the keyword "nano fertilizers," which is increased from earlier numbers as specified in the paper. Agrochemicals by increasing numbers indicate that active study is going on in this particular area and the field is growing as researchers are publishing work on this topic. This study also reflects the most often occurring searches in conjunction with "nano fertilizer" from 2013 to 2023 and those are sustainable agriculture, agriculture, and NPs and these are among the most addressed fields or domains. So, it can be implied that the growing interest in this field is a result of its potential benefits and its positive impact on sustainable agriculture. Despite the progress in science that paved the way for the development of NFs and their potential to increase agricultural sustainability and production, there are still safety concerns about the use of nanomaterials for people's health and the environment. Although there is a lot of promise for NFs, they still need strong regulation and risk assessment to address growing concerns and safe implementation (Yadav et al. 2023a).

In this chapter, NF's role in the sustainable development of plants and production will be highlighted. The following sections provide a brief overview of NF synthesis and its mechanism for improving soil fertility and agricultural productivity. NFs' benefits over traditional chemical fertilizers are also been reviewed along with the restrictions of NF utilization. Finally, this chapter discusses current limitations and potential future applications of NFs. The flow of the overall chapter is shown in Figure 3.1.

3.2 NFs, Its Types, and Synthesis Methods

NFs are processed NPs that can directly provide nutrients to plants and promote production as well as growth (Sun et al. 2022). In order for NPs to be employed as fertilizers, a synthesis method which is capable to form large-scale particles with regulated physicochemical qualities at a reasonable cost is needed (Raliya et al. 2018). NFs are synthesized via physical, chemical, and biological techniques. Physical methods are simpler and more scalable than chemical procedures, while chemical methods provide precise control over particle shape and size. Biological approaches, which make use of microbes and plant

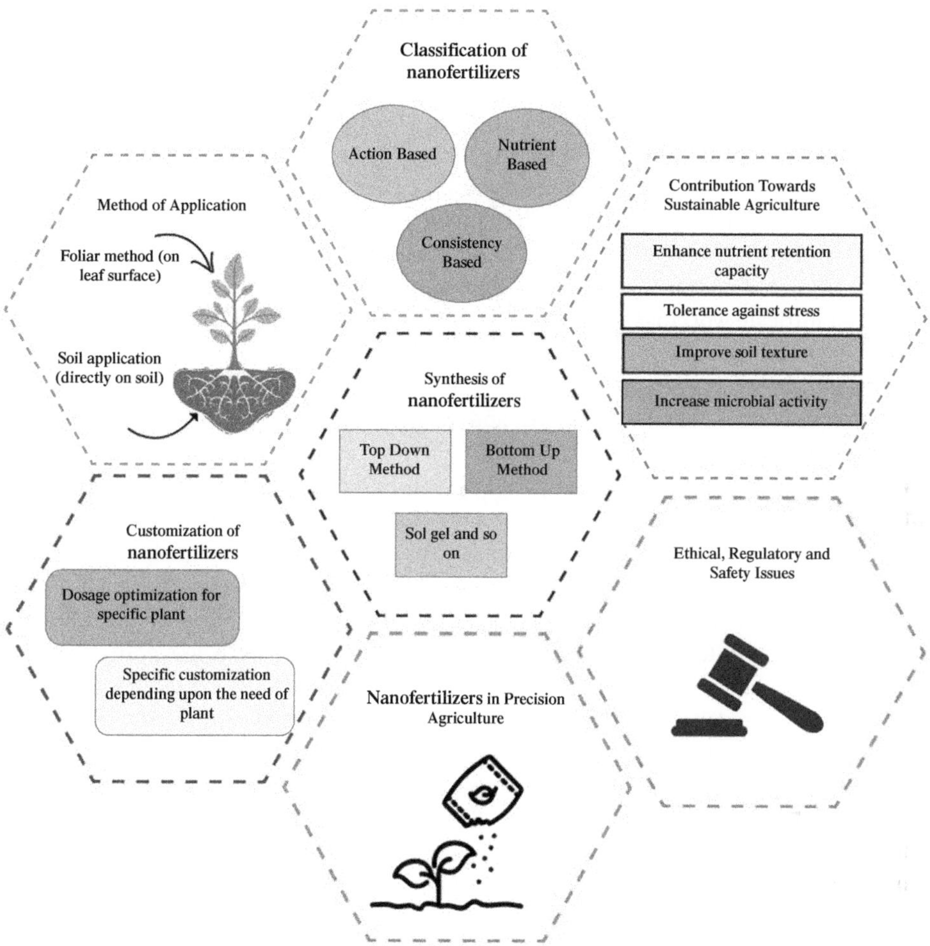

Figure 3.1 Role of nanofertilizers in sustainable agriculture.

extracts, provide ecologically beneficial options. NFs' dimensions, forms, surface characteristics, and interactions with plant cells all affect how they work (Padhan et al. 2024). Nanomaterial components, such as silica, Fe, ZnO, metal oxides, and other nanomaterials, along with crop type, significantly impact their effectiveness as NFs for plant growth. Moreover, addition of nutrients, either alone or in a form of combination along with absorbents that are nanosized, leads to the creation of NFs. When it comes to cationic nutrients, they are simply loaded and directed without any modification to the formation of NPs. However, in case of anionic nutrients, they are loaded after certain surface modifications to form NPs, which can be achieved by utilizing various chemical and physical methods (Nongbet et al. 2022).

High-energy ball milling is a top-down strategy for creating NFs, which is inexpensive, simple to operate, and suitable for large-scale manufacturing. Planetary ball mills are used to create most NPs, but little research has been done on NF production. Milling media, or process control agents, help grind particles by controlling the mill's internal temperature and minimizing cold welding. Wet milling is believed to be more efficient in comparison to salt-assisted milling and dry milling. A higher ball-to-powder ratio reduces particle, size but adds contaminants to the final product. Real-time ball milling is an impractical, time-consuming process, it is and ineffective; therefore, modeling the planetary ball milling process is crucial. The dynamic milling process is represented by mathematical, statistical, and analytical models. Statistical models provide most and least important parameter details, but dependable optimization requires a large amount of baseline data. Effective planetary ball milling can be simulated using numerical models, but they have many variables, and intricate workings that are challenging to understand and include numerically. Most models calculate the principal output, milling energy, or power and conclude that greater power or energy results in greater size reduction. A few models have connected the milling energy and kinematic equations, and the primary goal of creating NF is to achieve nanoscale size (Pohshna et al. 2020).

Particle size and content may be precisely controlled due to the chemical nature of this procedure. "Emulsion," "co-precipitation," "reverse micelle" creation, and micelle formation are a few controlled synthesis processes that can yield homogeneous NPs with little coagulation or aggregation. Once created, the NPs' physical and chemical characteristics, including their stability, dispersibility, and solubility, need to be evaluated using a variety of analytical techniques. In general, the top-down method is less costly than the bottom-up method, as it creates NPs by assembling smaller building pieces. The top-down method may create NPs with certain size distributions and take less time. It might not be as accurate as the bottom-up method, though, and it might cause the NPs to develop flaws. Furthermore, some materials may degrade as a result of the top-down approach, rendering them less appropriate for particular uses. The economy of a top-down method of NP synthesis generally relies on the specific utilization, the intended qualities of the nanomaterial, as well as the expense of necessary ingredients and equipment (Lin et al. 2014).

3.2.1 NF and Their Significance in Current Agriculture

The agricultural production system is dealing with various problems like reduction in crop yield, huge amounts of nutrient deficiencies, and a decrease in organic matter in the soil (IFFCO 2022). NFs are arising solution of sustainable agriculture that can overcome various challenges and also provide high effectiveness. Moreover, they are also capable of decreasing the harmful effects of chemical fertilizers on the environment (Yadav et al. 2023a). They are exclusive, eco-friendly, and less expensive fertilizers that are capable of performing a vital role in the field of agriculture. Various research has presented on the NF's role in soil and plants as they are capable of working effectively to enhance the production of agriculture. For example, a study conducted by Azam et al. stated that the NFs of ZnO can improve the yield and growth of maize. They can applied as NFs for cultivating crops that are grown in soil with zinc deficiency and have the potential to encourage the crop's qualitative and quantitative value (Azam et al. 2022). This can be supported by the research performed on NFs of nitrogen as they are a resource of quick plant development (root and shoot system of a plant) because of enhancement in chlorophyll level in leaves of a plant (Verma et al. 2023). The research performed by Tang and Fei demonstrated that the potassium NFs are capable of enhancing levels of chlorophyll and have a prominent improvement in plant parameters in a peanut plant (Tang and Fei 2021). Thus, the application of NFs has led to the creation of a significant step in the direction of sustainable agriculture in the entire world along with its amazing advantages, such as improving nutrient uptake and efficiency while being eco-friendly and acting as a sustainable substitute for conventional fertilizers (Table 3.1).

Table 3.1 Effect of various nanofertilizers application in sustainable growth of different crops.

Type of nanofertilizers	Crops	Application method and conditions	Effects	References
Nano-sulfur (NS)	Rice	200 mg/kg NS via soil method and application period: 35 days for seeding, 120 days for maturation of plant	• It enhances the 40% biomass of rice seeding • By 26% improves the maturation of plant seed growth	G Meselhy et al. (2021)
Nano-zeolite (NZs) and hybrid nanocomposite (HNCs)	Strawberry	Soil and foliar application method	• NZS increased the growth level in comparison to HNCF + NZs and HNCSF • However, HNCSF affects the fat, protein, and ash • HNCF + NZs enhances the energy and carbohydrates	Rahman, Hasan, and Khan (2021)
ZnO NPs	Squash plants	Foliar application	• Increases the plant yield and development	Shebl et al. (2019)
B and Zn NPs	*Punica granatum*	Foliar application	• Enhances the fruit yield in a plant	Davarpanah et al. (2016)
Nano-calcium fertilizers	Pomegranate	Foliar application	• Decreases the cracking of a fruit	Davarpanah et al. (2018)
Zn NPs	Wheat	Foliar application	• Enhances the nutrient uptake and yield of crop	Meena, Jat, and Jain (2021)
Magnesium oxide	Groundnut and green gram	Nano-priming	• Releases antioxidants which help in increasing the production of chlorophyll	Vijai Anand et al. (2020)

(Continued)

Table 3.1 (Continued)

Type of nanofertilizers	Crops	Application method and conditions	Effects	References
Chitosan–silicon	Maize	Foliar application	• Enhances the plant development and its yield	Mujtaba et al. (2020)
Urea along with nano-nitrogen chelate	Sugarcane	Soil treatment	• Increases the production of sugar in a crop	Alimohammadi et al. (2020)
Urea-HAP	Wheat, rice	Soil treatment	• Slow release of N, reduced leaching of N, and enhanced nutrient content	Sharma, B. et al. (2022)
Chitosan	Tomato	Foliar application	• Improved plant stress tolerance (mitigate biotic stress)	Chun and Chandrasekaran (2019)
Iron (III) oxide combined with cow dung or flower waste; NPK	Tomato	Soil treatment	• Enhanced growth, improved soil quality, and also maintained soil moisture	Yadav et al. (2023d)
Mo	Chickpea	Foliar application	• Improved dehydrogenase enzyme activity and increased plant root nutrition and yield	Thavaseelan and Priyadarshana (2021)
Silica	*Zea mays L.*	Foliar application	• Enhanced plant's resilience to salinity • Increased plant growth	Saxena et al. (2016)
CeO_2 and TiO_2	*Hordeum vulgare L.*	Soil treatment	• Impacts the seed germination and also enhances the oxidative stress in the root cells of a plant	Mattiello et al. (2015)
$CaCO_3$	Peanut	Foliar application	• Increases the plants seeding and growth • Enhances the plant's yield	El-temsah et al. (2023)
Mg-NP and Fe-NP	Black-eyed pea	Foliar application	• Increases the photosynthetic efficiency of a plant • Enhances the yield of plants	Delfani et al. (2014)

3.2.2 Classification of NFs

NFs are classified into three types: nanoscale fertilizers, nanoscale additives, and nanoscale coatings. Nanoscale fertilizers (nutrient-containing SiO_2, Fe, and other NPs). They are very small in size; they can easily diffuse through the root and leave pores resulting in efficient absorption of nutrients. Nanoscale additive fertilizers are called traditional fertilizers that utilize nanoscale additions like metal and metal oxide NPs or carbon-based nanomaterials, and nanoscale coating fertilizers are conventional fertilizers that are encapsulated or packed with nano-sized particles. The coating improves porosity on the surface, which helps in longer retention of some nutrients (Mikkelsen 2018). NFs have many types based on their varying parameters and characteristics. Therefore, its classification is inconsistent. It can be classified based on the material used, nutrients they carry, and consistency or action. As mentioned by Yadav et al. (2023), NFs are categorized into four main types based on their action, nutrient composition, consistency, and delivery mechanisms. Each type offers specific benefits for nutrient efficiency, plant growth, environmental sustainability, and promoting agricultural practices. An explanation of each type and its corresponding subtypes are given as follows.

3.2.2.1 Action-based NFs

Action-based NFs are based on the mode of action they deliver. They are made to carry out particular tasks that improve the development of plants and nutrient administration. This category includes four primary types: regulated release, targeted delivery, plant growth boosting, and water and nutrient loss regulating NFs. Examples of controlled-release NFs are nanocapsule-based NF type – in this type, fertilizers are delivered via nanocapsules, which are composed of biopolymers, metal oxides (M_2O_3), and carbon nanotubes. They decrease the danger of leaching by forming a barrier of defense, realizing nutrients gradually, and maintaining nutritional stability (Sinha et al. 2019), and example of targeted delivery is chitosan-based NF – in this, a positively charged polymer called chitosan comes in contact with negatively charged substances to create complexes with fertilizer particles that make them more accessible to plants. Chitosan improves nutrient absorption by trying to shield biomolecules from light, temperature changes, and pH shifts (Mujtaba et al. 2020).

3.2.2.2 Nutrient-based NFs

These are the fertilizers that improve the delivery and efficiency of nutrients needed for the growth of plants. Nutrient-based NFs are classified based on the type of nutrient they provide: inorganic, organic, hybrid, and nutrient-loaded NFs. Inorganic NFs include metallic and nonmetallic NPs, which can provide essential macronutrients like nitrogen (N), phosphorous (P), sulfur (S), and others to plants (Yadav et al. 2023d). In this type, one or many nutrients are encased within a specific nanomaterial. For example, for the localized delivery and slow release of N, hydroxyapatite (NP) coated with urea-based NF has been made. It is also a reservoir of calcium and phosphate. In contrast with pure urea, it lowers NH_3 volatilization. (Basavegowda and Baek 2021). Hybrid NFs are made up of NPs of a specific nutrient and they are covered with a layer that protects them and therefore allows them to increase their availability to plants. Hybrid materials are a combination of base materials like polymer and nanomaterials made from inorganic materials, which are dispersed throughout the polymer phase. This coupling provides them with improved physical and mechanical characteristics; hence, these materials perform better (Avila-Quezada et al. 2022).

3.2.2.3 Consistency-based NFs

Consistency-based NFs are categorized by their physical form or the type of coating they use, such as surface-coated NFs, synthetic polymer–coated NFs, and biological product–coated NFs. Artificial polymers like polyacrylamide help in the controlled release of nitrogen and reduce its loss through leaching, thereby increasing its efficiency. We can also use natural polymers like alginate acting as a biological coating, which can enhance the availability of essential macronutrient phosphorous and its uptake (Yadav et al. 2023d).

3.2.2.4 Nanocarrier-loaded NFs

Nanocarrier-loaded NFs utilize various nanocarriers to supply nutrients to crops: nanogels, nanocapsules, and nanoscale porous materials. Among the diverse classifications, this structured categorization focuses on the various strategies used to maximize the uptake of nutrients and improve plant development, which eventually supports sustainable farming methods.

3.2.2.5 Nano-biofertilizers

Nano-biofertilizers are a combination of biological fertilizers and nanostructures. Biofertilizers are made up of microorganisms that naturally increase soil productivity by nitrogen fixation, synthesizing growth-promoting substances, and solubilizing phosphorus (Nongbet et al. 2022a). These biofertilizers have been synthesized by using microorganisms like *Azotobacter, Anabaena, Rhizobium, and Pseudomonas* spp., which are cost-effective, renewable, and environmentally friendly. They are proven as good alternatives to synthetic fertilizers. Soil quality improvement, essential nutrient deposition, and generation of plant defense molecules are some significant advantages of biofertilizers. In one study, nitrogen-fixing bacteria were delivered to plants via chitosan-based NPs, which improved the bacteria's ability to enter plant roots and increased plant growth and yield (Yadav et al. 2023d).

3.3 Mode of Action

The efficient and maximum absorption of nutrients for plants depends upon the reactivity of nanomaterials. Moreover, they have improved the efficiency of applications that provide minimal loss of nutrients compared to chemical fertilizers (Fatima et al. 2021). The effective absorption, accumulation, and distribution of NFs are based on various factors, such as soil pH, the concentration of organic material in soil, its texture, and also factors related to the NPs like their coating and size (Ma et al. 2018). The absorption of NFs is via the plants' roots and leaves affecting a plant's bioavailability, action, and absorption. For example, research conducted by macronutrient NFs enhances plant growth by 19% compared to synthetic fertilizers. In addition, macronutrient NFs are superior by 18%, and NFs that carry macronutrients enhance the development of crops by 29% in comparison to various other fertilizers (Júnior et al. 2020).

3.3.1 Mechanism of Nutrient Release and Uptake by Plants

Typically, NFs involve small particles of polymers or metal oxides like zinc oxide, polyethylene, polystyrene, and titanium oxide. All these materials have various properties that make them great for various delivery fertilizers. As an example, zinc oxide and titanium oxide absorb maximum and are capable of providing nutrients directly to the cells of a plant. Whereas in the case of polymers, they have many versatile properties capable of designing the release of nutrients for a long period enabling them to be precise along with a controlled release of nutrients (Padhan et al. 2024). The NFs use site-specific and controlled nutrient release mechanisms that allow them to increase the delivery efficiency of nutrients, decrease nutrient loss along with the dose reduction, and reduce soil and water pollution, which helps to fulfill essential characteristics in various fields like economic, social, health, and environmental. These methods can reduce the requirement for the utilization of artificial fertilizers, pesticides, and irrigation on crops that help to conserve energy and resources (Azeem et al. 2023).

There are various mechanisms by which plants can uptake the micronutrients and macronutrients. Upon releasing NFs in the environment near the crop, the nutrients released by NF are absorbed by various routes in plants via leaf, root, endocytosis, and other parts of plants (Ioannou et al. 2020) as shown in Figure 3.2. The study conducted by Satori stated the comparison between two kinds of absorption methods of plants on orange crops. They have examined the method of nutrient uptake in plants via root and leaf by continued application of zinc for five years. The research demonstrated that the root canal method is more significant as proven. A further more crucial route of nutrient uptake is via a leaf and the application of fertilizers is conceivable as same as the pesticides and herbicides. Therefore, the application of fertilizers is done in the foliar method of application, which can reduce contamination and is highly effective (Sartori et al. 2008). This can be supported by the research conducted by Omer in 2019, in which the impact of chitosan NPs along with modified carbon nanotubes packed with nitrogen, potassium, and phosphorus in the form of fertilizers was evaluated, which were applied to the French bean plant. The mode of application was leaf application and seed priming. In the result of the experiment, they observed that the leaf application of fertilizers is more efficient than the seed priming method because it can decrease the days of harvesting without impacting the crop yield (Abdel-Aziz et al. 2019).

The primary methods of nutrient uptake in plants are leaf and root, but these routes are supported by various other methods like the endocytosis method. The process of endocytosis involves the transfer of components from the extracellular matrix to the cell by utilizing the vesicles provided by the cell membrane. Various parameters have a direct impact on the nutrient uptake of plants, such as the size of a particle, chemical composition, plant stage, concentration, environmental condition, and exposure period of an element. Thus, it is vital to conduct a deep study of every crop when the utilization of NFs is done as

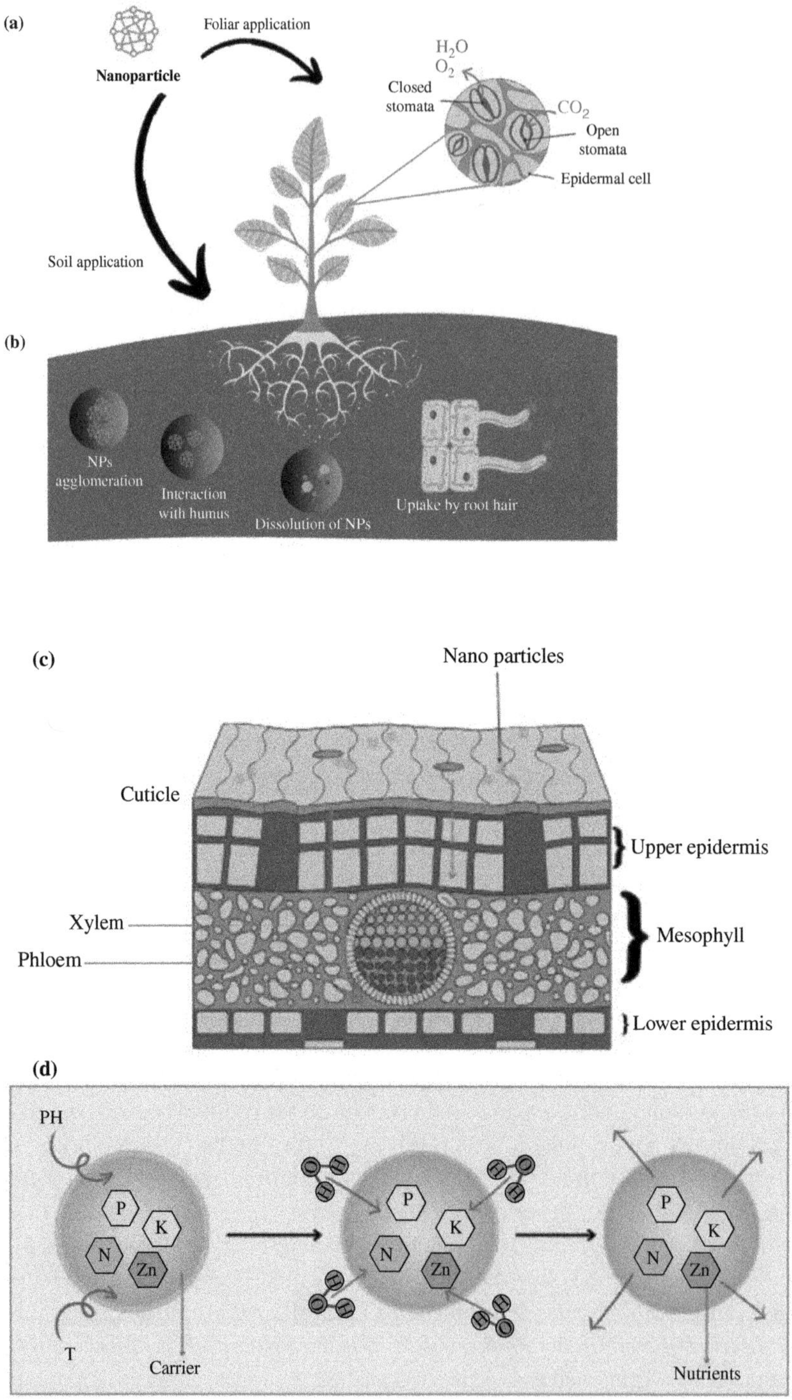

Figure 3.2 (a) Method of application of nanofertilizers, (b) NPs' release in soil and their uptake via the root of the plant, (c) uptake of NPs through the leaf, and (d) controlled release of agrochemicals from nanofertilizers.

a resource for plant nutrients (Júnior et al. 2020). For example, research performed by Apodaca et al. (2017) shows that when NPs of copper along with kinetin (a plant hormone) were applied to the bean plants, they observed a large concentration of copper in the tissues of the roots, which demonstrates the function of endocytosis in the plant's nutrient-absorbing mechanisms (Apodaca et al. 2017).

3.3.2 Increased Nutrition Uptake by Plants

There are various significant properties of NF particles, such as being smaller than the pore size of leaves and roots and having a large surface area that helps in easy penetration of them in a plant cell. These properties of NFs aid in enhancing the uptake of nutrients along with increasing nutrient use efficiency. Thus, decreasing the size of a particle can result in an enhancement in the surface area of a particle, whereas increasing the number of particles in a per unit area during fertilization can offer more chances to create contact with NFs, which results in a higher penetration along with increasing uptake of nutrient (Meena et al. 2017).

According to Tarafdar (2012), fertilizers that are encapsulated with NPs are capable of enhancing the accessibility of nutrients along with the uptake of nutrients in a crop. For example, in the research conducted by Abou-yuoseff et al. (2022), in this experiment, they evaluated the impact of conventional fertilizers and NPK NFs based on different parameters, such as plant growth, nutrition content, and yield of the crop in Faba Bean. As a result of an experiment, they have observed that NPK NFs when applied in the recommended dose of N, K, and P lead to an increase in crop yield and nutrient uptake that enhances the dry weight of a crop (Tarafdar 2012). Therefore, NFs act as important tools in the field of agriculture that help to improve crop yield and development by enhancing nutrient use efficiency (NUE) and also reduce expenses of cultivation and the loss of nutrients. Moreover, NFs have the potential to increase nutrient uptake from the soil, improve the NUE, and increase the productivity of a crop by reducing the loss of nutrients (Ain et al. 2023).

3.3.3 Improved Water and Nutrient Retention in Soil

NFs are composed of those NPs that can control the fertilizer's rate of discharge in soil, enabling less utilization of fertilization and even maintaining a similar outcome of crops. NFs are created in such a way that help to decrease water loss and also control the release of nutrients via using various methods. Among the several methods, one technique is that NFs are encapsulated by the porous medium which provides a slow nutrient release beyond the same period (Okey-Onyesolu et al. 2021). Another technique includes the modification in the surface of NFs that makes them hydrophilic, which leads to decreased water loss caused by evaporation and an enhancement in their water holding capacity (Sivarethinamohan and Sujatha 2021).

Nanobeads and nanoemulsions are certain examples of NFs that can control soil's nutrient and water loss. Nanobeads are composed of several elements like carbon, iron, and various other kinds of metals. These are small beads that can penetrate even in a tiny break of soil, which assists in fertilizing the plants. Nanoemulsions are created via a combination of surfactant with liquid by utilizing a high-energy mixing process that provides stable homogenized mixtures. They are small droplets created to have both insoluble and soluble nutrients. They help in crop health and yield enhancement. However, there is a need for more research and studies concerning developing less expensive and sustainable strategies for crop enhancement and soil health.

3.4 Contribution Toward Sustainable Agriculture

The implementation of an effective plant nutrition system that causes less environmental harm is crucial for sustainable agriculture, which calls for the use of agrochemicals to a minimum. Thus, by offering advanced and innovative solutions, nanotechnological approaches are being applied to tackle a global agricultural issue. Among these cutting-edge methods, NFs have become a pivotal technological advancement that is essential to advancing the sustainability of agriculture (Basavegowda and Baek 2021). NFs promote sustainable development in the following ways.

3.4.1 Enhanced Nutrient Retention Capacity

To ensure effective nutrient delivery and crop uptake, NFs release nutrients gradually over time (Avila-Quezada et al. 2022). Coating nutrients with nanomaterials allows for regulated release, prevents unfavorable nutrient loss, and allows nutrients to gently seep into the soil. By minimizing the loss of nutrients due to runoff and leaching, furthermore, by reducing rapid volatility and deterioration, the utilization of nanoscale fertilizers can enhance the fertility of soil along with nutrient quality over a certain period, thereby boosting crop output. The optimum places for soil remediation are those with strong nutrient retention capabilities and regular precipitation patterns (Nongbet et al. 2022a). An increase in crop yield and the need for agricultural labor may result from plants' enhanced capacity for nitrogen absorption (Yadav, Yadav, and Abd-Elsalam 2023b). NFs have a high surface-to-volume ratio, thanks to their nanostructure, which increases the number of biologically active sites, thereby absorbing nutrients properly and gently (Solanki et al. 2015).

3.4.2 Biotic and Abiotic Stress Tolerance by Plants

Plants respond to environmental stresses by undergoing genetic and physicochemical changes at the molecular level. NFs can penetrate cell walls and provide nutrients to the root zone and other sections of the plant rapidly, and they are capable of enhancing plant stress tolerance (Yadav et al. 2023d). For example, silica particles enhance plants' resilience to abiotic stresses like drought, heavy metal toxicity, and salinity by enhancing antioxidant enzyme induction, absorption processes, toxic metal ion immobilization, and metal ion categorization (Nongbet et al. 2022a). Research has demonstrated that silica nanoparticles (SNPs) enhance growth and yield in stressed plants because SNPs are absorbed by the root and form a thin layer in the cell wall to support the plant's ability to withstand stress and continue to produce. In a study conducted on maize plants, it was observed that maize seedlings absorb NPs synthesized from SiO_2 than microparticles. Efficient absorption meets the growth requirement of the plant and observed good growth. When these SiO_2 NPs were applied at higher concentrations in pear seedlings, no toxicity and adverse effects were observed on plant biology (Saxena et al. 2016).

In a recent case study, the use of chitosan NPs has demonstrated encouraging outcomes in reducing crop stress brought on by *Fusarium andiyazi*-induced tomato wilt disease. According to in vitro research, chitosan NPs at a dosage of 5.0 mg/mL inhibited radial mycelial development to the greatest extent among the tested concentrations—achieving an astounding 73.8%. The pathogenesis-related (PR) proteins and antioxidant genes, which are essential parts of the plant defense system against pathogen assaults, are upregulated in the underlying process. This finding emphasizes the potential of chitosan NPs as a strong biocontrol tool to boost plant resilience and lessen the effects of tomato crop wilt disease (Chun and Chandrasekaran 2019; Tortella et al. 2023).

3.4.3 Increase Microbial Activity

The interaction of NPs with the crop microbiome is one of the factors that influence plant growth (Yadav et al. 2023d). Microbes that produce bioactive chemicals, growth hormones, vitamins, and antioxidants, such as *Azotobacter, Rhizobium*, and a few species of *Pseudomonas*, help plants fix nitrogen and phosphate-solubilize it indirectly (Basavegowda and Baek 2021). They also protect against pathogens. NFs promote plant growth and the activity of soil microbes that break down organic materials. The improved dehydrogenase enzyme activity, which indicates the metabolic activity of a microorganism and creates organic acids, has been linked to the overall increased development of mung bean plants. This enhances the accessibility of phosphorus in the soil for future crop absorption. The rhizosphere's microbial community controls soil fertility by carrying out vital biogeochemical nutrient cycling (Ahmed et al. 2021). Mo-NPs, used as a micronutrient source in chickpeas, increased plant root nutrition, antioxidant enzyme activity, and resistance to plant pathogens, thereby enhancing the yield, performance, and disease resistance of legume crops in loamy soil (Thavaseelan and Priyadarshana 2021).

3.4.4 Lesser Environmental Pollution

Overutilization of fertilizers pollutes the ecosystem because residual and unutilized fertilizer particles contaminate the water bodies, soil, and atmosphere. However, it is believed that applying fertilizers with a slow and sustained release of nutrients will enhance nutrient use (Tarafder et al. 2020). Besides being nontoxic, NFs are secure for humans and the ecosystem. NFs are formed in a way to release nutrients gradually over a period in a controlled manner and lower the risk of ecological contamination by reducing the loss of nutrients (Basavegowda and Baek 2021). NFs show high solubility and diffusion rates. NFs can take nutrients efficiently, decreasing the quantity of fertilizer applied and, as a result, the toxicity of the soil environment (Avila-Quezada et al. 2022). By applying these special qualities of NPs, NFs can improve plant nutrient availability, raise crop yields, lower environmental impact caused by using fertilizers, and support sustainable farming methods.

3.5 Customization of NFs

Nanotechnology enables the precise engineering of NPs that provide a gradual delivery of nutrients over some time. This allows the farmer to develop customized fertilizers by utilizing specific methods designed according to the requirements demanded by specific types of plants or crops, the condition of local soil, and the development stage of plants (Yedoti et al. 2024). The customization of NFs is conducted because they can engage with a treatment that treats the lack of accessibility of nutrients in plants. The customization or designing of NF is easily conducted by the presence of the atom on the surface of NPs allowing ease in aligning them into desired attributes (Zahra et al. 2022). This can be supported by research performed on inorganic types of NFs, which provide a special benefit due to their capability to get customized, and they are customized based upon nutrient needs related to specific crops that provide particular utilization of nutrients that help in improving the yield of the crop. Inorganic fertilizers are composed of metalloids, non-metallic, and metallic NPs. These fertilizers can supply plants with vital nutrients like nitrogen, potassium, and phosphorus. Inorganic fertilizers are created in a way that can enhance nutrient uptake efficiency (NUE) and increase crop yields (Yadav et al. 2023).

Customizing the delivery system of the NF is a crucial step as it is cost-effective and provides an impactful delivery of nutrients to crops and plants. Moreover, NPs offer various advantageous compounds like minerals, enzymes, and vitamins, which permit the customization of NFs (Babu et al. 2022). NFs are designed to make slow delivery over time, which helps reduce nutrient loss and decreases environmental safety. The research conducted on NFs around the world is less. However, the literature on fertilizers has shown that customizing fertilizers can make sustainable farming productive (Subramanian et al. 2015).

3.5.1 Dosage Optimization

A wide range of studies is needed to identify various traits of NFs like their mode of action, method of application, types of NFs, and their effect on the crop. A slight change in the concentration of NFs can lead to a harmful impact on plant yield and growth. Therefore, the optimization of the amount of dosage of various NFs for distinct plant species is necessary (Gade et al. 2023). For example, when an in-depth study was conducted on micronutrient NFs to evaluate elements that affect their accessibility in a field, their values are compared to conventional fertilizers in the case of biofortification when the application method is fertigation with commonly accessible micronutrient fertilizers acting as one option. This happened due to the nanotoxicity produced by these kinds of NFs. Thus, to overcome the nanotoxicity of micronutrient NFs, it is vital to know the optimal dose of them for every crop (Nongbet et al. 2022). NF encourages the crops to have better yield and development through a suitable concentration and dosage. Optimizing the dosage for various NFs for distinct plants can lead to ecologically friendly and higher productivity of a production system (Al-Juthery et al. 2021). This can be supported by the study conducted by Tarar et al., in which they synthesized the NFs from banana biochar (BNF), then evaluated their surface morphology, and also determined their impact on the first stage of development parameters of broccoli, red okra, and cucumber. As a result, they observed that BNF has increased the 50% of germination of cucumber, okra, and broccoli. Moreover, in comparison to traditional fertilizers, they have increased the root and shoot length of every individual vegetable. They have also optimized the dose of BNF, which is 37.5 kg/ha, because at this dose, the improvement in soil quality and slow-release of nutrients is achieved by providing a further supply of organic matter (Tarar et al. 2023). Thus, optimizing the dosage of NFs can protect the plant from their toxicity.

3.5.2 Method of NFs' Application

Majorly, there are three kinds of NF application methods: foliar, soil treatment, and seed nanopriming. In the foliar application method, NFs are applied in a spray form on a leaf surface, offering immediate absorption by the leaf surface (Liu and Lal 2015). Whereas in the case of seed nanopriming, seeds are precoated or soaked in a solution that already has NFs in it, and this process is done before the plantation of seeds (Adisa et al. 2019). The soil treatment involves the direct application of NFs in the soil through banding, localized placement, and broadcasting of NFs (Subramanian et al. 2015). The choice of an ideal mode for the utilization of NFs is vital because it helps to obtain the optimal development of a plant. The choice of mode is based on the quality of the soil, climate, and nutrient accessibility that have an impact on the uptake and use of nutrients. Comprehension of these variables based on this selection of the ideal mode of application is done, which helps in decreasing ecological effects, enhancing the crop yields, and leading to creation of additional sustainable procedures of agriculture (Yadav et al. 2023).

3.5.2.1 Foliar Spray

It is a modernized and advanced mode of application of NFs in which fertilizers are directly sprayed on the plant's leaves in liquid form, which allows quick absorption of the nutrient via an exterior part of a leaf. This method provides a specific, ideal, quick delivery of NFs and also involves the precise translocation of NFs in plants. This technique can control and delay the release of nutrients to achieve increased efficiency of substance. The process of absorption in this method occurs directly or via stomata and endocytosis; however, this process is primarily based on the size of the particles. After the process of absorption is completed, these particles accumulate in the vacuoles of plant cells. Different variables affect the process of absorption and transfer of NPs such as traits of plants, ecological conditions, and physical properties of an NP (Ding et al. 2023).

The foliar spray provides various applications in comparison to the soil treatment; it has a quick response, decreases the leaching of nutrients, and enhances the use of nutrients. This can be supported by various studies related to this method. For example, a study conducted by Lopez-Lima et al. (2021) stated that when Cu NPs are applied via a foliar spray method in a *Solanum lycopersicum* plant, the yield of the plant was enhanced by 80% along with a 30% decrease in the need for Cu concentration in comparison to traditional copper fungicides. Yet there are certain disadvantages of this technique, such as nutrient penetration and movement occurring via cuticles of the leaf (Júnior et al. 2020).

3.5.2.2 Nanopriming

Nanopriming means seed priming with NPs or nanoagents which are capable of protecting the seed, enhancing the germination of the seed, boosting growth of a seed, and improving the yield of the seed by offering resistance to various kinds of plant stress (Nile et al. 2022). They can control the biochemical procedure despite maintaining a balance between the various plant hormones and reactive oxygen compounds. They are utilized to encourage the plant to develop, and its metabolism provides resistance to abiotic stress and enhances the germination of a seed (Salam et al. 2022). In this method, the seed is soaked in the solution having NFs, which helps in decreasing the application of fertilizer to half and provides better results (do Espirito Santo Pereira et al. 2021). The research performed on bean seeds that are coated or primed with NPs of chitosan of different concentrations for a period of three hours, and they also have treatment of NaCl of 100 mM, which results in increased length of radicle and germination of seed (Zayed et al. 2017). By utilizing this method, NFs are capable of providing power to antioxidants and decreasing the formation of reactive oxygen species (ROS; Sharma et al. 2023).

3.5.2.3 Soil Treatment

In this method, the NFs are applied directly to the soil by utilizing traditional ways like side-dressing, fertigation, and broadcasting techniques. After the application of NFs via the soil treatment method, the NPs are absorbed by the roots of a plant or they can penetrate root cells through endocytosis (Ahmed et al. 2021; Miralles et al. 2012). In soil treatment, NFs are capable of interacting with the particles of soil and plants and also influence microbes present in soil. These influences can lead to alteration in their function and actions. The controlled release of nutrients via NPs led to improvement in plant development and yields (Madzokere et al. 2021). However, there are some disadvantages of this technique, such as they are expensive, have some long-term harmful effects on NPs, and have issues in regulation (Mishra et al. 2019).

3.6 NFs' Integration with Precision Agriculture

A type of agricultural production management known as "precision farming" makes use of information technology to monitor and control crop production inputs, including water, herbicides, and fertilizers. Precision farming techniques might be enhanced by the use of NFs, which have been made possible by recent developments in nanotechnology (Raliya et al. 2018). With the rising growth of computer capabilities and the realization of data's worth as exploitable information, artificial intelligence (AI) and machine learning techniques have been rising more popularly to be a way to find trends in big datasets that indicate forthcoming results. Neural network algorithms are one of the most popular methods. They create a model that produces estimations using the remaining data by using an unaltered subset of the total data that is accessible as the training set; the predictions' validity is assessed to make sure they could not have arisen at random. There are primary categories of machine learning algorithms and instances of their use in the creation of nanomaterials and/or agriculture, which are as follows – supervised learning, unsupervised learning, and reinforcement learning (Baştanlar and Özuysal 2014).

Supervised learning is an algorithm that predicts a target outcome from a set of independent variables. Examples of supervised learning methods include decision trees, regression, random forest algorithms, logistic regression, and k-nearest neighbors (KNN). KNN algorithms find applications in the fields of agriculture, agronomy, nanomaterials design, safety, and interactions. They have been used to predict water retention, develop quantitative structure–activity relationship (QSAR) models for nanomaterials' cellular association, and predict biomolecule coronas and their evolution during nanomaterials' uptake into plants. These algorithms can also be incorporated into the International Air Transport Association (IATA) by utilizing the water-retention model to project the mobility of NPs in soil (Afantitis et al. 2018).

Unsupervised learning is a machine learning algorithm, where it does not require any prior instructions to predict the outcome. It does it by analyzing the patterns or structures in the data given and then it interprets the data by clustering the same kind of data in one. "K-means clustering and a priori algorithms" are two examples. Segmentation technique has been applied in agriculture and agronomy; agricultural lands may be divided into several control units using unsupervised learning methods. Then, by regulating these units uniquely according to their features, farming operations may be optimized and made more specialized (Pedroso et al. 2010). K-means clustering has been applied to process the raw data from mass spectrometry (spICP-MS) to quantify the disintegration and uptake of nanomaterials. It can predict or customize the NPs in different soil and climate conditions. This involves finding out the effectiveness of nanoagrichemicals and their net yield. This is important for interpreting the impact they have on the environment and utilizing their positive potential in agriculture (Bi et al. 2014). Reinforcement learning is also a type of machine learning where software optimizes its action based on a trial-and-error approach and is reliant on past experiences to make decisions. "Markov decision process" is an exemplification of reinforcement learning.

3.7 Ethical, Regulatory, and Safety Issues

Nowadays, the application of NFs in the agricultural field is excessively enhanced due to their ability to increase crop productivity and enhance soil quality. However, using agricultural products produced by nanotechnology creates concerns about their potential to harm human health and the environment (Khot et al. 2012). To resolve these problems, regulatory organizations at a global level have designed various rules and regulations that ensure the safe utilization of NFs. Certain critical guidelines and rules in terms of nanotechnology-based agricultural products are as follows (Kumari et al. 2023):

- *Regulatory control:* It should be made by each country. They should create a regulatory framework for using nanotechnology in agriculture by themselves. For example, in the United States, all regulations related to nanotechnology-based agri products are created and maintained by government organizations like the US Food and Drug Administration (FDA) and the United States Department of Agriculture (USDA).
- *Risk screening:* This is an essential step conducted by the regulatory organization to evaluate the risk-causing NP in nanotechnology-based agricultural products. The assessment is performed on properties of an NP like the toxicity caused by NPs when they are released into the environment and their effect on human health.
- *Labeling:* The regulatory organization critically requires the labeling of the nanomaterial present in agricultural production. The knowledge about the nanomaterial present in the agri product is essential to consumers so that they can decide about the utilization and purchase of the goods.

- *Standardization at the international level:* The formation of international standards is essential to ensure about quality and safety of nanotechnology-based agricultural products. International Organizations form various standards linked to nanotechnology like ISO/TS 80004-1, and these standards provide information about the nanomaterial's definition and terminology.
- *Research and development:* The regulatory organization provides motivation and support to the research and development related to nanotechnology-based agricultural products and also helps to provide information about the safety issues of the products related to the environment and human health. For example, the government of the United States has established a program called Small Business Innovation Research (SBIR) to encourage the formation of new technologies and small businesses. The US Department of Agriculture used this program to encourage the development and formation of NFs. The government of Canada also provides credit for tax in business, which has funded the research and development in experiments' development and scientific programs.

In India, the development of all rules and regulations related to nanotechnology-based agricultural products is done by various bodies in the Department of Biotechnology (DBT). For example, nano-agri-input products (NAIPs) maintain all regulatory frameworks related to NF, which comes under the DBT (Adholeya et al. 2020).

Figure 3.3 demonstrates various factors of a regulatory framework related to the commercialization of NFs (Yadav, Yadav, and Abd-Elsalam 2023). Despite such amazing developments in nanotechnology research, it still lacks the standards related to the regulatory framework linked with the commercialization of nano-enabled products, which leads to a slowdown of their development in various countries (Adholeya et al. 2017).

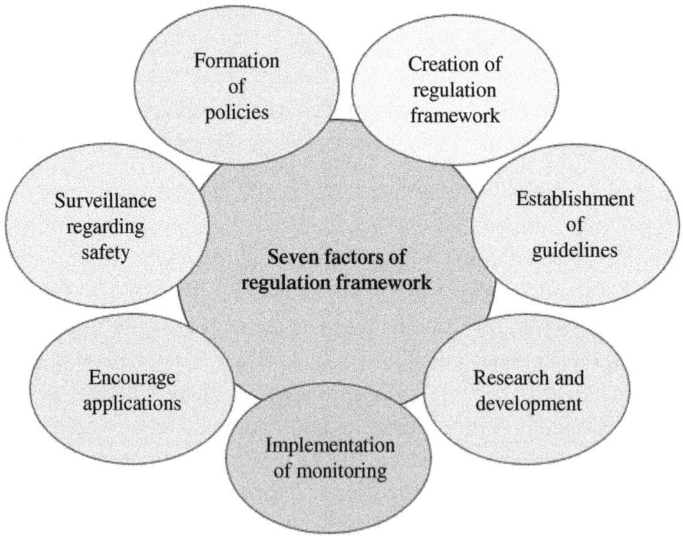

Figure 3.3 Factors of regulation framework involved in the commercialization of nanofertilizers.

However, there are still a lot of developments required by the rules and regulations related to NFs and nanotechnology-based agriculture products. This development will help in the faster application and development of NFs in various countries, leading to the formation of sustainable agriculture.

3.8 Advantages and Limitations

There are various advantages of NFs compared to conventional fertilizers, which are mentioned below. However, there are certain disadvantages of NFs and it is also vital to discuss them (Nichepom 2023).

3.8.1 Advantages of NFs

In comparison to traditional fertilizers, NFs provide a variety of advantages for environmentally responsible and sustainable crop development (Avila-Quezada et al. 2022). This includes improved efficiency and reduced environmental impact through direct nutrient delivery to plants. This kind of innovation has the potential to increase crop yields while reducing fertilizer use (Colipano and Cagasan 2022).

- *Higher solubility:* NFs are more soluble and dispersed than traditional synthetic fertilizers, allowing plants to absorb more nutrients (Thavaseelan and Priyadarshana 2021). They are more soluble because they dissolve more readily in solvents like water because of their smaller particle size and greater surface area. In comparison to traditional fertilizers, nano-hydroxyapatite and nano-zinc oxide have demonstrated increased solubility.
- *Environmental benefits:* One of the main causes of environmental degradation and climate change is nutrient loss from agricultural fields as a result of leaching and gaseous emissions (Kumar et al. 2021). NFs offer environmental benefits like improved efficiency and reduced environmental impact, for example, when applied on the surface, nano-nitrogen can improve nutrient absorption efficiency, decrease urea losses, and mitigate contamination of the soil, air, and water (Kumar et al. 2021).
- *Efficient nutrient delivery to plants and increased crop yield:* Plant nutrients are delivered as nano-sized emulsions using fertilizers based on nanotechnology, which encapsulates the nutrients in nano forms. Crop output is increased as an outcome of increased fertilizer usage efficiency, improved crop production, and a reduced environmental impact (Kumar et al. 2021).
- *Increased plant growth:* NFs maximize plant growth and development, providing the right nutrients at the right time and in the right amounts. Zinc oxide, a NP with NF potential, increased tomato plant growth and biomass output when applied to tomato leaves (Avila-Quezada et al. 2022).
- *High bioavailability:* NFs have high bioavailability compared to conventional fertilizers. By activating different pathways and enzymes within the plant system, the application of NFs improves the bioavailability of nutrients and raises the biomass of roots and rhizosphere microbial population.

- *Controlled or slow release:* Common ways of distribution include coating and drenching. These can guarantee that fertilizer is applied uniformly throughout the soil and regulate the rate of nutrient discharge over time (Yadav et al. 2023d). Plants receive a consistent supply of nutrients from NFs, which promotes plant development and production. They lessen the requirement for regular fertilizer reapplication and the amount of nutrients that leak into the environment (Kumar et al. 2021).

3.8.2 Limitation of NFs

Even though there are many advantages to using NF in agriculture, there are also certain limitations in utilizing nanofertilization (Iqbal et al. 2019). Mainly, studies that are conducted related to NFs are either at a small scale or in a lab because conducting them on a large scale (computerization of NFs) involves many challenges such as high cost, issues in standardization, and lack of knowledge in the formation of nanofertilization (Kah et al. 2018; Yadav, Yadav, and Abd-Elsalam 2023a). For example, the limitation involved with foliar application of NF such as if a spray concentration is high, they are flammable and require an ample area of a functional leaf (Achari and Kowshik 2018). The effect caused by applying nutrients when supplied in a molecule form in a field via the process of transformation inside a plant and their effect on ecology needs to be investigated. For example, an investigation is needed to understand whether the NFs that are applied to plants get converted into an ionic form within a plant and get involved with a plant's protein and metabolites or whether they do not get converted but remain in intact form and even can reach the consumers via a food chain process. There are certain risks that the utilization of NF includes, such as risks caused to human health, risks caused to the environment, and risks caused to the ecological system (Biswal 2023).

- *Risks caused to human health:* NFs' capacity to affect human health is a significant concern. The capacity to be toxic is because of their properties like their small size, which leads to easy absorption by an organism. According to Gupta et al., research demonstrated that ingestion of NF can cause harm to the liver, kidneys, and gastrointestinal tract in experimental animals (Gupta and Xie 2018). Moreover, NPs can surpass the biological barrier like a blood–brain barrier, leading to neurological harm (Zulfiqar et al. 2019).
- *Risks caused to the environment:* The discharge of NPs in the environment causes contamination in water, air, and soil. These accumulations of NPs are capable of causing damage to soil ecology and also able to reduce its fertility. Furthermore, it is possible that the leakage of NPs from soil can cause damage to aquatic animals, which eventually can harm the food chain (Gade et al. 2023).
- *Risks caused to the ecological system:* A further crucial concern is NFs' capacity that affects the nontargeted organisms. NFs can tamper with the organism's proliferation, growth, and reproductivity, leading to a population reduction. However, their utilization benefits certain microbes like mycorrhizal fungi and nitrogen-fixing bacteria (Yadav et al. 2023).

Therefore, these limitations related to NFs should be addressed and various investigations should be conducted that help in different ways to overcome these limitations of NFs.

3.9 Conclusion and Future Perspective

The advent of NFs represents a significant breakthrough in the field of sustainable agriculture. NFs act as a modern technique for the fertilization of crops and enhance their yield. These are not only cost-effective and eco-friendly but also can act as one of the alternatives to chemical fertilizers. The reason for this is due to their outstanding properties such as providing site-specific delivery and also having controlled release of agrochemical components. These types of fertilizers can also be customized to release nutrients in a controlled manner, reducing waste and environmental impact while enhancing crop yields. Thus, these properties of NFs make them capable of playing a vital role in sustainable agriculture. In the agricultural sector, a revolution of precise agriculture and the integration of AI to design NFs provide pollution reduction and enhance the NUE of plants. They are also capable of enhancing the quality of soil without reducing the microbial activity of the soil. Customization of NF leads to the development of fertilizers that are developed according to the soil condition and requirements of plants that they need for their growth and proliferation. Dosage optimization also minimizes the adverse effect of over-fertilization in soil and water systems.

Looking into the future, continued research and development in NF technologies promise even greater advancements. Despite having so many advantages, they also have some limitations, which can harm human health, ecological systems, and the environment. Due to this, there are many rules, regulations, and guidelines have been developed for the implementation of NF. Establishing robust regulatory frameworks will ensure that NFs are used safely and responsibly, preventing unintended consequences. Through careful management, NFs can pave the way for a more sustainable and equitable agricultural future.

References

Abdel-Aziz, H.M.M., Hasaneen, M.N.A., and Omer, A.M. (2019). Impact of engineered nanomaterials either alone or loaded with NPK on growth and productivity of French bean plants: seed priming vs foliar application. *South African Journal of Botany* 125: 102–108. doi: 10.1016/j.sajb.2019.07.005.

Abou-yuoseff, A.M., El–Khair, A., El–Mohtasem, M.O. et al. (2022). Impacts of nano-fertilizers and chemical fertilizers on plant growth and nutrient uptake by Faba Bean (*Vicia faba* L.) plant. *Al-Azhar Journal of Agricultural Research* 47 (1): 216–229. doi: 10.21608/ajar.2022.266504.

Achari, G.A. and Kowshik, M. (2018). Recent developments on nanotechnology in agriculture: plant mineral nutrition, health, and interactions with soil microflora. *Journal of Agricultural and Food Chemistry* 66 (33): 8647–8661. doi: 10.1021/acs.jafc.8b00691.

Adam, D. (2021). How far will global population rise? Researchers can't agree. *Nature* 597 (7877): 462–465. doi: 10.1038/d41586-021-02522-6.

Adholeya, A., Das, R.K., Dubey, M.K. et al. (2017). "Regulation of nanoproducts in agriculture" Zero Draft Policy on regulation of nanoproducts in agriculture. *The Energy and Resources Institute at teriin*. https://www.teriin.org/sites/default/files/2018-03/zero-draft-policy.pdf.

Adholeya, A., Dinda, A., Subramanian, K.S. et al. (2020). Guideline for evaluation of nano-based agri-input and food products in India, Department of Biotechnology, Ministry of Science & Technology, Ministry of Agriculture and Farmers' Welfare. https://dbtindia.gov.in/sites/default/files/uploadfiles/Guidlines%20for%20Evaluation%20of%20Nano%20Based%20Agri%20Input%20and%20Food%20Products%20in%20India.pdf.

Adisa, I.O., Pullagurala, V.L.R., Peralta-Videa, J.R. et al. (2019). Recent advances in nano-enabled fertilizers and pesticides: a critical review of mechanisms of action. *Environmental Science: Nano* 6 (7): 2002–2030. doi: 10.1039/C9EN00265K.

Afantitis, A., Melagraki, G., Tsoumanis, A. et al. (2018). A nanoinformatics decision support tool for the virtual screening of gold nanoparticle cellular association using protein corona fingerprints. *Nanotoxicology* 12 (10): 1148–1165. doi: 10.1080/17435390.2018.1504998.

Ahmed, B., Rizvi, A., Ali, K. et al. (2021). Nanoparticles in the soil–plant system: a review. *Environmental Chemistry Letters* 19 (2): 1545–1609. doi: 10.1007/s10311-020-01138-y.

Ain, Q.U., Hussain, H.A., Zhang, Q. et al. (2023). Use of nano-fertilizers to improve the nutrient use efficiencies in plants. In: *Sustainable Plant Nutrition*, 299–321. Elsevier. doi: 10.1016/B978-0-443-18675-2.00013-4.

Alimohammadi, M., Panahpour, E., and Naseri, A. (2020). Assessing the effects of urea and nano-nitrogen chelate fertilizers on sugarcane yield and dynamic of nitrate in soil. *Soil Science and Plant Nutrition* 66 (2): 352–359. doi: 10.1080/00380768.2020.1727298.

Al-Juthery, H.W.A., Lahmod, N.R., and Al-Taee, R.A.H.G. (2021). Intelligent, nano-fertilizers: a new technology for improvement nutrient use efficiency (Article Review). *IOP Conference Series: Earth and Environmental Science* 735 (1): 012086. doi: 10.1088/1755-1315/735/1/012086.

Apodaca, S.A., Tan, W., Dominguez, O.E. et al. (2017). Physiological and biochemical effects of nanoparticulate copper, bulk copper, copper chloride, and kinetin in kidney bean (*Phaseolus vulgaris*) plants. *Science of the Total Environment* 599–600: 2085–2094. doi: 10.1016/j.scitotenv.2017.05.095.

Avila-Quezada, G.D., Ingle, A.P., Golinska, P. et al. (2022). Strategic applications of nano-fertilizers for sustainable agriculture: benefits and bottlenecks. *Nanotechnology Reviews* 11 (1): 2123–2140. doi: 10.1515/ntrev-2022-0126.

Azam, M., Bhatti, H.N., Khan, A. et al. (2022). Zinc oxide nano-fertilizer application (foliar and soil) effect on the growth, photosynthetic pigments and antioxidant system of maize cultivar. *Biocatalysis and Agricultural Biotechnology* 42: 102343. doi: 10.1016/j.bcab.2022.102343.

Azeem, A., Abbas, N., Azeem, S. et al. (2023). Physiological and molecular mechanism of nanoparticles induced tolerance in plants. In: *Emerging Contaminants and Plants: Interactions, Adaptations and Remediation Technologies* (ed. T. Aftab), 233–248. Cham: Springer International Publishing. doi: 10.1007/978-3-031-22269-6_9.

Babu, S., Singh, R., Yadav, D. et al. (2022). Nanofertilizers for agricultural and environmental sustainability. *Chemosphere* 292: 133451. doi: 10.1016/j.chemosphere.2021.133451.

Basavegowda, N. and Baek, K.-H. (2021). Current and future perspectives on the use of nanofertilizers for sustainable agriculture: the case of phosphorus nanofertilizer. *3 Biotech* 11 (7): 357. doi: 10.1007/s13205-021-02907-4.

Baştanlar, Y. and Özuysal, M. (2014). Introduction to machine learning. In: *miRNomics: MicroRNA Biology and Computational Analysis* (ed. M. Yousef and J. Allmer), 105–128.

Totowa, NJ: Humana Press (Methods in Molecular Biology).doi: 10.1007/978-1-62703-748-8_7.

Bi, X., Lee, S., Ranville, J.F. et al. (2014). Quantitative resolution of nanoparticle sizes using single particle inductively coupled plasma mass spectrometry with the K-means clustering algorithm. *Journal of Analytical Atomic Spectrometry* 29 (9): 1630. doi: 10.1039/C4JA00109E.

Biswal, D. (2023). Use of nanofertilizers in agriculture: advantages, disadvantages, and future implications. In: *Advances in Environmental Engineering and Green Technologies* (ed. R. Lone and J.A. Malik), 102–133. IGI Global. doi: 10.4018/978-1-6684-5533-3.ch006.

Chandini, R.K., Kumar, R., and Om, P. (2019) The impact of chemical fertilizers on our environment and ecosystem. In: Research Trends in Environmental Sciences, 2e, 71–86. https://www.researchgate.net/publication/331132826.

Chun, S.-C. and Chandrasekaran, M. (2019). Chitosan and chitosan nanoparticles induced expression of pathogenesis-related proteins genes enhances biotic stress tolerance in tomato. *International Journal of Biological Macromolecules* 125: 948–954. doi: 10.1016/j.ijbiomac.2018.12.167.

Colipano, J. M. and Cagasan, U.A. (2022). A review on the impact of organic, conventional and nano fertilizer application in crop production. *Eurasian Journal of Agricultural Research* 6 (2): 101–109.

Davarpanah, S., Tehranifar, A., Abadía, J. et al. (2018). Foliar calcium fertilization reduces fruit cracking in pomegranate (*Punica granatum* cv. Ardestani). *Scientia Horticulturae* 230: 86–91. doi: 10.1016/j.scienta.2017.11.023.

Davarpanah, S., Tehranifar, A., Davarynejad, G. et al. (2016). Effects of foliar applications of zinc and boron nano-fertilizers on pomegranate (*Punica granatum* cv. Ardestani) fruit yield and quality. *Scientia Horticulturae* 210: 57–64. doi: 10.1016/j.scienta.2016.07.003.

Delfani, M., Baradarn Firouzabadi, M., Farrokhi, N. et al. (2014). Some physiological responses of black-eyed pea to iron and magnesium nanofertilizers. *Communications in Soil Science and Plant Analysis* 45 (4): 530–540. doi: 10.1080/00103624.2013.863911.

Ding, Y., Zhao, W., Zhu, G. et al. (2023). Recent trends in foliar nanofertilizers: a review. *Nanomaterials* 13 (21): 2906. doi: 10.3390/nano13212906.

do Espirito Santo Pereira, A., Caixeta Oliveira, H., Fernandes Fraceto, L. et al. (2021). Nanotechnology potential in seed priming for sustainable agriculture. *Nanomaterials* 11 (2): 267. doi: 10.3390/nano11020267.

El-temsah, M.E., Abd-Elkrem, Y.M., El-Gabry, Y.A. et al. (2023). Response of diverse peanut cultivars to nano and conventional calcium forms under alkaline sandy soil. *Plants* 12 (14): 2598. doi: 10.3390/plants12142598.

Fatima, F., Hashim, A., and Anees, S. (2021). Efficacy of nanoparticles as nanofertilizer production: a review. *Environmental Science and Pollution Research* 28 (2): 1292–1303. doi: 10.1007/s11356-020-11218-9.

G Meselhy, A., Sharma, S., Guo, Z. et al. (2021). Nanoscale sulfur improves plant growth and reduces arsenic toxicity and accumulation in rice (*Oryza sativa* L.). *Environmental Science & Technology* 55 (20): 13490–13503. doi: 10.1021/acs.est.1c05495.

Gade, A., Ingle, P., Nimbalkar, U. et al. (2023). Nanofertilizers: the next generation of agrochemicals for long-term impact on sustainability in farming systems. *Agrochemicals* 2 (2): 257–278. doi: 10.3390/agrochemicals2020017.

Gupta, R. and Xie, H. (2018). Nanoparticles in daily life: applications, toxicity and regulations. *Journal of Environmental Pathology, Toxicology and Oncology* 37 (3): 209–230. doi: 10.1615/JEnvironPatholToxicolOncol.2018026009.

IFFCO (2022). Nano urea fertilizer for sustainable farming (no date) IFFCO nano urea. https://nanourea.in/en/nano-urea (accessed 22 July 2024).

Ioannou, A., Gohari, G., Papaphilippou, P. et al. (2020). Advanced nanomaterials in agriculture under a changing climate: the way to the future? *Environmental and Experimental Botany* 176: 104048. doi: 10.1016/j.envexpbot.2020.104048.

Iqbal, M., Umar, S., and Mahmooduzzafar (2019). Nano-fertilization to enhance nutrient use efficiency and productivity of crop plants. In: *Nanomaterials and Plant Potential* (eds. A. Husen and M. Iqbal), 473–505. Cham: Springer International Publishing. doi: 10.1007/978-3-030-05569-1_19.

Júnior, A., Mulinari, J., De Oliveira, C. et al. (2020) Nanofertilizers: an overview. In: *Ciência, Tecnologia e Inovação: do campo à mesa. Congresso Internacional Da Agroindústria*. Instituto Internacional Despertando Vocações. doi: 10.31692/ICIAGRO.2020.0041.

Kah, M., Kookana, R.S., Gogos, A. et al. (2018). A critical evaluation of nanopesticides and nanofertilizers against their conventional analogues. *Nature Nanotechnology* 13 (8): 677–684. doi: 10.1038/s41565-018-0131-1.

Kashyap, S., Kumar, R., Ram, H. et al. (2023). Quantitative and qualitative response of fodder maize to use of bulk and nano-fertilizers in north western plains of India. *Agronomy* 13 (7): 1889. doi: 10.3390/agronomy13071889.

Khot, L.R., Sankaran, S., Maja, J.M. et al. (2012). Applications of nanomaterials in agricultural production and crop protection: a review. *Crop Protection* 35: 64–70. doi: 10.1016/j.cropro.2012.01.007.

Kumar, Y., Singh, T., Raliya, R. et al. (2021). Nano fertilizers for sustainable crop production, higher nutrient use efficiency and enhanced profitability. *Indian Journal of Fertilizers (IJF)* 17: 1206–1214.

Kumari, R., Suman, K., Karmakar, S. et al. (2023). Regulation and safety measures for nanotechnology-based agri-products. *Frontiers in Genome Editing* 5: 1200987. doi: 10.3389/fgeed.2023.1200987.

Lin, P.-C., Lin, S., Wang, P.C. et al. (2014). Techniques for physicochemical characterization of nanomaterials. *Biotechnology Advances* 32 (4): 711–726. doi: 10.1016/j.biotechadv.2013.11.006.

Liu, R. and Lal, R. (2015). Potentials of engineered nanoparticles as fertilizers for increasing agronomic productions. *Science of the Total Environment* 514: 131–139. doi: 10.1016/j.scitotenv.2015.01.104.

Lopez-Lima, D., Mtz-Enriquez, A.I., Carrión, G. et al. (2021). The bifunctional role of copper nanoparticles in tomato: effective treatment for *Fusarium* wilt and plant growth promoter. *Scientia Horticulturae* 277: 109810. doi: 10.1016/j.scienta.2020.109810.

Ma, C., White, J.C., Zhao, J. et al. (2018). Uptake of engineered nanoparticles by food crops: characterization, mechanisms, and implications. *Annual Review of Food Science and Technology* 9 (1): 129–153. doi: 10.1146/annurev-food-030117-012657.

Madzokere, T.C., Murombo, L.T., and Chiririwa, H. (2021). Nano-based slow releasing fertilizers for enhanced agricultural productivity. *Materials Today: Proceedings* 45: 3709–3715. doi: 10.1016/j.matpr.2020.12.674.

Mattiello, A., Filippi, A., Pošćić, F. et al. (2015). Evidence of phytotoxicity and genotoxicity in *Hordeum vulgare* L. exposed to CeO_2 and TiO_2 nanoparticles. *Frontiers in Plant Science* 6: 1043. doi: 10.3389/fpls.2015.01043.

Meena, R.H., Jat, G., and Jain, D. (2021). Impact of foliar application of different nano-fertilizers on soil microbial properties and yield of wheat. *Journal of Environmental Biology* 42: 302–308. doi: 10.22438/JEB/42/2/MRN-1465.

Mikkelsen, R. (2018). Nanofertilizer and nanotechnology: a quick look. *Better Crops with Plant Food* 102 (3): 18–19. doi: 10.24047/BC102318.

Miralles, P., Church, T.L., and Harris, A.T. (2012). Toxicity, uptake, and translocation of engineered nanomaterials in vascular plants. *Environmental Science and Technology* 46 (17): 9224–9239. doi: 10.1021/es202995d.

Mishra, M., Dashora, K., Srivastava, A. et al. (2019). Prospects, challenges and need for regulation of nanotechnology with special reference to India. *Ecotoxicology and Environmental Safety* 171: 677–682. doi: 10.1016/j.ecoenv.2018.12.085.

Mujtaba, M., Khawar, K.M., Camara, M.C. et al. (2020). Chitosan-based delivery systems for plants: a brief overview of recent advances and future directions. *International Journal of Biological Macromolecules* 154: 683–697. doi: 10.1016/j.ijbiomac.2020.03.128.

Nichepom. (2023). Fertilizers of the future: nano vs. traditional – which way to go? 8 June. https://www.nicheagriculture.com/nano-fertilizers-vs-traditional-fertilizers/ (accessed 28 July 2024).

Nile, S., Thiruvengadam, M., Wang, Y. et al. (2022). Nano-priming as emerging seed priming technology for sustainable agriculture – recent developments and future perspectives. *Nanobiotechnology* 20: 254. doi: 10.1186/s12951-022-01423-8.

Nongbet, A., Mishra, A.K., Mohanta, Y.K. et al. (2022). Nanofertilizers: a smart and sustainable attribute to modern agriculture. *Plants* 11 (19): 2587. doi: 10.3390/plants11192587.

Okey-Onyesolu, C.F., Hassanisaadi, M., Bilal, M. et al. (2021). Nanomaterials as nanofertilizers and nanopesticides: an overview. *ChemistrySelect* 6 (33): 8645–8663. doi: 10.1002/slct.202102379.

Padhan, S.R., Kar, I., Mohanty, A. et al. (2024). Nanofertilizers: types, synthesis, methods, and mechanisms. In: *Nanofertilizers for Sustainable Agroecosystems* (ed. K.A. Abd-Elsalam and M.A. Alghuthaymi), 61–98. Cham: Springer Nature Switzerland (Nanotechnology in the Life Sciences). doi: 10.1007/978-3-031-41329-2_3.

Pedroso, M., Taylor, J., Tisseyre, B. et al. (2010). A segmentation algorithm for the delineation of agricultural management zones. *Computers and Electronics in Agriculture* 70 (1): 199–208. doi: 10.1016/j.compag.2009.10.007.

Pohshna, C., Mailapalli, D.R., and Laha, T. (2020). Synthesis of nanofertilizers by planetary ball milling. In: *Sustainable Agriculture Reviews 40* (ed. E. Lichtfouse), 75–112. Cham: Springer International Publishing (Sustainable Agriculture Reviews). doi: 10.1007/978-3-030-33281-5_3.

Rahman, Md. H., Hasan, Md. N., and Khan, Md. Z.H. (2021). Study on different nano fertilizers influencing the growth, proximate composition and antioxidant properties of strawberry fruits. *Journal of Agriculture and Food Research* 6: 100246. doi: 10.1016/j.jafr.2021.100246.

Rajput, V.D., Singh, A., Minkina, T. et al. (2021). Nano-enabled products: challenges and opportunities for sustainable agriculture. *Plants* 10 (12): 2727. doi: 10.3390/plants10122727.

Raliya, R., Saharan, V., Dimkpa, C. et al. (2018). Nanofertilizer for precision and sustainable agriculture: current state and future perspectives. *Journal of Agricultural and Food Chemistry* 66 (26): 6487–6503. doi: 10.1021/acs.jafc.7b02178.

Salam, A., Afridi, M.S., Javed, M.A. et al. (2022). Nano-priming against abiotic stress: a way forward towards sustainable agriculture. *Sustainability* 14 (22): 14880. doi: 10.3390/su14221 4880.

Saraiva, R., Ferreira, Q., Rodrigues, G.C. et al. (2023). Nanofertilizer use for adaptation and mitigation of the agriculture/climate change dichotomy effects. *Climate* 11 (6): 129. doi: 10.3390/cli11060129.

Sartori, R.H., Boaretto, A.E., Villanueva, F.C.A. et al. (2008). Absorção radicular e foliar de 65Zn e sua redistribuição em laranjeiras. *Revista Brasileira de Fruticultura* 30: 523–527. doi: 10.1590/S0100-29452008000200044.

Saxena, R., Tomar, R.S., and Kumar, M. (2016). Exploring nanobiotechnology to mitigate abiotic stress in crop plants. *Journal of Pharmaceutical Sciences and Research* 8 (9): 974.

Sharma, B., Shrivastava, M., Afonso, L.O. et al. (2022). Metal doped nitrogenous hydroxyapatite nanohybrids slowly release nitrogen to crops and mitigate ammonia volatilization: an impact assessment. *NanoImpact* 28: 100424. doi: 10.1016/j.impact.2022.100424.

Sharma, B., Tiwari, S., Kumawat, K.C. et al. (2023). Nano-biofertilizers as bio-emerging strategies for sustainable agriculture development: potentiality and their limitations. *Science of the Total Environment* 860: 160476. doi: 10.1016/j.scitotenv.2022.160476.

Shebl, A., Hassan, A.A., Salama, D.M. et al. (2019). Green synthesis of nanofertilizers and their application as a foliar for *Cucurbita pepo* L. *Journal of Nanomaterials* 2019 (1): 3476347. doi: 10.1155/2019/3476347.

Shukla, P., Chaurasia, P., Younis, K. et al. (2019). Nanotechnology in sustainable agriculture: studies from seed priming to post-harvest management. *Nanotechnology for Environmental Engineering* 4 (1): 11. doi: 10.1007/s41204-019-0058-2.

Singh, M.D., Chirag, G., Prakash, P.O. et al. (2017). Nano-fertilizers is a new way to increase nutrients use efficiency in crop production. *International Journal of Agricultural Sciences* 9: 3831–3833.

Sinha, T., Bhagwatwar, P., Krishnamoorthy, C. et al. (2019). Polymer based micro- and nanoencapsulation of agrochemicals. In: *Polymers for Agri-Food Applications* (ed. T.J. Gutiérrez), 5–28. Cham: Springer International Publishing. doi: 10.1007/978-3-030-19416-1_2.

Sivarethinamohan, R. and Sujatha, S. (2021). Unlocking the potentials of using nanotechnology to stabilize agriculture and food production. *AIP Conference Proceedings* 2327: 020022. doi: 10.1063/5.0039418.

Solanki, P., Bhargava, A., Chhipa, H. et al. (2015). Nano-fertilizers and their smart delivery system. In: *Nanotechnologies in Food and Agriculture* (ed. M. Rai, C. Ribeiro, L. Mattoso et al.), 81–101. Springer International Publishing. doi: 10.1007/978-3-319-14024-7_4.

Subramanian, K.S., Manikandan, A., Thirunavukkarasu, M. et al. (2015). Nano-fertilizers for balanced crop nutrition. In: (ed. M. Rai, C. Ribeiro, L. Mattoso et al.) *Nanotechnologies in Food and Agriculture*, 69–80. Cham: Springer International Publishing. doi: 10.1007/978-3-3 19-14024-7_3.

Sun, B. and Barnard, A.S. (2019). Visualising multi-dimensional structure/property relationships with machine learning. *Journal of Physics: Materials* 2 (3): 034003. doi: 10.1088/2515-7639/ab0faa.

Sun, Y., Zhu, G., Zhao, W, et al. (2022). Engineered nanomaterials for improving the nutritional quality of agricultural products: a review. *Nanomaterials (Basel)* 12 (23): 4219. doi: 10.3390/nano12234219.

Tang, S. and Fei, X. (2021). Refractory calcium phosphate-derived phosphorus fertilizer based on hydroxyapatite nanoparticles for nutrient delivery. *ACS Applied Nano Materials* 4 (2): 1364–1376. doi: 10.1021/acsanm.0c02921.

Tarafdar, J. (2012). Nanoparticle synthesis characterization and application to solve some chronic agricultural problems. https://www.researchgate.net/publication/262375588.

Tarafder, C., Daizy, M., Alam, M.M. et al. (2020). Formulation of a hybrid nanofertilizer for slow and sustainable release of micronutrients. *ACS Omega* 5 (37): 23960–23966. doi: 10.1021/acsomega.0c03233.

Tarar, O.F., Asghar, A., Qayyum, S.A. et al. (2023). Synthesis and surface morphology of banana biochar-based nano-fertilizer and its effect on first stages of growth parameters of cucumber, broccoli, and red okra. *Journal of the Saudi Society of Agricultural Sciences* 22 (8): 535–545. doi: 10.1016/j.jssas.2023.06.002.

Thavaseelan, D. and Priyadarshana, G. (2021). Nanofertilizer use for sustainable agriculture. *Journal of Research Technology & Engineering* 2: 19.

Tortella, G., Rubilar, O., Pieretti, J.C. et al. (2023). Nanoparticles as a promising strategy to mitigate biotic stress in agriculture. *Antibiotics* 12 (2): 338. doi: 10.3390/antibiotics12020338.

Verma, K.K., Song, X.P., Degu, H.D. et al. (2023). Recent advances in nitrogen and nano-nitrogen fertilizers for sustainable crop production: a mini-review. *Chemical and Biological Technologies in Agriculture* 10 (1): 111. doi: 10.1186/s40538-023-00488-3.

Vijai Anand, K., Anugraga, A.R., Kannan, M. et al. (2020). Bio-engineered magnesium oxide nanoparticles as nano-priming agent for enhancing seed germination and seedling vigour of green gram (*Vigna radiata* L.). *Materials Letters* 271: 127792. doi: 10.1016/j.matlet.2020.127792.

Yadav, A., Yadav, K., and Abd-Elsalam, K.A. (2023a). Exploring the potential of nanofertilizers for a sustainable agriculture. *Plant Nano Biology* 5: 100044. doi: 10.1016/j.plana.2023.100044.

Yadav, A., Yadav, K., and Abd-Elsalam, K.A. (2023b). Nanofertilizers: types, delivery and advantages in agricultural sustainability. *Agrochemicals* 2 (2), 296–336. doi: 10.3390/agrochemicals2020019.

Yedoti, V., Supraja, N., and Chadipiralla, K. (2024). Review on nanotechnology in organic fertilizers: a sustainable agricultural revolution. *International Journal of Plant and Soil Science* 36 (6): 10–17. doi: 10.9734/ijpss/2024/v36i64601.

Zahra, Z., Habib, Z., Hyun, H. et al. (2022). Overview on recent developments in the design, application, and impacts of nanofertilizers in agriculture. *Sustainability* 14 (15): 9397. doi: 10.3390/su14159397.

Zayed, M.M., Elkafafi, S.H., Zedan, A.M. et al. (2017). Effect of nano chitosan on growth, physiological and biochemical parameters of *Phaseolus vulgaris* under salt stress. *Journal of Plant Production* 8 (5): 577–585. doi: 10.21608/jpp.2017.40468.

Zulfiqar, F., Navarro, M., Ashraf, M. et al. (2019). Nanofertilizer use for sustainable agriculture: advantages and limitations. *Plant Science* 289: 110270. doi: 10.1016/j.plantsci.2019.110270.

4

Nanotechnology is an Emerging Tool for Stress Management in Crop Plants

Mohd Anas[1], Mohammad Umar[2], and Abdul Razzak[3]

[1] *Interdisciplinary Nanotechnology Centre, Aligarh Muslim University, Aligarh, India*
[2] *Institute of Plant Genetics and Biotechnology, Plant Science and Biodiversity Centre, Slovak Academy of Sciences, Nitra, Slovakia*
[3] *Institute of Forest Ecology, Slovak Academy of Sciences, Technical University in Zvolen, Zvolen, Slovakia*

4.1 Introduction

The environmental challenge to agriculture is multifaceted, and it includes the stress factors emanating from biotic factors involving pests, diseases, and invasive organisms, while abiotic stresses are related to drought, salinity, temperature extremes, and nutritional deficiencies. Environmental stress interferes with plant development and crop productivity and ultimately causes severe food insecurity globally. In this respect, Fisher et al. (2018) highlighted that such unfavorable conditions limit agricultural productivity and deteriorate the situation with food security, hence posing a major issue to the stability and sustainability of the supplies of food across different regions. To achieve more sustainable agricultural production and enable food security, the tolerance of plants to various types of stressors should be enhanced. Research indicates that nearly 800 million individuals are affected by chronic hunger, while approximately 2 billion people endure micronutrient deficiencies. According to projections from the FAO (2017), it is anticipated that by 2030, around 653 million individuals will still experience undernutrition. As a result, food security is poised to persist as a considerable challenge, particularly given that the global human population is projected to approach 10 billion by 2050 (Ashraf and Javed 2023). The FAO estimated in 2011 that almost a third of the food grown worldwide is discarded, equivalent to nearly 1.3 billion tons per year (Gustavsson et al. 2011). Preharvest crop losses account for approximately 35% of the total losses brought about by diseases, animal pests, weeds, and abiotic stresses. This represents about 1051.5 million tons. Extra losses post harvesting and storage amount to around 690 million tons, as reported by Mesterházy, Oláh and Popp (2020). The biotic and abiotic stresses contributed to tremendous crop yield loss. Insect pests have been known to be responsible for about 15–20% yield losses in main food and cash crops alone (Rathee and Dalal 2018). Elsewhere, estimated global yield losses by pathogenic diseases were in the range of 11–30% (Savary et al. 2019; Van Esse, Reuber, and

Nanotechnology-based Sustainable Agriculture, First Edition. Edited by Pardeep Singh et al.

van der Does 2020). Besides, a wide range of morphological, physiological, biochemical, and molecular changes are induced in plants by abiotic stresses that include drought, water logging, extreme temperature variability, salinity, heavy metal toxicity, and mineral toxicity. These changes in turn lead to a negative impact on the growth of plants, their development, and yield outcomes from this system (Atkinson and Urwin 2012). It was also observed that these stresses all contribute to huge losses in growth and yield, ranging from 50 to 70% (Zahra et al. 2021). An excess of reactive oxygen species (ROS) overproduction is a major cause of such losses in crops due to abiotic stress (Farooq et al. 2019). Global models project an increase in CO_2 concentrations from 400 to 800 ppm in the next century (Hofmann et al. 2009). Besides, more than 45% of the cultivable land is under threat of drought (Thornton et al. 2014). More than 27% of global land area faces aridity and under most circumstances, most of the cultivated crop species are sensitive to salinity stress in the range of 1.0–1.8 d S/m. As reported by Panta et al. (2014), these abiotic stressors could create a 10–50% yield loss. Similarly, pollution by heavy metals such as Cr, Cd, As, Pb, Cu, and Hg causes adverse effects on seed germination, photosynthesis, respiration, and transpiration. It also reduces the growth of the plant species, crop yield, and quality of fruits and vegetables (Manzoor, Sharma, and Wani 2018). Synthetic chemicals, particularly pesticides, have been widely used in the recent past as a technique to minimize abiotic and biotic stressors from crop species. However, the major use of these chemicals has also pointed out significant health risks and environmental impacts. Studies have shown alarming trends: a 35% reduction in the rate of soil respiration, nearly ubiquitous contamination of water, affecting up to 90% of all cropland, an astonishing decline of insect biomass by 70%, and a decline of bird populations by 50% within farmland. That is besides the very alarming trend of a 30% decline in the honey bee population and a 42% decline in species richness in parts of the world such as Europe, Australia, and the America. Other than causing environmental effects, pesticide exposure has posed serious health risks with a 25–30% increased risk of cancers and mental disorders and an alarming 50% increase in leukemia, lymphoma, and brain cancers among children. The results underscore the urgent need for sustainable agricultural practices that alleviate the use of harmful chemicals, to focus on long-term ecological health and human well-being (Ali et al. 2021). Moreover, diverse stresses acting cumulatively increase the complexity of the plant responses, pinpointing the requirement of a renewed effort akin to a second green revolution to meet global food demands. The very reason has, therefore, evoked an urgent requirement of means that, at the same time as public health, would underscore ecological sustainability. As a critical "Key Enabling Technology," nanotechnology holds promising opportunities for the enhancement of agricultural productivity and the reduction of ecological impacts (Parisi, Vigani, and Rodríguez-Cerezo 2015). This new approach has therefore targeted the challenges of both biotic and abiotic stressors, hence ensuring long-term agricultural sustainability. Hossain, Yasmeen, and Komatsu (2020) and Baazaoui et al. (2023) studies provided important evidence on the potential of nanotechnology, which promises to improve plant tolerance to stresses and therefore presents a viable strategy for meeting these challenges and enhancing the stability of food systems. Indeed, nanoparticles (NPs) have shown tremendous potential in regulating plant growth and development by enhancing tolerance to various biotic and abiotic factors. This is achieved through increased activities in antioxidant activity as reported by Ahmed et al. (2021) and photosynthetic efficiency. Application of nanotechnology in agriculture enhances the ecological footprint of the said agrochemicals, such as pesticides, herbicides,

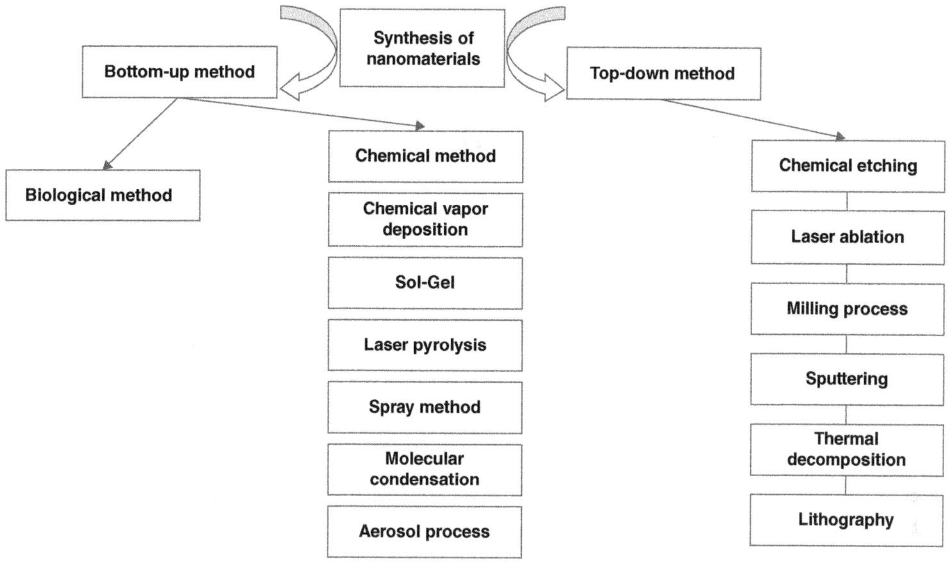

Figure 4.1 Illustration of different approaches for the synthesis of nanoparticles.

and fertilizers (Adisa et al. 2019; Ali et al. 2021). As a result of this reduction, input costs related to agricultural produce decrease while pollution is minimized at the same time. NPs are synthesized through various procedures including physical, chemical, and biological methods as illustrated in Figure 4.1 using metals or metal oxides (Singh 2016). Formidable effectiveness against phytopathogens and insect pests has been noted in pesticides, insecticides, and repellents fabricated through nanotechnology incorporating metals (Srujana and Bhagat 2022). However, biological synthesis by using NPs with plants or plant extracts, entomopathogens, and other biomaterials is believed to carry a number of benefits over chemical or physical methods. Plants and their parts are a repository of various enzymes, sugars, and a number of phytochemicals including flavonoids, latex, phenolics, terpenoids, alcohols, amines, and cofactors. These phytochemicals serve in place of natural reducing and stabilizing agents in metal NP synthesis. This approach enables not only the fabrication of NPs with well-defined sizes and shapes but also yields less environmental pollution and hence proved to be a more friendly and eco-friendly method by Kumar and Yadav (2009) and Siddiqui et al. (2014). NPs are also applied gradually as part of their effectiveness in mitigating the adverse effects that various abiotic stressors have on plant systems. Notably, the metallic NPs have proved to have manifold uses in these biological systems. In fact, Wang et al. (2022) reported that silica NPs improved growth in plants and further developed the potential of those plants to bear up with the biotic stressors. In this connection, copper NPs, zinc oxide, and selenium also came forward as promising nanofertilizers for agricultural sectors. According to Kohatsu et al. (2021) and Ghani et al. (2022), NPs generally induce the synthesis of phytohormones in plants to enhance growth and metabolic activities under certain abiotic stressors, such as drought and salinity. They move towards sustainable agriculture because their adoption contributes to reducing dependence on conventional chemical use by inducing environmentally friendly modes of farming. In this view, chitosan NPs, which release nitric oxide, have been indicated to ameliorate the negative impacts triggered by saline stress (Oliveira et al. 2016). The application of these nitric oxide releasing chitosan NPs in soils was demonstrated to maintain the root network of soybeans and promote overall development under copper stress conditions (Gomes et al. 2022). NPs, as they have made their way in, have revolutionized pest control applications in agriculture by decreasing the extent of conventional pesticide use to a great extent. Notably, polymeric, metal and metal-oxide NPs show very high efficiency in alleviating biotic stress in crops (Tripathi, Singh, and Pandey-Rai 2022). With high antimicrobial potential, silver and copper NPs receive much attention in biotechnological research and could be used as the leading agents to apply against microbial threats (Bruna et al. 2021). However, considering their antimicrobial efficacy, biocompatibility, and biodegradability, chitosan NPs have emerged as a pivotal tool in plant biotic stress control studies, which would prefer them over other sustainable agricultural applications (Singh et al. 2022). Polymeric NPs are valued not only for their antimicrobial properties but also for being able, through nano-encapsulation, to act as delivery systems for biocontrol agents that effectively reduce biotic stress in plants (Riseh et al. 2022; Sobral, Martins, and Sobral 2022). A number of recent reports have shown that, in plant systems, NPs can be used to alter metabolic pathways aimed at promoting ameliorations in biotic stress in plants (Kumari et al. 2020). Figure 4.2 depicts the different applications involved in a NP-mediated plant expressing

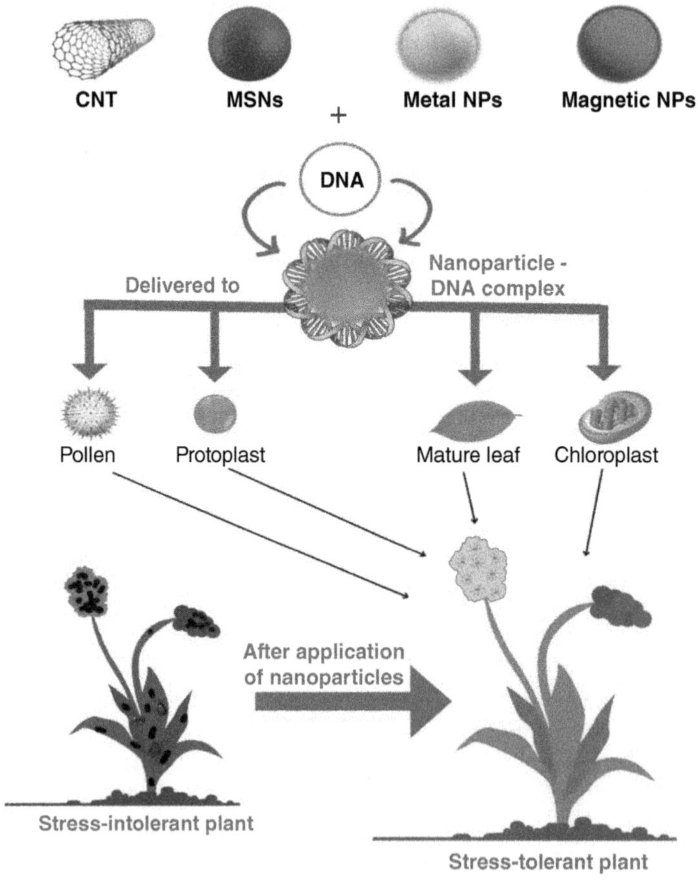

Figure 4.2 Illustration of different nanoparticle applications for the mitigation of stress.

stress tolerance. This chapter focuses on the progress of nanotechnology in agriculture and the further efficiency of biosynthesized NPs in mitigating various abiotic and biotic stresses linked with crop production.

4.2 Synthesis and Characterization of Nanomaterials

The various techniques used in NP synthesis are discussed briefly and well covered. This includes physical, chemical, and biological processes that can be bottom-up or top-down techniques. The way of synthesis has a significant influence on the physical and chemical characteristics when it comes to creating nanomaterial based products that interfere directly or indirectly with living cells. The effect of nano-based substances can be fulfilled by choosing an adequate synthesis method according to the utilization.

4.2.1 Bottom-up Method

With this technique, scientists grow NPs with the right kinds of defects and a consistent level of chemicals inside by using atoms and molecules as building blocks to purposely form them through chemical reactions. This is represented by a number of methods and measures.

4.2.2 Chemical Method

So, the chemical approach as a synthetic pathway to NP is still top among other methods due to its high degree of versatility and ease. This method offers the possibility of controlling several parameters including precursors, reductants, and surfactants or concentrations, which determine the size and shape composition structure of NPs. These factors can be tuned to obtain distinct NP morphologies. The various chemical methods are the co-precipitation method, thermal decomposition, solvothermal decomposition, sonochemical method, self-assembly using a template pyrolysis, electrospinning chemical method, vapor deposition method, atomic or molecular condensation method and photochemical method. There is a reaction in which NPs form at the electrode electrolyte interface because of electrodes located far apart combining during electrochemical synthesis. Control over the electrochemical parameters allows them to adjust the shape and size of the particles produced in particular (Singh et al. 2022).

4.2.3 Biological Method

The synthesis of NPs can be carried out using various chemicals for the reduction of metal precursors, such as hydrazine hydrate, lithium aluminum hydride, or sodium borohydride. But they are toxic and ecotoxic in nature. To solve this, an eco-friendly and safe method using plant extracts and microorganisms, i.e. bacteria, fungi, algae, and viruses, are used as the only option for these harmful chemicals (Oza et al. 2020). Extracts of plant leaves, seeds, and algae have been among the sources capable of synthesizing different

NPs such as metal oxides and zero-valent NPs. These extracts naturally contain sugars, proteins, flavonoids, and polyphenols as active ingredients that act as reducing agents and capping agents during the synthesis of NPs, making these methods economical and environmentally friendly (Öztürk, Gürsu, and Dağ 2020; Sur et al. 2018). Then, the precursor salts and different reductase enzymes are also absorbed by microbes, through which reduction and detoxification occur. They contain many enzymes, proteins, and metal-resistance genes that act as reducing (i.e. counter ion) or capping agents against metals. The microorganism based NP is synthesized by cell bound or extracellular synthesis (Molnár et al. 2018).

4.2.4 Top-down Method (Physical Approach)

Starting with bulk-sized materials that you reduce to smaller, nanometer-sized pieces, the simplest and oldest method is the mechanical method, which consists of grinding fragments of carbon steel with tungsten carbide or iron balls. It also includes methods such as lithography, laser ablation, sputtering and electrospray (Baig, Kammakakam, and Falath 2021).

4.3 Characterization of Nanomaterials

Characteristics of reaction like the quality of precursor, type and amount of the reactants involved, pH of the reaction solution, heating rates, and stirring conditions are influencing factors that promote a variety in shapes and sizes. Control over the morphology of NPs is a key aspect, and it can often be achieved by modulation of synthesis methods or reducing agents/reagents/stabilizers used during their production process (Baig, Kammakakam, and Falath 2021). The performance of NPs depends on their physical and chemical properties, which are contingent upon the size of the NPs. This is because smaller NPs, for example, have higher reactivity and crystalline structures in comparison to their larger counterparts. Size-dependent properties are exploited in a variety of uses such as catalysis and magnetic phenomena where particular sizes of larvae-NP synthesis can optimize the required features (Ozel, Kockar, and Karaagac 2015). The structure of NPs also influences their properties. Nanoparticle shapes, such as rods, spheres, or cubes, can also induce properties and effects that are not seen through a conventional raw form of material. This way, the shape change in NPs can become selectively controlled and suitable for specific applications such as in drug delivery systems where the aspect ratio plays a critical role in their biological interactions and stability (Wu et al. 2016). One of the characteristics of NPs is their high surface area, facilitating pronounced interactions with the environmental matrix. NPs can adhere to and trap contaminants, which makes them especially useful for things like environmental remediation. Owing to their small size, the high-surface curvature of NPs increases catalytic activity and they are employed as effective catalysts in various chemical reactions (Feng et al. 2012). NPs exhibit peculiar optical properties, which are primarily governed by size and phenomenon of surface plasmon resonance, especially in metal NPs like gold (Au). This effect meditates several optical fields such as Mie scattering, surface-enhanced luminescence, and Raman scattering, which can be used for different applications from sensing/imaging to photothermal therapies (Huang et al. 2007). Some NPs, such as iron oxides, have magnetic properties and can be manipulated using external magnetic fields. In application, this is an important property for magnetic resonance imaging (MRI) and targeted drug delivery systems that contain NPs as they need to be manipulated in a specific direction under external magnetic fields. Certain NPs, such as silver (Ag)-based particles, are known to have strong antimicrobial activities over a range of different bacterial strains (Lok et al. 2007). As these properties depend on the size and shape of NPs, they can be used in biomedicine for antibacterial, antifungal or antiviral activity (Rai, Yadav, and Gade 2009).

4.4 Applications of Nanotechnology in Managing Abiotic Stress

4.4.1 Drought Stress

Drought presents a considerable risk to crop development and productivity. There are studies by Farooq et al. (2009) and Gupta and Sidana (2020) that provide evidence of the detrimental effects of drought on crops. A comprehensive review of hazardous weather events in the United States between 1980 and 2012 indicated that droughts and heat stress had resulted in damages to agriculture costing approximately US$200 billion. The monetary cost resulting from droughts was estimated at US$50 billion (Suzuki et al. 2014). Drought induces changes in both plants and their molecular processes. Changes are imperative for plants to persist and survive the drought-induced water deficit conditions. Plants employ methods of adaptation under stress such as leaf contraction, altering hormone concentrations, and increasing stress membrane gene expression to achieve tolerating stressful conditions (Gupta et al. 2020). In recent years, scientists began to study the use of NPs through methods as a sustainable technique to lessen drought stress imposed on plants. For instance, Ali et al. (2019) examined whether NPs produced by methods had an effect on *Tephrosia apollinea* plants under drought stress. In this examination, silver NPs facilitated protection against bacteria, particularly *Escherichia coli* and *Staphylococcus aureus*, and protected the membranous cells in *Tephrosia apollinea*. Moreover, under drought conditions, NPs increased hydrogen peroxide levels in the roots of *T. apollinea*, and decreased membrane damage caused by drought stress (Ali et al. 2019). Moreover, green NPs synthesized from materials had a positive impact on the growth of seeds or seedlings since they germinated at 89%, which was 6% faster than the control in a germination study with lentil plants under drought conditions (Hojjat and Ganjali 2016). The observed positive impact on growth was because there was an altered level of water present in the living tissues of the live seedlings. To summarize, NPs sourced from materials may positively influence the reducing effects of drought on plant growth. However, in order to determine proper physiological mechanisms, we need to study the biochemical and genetic mechanisms of the plants so that we enhance the crop's ability to limit drought stresses. Based on Lowry, Avellan, and Gilbertson (2019), there is an opportunity to improve the photosynthesis process by enhancing enzyme activity in the C3 cycle and promoting the synthesis of pigments that promote plant growth through nanotechnology. Harmful effects of oxidative stress from NPs include toxicity, which may induce cell signals such as cell cycle disruption and antioxidant enzymes (Jalil and Ansari 2019). Some micro elements such as Zn and Cu enhance the plant's tolerance to drought stress and at the same time influences root growth. Cu and Zn NPs lead to increased root hair formation and lateral root formation while root length is somewhat inhibited, thereby restricting water access. Copper NPs enhance lignification affecting the water transport and cell wall looseness (Nair and Chung 2015). Results for grown-up plants with Cu NPs involve an increase of effective solutes such as anthocyanins and a pronounced ability of plants to tolerate drought stress through higher proline levels. In wheat, Cu NPs result in the formation of ROS largely due to abscisic acid (ABA)-induced stress, which offers protection against drought (Dimkpa et al. 2012). This may be achieved by delivering more water in the form of fullerenol NPs (FNP), with the chemical formula $C_{60}(OH)_{24}$. FNPs can penetrate root and leaf tissues and interact with different compartments firmly attaching water molecules that might enhance the mechanisms of drought tolerance (Borišev et al. 2016; Verma et al. 2019). Experiments on sugar beet plants revealed that FNPs enhanced the accumulation of proline and antioxidant enzymes to improve the plant's resistance to water deficit. Despite the need for more detailed investigations of the FNPs' action, the information in the research of Liu et al. (2016) suggests that they can enhance the process of water management in the sphere of agriculture. Silver nanoparticles (AgNPs) are widely used in research. The antibacterial effects of AgNPs have been well documented (Beyene et al. 2017). Although the AgNPs can alter plant hydraulic conductivity, they can also hamper cell division by penetrating through plant tissues (Tripathi et al. 2017). As confirmed by Hojjat and Ganjali (2016), the use of AgNPs enhances the germination index of lentils under conditions of water-deficit stress as well as root length and biomass. But, AgNPs may decrease the antioxidant enzymes in the plants such as *Tanacetum parthenium* because plants may have lesser requirements to protect from oxidative damage. Investigation on *Carum copticum* under drought stress indicated that the magnetic field has an impressive enhancement of yield comparative to AgNPs (Seghatoleslami et al. 2015). NPs help increase the ability of plants to tolerate drought stress through increased antioxidant capacity, increased nutrient content and efficiency of the photosynthesis process, decreased production of ROS, as well as the regulation of proteins and signaling. Across the use of different species of plants, application methods, size of the particles, and concentration, the effects differ. Further work detailing its functions, methods of interaction with biomolecules, and operation on the genetic level when water is scarce is necessary. However, certain works describe the possible environmental hazards, which is why more focus has been placed on studying the interaction between plants and NPs and their influence on the environment.

4.4.2 Salt Stress

Salinity stress affects over 20% of the cultivable lands worldwide and represents a major threat to plant growth due to its current costs to our environment. This stress presents itself in the form of osmotic and particular ion toxicities and imbalance of minerals (Atta et al. 2023). Application of NPs has been reported to positively influence plant growth under such a situation (Rubilar et al. 2013). NPs stimulate the production of carbohydrates and proteins and increase enzyme activity such as catalase, guaiacol peroxidase (POD), and ascorbate peroxidases (APX). This is most likely due to important enzymatic functions of the enzymes for salt stress alleviation through scavenging ROS and hormonal regulation (Ghosh et al. 2016). The nanofertilizer's application of Se and Cu as foliar spray on other hand improved the concentration (but less than that by NPs), so decreased detrimental effects of salinity stress and consequently growth rate were significantly enhanced in tomato plants. Nevertheless, the use of saline water affected soil properties negatively (Saffan et al. 2022). Under NaCl stress, extreme uptake of sodium (Na^+) and chloride (Cl^-) ions occurs, resulting in oxidative damage due to increased influx of ROS and reduced availability of K^+ and Ca^{2+} ions in plants (Gul et al. 2019). NPs as a bio-modifier can prevent the Na^+ and Cl^- ions that penetrate into plants. For example, zinc oxide nanoparticles (ZnO-NPs) were found to increase K^+ and Ca^{2+} uptake, while it diminishes Na^+ and Cl^- accumulation. Key antioxidant enzymes are supported by diverse breeds of NPs. The catalase enzyme is also supported by iron (Fe), cesium (Cs), and cobalt NPs, as well as the peroxidase (POD) with those included in copper. It is shown that low levels of silver NPs (Ag-NPs) enhance the activity of antioxidant enzymes and promote the growth of plants in saline conditions. For example, titanium dioxide NPs (TiO_2-NPs) have the capability to relieve salinity stress in *Dracocephalum moldavica* by activating antioxidant defense systems, reducing oxidative damage caused by low-quality water, and promoting plant growth in saline conditions (Gohari et al. 2020). Finally, it may be concluded that the introduction of NPs can be relied upon to enhance salinity tolerance of plants by the induction of antioxidant system, retention of tissue water content, and maintenance of ion equilibrium. Regardless of these auspicious results, more progress has to be made in the optimization of the application of NPs for field application to help enhance plant salt tolerance.

4.4.3 Thermal Stress

Both elevated and lowered temperature extremes are important aspects affecting plant growth development and yield (Choudhary et al. 2022). Heat stress is defined as an excessive temperature due to which plant growth development or yield becomes impossible to recover for a duration more than the respective plant's critical temperature (Haider et al. 2022). This stress causes extreme generation of ROS leading to membrane and other essential cell molecules being exposed to oxidative injury (Mushtaq et al. 2022; Ullah et al. 2022) and drastically affects the levels of photosynthetic pigments and the processes of carbon uptake especially in crops such as wheat and chickpea. Interactions with low-temperature stress include either chilling (0–15 °C) or freezing (less than 0 °C) temperatures. Chilling stress can induce damage without the formation of ice, while freezing stress can lead to impairment because of the formation of ice (Arshad et al. 2017). These low-temperature stresses inhibit the development of plants due to the occurrence of photoinhibition, oxidative injury as a result of ROS, inefficient utilization of nutrients, the activity of various enzymes which get impaired, as well as a decline in carbon fixing ability (Farooq et al. 2009). The nanomaterials' usage has efficiently ameliorated the effect of heat stress in plants. For instance, Iqbal et al. (2019) proved that the use of AgNPs made up of the extract of *Moringa* enhanced biomass and the area of leaf in wheat plants exposed to high-temperature stress. The green-synthesized AgNPs were able to maintain tissue water contents and boost chlorophyll contents over the controls in the case of wheat plants (Iqbal et al. 2019). The use of *Moringa* extract-synthesized AgNPs conferred the plants with the ability to tolerate heat stress by enhancing the antioxidant defense mechanism through the modulation of the antioxidant genes *MeCu/ZnSOD* and *MeAPX2* conducted in *Arabidopsis thaliana* (Kohan-Baghkheirati and Geisler-Lee 2015). Selenium (Se) is an important and beneficial component that is essential for the performance of antioxidant protection mode of action in plants. Selenium nanoparticles (SeNPs) at a low dose have been reported to enhance the antioxidant protection system and the activity of some antioxidant enzymes and enhance the levels of photosynthetic pigments as well as growth in tomatoes (Haghighi, Abolghasemi, and da Silva 2014) and grain sorghum (Djanaguiraman et al. 2018) crops. SeNPs with the application of the microbial strain *Lactobacillus casei* improved high-temperature stress tolerance in *Chrysanthemum* due to the start of the antioxidant protection system (Seliem et al. 2020). The usage of TiO_2 NPs also showed improved heat tolerance in tomatoes by modulating stomatal oscillations (Qi, Liu, and Li 2013). Heat shock proteins (HSPs) are a class of chaperone proteins in plants induced under variable stressful heat (Reinle et al. 2022). HSPs ultimately play a crucial role in stabilizing protein structure and inducing heat tolerance in plants. Several nanomaterials exhibited excellent potential to increase levels of heat stress tolerance following the induction of HSPs. In one study, Zhao et al. (2012) demonstrated that CeO_2 NPs can protect maize from heat stress by strengthening the expression of HSP70 and promoting growth. Multiwalled carbon nanotubes (MWCNTs) improved heat stress tolerance in tomatoes through induction of similar stress-related genes including HSP90 and promoting better growth (Khodakovskaya et al. 2011). Nanomaterials have also produced significant freezing stress tolerance for a range of plant species. For instance, SeNPs positively influenced the status of pigments related to photosynthesis and the system of antioxidant protection while increasing the proliferation of tomato plants in low-temperature stress (Haghighi, Abolghasemi, and da Silva 2014). TiO_2 NPs increased the cold tolerance of chickpeas by enhancing antioxidant enzyme activities and chlorophyll contents and lowering ROS and oxidative stress more effectively than cold stress controls. Additionally, TiO_2 NPs stimulated cold tolerance gene expression. Photosynthesis, which is crucial for food production, is dramatically impaired by cold stress in plants (Farooq et al. 2019). Photosynthetic pigments and active ingredients like RUBISCO, which are central to the photosynthetic processes, are reduced due to cold stress, and so is the photosynthetic rate (Stahl-Rommel et al. 2022). Advanced materials such as nanomaterials have been shown to be effective in protecting the thylakoid membrane from ROS damage (Ranjan et al. 2022) and optimizing the electron transport and carbon fixation by chloroplasts (Husen and Siddiqi 2014). Using TiO_2 NPs in chickpea examples, their antioxidant activities including superoxide dismutase, catalase, ascorbate peroxidase, glutathione peroxidase, guaiacol peroxidase, polyphenol oxidase, lipoxygenase and allene oxide synthase were on the rise. There was also an increase in the chlorophyll content and a rise in the gene expression related to the chlorophyll-binding protein and RUBISCO (Hasanpour, Maali-Amir, and Zeinali 2015) when plants were subjected to cold stress, which ultimately enhanced the plant's competence to tolerate cold conditions.

4.4.4 Toxic Metal Stress

Contamination in soil with toxic metals has a detrimental effect on the growth of plants, soil ecosystems, and human health. Based on metal concentration in soil contamination, metals can be categorized into two groups: (i) essential and (ii) non-essential. Essential metal are one of the specific nutrient categories for plant growth and maturation as micronutrients: cobalt (Co), copper (Cu), iron (Fe), manganese (Mn), molybdenum (Mo), nickel (Ni), selenium (Se), and zinc (Zn). Although necessary for plant growth, essential metals can exert toxicity on cellular growth and health when concentrations exceed a given threshold of an essential micronutrient(s) for plants (Angulo-Bejarano, Puente-Rivera, and Cruz-Ortega 2021). Non-essential metals, however, include toxicant metals such as arsenic (As), cadmium (Cd), chromium (Cr), lead (Pb), mercury (Hg), and silver (Ag). Non-essential metals do not serve any biological functions in plants, yet are harmful even at low levels, and affect plant growth and/or health (Singh et al. 2020). Toxic metals exert a form of stress on the plant, creating an undesirable condition in the plant mainly because of (i) metabolic changes through modified protein structures, (ii) functional changes (interference) in functional groups of essential compounds, and (iii) evidence of oxidative damage to biological membranes (Dubey et al. 2018; Hamid et al. 2019). When applied simultaneously or sequentially with microbes, the implementation of the assigned nanomaterials and treatments allows toxic metal effects that would typically cause plant stress to be ameliorated. For example, inoculating arbuscular mycorrhizal fungi in combination with nanoscale zero-valent iron (nZVI), even at lower doses (100 mg/kg), has been shown to improve the Cd immobilization, thereby decreasing its absorption in sweet sorghum grown in acidic contaminated soil (Cheng et al. 2021). Likewise, usage of the polyvinylpyrrolidone-coated iron oxide (Fe_2O_3) NPs together with the bacterium which is Gram-negative *Halomonas* sp. effectively eliminates Cd and Pb, significantly decreasing the time of repairment. When used alone, these NPs eliminated about 66% of Cd and 82% of Pb, while the bacterium alone achieved approximately 84% removal of Cd and 81% of Pb. However, the combination of these NPs and the bacterium led to the complete removal of both Cd and Pb (Cao et al. 2020). Nanomaterials can also enhance the effectiveness of phytoextraction for the remediation of soil. For instance, in nZVI presence at a concentration of 100 mg/kg, *Lolium perenne* plants demonstrated the maximum build-up of lead (Pb), reaching 1175 μg/pot. At elevated applications of nZVI, the uptake of lead was reduced once again at 2000 mg/kg due to higher levels of nZVI, and induced stress was attributed (Huang et al. 2018). The type of stress induced was oxidative stress. The closer and more often that plants experience oxidative stress, the greater the potential for negative impacts on plant health will be increased. Nanotechnology can use biologically synthesized materials as a mitigable pathway of plant responses to toxic metal stress imposed on the soil. These biologically derived nanomaterials provide an environment-friendly process for mitigating the adverse effects of toxic metals, improving soil health and supporting sustainable agricultural practices (Ahmed et al. 2021). NPs with biogenic copper created by a bacterium that is copper-resistant (Shigella flexneri SNT22) have been shown to cut Cd uptake by 50% in wheat plants grown in contaminated soil (Noman et al. 2020). This treatment also showed a 28% increase in shoot dry weight, a 44% rise in plant height, and improvements in nitrogen and phosphorus levels by 41 and 58%, respectively. It is believed that these biogenic copper NPs adsorb cadmium (Cd) on their surfaces, thereby preventing its uptake by the plants (Noman et al. 2020). In addition, these tiny particles might hinder the uptake of metals by plants by competing for entry points in their membranes. Once they enter the plant, they trigger the activity of enzymes related to defense mechanisms that help limit the spread of metals within the plant. While the idea of using nanotechnology to alleviate metal stress is appealing, research in this developing area is still quite limited. Nanotechnology blended with bioremediation enhances the advantages of both methods and presents the opportunity to speed up the creation of environmentally friendly cleanup systems. Nevertheless, the chemical composition of contaminated soils can greatly impact how effectively NPs interact with pollutants and microorganisms (Cecchin et al. 2017). This strategy needs further exploration for field application, especially concerning the fate and environmental impacts of nanomaterials. Understanding how these materials operate in real-world settings and their potential effects on ecosystems is crucial for developing effective and sustainable remediation strategies.

4.4.5 Organic Pollutants Stress

In the environment, we can commonly find organic pollutants (OPs) and they come in various forms with more than 20 different types recognized (Geissen et al. 2015). Typically, these pollutants can be divided into three main categories: industrial chemicals, pesticides, and pharmaceuticals. Within these groups, some OPs are particularly prevalent such as polycyclic aromatic hydrocarbons (PAHs), organochlorine pesticides, polychlorinated dichlorodiphenyltrichloroethane (DDT), polybrominated diphenyl ethers (PBDEs), and hexachlorocyclohexane (HCH, Han and Currell 2017). Each of these classes has distinct environmental impacts and presents different threats to ecosystems and human beings. These organic contaminants are extremely harmful and can be mutagenic or cancer causing even in doses. Their durability and persistence in the environment can be attributed to their resistant and water-repellent characteristics enabling them to endure for many years. OPs have the potential to accumulate in the food chain jeopardizing human health and causing toxicity in plants. At levels, OPs negatively influence plant germination and growth by disrupting processes, such as enzyme activity, carbon absorption, and various metabolic functions. Similar to other stresses, OPs trigger stress in plants by the production of ROS (Zhang et al. 2007). This oxidative stress results in peroxidation (LPO), which has the potential to harm DNA/RNA and cell membranes. Phytoremediation, using plants to clean up OPs, is an economically viable and environmentally friendly approach. However, its efficiency decreases, with OP concentrations due to their harmful effects on plants. To tackle this issue using nanomaterials could potentially lessen the harmful effects of OPs on vegetation. For instance, a study on the effect of bimetallic NPs Ni/Fe on the toxicity of PBDEs in Chinese cabbage found that these nanomaterials demonstrated significant environmental benefits. Among the various plants, the phosphorylation of nanomaterials can remove negative influences of BDPs from the plants (Wu et al. 2016). Contrastingly, research in another arena showed how the efficiency of the adsorptive process of four insecticides including chlordane by lettuce was improved by carbon nanotubes (CNTs). The use of CNTs prominently lowers the absorption of chlordane components and *p,p*”-DDE in lettuce plantlets. The CNTs seem to capture the organic pollutants (OPs); thus, the avoidance is achieved by limiting the transfer of the organic substances into the plant tissues (Hamdi et al. 2015). AgNPs also have a developed process of reducing the uptake and accumulation of the substance *p,p*”-DDE in both zucchini and soybeans (De et al. 2013) which has been observed lately. MWCNTs have also shown promise in the reduction of pyrene and 1-methylpyrene levels in the roots and shoots of maize while also decreasing their mobility within the plant (Zhang et al. 2017). In addition, some of the NPs such as ZnO, CNTs, nZVI particles, graphene QDs, and AgNPs have been demonstrated to promote plant growth under stress by different processes. For instance, ZnO and the QDs of graphene may work as nanofertilizers, which will improve the health and growth of the plants (Chakravarty, Erande, and Late 2015). Praveen et al. (2018) found that the use of ZnO-NPs was responsible for the increased tolerance to Fe_2O_3 stress in Indian mustard by altering the activities of the enzyme. Besides that, NPs can regulate the growth of plants by absorbing more water and nutrients and affecting the community of the microbes in the soil. Biosynthesized nanomaterials are also capable of degrading different types of OPs in the environment. Biosynthesized palladium NPs (PdNPs) have been shown to be considerably more effective than commercial palladium catalysts in the dehalogenation of various toxic congeners of polychlorinated biphenyls (PCBs) in sediments and water. In this case, pelletized cells of *Desulfovibrio desulfuricans* had five times more catalytic dehalogenation of the flame-retardant tris-(chloroisopropyl)-phosphate (TCPP) than chemically reduced palladium powder (Deplanche et al. 2009). The trichloroethane (TCE) dehalogenation and diatrizoate deiodination rates were increased by a lot due to the usage of biosynthesized PdNPs on a graphite cathode (De et al. 2011). In this respect, these results reaffirm the ability of biosynthesized NPs to environmental decontamination. However, there is still a lack of research regarding their potential to decrease the phytotoxicity of OPs. The method of biosynthesized NPs used to secure the bad effects of OPs on plant health should be further investigated.

4.4.6 Hypoxia and Anoxia Stresses

Climate change leads to more environmental stresses on plants, like submersion and waterlogging (Raza et al. 2019). The latter factors are one of the reasons why there is a difference in the availability of oxygen (O_2) around plant roots, thus making it possible to have oxygen levels that vary. Under submerged conditions, where plants are either fully or partially submerged, oxygen can be found in normal levels (normoxia), hypoxia (partial O_2 deficiency), and anoxia (complete O_2 deficiency). This is an example of why this phenomenon is happening as oxygen dissolves in water at a much slower rate than it does in air (Zahra et al. 2021). Thus, it is the case that the plants in the waterlogged regions have to face the problems of getting the right amount of oxygen, which turns out to be detrimental to the respiratory system of the plants and the vitality of the plants. Flooding and hypoxia have a significant impact on plant life by preventing them from performing essential functions such as breathing, producing seeds, and developing normally (Komatsu, Hiraga, and Yanagawa 2012). Hypoxia is a condition where oxygen is deficient, which is why under such conditions, the production of adenosine triphosphate (ATP) through phosphorylation greatly decreases. In addition, there is a resultant timeout of material metabolic by-products like lactate and ethanol, which are in high concentrations and thus contributed to the cytosol acidification reaction (Drew et al. 1997). The two relative hypoxia and anoxia stresses affect the plant in various ways. They lead to a reduction of the level in the green pigment, they obstruct the gas exchange through the stomata, and they reduce the photosynthetic activity which is very harmful to plants; they can also be a source of oxidative damage through the increased production of lipid peroxidase as a result of the overproduction of ROS in the leaf and root tissues (Iqbal et al. 2021). With assisted or unassisted selection, upregulating genes and, molecular and breeding strategies have all been used to improve the plant along with the hypoxic/aerobic environment. These strategies are intended to enhance the plant's ability to live in oxygen-limited situations, while not limiting its growth or productivity (Vartapetian et al. 2014). New strategies and the use of tools like nanobiotechnology have become indispensable in the joint implementation of traditional practices and sustainable plant growth (Vartapetian et al. 2014). Nanotechnology, which is green and bioprocesses, is a green option that helps limit some abiotic stressors such as hypoxia and anoxia (Bartolucci et al. 2020). SiNPs-treated plants have revealed more capacity to cope with hypoxia stress than the plants treated with conventional silicon (Iqbal et al. 2021). This increase in resilience is undoubtedly due to elevated antioxidant activities, the gain of structural osmoprotectants, and more competent regulation of micronutrients (Iqbal et al. 2021). Almutairi et al. (2019) researched the usage of silver (AgNPs) and aluminum oxide (Al_2O_3) NPs to surpass the bad effects of flooding stress. The same methods were used by Mustafa, Sakata, and Komatsu (2015) to carry out the infection of soybeans with Al_2O_3NPs, ZnONPs, and AgNPs which are the major agents of flooding stress. Indeed, analysis of proteins showed that the stress of flooding altered the presence of proteins in the energy metabolism pathway, but the application of Al_2O_3NPs decreased these changes. Mustafa and Komatsu (2016) learned that the size paradigm of NPs is more critical in soybean anoxia tolerance than the type and amount of NPs. In addition, nanomaterials have also been found to reduce flooding (both hypoxia and anoxia) damages via repressing ethylene biosynthesis in *Arabidopsis* plantlets by Syu et al. (2014). Further investigation is required to appreciate the potential of nanomaterials in enhancing hypoxia and anoxia tolerances, which would be useful for providing better defense tools over these harsh environmental conditions against disaster effects.

4.5 Environmental Implications: Case Studies and Recent Plant Research

The tiny size of NPs is advantageous, but it can also lead to toxicity risk and other potential adverse effects. Specifically, the adverse effects include environmentally related hazards along with human and/or animal health impacts; hence, the need to evaluate the safety of newly developed nanomaterials will increase. Even though nanoproducts are being rapidly introduced into the market, there is still a lack of comprehensive guidelines to address these concerns. Regulatory agencies are working to gather information from published studies, but the gap remains due to insufficient legal and scientific resources. In 2006, the United States Food and Drug Administration (FDA) established the Nanotechnology Task Force to serve consumers (Acharya and Pal. 2020). Furthermore, the European Commission is currently reviewing the influence of nanomaterials on human health. The competent body carefully considers processes and classification methods, and may propose changes to relevant sectors under the auspices of EU guidelines, whereby nano-products used for pest control have multiple environmental risks, whereby the highly current feature of NPs may engage in unpredicted reactions with environmental constituents, resulting in potentially unforeseen environmental consequences. Various physical forms of nanopesticides (emulsifiable concentrates, nano-dispersers, suspension concentrates, etc.) have diverse environmental interactions due in part to chemical composition and preparation methods used. It is imperative to gain a comprehensive knowledge of the long-term safety and environmental impact of nanopesticides and their active components prior to them being used commercially. Recommendations for nanotechnology research and development have been made by the Institute for Food and Agricultural Standards (http://www.msu.edu/~ifas/). In this context, extra NGOs and citizen organizations must actively participate (Acharya and Pal. 2020).

4.5.1 NPs as Phytoregulators

To provide sustenance and nourish an increasing number of people in the face of challenges such as climate change, negligent agrochemical use, reduced land for agriculture, unavoidable pathogen infections, and a scarcity of water resources, advancements in technology that will increase productivity in farming and speed up environmentally friendly agricultural growth are critical. As a result, developing nanotechnology may provide a basis for increased production and plant growth. Over half of global agricultural productivity and production would have been lost in the absence of nanotechnology, according to reports, resulting in an unimaginable financial loss (Kah and Kookana 2020). In agricultural production, promising results have been shown in various plant growth parameters due to the application of nanomaterials (NMs). A noticeable trend of a "low-promoting and high-inhibiting" response between higher plants and NMs at different concentrations has been observed. Immersion of pre-germinated watermelon seeds in a solution of magnetic Fe_2O_3 NPs resulted in accelerated root development and increased vegetative biomass (Farooq et al. 2022). Srivastava et al. (2014) demonstrated that spinach seeds treated with an aqueous suspension of iron pyrite NPs (FeS_2) for 14 hours produced a significant increase in spinach yield. Similarly, Acharya et al. (2020) demonstrated that the applied silver (Ag) NPs suspension to watermelon (*Citrullus lanatus*) seeds for 12 hours led to a notable enhancement in germination, resulting in increased growth, fruit yield, and improved quality. Additionally, treatment of tomato seeds with low concentrations of silicon dioxide (SiO_2) NPs, particularly with the highest values for seed germination index, seedling vigor index, germination time, and percentage of seed germination, was obtained with 8g/L of SiO_2 NPs (Siddiqui et al. 2014). Spraying low concentrations of two specific iron-based NPs, Fe and Fe_3O_4, on maize (*Zea mays*) leaves significantly boosts their growth and enhances photosynthesis (Li et al. 2021). In maize, using Fe_3O_4 NPs increases ferritin levels, decreases lipid peroxidation, and helps maintain consistent iron levels (Jalali, Ghanati, and Modarres-Sanavi 2016). According to a study by Tighe-Neira et al. (2020), both TiO_2 NPs and microparticles (μPs) provide a short-term improvement in photosynthesis, possibly through their influence on diffusion, which also increases water use efficiency and slightly boosts the plant's secondary metabolic processes. Moreover, Raliya et al. (2015) proved that using TiO_2 NPs has been shown to raise tomato plants' levels of chlorophyll The treatment of NPs has been reported to enhance plant growth and development through an increase in mRNA, thus seems under-regulated by protein coding genes or miRNA expressions. Fe_2O_3 NPs enhance the expression of ferric-chelate reductase (FRO2) and iron-regulated ferrous iron transporter (IRT1), leading to increased root-to-shoot transport process mobility for both FER- and/or IRT1-like gene homologs. Furthermore, these impacts promoted the efficiency of uptake by enhancing specific processes related to the chelation-mediated pathway. However, most soils contain low selenium (Se) levels and several limiting factors reduce its absorption. The plant leaves are an essential part of higher plants, which regularly conduct the metabolism of antioxidants, photosynthesis, secondary metabolites, and carbohydrate synthesis. Se is an effective environmental element for the high-level metabolic processes in a living body, with obvious biological functions (Lanza and Dos Rei 2021). The interest in Se NPs is inferred from these facts, and its synthesis can be potentially useful for applications as diverse uses. Higher bioavailability, lower toxicity, and higher biological activity of Se NPs have been reported in comparison to either inorganic or organic selenium species. Se NPs were the most effective for tobacco organogenesis and root growth, while Na_2SeO_3 completely inhibited callus proliferation and rooting. Hence, nanostructure substances have the potential to deliver new plant features and great performance that act as growth regulators for increasing crop yield (Zhao et al. 2020).

4.5.2 NPs for Preserving Soil Integrity and Functionality

To fully tap into the potential of this new family of tools, with a particular focus on agriculture, it is essential to mimic natural processes through cutting-edge scientific means. In general terms, the engineered nanomaterials introduced into soil systems pursue three main goals: enhancing the activity of soil enzymes, curbing the expansion of diseases instigated by insects or microorganisms, and increasing the availability of nutrients in the soil environment. Engineered NMs in the soils help to enhance the soil efficiency and, in so doing, optimize the utilization of fertilizers, which, in turn, boosts plant growth and supports environmental conservation. This selective introduction in soils is almost always seen as complementary to conventional sources of fertilization, often obtaining crop yields on par or even higher than usual while reducing the volume of applied products and minimizing nutrient leaching. The type and activity of soil enzymes are closely related to the process of material transformation in various environments, which together influences the level of soil fertility, a critical driving factor facilitating plant growth and nutrient utilization. The most sensitive indicators of the soil's biochemical status are enzymes; hence, enzyme-catalyzed reactions that drive immobilization and mineralization processes associated with organic matter transformation within the soil, through microbial activity, work in concert as a result. In addition, Ben-Moshe et al. (2013) studied the NPs on soil micro-properties present with structural and compositional changes in organic/inorganic materials by means of Fourier transform infrared (FTIR) and fluorescent techniques. It is worth mentioning that the applications of metallic NPs in soil stimulate the upregulation of some basic biogeochemical cycles (nitrogen, phosphorus, and carbon), macronutrient nitrogen (N), and phosphorous (P) bioavailability through altering microbial activity (Asadishad et al. 2017). Gold was incorporated into the soil system by Asadishad et al. (2017) through the enhancement of activity enzymes such as phosphatase while only micro-glucosidase also increased quite significantly. They are key to the health and quality of soil as they produce critical enzymes (Das and Verma 2011). Applying iron oxide NPs (Fe_3O_4 and Fe_2O_3) in concentrations of 0.42, 0.84, and 1.26 mg per gram of soil significantly enhances the activity of soil enzymes, such as invertase and urease, respectively (He et al. 2011). TiO_2 NPs were reported to have also increased urease activity in the soil by 67% of upregulation (Du et al. 2011). The activity of soil invertase is related to the dynamic variation in soil organic carbon (SOC) pools; it directly reflects the accumulation and decomposition processes of organic matter. This activity serves as a crucial marker of the soil carbon cycle (Zhong et al. 2019). Urease, however, plays a vital role in breaking down urea into ammonia (NH_3) and carbon dioxide (CO_2), pivotal elements in nutrient cycling. The positive correlation between soil enzyme activities and the availability of nutrients highlights their importance in enhancing soil fertility. As soil fertility improves, plants experience more robust growth, vigorous root development, and enhanced nutrient absorption (Verma et al. 2019). Chelated Fe resulted in decreased growth and productivity compared to the other applications of iron, while sandy soil had improved availability of Fe because it was treated with iron oxide NPs (Fe_2O_3 NPs), and after that, they were found by peanut plants (Rui et al. 2016). Because of its large number of planar structures from sp^2 and nonplanar (graphitic sheets and edges) C atoms, graphene oxide (GO) has special characteristics as water conduits for soil. Even at the final size, a minor dosage of GO was explored to significantly promote spinach and chives. The oxygen functional groups of GO acted as polar regions, resulting in the adsorption of water, and the hydrophobic sp^2 domains worked like channels to carry out oils due to this unique property. It could be applied for promoting seed germination since it allowed a fast transportation rate. Interaction of GO with soil grain surfaces served to retain GO within the planted treatment, thus preventing leaching and dissipation (He et al. 2018). Hence, more focus is needed in this aspect playing around with formulations for optimized dosage for the appropriate growth, synthesis, and generation of NPs from readily biodegradable organic/natural sources that are utilized as nutrients. Depending on their designs, they have indeed validated a plethora of useful impacts on crop improvement; however, the design of the soil is complex, and the mechanisms by which earth sprays on nitrogen-inductive nanomaterial soil give rise to beneficial, rather than detrimental, effects remain unknown and are a subject of ongoing research (Farooq et al. 2022).

4.5.3 Utilizing Nanopesticides in Plant Defense

The biotic and abiotic stresses lead to the activation of different defense mechanisms in plants (Ali et al. 2024). Pests mainly the fields, destroying all agricultural production. To kill pathogens, people often use pesticides that harm harmless or beneficial insects like bees. It aids in the eradication of hazardous pests that breed in soil, hence improving yield and the nutritional value of food. It is estimated that more than 90% of pesticides applied are absorbed into the environment, making it physically impossible to reach the disease control sites. Additionally, pesticide residues have the ability to damage groundwater and topsoil in the ecosystem. Recent advancements in the field like the development of nanoformulations have lightened up a new path for efficient usage with fewer toxic effects (Nuruzzaman et al. 2016). Kah and Kookana (2020) show that nanopesticides are different from the traditional ones in behavior consistent with the increased efficiency of the latter. There are particular advantages to using pesticides based on NPs over conventional pesticides, such as increased solubilization in water, enhanced formulation stability, elimination of toxic and volatile organic solvents through the AI's slow release, and improved mobility and insecticidal action (Sasson et al. 2007). As recent research has shown, to design NPs, substances synthesized as materials include C, silicates, semiconducting polymers, polymers, polysaccharides, metals, oxides, lipids, semiconductor quantum dots, ceramics, emulsions, proteins, and dendrimers. The fungicide of *Fusarium oxysporum* wilt is watermelon; mature CuO nanopesticides represent strong antifungal activity (Elmer et al. 2018). Silica NP application to squash plants reduces the infestation of *Spodoptera littoralis* larvae, corroborating with their results showing that silica NPs suppress *S. littoralis* on tomato plants (El-Helaly et al. 2016). Plants are more protected against the bacterial disease *Pseudomonas* by SiO_2 NPs with pesticide action than by their dissolved counterpart $Si(OH)_4$. According to unpublished data, the protective mechanism of SaNPs is based on the enhancement of SA-dependent plant immunities caused by partially released $Si(OH)_4$ from NPs entering stomata and subsequent distribution within spongy mesophyll. Additionally, solid nanoparticle-induced fully intact particles deposited in flowers are likely partially responsible for this, as they can also induce systemic acquired resistance (El-Shetehy et al. 2021). Nanopesticides were applied as TiO_2 and ZnO NPs in an experiment. The way these nanopesticides function is directly related to the production of electrons and protons under the setting of UV light. This process starts with the production of ROS, which can decompose complicated chemical substances and inactivate a vast variety of biological bacteria. ROS also generates superoxide anions, hydrogen peroxide, and hydroxyl radicals, which are complementary to ROS. There could be a different mechanism for the action of TiO_2 and ZnO NPs as pesticides; however, the primary mode was proposed to relate by UV light such that ROS are formed through photocatalysis (Zhang et al. 2018). According to Hao et al. (2018), Fe_2O_3 and TiO_2 not only severely suppress the growth of turnip mosaic virus in tobacco at both the molecular and protein levels but they also significantly stimulate biomass. Moreover, the nanostructures of copper exhibited much higher efficacy in reducing the risk of bacterial spot development in comparison with already existing micron-sized copper bactericides when considering both greenhouse and field experimental in situ conditions (Strayer-Scherer et al. 2018). Besides, the in vivo-application of external CuO-CF NPs in various concentrations from 5 to 350g/mL on diseased tomato seedlings infected with *Fusarium oxysporum* resulted in a reduction or absence of disease and a positive impact on plant growth parameters and fruit quality (Ashraf et al. 2021). Therefore, it was essential to use a special stimuli-responsive mechanism for the controlled release of pesticides based on plant demand and hence minimal degradation with an extended effective life span without causing eco-toxicity. Hence, future studies should emphasize the design of complete technologies for a massive scale production of any nanopesticide at an affordable price (Farooq et al. 2022).

4.5.4 Antimicrobial Action of NPs

Throughout the years, losses of large crops ensue from microbial diseases induced by viruses, fungi, and bacteria. Existing fungicides and antibacterial agents to combat these diseases have presented several issues concerning health problems and environmental pollution. In recent years, nanomaterials have been employed as an eco-friendly antibacterial in plant defense to lower infection risks due to lower toxicity and administration dosage (Fu et al. 2020). The core mechanism involves the production of ROS upon light radiation, which destroys fungi and microorganisms such as bacteria. NPs also release metal ions such as Ag^+ and Zn^{2+}, which cause disruption in bacterial processes and cell wall degradation, for instance, Ag NPs biosynthesized at a concentration 100μg/mL reduced chickpea fusarium wilt incidence by 73.33% compared to a traditional fungicide (Kaur et al. 2018). According to Fernández et al. (2016), when Ag NPs are applied at low concentrations (3 ppm), the type of silver significantly inhibits the growth of plant pathogenic fungi inoculated in potatoes. Antibacterial properties of magnesium oxide (MgO) NPs were more as compared to bulk MgO against *Ralstonia solanacearum*. This resulted from the direct binding of MgO NPs with bacterial surfaces, which consequently inflicted physical injuries to cell membranes, leading to a decline in motility and biofilm formation potential in *R. solanacearum* cells (Cai et al. 2018). Compared with commercial copper compounds, Cu-NPs possessed much greater bactericidal activities than important plant pathogens (Varympopi et al. 2020). Similarly, Majumdar, Ghosh, and Mukherjee (2021) reported Cu-NPs as a potent antibacterial agent against *Xanthomonas oryzae pv oryzae* (Xoo), an etiological pathogen of the blast. In addition to the promising efficacy, an important new observation was that 80% less metallic Cu was leached into the environment from a formulation of an advanced Co-nanocomposite compared to a commonly used commercial product, copper-mancozeb when examined for the cost-benefit arising out of such nanoscale antimicrobial agents in plant protection. These have the potential to reduce dependency and energy costs of producing synthetic chemicals reducing environmental hazards due to agrochemicals, while also offering a very sustainable solution for maintaining plant health effectively (Young et al. 2017; Table 4.1).

Table 4.1 Different nanoparticles and their impact.

Nanoparticles	Impact	References
Iron oxide (Fe_2O_3) NPs	Boost root growth and biomass in watermelon	Farooq et al. (2022)
Iron pyrite (FeS_2) NPs	Increase spinach yield significantly	Srivastava et al. (2014)
Silver (Ag) NPs	Enhance watermelon germination, growth, and fruit quality	Acharya et al. (2020)
Silicon dioxide (SiO_2) NPs	Improve tomato seed germination and vigor	Siddiqui et al. (2014)
Iron-based (Fe and Fe_3O_4) NPs	Promote maize growth and photosynthesis	Li et al. (2021)
Titanium dioxide (TiO_2) NPs	Boost photosynthesis and chlorophyll content in tomatoes	Tighe-Neira et al. (2020); Raliya et al. (2015)

(Continued)

Table 4.1 (Continued)

Nanoparticles	Impact	References
Selenium (Se) NPs	Promote organogenesis and root growth in tobacco	Lanza and Dos Rei (2021); Zhao et al. (2020)
Iron oxide (Fe_2O_4 and Fe_2O_3) NPs	Enhance soil enzyme activity, improving fertility	He et al. (2011)
Titanium dioxide (TiO_2) NPs	Increase soil urease activity, aiding nutrient cycling	Du et al. (2011)
Gold NPs	Improve soil enzyme activity linked to phosphorus availability	Asadishad et al. (2017)
Graphene oxide (GO)	Enhances water adsorption and nutrient delivery in soil	He et al. (2018)
Silica (SiO_2) NPs	Protect plants against bacterial pathogens via the controlled release	El-Shetehy et al. (2021)
Zinc oxide (ZnO) NPs	Act as a nanopesticide, generating reactive oxygen species to combat pathogens	Zhang et al. (2018)
Copper oxide (CuO) NPs	Exhibit strong antifungal activity and improves plant health	Elmer et al. (2018); Ashraf et al. (2021)

4.6 Conclusion and Future Perspectives

Consequently, the use of nanotechnology is a seminal solution to the complex quandaries of abiotic stress management among crop plants. Nanotechnology holds promise in adventitious plant research and considerable attention has been paid to the application of nanomaterials in the agriculture and food production sectors. The efficiency to increase nutrient uptake, water retention, and modulate plant growth, as well as defense mechanisms, is among the few ways that nanomaterials improve stress tolerance. Research has demonstrated that at the nanoscale, AgNPs, titanium dioxide (TiO_2), and selenium NPs can remarkably promote the resistance of plants to various stressful situations through enhanced antioxidant activities, photosynthetic efficiency, and plant growth. The application of nanotechnology in agriculture envisages not only potentially economically improved yields and crop quality but also improved new environmentally friendly technology based on the reduction of traditional chemical elements, which are produced when assimilating fertilizers that lead to major human effects. However, we still do not know enough about the long-term environmental and health impacts of these products requiring more extensive research in this area as well as appropriate regulatory provisions for their safe use in agricultural practices. However, the main scope of nanotechnology in agriculture is yet to come and that lies in developing multifunctional NPs according to crop requirements and environmental conditions surrounding it. Future studies should aim to unravel the physicochemical and molecular mechanisms that govern how nanomaterials are capable of boosting plant tolerance towards abiotic stresses. Similarly, research in other metal NPs is also essential to model an appropriate use in plant growth and stress tolerance. So, achieving ecologically friendly and biodegradable NPs are designed to lower the ecological footprints while improving soil health. In addition, we need field trials on a large scale, and also the discoveries made in the laboratory should be applied as an effective technology for manufacturing of nanopesticides and the production of nano fertilizers that can be applied to agriculture practices. This will require interdisciplinary collaboration between plant scientists, chemists (material science), environmentalists, and policymakers to exploit the full potential of nanotechnology. Implementation of appropriate guidelines and regulatory frameworks on nanomaterial use will guarantee that their contribution to food security is continued through ways that are safe and environmentally friendly.

References

Acharya, A. and Pal, P.K. (2020). Agriculture nanotechnology: translating research outcome to field applications by influencing environmental sustainability. *NanoImpact* 19: 100232.

Acharya, P., Jayaprakasham G.K., Crosby, K.M. et al. (2020). Nanoparticle-mediated seed priming improves germination, growth, yield, and quality of watermelons (*Citrullus lanatus*) at multi-locations in Texas. *Scientific Reports* 10 (1): 5037.

Adisa, I.O., Pullagurala, V.L., Peralta-Videa, J.R. et al. (2019). Recent advances in nano-enabled fertilizers and pesticides: a critical review of mechanisms of action. *Environmental Science: Nano* 6 (7): 2002–2030.

Ahmed, T., Noman, M., Ijaz, M. et al. (2021). Current trends and future prospective in nanoremediation of heavy metals contaminated soils: a way forward towards sustainable agriculture. *Ecotoxicology and Environmental Safety* 227: 112888.

Ahmed, T., Noman, M., Manzoor, N. et al. (2021). Nanoparticle-based amelioration of drought stress and cadmium toxicity in rice via triggering the stress responsive genetic mechanisms and nutrient acquisition. *Ecotoxicology and Environmental Safety* 209: 111829.

Ali, J., Mukarram, M., Abbas, A. et al. (2024). Wound to survive: mechanical damage suppresses aphid performance on brassica. *Journal of Plant Diseases and Protection* 131: 781–792. doi: 10.1007/s41348-024-00871-8.

Ali, M., Afzal, S., Parveen, A. et al. (2021). Silicon mediated improvement in the growth and ion homeostasis by decreasing Na^+ uptake in maize (*Zea mays* L.) cultivars exposed to salinity stress. *Plant Physiology and Biochemistry* 158: 208–218.

Ali, M.A., Mosa, K.A., El-Keblawy, A. et al. (2019). Exogenous production of silver nanoparticles by *Tephrosia apollinea* living plants under drought stress and their antimicrobial activities. *Nanomaterials* 9 (12): 1716.

Ali, S., Ullah, M.I., Sajjad, A. et al. (2021). Environmental and health effects of pesticide residues. In: *Sustainable Agriculture Reviews 48: Pesticide Occurrence, Analysis and Remediation* Vol. 2 Analysis, 311–336. Cham: Springer.

Almutairi, Z. M. (2019). Plant molecular defense mechanisms promoted by nanoparticles against environmental stresses. *International Journal of Agriculture & Biology* 21: 259–270.

Angulo-Bejarano, P.I., Puente-Rivera, J., and Cruz-Ortega, R. (2021). Metal and metalloid toxicity in plants: an overview on molecular aspects. *Plants* 10 (4): 635.

Arshad, M.S., Farooq, M., Asch, F. et al. (2017). Thermal stress impacts reproductive development and grain yield in rice. *Plant Physiology and Biochemistry* 115: 57–72.

Asadishad, B., Chahal, S., Cianciarelli, V. et al. (2017). Effect of gold nanoparticles on extracellular nutrient-cycling enzyme activity and bacterial community in soil slurries: role of nanoparticle size and surface coating. *Environmental Science: Nano* 4 (4): 907–918.

Ashraf, H., Anjum, T., Riaz, S. et al. (2021). Inhibition mechanism of green-synthesized copper oxide nanoparticles from *Cassia fistula* towards *Fusarium oxysporum* by boosting growth and defense response in tomatoes. *Environmental Science: Nano* 8 (6): 1729–1748.

Ashraf, J. and Javed, A. (2023). Food security and environmental degradation: do institutional quality and human capital make a difference? *Journal of Environmental Management* 331: 117330.

Atkinson, N.J. and Urwin, P.E. (2012). The interaction of plant biotic and abiotic stresses: from genes to the field. *Journal of Experimental Botany* 63 (10): 3523–3543.

Atta, K., Mondal, S., Gorai, S. et al. (2023). Impacts of salinity stress on crop plants: improving salt tolerance through genetic and molecular dissection. *Frontiers in Plant Science* 14: 1241736.

Baazaoui, N., Bellili, K., Messaoud, M. et al. (2023). Bio-nano-remediation of olive oil mill wastewater using silicon dioxide nanoparticles for its potential use as biofertilizer for young olive plants. *Silicon* 15 (17): 7395–7411.

Baig, N., Kammakakam, I., and Falath, W. (2021). Nanomaterials: a review of synthesis methods, properties, recent progress, and challenges. *Materials Advances* 2 (6): 1821–1871.

Bartolucci, C., Antonacci, A., Arduini, F. et al. (2020). Green nanomaterials fostering agrifood sustainability. *TrAC Trends in Analytical Chemistry* 125: 115840.

Ben-Moshe, T., Frenk, S., Dror, I. et al. (2013). Effects of metal oxide nanoparticles on soil properties. *Chemosphere* 90 (2): 640–646.

Beyene, H.D., Werkneh, A.A., Bezabh, H.K. et al. (2017). Synthesis paradigm and applications of silver nanoparticles (AgNPs), a review. *Sustainable Materials and Technologies* 13: 18–23.

Borišev, M., Borišev, I., Župunski, M. et al. (2016). Drought impact is alleviated in sugar beets (*Beta vulgaris* L.) by foliar application of fullerenol nanoparticles. *PLoS One* 11 (11): e0166248.

Bruna, T., Maldonado-Bravo, F., Jara, P. et al. (2021). Silver nanoparticles and their antibacterial applications. *International Journal of Molecular Sciences* 22 (13): 7202.

Cai, L., Chen, J., Liu, Z. et al. (2018). Magnesium oxide nanoparticles: effective agricultural antibacterial agent against *Ralstonia solanacearum*. *Frontiers in Microbiology* 9: 790.

Cao, X., Alabresm, A., Chen, Y.P. et al. (2020). Improved metal remediation using a combined bacterial and nanoscience approach. *Science of the Total Environment* 704: 135378.

Cecchin, I., Reddy, K.R., Thomé, A. et al. (2017). Nanobioremediation: integration of nanoparticles and bioremediation for sustainable remediation of chlorinated organic contaminants in soils. *International Biodeterioration & Biodegradation* 119: 419–428.

Chakravarty, D., Erande, M.B., and Late, D.J. (2015). Graphene quantum dots as enhanced plant growth regulators: effects on coriander and garlic plants. *Journal of the Science of Food and Agriculture* 95 (13): 2772–2778.

Cheng, P., Zhang, S., Wang, Q. et al. (2021). Contribution of nano-zero-valent iron and arbuscular mycorrhizal fungi to phytoremediation of heavy metal-contaminated soil. *Nanomaterials* 11 (5): 1264.

Choudhary, M., Yan, G., Siddique, K.H. et al. (2022). Heat stress during meiosis has lasting impacts on plant growth and reproduction in wheat (*Triticum aestivum* L.). *Agronomy* 12 (5): 987.

Das, S.K. and Varma, A. (2011). Role of enzymes in maintaining soil health. In: *Soil Enzymology*, 25–42. Berlin: Springer.

De Gusseme, B., Hennebel, T., Vanhaecke, L. et al. (2011). Biogenic palladium enhances diatrizoate removal from hospital wastewater in a microbial electrolysis cell. *Environmental Science & Technology* 45 (13): 5737–5745.

De La Torre-Roche, R., Hawthorne, J., Deng, Y. et al. (2013). Multiwalled carbon nanotubes and C60 fullerenes differentially impact the accumulation of weathered pesticides in four agricultural plants. *Environmental Science & Technology* 47 (21): 12539–12547.

Deplanche, K., Snape, T.J., Hazrati, S. et al. (2009). Versatility of a new bioinorganic catalyst: palladized cells of *Desulfovibrio desulfuricans* and application to dehalogenation of flame retardant materials. *Environmental Technology* 30 (7): 681–692.

Dimkpa, C.O., McLean, J.E., Latta, D.E. et al. (2012). CuO and ZnO nanoparticles: phytotoxicity, metal speciation, and induction of oxidative stress in sand-grown wheat. *Journal of Nanoparticle Research* 14: 1–15.

Djanaguiraman, M., Belliraj, N., Bossmann, S.H. et al. (2018). High-temperature stress alleviation by selenium nanoparticle treatment in grain sorghum. *ACS Omega* 3 (3): 2479–2491.

Drew, M.C. (1997). Oxygen deficiency and root metabolism: injury and acclimation under hypoxia and anoxia. *Annual Review of Plant Biology* 48 (1): 223–250.

Du, W., Sun, Y., Ji, R. et al. (2011). TiO2 and ZnO nanoparticles negatively affect wheat growth and soil enzyme activities in agricultural soil. *Journal of Environmental Monitoring* 13 (4): 822–828.

Dubey, S., Shri, M., Gupta, A. et al. (2018). Toxicity and detoxification of heavy metals during plant growth and metabolism. *Environmental Chemistry Letters* 16: 1169–1192.

El-Helaly, A.A., El-Bendary, H.M., Abdel-Wahab, A.S. et al. (2016). The silica-nano particles treatment of squash foliage and survival and development of *Spodoptera littoralis* (Bosid.) larvae. *Journal of Entomology and Zoology Studies* 4 (1): 175–180.

El-Shetehy, M., Moradi, A., Maceroni, M. et al. (2021). Silica nanoparticles enhance disease resistance in *Arabidopsis* plants. *Nature Nanotechnology* 16: 344–353.

Elmer, W., De La Torre-Roche, R., Pagano, L. et al. (2018). Effect of metalloid and metal oxide nanoparticles on *Fusarium* wilt of watermelon. *Plant Disease* 102: 1394–1401.

FAO F. (2017). *Agriculture Organization of the United Nations the Future of Food and Agriculture*. Rome: Trends and Challenges, FAO.

Farooq, M., Wahid, A., Kobayashi, N.S.M.A. et al. (2009). Plant drought stress: effects, mechanisms and management. In: *Sustainable Agriculture*, 153–188. Dordrecht: Springer.

Farooq, M.A., Hannan, F., Islam, F. et al. (2022). The potential of nanomaterials for sustainable modern agriculture: present findings and future perspectives. *Environmental Science: Nano* 9 (6): 1926–1951.

Farooq, M.A., Niazi, A.K., Akhtar, J. et al. (2019). Acquiring control: the evolution of ROS-induced oxidative stress and redox signaling pathways in plant stress responses. *Plant Physiology and Biochemistry* 141: 353–369.

Feng, L., Cao, M., Ma, X. et al. (2012). Superparamagnetic high-surface-area Fe_3O_4 nanoparticles as adsorbents for arsenic removal. *Journal of Hazardous Materials* 217: 439–446.

Fernández, J.G., Fernández-Baldo, M.A., Berni, E. et al. (2016). Production of silver nanoparticles using yeasts and evaluation of their antifungal activity against phytopathogenic fungi. *Process Biochemistry* 51: 1306–1313.

Fisher, M.C., Hawkins, N.J., Sanglard, D. et al. (2018). Worldwide emergence of resistance to antifungal drugs challenges human health and food security. *Science* 360 (6390): 739–742.

Fu, L., Wang, Z., Dhankher, O.P. et al. (2020). Nanotechnology as a new sustainable approach for controlling crop diseases and increasing agricultural production. *Journal of Experimental Botany* 71: 507–519.

Geissen, V., Mol, H., Klumpp, E. et al. (2015). Emerging pollutants in the environment: a challenge for water resource management. *International Soil and Water Conservation Research* 3 (1): 57–65.

Ghani, M.I., Saleem, S., Rather, S.A. et al. (2022). Foliar application of zinc oxide nanoparticles: an effective strategy to mitigate drought stress in cucumber seedling by modulating antioxidant defense system and osmolytes accumulation. *Chemosphere* 289: 133202.

Ghosh, M., Jana, A., Sinha, S. et al. (2016). Effects of ZnO nanoparticles in plants: cytotoxicity, genotoxicity, deregulation of antioxidant defenses, and cell-cycle arrest. *Mutation Research/ Genetic Toxicology and Environmental Mutagenesis* 807: 25–32.

Gohari, G., Mohammadi, A., Akbari, A. et al. (2020). Titanium dioxide nanoparticles (TiO_2 NPs) promote growth and ameliorate salinity stress effects on essential oil profile and biochemical attributes of *Dracocephalum moldavica*. *Scientific Reports* 10 (1): 912.

Gomes, D.G., Debiasi, T.V., Pelegrino, M.T. et al. (2022). Soil treatment with nitric oxide-releasing chitosan nanoparticles protects the root system and promotes the growth of soybean plants under copper stress. *Plants* 11 (23): 3245.

Gul, M., Wakeel, A., Steffens, D. et al. (2019). Potassium-induced decrease in cytosolic Na^+ alleviates deleterious effects of salt stress on wheat (*Triticum aestivum* L.). *Plant Biology* 21 (5): 825–831.

Gupta, A. and Sidana, A. (2020). Clinical hypnotherapy in grief resolution-a case report. *Indian Journal of Psychological Medicine* 42 (2): 193–197.

Gustavsson, J., Cederberg, C., Sonesson, U., Van Otterdijk, R., & Meybeck, A. (2011). *Global Food Losses and Food Waste*. Rome: FAO.

Haghighi, M., Abolghasemi, R., and da Silva, J.A. (2014). Low and high temperature stress affect the growth characteristics of tomato in hydroponic culture with Se and nano-Se amendment. *Scientia Horticulturae* 178: 231–240.

Haider, S., Raza, A., Iqbal, J. et al. (2022). Analyzing the regulatory role of heat shock transcription factors in plant heat stress tolerance: a brief appraisal. *Molecular Biology Reports* 49 (6): 5771–5785.

Hamdi, H., De La Torre-Roche, R., Hawthorne, J. et al. (2015). Impact of non-functionalized and amino-functionalized multiwall carbon nanotubes on pesticide uptake by lettuce (*Lactuca sativa* L.). *Nanotoxicology* 9 (2): 172–180.

Hamid, Y., Tang, L., Sohail, M.I. et al. (2019). An explanation of soil amendments to reduce cadmium phytoavailability and transfer to food chain. *Science of the Total Environment* 660: 80–96.

Han, D. and Currell, M.J. (2017). Persistent organic pollutants in China's surface water systems. *Science of the Total Environment* 580: 602–625.

Hao, Y., Yuan, W., Ma, C. et al. (2018). Engineered nanomaterials suppress turnip mosaic virus infection in tobacco (*Nicotiana benthamiana*). *Environmental Science: Nano* 5: 1685–1693.

Hasanpour, H., Maali-Amir, R., and Zeinali, H. (2015). Effect of TiO2 nanoparticles on metabolic limitations to photosynthesis under cold in chickpea. *Russian Journal of Plant Physiology* 62: 779–787.

He, S., Feng, Y., Ren, H. et al. (2011). The impact of iron oxide magnetic nanoparticles on the soil bacterial community. *Journal of Soils and Sediments* 11: 1408–1417.

He, Y., Hu, R., Zhong, Y. et al. (2018). Graphene oxide as a water transporter promoting germination of plants in soil. *Nano Research* 11: 1928–1937.

Hofmann, D.J., Butler, J.H., and Tans, P.P. (2009). A new look at atmospheric carbon dioxide. *Atmospheric Environment* 43 (12): 2084–2086.

Hojjat, S.S. and Ganjali, A. (2016). The effect of silver nanoparticles on lentil seed germination under drought stress. *International Journal of Farming and Allied Sciences* 5 (3): 208–212.

Hossain, Z., Yasmeen, F., and Komatsu, S. (2020). Nanoparticles: synthesis, morphophysiological effects, and proteomic responses of crop plants. *International Journal of Molecular Sciences* 21 (9): 3056.

Huang, D., Qin, X., Peng, Z. et al. (2018). Nanoscale zero-valent iron assisted phytoremediation of Pb in sediment: impacts on metal accumulation and antioxidative system of *Lolium perenne*. *Ecotoxicology and Environmental Safety* 153: 229–237.

Huang, X., Jain, P.K., El-Sayed, I.H. et al. (2007). Gold nanoparticles: interesting optical properties and recent applications in cancer diagnostics and therapy. *Nanomedicine* 2 (5): 681–693.

Husen, A. and Siddiqi, K.S. (2014). The Synthesis of nanoparticles: concept, controversy and application. *Nanoscale Research Letters* 9: 1–24.

Iqbal, M., Raja, N.I., Mashwani, Z.U. et al. (2019). Effect of silver nanoparticles on growth of wheat under heat stress. *Iranian Journal of Science and Technology, Transactions A: Science* 43: 387–395.

Iqbal, Z., Iqbal, M.S., Hashem, A. et al. (2021). Plant defense responses to biotic stress and its interplay with fluctuating dark/light conditions. *Frontiers in Plant Science* 12: 631810.

Jalali, M., Ghanati, F., and Modarres-Sanavi, A.M. (2016). Effect of Fe_3O_4 nanoparticles and iron chelate on the antioxidant capacity and nutritional value of soil-cultivated maize (*Zea mays*) plants. *Crop and Pasture Science* 67 (6): 621–628.

Jalil, S.U. and Ansari, M.I. (2019). Nanoparticles and abiotic stress tolerance in plants: synthesis, action, and signaling mechanisms. In: *Plant Signaling Molecules*, 549–561. Elsevier.

Kah, M. and Kookana, R. (2020). Emerging investigator series: nanotechnology to develop novel agrochemicals: critical issues to consider in the global agricultural context. *Environmental Science: Nano* 7 (7): 1867–1873.

Kaur, P., Thakur, R., Duhan, J.S. et al. (2018). Management of wilt disease of chickpea in vivo by silver nanoparticles biosynthesized by rhizospheric microflora of chickpea (*Cicer arietinum*). *Journal of Chemical Technology & Biotechnology* 93: 3233–3243.

Khodakovskaya, M.V., de Silva, K., Nedosekin, D.A. et al. (2011). Complex genetic, photothermal, and photoacoustic analysis of nanoparticle-plant interactions. *Proceedings of the National Academy of Sciences of the United States of America* 108 (3): 1028–1033.

Kohan-Baghkheirati, E. and Geisler-Lee, J. (2015). Gene expression, protein function and pathways of *Arabidopsis thaliana* responding to silver nanoparticles in comparison to silver ions, cold, salt, drought, and heat. *Nanomaterials* 5 (2): 436–467.

Kohatsu, M.Y., Lange, C.N., Pelegrino, M.T. et al. (2021). Foliar spraying of biogenic CuO nanoparticles protects the defence system and photosynthetic pigments of lettuce (*Lactuca sativa*). *Journal of Cleaner Production* 324: 129264.

Komatsu, S., Hiraga, S., and Yanagawa, Y. (2012). Proteomics techniques for the development of flood tolerant crops. *Journal of Proteome Research* 11 (1): 68–78.

Kumar, V. and Yadav, S.K. (2009). Plant-mediated synthesis of silver and gold nanoparticles and their applications. *Journal of Chemical Technology & Biotechnology: International Research in Process, Environmental & Clean Technology* 84 (2): 151–157.

Kumari, M., Pandey, S., Mishra, S.K. et al. (2020). Omics-based mechanistic insight into the role of bioengineered nanoparticles for biotic stress amelioration by modulating plant metabolic pathways. *Frontiers in Bioengineering and Biotechnology* 8: 242.

Lanza, M.G. and Dos Reis, A.R. (2021). Roles of selenium in mineral plant nutrition: ROS scavenging responses against abiotic stresses. *Plant Physiology and Biochemistry* 164: 27–43.

Li, M., Zhang, P., Adeel, M. et al. (2021). Physiological impacts of zero valent iron, Fe_3O_4 and Fe_2O_3 nanoparticles in rice plants and their potential as Fe fertilizers. *Environmental Pollution* 269: 116134.

Liu, F.Y., Xiong, F.X., Fan, Y.K. et al. (2016). Facile and scalable fabrication engineering of fullerenol nanoparticles by improved alkaline-oxidation approach and its antioxidant potential in maize. *Journal of Nanoparticle Research* 18: 1–13.

Lok, C.N., Ho, C.M., Chen, R. et al. (2007). Silver nanoparticles: partial oxidation and antibacterial activities. *JBIC Journal of Biological Inorganic Chemistry* 12: 527–534.

Lowry, G.V., Avellan, A., and Gilbertson, L.M. (2019). Opportunities and challenges for nanotechnology in the agri-tech revolution. *Nature Nanotechnology* 14 (6): 517–522.

Majumdar, T.D., Ghosh, C.K., and Mukherjee, A. (2021). Dual role of copper nanoparticles in bacterial leaf blight-infected rice: a therapeutic and metabolic approach. *ACS Agricultural Science & Technology* 1: 160–172.

Manzoor, J., Sharma, M., and Wani, K.A. (2018). Heavy metals in vegetables and their impact on the nutrient quality of vegetables: a review. *Journal of Plant Nutrition* 41 (13): 1744–1763.

Mesterházy, Á., Oláh, J., and Popp, J. (2020). Losses in the grain supply chain: causes and solutions. *Sustainability* 12 (6): 2342.

Molnár, Z., Bódai, V., Szakacs, G. et al. (2018). Green synthesis of gold nanoparticles by thermophilic filamentous fungi. *Scientific Reports* 8 (1): 3943.

Mushtaq, Z., Faizan, S., Gulzar, B. et al. (2022). Changes in growth, photosynthetic pigments, cell viability, lipid peroxidation and antioxidant defense system in two varieties of chickpea (*Cicer arietinum* L.) subjected to salinity stress. *Phyton* 91 (1): 149.

Mustafa, G., Sakata, K., and Komatsu, S. (2015). Proteomic analysis of flooded soybean root exposed to aluminum oxide nanoparticles. *Journal of Proteomics* 128: 280–297.

Mustafa, G. and Komatsu, S. (2016). Insights into the response of soybean mitochondrial proteins to various sizes of aluminum oxide nanoparticles under flooding stress. *Journal of Proteome Research* 15 (12): 4464–4475.

Nair, P.M.G. and Chung, I.M. (2015). Study on the correlation between copper oxide nanoparticles induced growth suppression and enhanced lignification in Indian mustard (*Brassica juncea L.*). *Ecotoxicology and Environmental Safety* 113: 302–313.

Noman, M., Ahmed, T., Hussain, S. et al. (2020). Biogenic copper nanoparticles synthesized by using a copper-resistant strain Shigella flexneri SNT22 reduced the translocation of cadmium from soil to wheat plants. *Journal of Hazardous Materials* 398: 123175.

Nuruzzaman, M., Rahman, M.M., Liu, Y. et al. (2016). Nanoencapsulation, nano-guard for pesticides: a new window for safe application. *Journal of Agricultural and Food Chemistry* 64: 1447–1483.

Oliveira, H.C., Gomes, B.C., Pelegrino, M.T. et al. (2016). Nitric oxide-releasing chitosan nanoparticles alleviate the effects of salt stress in maize plants. *Nitric Oxide* 61: 10–19.

Oza, G., Reyes-Calderón, A., Mewada, A. et al. (2020). Plant-based metal and metal alloy nanoparticle synthesis: a comprehensive mechanistic approach. *Journal of Materials Science* 55: 1309–1330.

Ozel, F., Kockar, H., and Karaagac, O. (2015). Growth of iron oxide nanoparticles by hydrothermal process: effect of reaction parameters on the nanoparticle size. *Journal of Superconductivity and Novel Magnetism* 28: 823–829.

Öztürk, B.Y., Gürsu, B.Y., and Dağ, İ. (2020). Antibiofilm and antimicrobial activities of green synthesized silver nanoparticles using marine red algae *Gelidium corneum*. *Process Biochemistry* 89: 208–219.

Panta, S., Flowers, T., Lane, P. et al. (2014). Halophyte agriculture: success stories. *Environmental and Experimental Botany* 107: 71–83.

Parisi, C., Vigani, M., and Rodríguez-Cerezo, E. (2015). Agricultural nanotechnologies: what are the current possibilities? *Nano Today* 10 (2): 124–127.

Praveen, A., Khan, E., Ngiimei, D.S. et al. (2018). Iron oxide nanoparticles as nano-adsorbents: a possible way to reduce arsenic phytotoxicity in Indian mustard plant (*Brassica juncea* L.). *Journal of Plant Growth Regulation* 37: 612–624.

Qi, M., Liu, Y., and Li, T. (2013). Nano-TiO2 improves the photosynthesis of tomato leaves under mild heat stress. *Biological Trace Element Research* 156: 323–328.

Rai, M., Yadav, A., and Gade, A. (2009). Silver nanoparticles as a new generation of antimicrobials. *Biotechnology Advances* 27 (1): 76–83. doi: 10.1016/J.BIOTECHADV.2008.09.002.

Raliya, R., Nair, R., Chavalmane, S. et al. (2015). Mechanistic evaluation of translocation and physiological impact of titanium dioxide and zinc oxide nanoparticles on the tomato (*Solanum lycopersicum* L.) plant. *Metallomics* 7 (12): 1584–1594.

Ranjan, A., Rajput, V.D., Kumari, A. et al. (2022). Nanobionics in crop production: an emerging approach to modulate plant functionalities. *Plants* 11 (5): 692.

Rathee, M. and Dalal, P. (2018). Emerging insect pests in Indian agriculture. *Indian Journal of Entomology* 80 (2): 267–281.

Raza, A., Razzaq, A., Mehmood, S.S. et al. (2019). Impact of climate change on crops adaptation and strategies to tackle its outcome: a review. *Plants* 8 (2): 34.

Reinle, K., Mogk, A., and Bukau, B. (2022). The diverse functions of small heat shock proteins in the proteostasis network. *Journal of Molecular Biology* 434 (1): 167157.

Riseh, R.S., Hassanisaadi, M., Vatankhah, M. et al. (2022). Nano/microencapsulation of plant biocontrol agents by chitosan, alginate, and other important biopolymers as a novel strategy for alleviating plant biotic stresses. *International Journal of Biological Macromolecules* 222: 1589–1604.

Rubilar, O., Rai, M., Tortella, G. et al. (2013). Biogenic nanoparticles: copper, copper oxides, copper sulphides, complex copper nanostructures and their applications. *Biotechnology Letters* 35: 1365–1375.

Rui, M., Ma, C., Hao, Y. et al. (2016). Iron oxide Nanoparticles as a potential iron fertilizer for peanut (*Arachis hypogaea*). *Frontiers in Plant Science* 7: 815.

Saffan, M.M., Koriem, M.A., El-Henawy, A. et al. (2022). Sustainable production of tomato plants (*Solanum lycopersicum* L.) under low-quality irrigation water as affected by bio-nanofertilizers of selenium and copper. *Sustainability* 14 (6): 3236.

Sasson, Y., Levy-Ruso, G., Toledano, O. et al. (2007). Nanosuspensions: emerging novel agrochemical formulations. In: *Insecticides Design Using Advanced Technologies* (ed. I. Ishaaya, R. Nauen, and A.R. Horowitz), 1–32. Berlin: Springer.

Savary, S., Willocquet, L., Pethybridge, S.J. et al. (2019). The global burden of pathogens and pests on major food crops. *Nature Ecology & Evolution* 3 (3): 430–439.

Seghatoleslami, M., Feizi, H., Mousavi, G. et al. (2015). Effect of magnetic field and silver nanoparticles on yield and water use efficiency of *Carum copticum* under water stress conditions. *Polish Journal of Chemical Technology* 17 (1): 110–114.

Seliem, M.K., Hafez, Y., and El-Ramady, H. (2020). Using nano-selenium in reducing the negative effects of high temperature stress on *Chrysanthemum morifolium* Ramat. *Journal of Sustainable Agricultural Sciences* 46 (3): 47–60.

Siddiqui, M.H., Al-Whaibi, M.H., Faisal, M. et al. (2014). Nano-silicon dioxide mitigates the adverse effects of salt stress on *Cucurbita pepo* L. *Environmental Toxicology and Chemistry* 33 (11): 2429–2437.

Singh, A. (2016). Scope of nanotechnology in crop science: profit or loss. *Research & Reviews: Journal of Botanical Sciences* 5.

Singh, A., Kumar, H., Kumar, S. et al. (2022). Role of chitosan and chitosan-based nanoparticles in antioxidant regulation of plants. In: *Role of Chitosan and Chitosan-Based Nanomaterials in Plant Sciences*, 321–341. Elsevier.

Singh, S., Yadav, V., Arif, N. et al. (2020). Heavy metal stress and plant life: uptake mechanisms, toxicity, and alleviation. In: *Plant Life Under Changing Environment*, 271–287. Academic Press.

Sobral, M.C., Martins, I.M., and Sobral, A.J. (2022). Role of chitosan and chitosan-based nanoparticles against heavy metal stress in plants. In: *Role of Chitosan and Chitosan-Based Nanomaterials in Plant Sciences*, 273–296. Academic Press.

Srivastava, G., Das, C.K., Das, A. et al. (2014). Seed treatment with iron pyrite (FeS_2) nanoparticles increases the production of spinach. *RSC Advances* 4 (102): 58495–58504.

Srujana, S. and Bhagat, D. (2022). Sensors for agricultural pest management using nanotechnology/nanomaterials. *Nanotechnology for Environmental Engineering* 7 (1): 319–323.

Stahl-Rommel, S., Kalra, I., D'Silva, S. et al. (2022). Cyclic electron flow (CEF) and ascorbate pathway activity provide constitutive photoprotection for the photopsychrophile, *Chlamydomonas* sp. UWO 241 (renamed *Chlamydomonas priscuii*). *Photosynthesis Research* 1: 1–6.

Strayer-Scherer, A., Liao, Y.Y., Young, M. et al. (2018). Advanced copper composites against copper-tolerant *Xanthomonas perforans* and tomato bacterial spot. *Phytopathology* 108: 196–205.

Sur, U.K., Ankamwar, B., Karmakar, S. et al. (2018). Green synthesis of silver nanoparticles using the plant extract of Shikakai and Reetha. *Materials Today: Proceedings* 5 (1): 2321–2329.

Suzuki, N., Rivero, R.M., Shulaev, V. et al. (2014). Abiotic and biotic stress combinations. *New Phytologist* 203 (1): 32–43.

Syu, Y.Y., Hung, J.H., Chen, J.C. et al. (2014). Impacts of size and shape of silver nanoparticles on Arabidopsis plant growth and gene expression. *Plant Physiology and Biochemistry* 83: 57–64.

Thornton, P.K., Ericksen, P.J., Herrero, M. et al. (2014). Climate variability and vulnerability to climate change: a review. *Global Change in Biology* 20 (11): 3313–3328.

Tighe-Neira, R., Reyes-Díaz, M., Nunes-Nesi, A. et al. (2020). Titanium dioxide nanoparticles provoke transient increase in photosynthetic performance and differential response in antioxidant system in *Raphanus sativus* L. *Scientia Horticulturae* 269: 109418.

Tripathi, D., Singh, M., and Pandey-Rai, S. (2022). Crosstalk of nanoparticles and phytohormones regulate plant growth and metabolism under abiotic and biotic stress. *Plant Stress* 6: 100107.

Tripathi, D.K., Singh, S., Singh, S. et al. (2017). An overview on manufactured nanoparticles in plants: uptake, translocation, accumulation and phytotoxicity. *Plant Physiology and Biochemistry* 110: 2–12.

Ullah, A., Nadeem, F., Nawaz, A. et al. (2022). Heat stress effects on the reproductive physiology and yield of wheat. *Journal of Agronomy and Crop Science* 208 (1): 1–7.

Varympopi, A., Dimopoulou, A., Theologidis, I. et al. (2020). Bactericides based on copper nanoparticles restrain growth of important plant pathogens. *Pathogens* 9: 1024.

Van Esse, H.P., Reuber, T.L., and van der Does, D. (2020). Genetic modification to improve disease resistance in crops. *New Phytologist* 225 (1): 70–86.

Vartapetian, B.B., Dolgikh, Y.I., Polyakova, L.I. et al. (2014). Biotechnological approaches to creation of hypoxia and anoxia tolerant plants. *Acta Naturae* (англоязычная версия) 6 (2 (21)): 19–30.

Verma, S.K., Das, A.K., Gantait, S. et al. (2019). Applications of carbon nanomaterials in the plant system: a perspective view on the pros and cons. *Science of the Total Environment* 667: 485–499.

Wang, L., Ning, C., Pan, T. et al. (2022). Role of silica nanoparticles in abiotic and biotic stress tolerance in plants: a review. *International Journal of Molecular Sciences* 23 (4): 1947.

Wu, J., Xie, Y., Fang, Z. et al. (2016). Effects of Ni/Fe bimetallic nanoparticles on phytotoxicity and translocation of polybrominated diphenyl ethers in contaminated soil. *Chemosphere* 162: 235–242.

Young, M., Ozcan, A., Myers, M.E. et al. (2017). Multimodal generally recognized as safe zno/nanocopper composite: a novel antimicrobial material for the management of citrus phytopathogens. *Journal of Agricultural and Food Chemistry* 66: 6604–6608.

Zahra, N., Hafeez, M.B., Shaukat, K. et al. (2021). Hypoxia and anoxia stress: plant responses and tolerance mechanisms. *Journal of Agronomy and Crop Science* 207 (2): 249–284.

Zhang, H., Liu, Y., Shen, X. et al. (2017). Influence of multiwalled carbon nanotubes and sodium dodecyl benzene sulfonate on bioaccumulation and translocation of pyrene and 1-methylpyrene in maize (Zea mays) seedlings. *Environmental Pollution* 220: 1409–1417.

Zhang, L., Jia, H., Liu, C. et al. (2018). Enhanced generation of reactive oxygen species and photocatalytic activity by PT-based metallic nanostructures: the composition matters. *Journal of Environmental Science and Health, Part C* 37: 1–13.

Zhang, L.Z., Wei, N., Wu, Q.X. et al. (2007). Anti-oxidant response of *Cucumis sativus* L. to fungicide carbendazim. *Pesticide Biochemistry and Physiology* 89 (1): 54–59.

Zhao, J., Liang, X., Zhu, N. et al. (2020). Immobilization of mercury by nano-elemental selenium and the underlying mechanisms in hydroponic-cultured garlic plant. *Environmental Science: Nano* 7 (4): 1115–1125.

Zhao, L., Peng, B., Hernandez-Viezcas, J.A. et al. (2012). Stress response and tolerance of *Zea mays* to CeO_2 nanoparticles: cross talk among H_2O_2, heat shock protein, and lipid peroxidation. *ACS Nano* 6 (11): 9615–9622.

Zhong, Q., Wang, K., Nie, M. et al. (2019). Responses of wetland soil carbon and nutrient pools and microbial activities after 7 years of experimental warming in the Yangtze Estuary. *Ecological Engineering* 136: 68–78.

5

Impacts of Nanomaterials on Soil Microbial Communities

Nisha Kumari[1]*, Abhishek Tiwari*[1]*, Ingle Sagar Nandulal*[1]*, Sai Parasar Das*[1]*, Bhabani Prasad Mondal*[1]*, Bipin Bihari*[1]*, Pritam Ganguly*[1]*, Chandini*[2]*, and Randeep Kumar*[1]

[1] *Department of Soil Science and Agricultural Chemistry, Bihar Agricultural University Sabour, Bhagalpur, Bihar, India*
[2] *Department of Agronomy, Bihar Agricultural University Sabour, Bhagalpur, Bihar, India*

5.1 Introduction

The study and manipulation of matter at scales between about 1 and 100 nm are topics covered by nanotechnology. According to Klaine et al. (2008), materials having at least one dimension falling within this size range are categorized as nanoparticles (NPs). In a 1974 publication, University of Tokyo researcher Norio Taniguchi coined the term "nanotechnology" (Taniguchi 1974). Later, the idea became well-known due to Eric Drexler's (1986) book *Engines of Creation*. The distinct characteristics of NPs, which set them apart from bulk materials and individual molecules, have attracted a lot of attention. Depending on the material that makes up their core, these NPs can be either synthetic or natural. Inorganic NPs can be divided into metal oxides (like aluminum oxide, cerium oxide, copper oxide, indium oxide, magnesium oxide, nickel oxide, silicon dioxide, titanium dioxide, zinc oxide, and zirconium dioxide), metal nanoparticles (MNPs; like aluminum, bismuth, cobalt, copper, gold, iron, indium, molybdenum, nickel, silicon, silver, tin, titanium, tungsten, and zinc), and quantum dots. Carbon nanotubes (CNTs) and fullerenes are examples of organic NPs. The potential toxic effects of metal-based NPs on the diversity, activity, and abundance of plant and animal life have been extensively studied and applied to a variety of situations (Ebbs et al. 2016; Ebrahimi et al. 2016; Olkhovych et al. 2016; Singh and Kumar 2016; Van et al. 2016a, 2016b; Cvjetko et al. 2017; McGee et al. 2017; Priester et al. 2017; You et al. 2018).

Owing to their wide range of applications, NPs are used in a wide range of industries, such as wastewater treatment, paints, plastics, coatings, consumer goods, agriculture, chemicals, electronics, optics, environmental remediation, food, fuel additives, textiles, and medicine (Salata 2004; Keller and Lazareva 2014; Yadav, Mungray, and Mungray 2014; Servin et al. 2017). The goal of creating new NP kinds with unique properties and uses is

Nanotechnology-based Sustainable Agriculture, First Edition. Edited by Pardeep Singh et al.

propelling the fast advancement of nanotechnology research, leading to higher production and usage. Currently, there is a list of approximately 1600 consumer products based on nanotechnology (Vance et al. 2015), and 2870 products based on or containing NPs are listed on www.nanodb.dk, an online database (Hansen et al. 2016). Global NP consumption was predicted to increase from 225 060 tons in 2014 to around 584 984 tons by 2019 (BCC Research 2024, while global NP production was estimated at 260 000–309 000 tons in 2010 (Yadav, Mungray, and Mungray 2014). The precise numbers for the production of NP worldwide are still unknown.

5.2 Types of Nanomaterials and Their Agricultural Applications

- *Metallic NPs*: Gold, silver, and iron oxide NPs are used in agriculture for purposes, such as pest control, soil remediation, and plant disease management. Silver nanoparticles (Ag-NPs), for example, exhibit antimicrobial properties that can be utilized to control soil-borne pathogens (Joudeh and Linke 2022).
- *Quantum Dots*: Quantum dots can be employed in precision agriculture for monitoring soil conditions and plant health through advanced sensing technologies. Their fluorescence properties allow for highly sensitive detection of soil nutrients and contaminants (Alivisatos et al. 2005).
- *Carbon-based Nanomaterials (NMs)*: Graphene and CNTs are used to enhance soil properties and improve plant growth. Graphene-based nanocomposites can increase soil water retention, while CNTs can enhance the delivery of nutrients and pesticides to plants (Dresselhaus et al. 2004).

5.3 Soil Microbial Communities: Role in Agriculture

5.3.1 Composition and Functions

Soil microbial communities, comprising bacteria, fungi, archaea, and other microorganisms, are crucial for maintaining soil health and fertility (Van Der Heijden, Bardgett, and Van Straalen 2008; Smith and Read 2008; Glick 2012). They play several key roles:

- *Nutrient Cycling*: Plants can get vital nutrients like phosphorus and nitrogen that are released by microbes during the breakdown of organic matter.
- *Disease Suppression*: Plant diseases can be inhibited by specific soil bacteria by means of competition, hostility, or the generation of antimicrobial chemicals.
- *Soil Structure*: Fungal hyphae and microbial exudates aid in soil aggregation, which enhances soil aeration and structure.
- *Symbiotic Relationships*: Beneficial relationships between mycorrhizal fungi and plant roots improve nutrient uptake and stimulate plant growth.
- *Plant Growth Promotion*: Plant growth–promoting rhizobacteria (PGPR) are bacteria that produce growth hormones or increase nutrient availability to promote plant growth.

- *Influences of Agricultural Practices*: Agricultural practices, such as fertilization, irrigation, and pesticide application, can significantly impact soil microbial communities. Practices that alter soil pH, moisture, and organic matter content can shift microbial populations and affect their functionality.

5.4 Effect of NPs on Microbial Diversity

Beneficial microbial groups, including bacteria and fungi, may face serious risks from the introduction of NPs into natural environments (Table 5.1). Many human activities cause these NPs to accumulate in soil: deliberate releases through technologies for remediating soil and water, use of these particles as pesticides and fertilizers in agriculture, and unintentional dispersal through sewage, water, air, and sludge (Tourinho et al. 2012; Tripathi, Champagne, and Tufenkji 2012; Maurer-Jones et al. 2013; Shandilya et al. 2015; Coll et al. 2016). Soil respiration and enzymatic activity measurements can be used to evaluate the effect of NPs on microbial activity (Simonin and Richaume 2015). For example, Xu et al. (2015) found that in flooded paddy soil, TiO_2- and CuO-NPs changed the community structures and decreased soil microbial biomass and enzymatic activities. Similar to this, You et al. (2018) found that in saline-alkali and black soils, exposure to ZnO, TiO_2, CeO_2, and Fe_3O_4 NPs affected bacterial communities and soil enzyme activities (such as invertase, urease, catalase, and phosphatase), which in turn caused changes in soil enzyme activity and disruptions to biological nitrogen fixation. According to research by Jiling et al. (2016), soil bacterial content was considerably decreased by high concentrations of Fe_3O_4 NPs. Additionally, it was discovered that ZnO and CeO_2 NPs hindered enzymatic activities and decreased the plate counts of *Azotobacter*, P-solubilizing, and K-solubilizing bacteria (Chai et al. 2015). Additionally, it has been demonstrated that titanium dioxide NPs decrease the number of active soil bacteria and enzymatic activity, which has a detrimental impact on microbial diversity, abundance, and activity (Buzea, Pacheco, and Robbie 2007; Solanki, John, and Ki-Bum 2008). On the other hand, research by Maliszewska (2016) showed that biogenic Au NPs at concentrations up to 33 mg/kg might be regarded as non-harmful because they did not significantly alter soil processes. However, ZnO-NPs generated by both biological and chemical means showed antibacterial efficacy against clinical isolates, including *Bacillus subtilis, Escherichia coli, Klebsiella pneumoniae, Pseudomonas aeruginosa, Salmonella typhi*, and *Staphylococcus aureus* (Lakshmi et al. 2012). According to Lofts et al. (2013), the main factor that renders both pathogenic and beneficial bacteria fatal is the release of copper ions from Cu NPs. The study by Concha-Guerrero, Brito, and Piñón-Castillo (2014), which examined the development of cavities, holes, membrane breakdown, blebs, cellular collapse, and lysis in soil bacterial isolates, provided more evidence of the significant toxicity of CuO-NPs to native soil bacteria. The impact of CuO-NPs on the soil microbial community is still not well understood, despite these observations. According to Frenk et al. (2013), inorganic NPs often represent a higher toxicity risk to soil microorganisms than organic NPs. For instance, Pradhan et al. (2011) discovered that by altering the related microbial populations, exposure to CuO and Ag-NPs reduced the rate of leaf degradation.

5.4.1 Changes in Microbial Community Structure

It has been shown that the relationship between microorganisms and plants greatly increases soil health, increases natural floral biodiversity, and improves agricultural yield and sustainability. Mohanty et al. (2014) state that depending on how soil microorganisms travel through soil layers, engineered nanomaterials (ENMs) can have different effects on them. The properties of the soil have a big impact on these impacts. For instance, studies reveal that the reactions of clay and sandy soils to magnetite and copper oxide (CuO) NPs are different (Frenk et al. 2013). Pawlett et al. (2013) also observed that more negatively than positively, microorganisms in sandy soils were impacted by zero-valent iron NPs. Furthermore, it was shown that elements such as salinity and organic carbon content influenced how ENMs affected microbial communities.

In freshwater and marine environments, ENMs have an effect on both planktonic organisms and biofilms, which are microorganisms found in aquatic ecosystems. For example, in controlled laboratory environments and in river water, carbon-based ENMs have been demonstrated to kill bacterial cells (Kang et al. 2009). In the same way, freshwater environments' microbial composition was changed by Ag-NPs (Das et al. 2012). Planktonic bacteria are less resistant to environmental stress than biofilms, which are communities of microorganisms contained in a polymer matrix; however, biofilms can still affect the toxicity of ENMs. Ag-NPs have been demonstrated to have an impact on the early development and subsequent succession of marine biofilm ecosystems (Cao et al. 2013).

Some research has revealed no substantial impact on community structure, despite the fact that many studies show a drop in microbial diversity following ENM exposure (Bradford et al. 2009). The main cause of ENMs' antibacterial qualities is their tiny size and high surface area, which allow them to interact directly with microbial cells and subcellular structures (Ng et al. 2013). According to Li et al. (2010), this contact may result in the generation of reactive oxygen species (ROS), damage to cell membranes, and inactivation of enzymes, all of which increase the antibacterial activity of the NPs.

5.4.2 Impacts of NMs on Microbial Function and Soil Health

An increasing amount of research is being conducted on the effects of ENMs, particularly NPs, on soil health and microbiological activity. Numerous important elements have been investigated, such as the roles of soil, methanogenesis, nutrient removal procedures, the effectiveness of treating industrial wastewater that contains ENMs, and evaluations of environmental toxicity. For instance, studies have revealed that Al_2O_3 ENMs lower the formation of methane, but when humic acids are absorbed by the NPs, their toxicity decreases (Alvarez and Cervantes 2012). In contrast, it has been discovered that Ag-NPs decrease methanogenesis in landfill bioreactors in a dose-dependent way. At 1 mg/kg, biogas production is constant, but at 10 mg/kg, methane output drastically decreases (Yang et al. 2012).

It's interesting to note that fullerene (C_{60}) treatment of biosolids from anaerobic sludge did not significantly reduce methanogenesis as observed in a subsequent study by Nyberg et al. (2008). Other ENMs that have been shown to impede nitrogen removal processes include Ag-NPs (Masrahi et al. 2014) and TiO_2 (Zheng et al. 2012). It has been demonstrated that graphene oxide worsens water turbidity and hinders the removal of nutrients, especially nitrogen and phosphorus, which results in lower effluent quality (Ahmed and Rodrigues 2013). Furthermore, it was discovered that Ag-NPs inhibited nitrification in wastewater treatment systems as well as soil microbial populations (Masrahi et al. 2014). According to Zheng et al. (2011), TiO_2 also reduced the effectiveness of nitrogen and phosphorus removal in activated sludge processes. In contrast, CuO-NPs increased the development of denitrifying bacteria, particularly those that reduce N_2O, which improved the removal of nitrogen overall (Chen et al. 2012). The reasons behind these effects are probably related to the unique toxicities of ENMs, since various bacteria react differently to NPs. For example, against some bacteria, Ag-NPs show antibacterial capabilities, but not against others (Sun et al. 2013).

Although ENMs raise some environmental problems, they also hold great promise for enhancing soil quality and promoting sustainable agriculture. ENMs may be able to help mitigate problems like erosion, compaction, acidification, salinization, loss of organic matter, and heavy metal contamination by promoting processes like mineralization, nutrient cycling, and microbial diversity. This is because of their small size, large surface area, slow-release behavior, and capacity to address soil health issues (Bandala and Berli 2018; Suppan 2013). According to recent studies, adding ENMs to soil can improve its physicochemical qualities and increase environmental resilience (Rafique et al. 2017). According to recent studies, adding ENMs to soil can improve its physicochemical qualities and increase environmental resilience (Rafiq et al. 2017). For example, it has been demonstrated that the physical and chemical characteristics of alluvial soils can be markedly enhanced by oxidized orthorhombic iron oxalate-capped Fe_3O_4 NPs (OCIO; Das et al. 2016). In one study, adding 10 mg/kg of OCIO reduced the bulk density of the soil by 7.8%, increased the amount of total organic carbon by 64.4%, increased the water-holding capacity by 11.5%, and increased the pH of the soil by almost 1.04 times. On the other hand, treatments employing Fe-EDTA or $FeSO_4$ resulted in a decrease in pH levels rather than an increase.

5.5 Ecotoxicology of NPs on Soil Microbial Community

5.5.1 Effects of Zinc NPs

Inhibition of critical soil functions, like respiration, ammonification, and dehydrogenase activity, has been reported for ZnO-NPs. Chai et al. reported that the growth of major fertility-enhancing soil bacteria, such as *Azotobacter*, phosphorus-solubilizing bacteria, and potassium-solubilizing microorganisms, was inhibited by the ZnO-NPs. The effect depends on the concentration of these NPs on nano and bulk forms of metal oxides as well as specific microorganisms in the soil bacterial community and is thus dose-dependent. The relationship does not follow a simple linear or sigmoidal pattern, but rather a characteristic hump-shaped curve (Kasemets et al. 2009; Simonin et al. 2017). Some of the most essential microbial phyla, namely *Proteobacteria, Bacteroidetes, Acidobacteria, Actinobacteria*, and *Firmicutes*, regulate the carbon, nitrogen, sulfur, and phosphorus biogeochemical cycles. They assist plants in nutrient uptake through providing nutritional facilities that enhance plant growth and inhibit pathogens of the plant (Mubeen et al. 2023; Giovannetti et al. 2023; Wu et al. 2021; Lugtenberg and Kamilova 2009; Yadav et al. 2017). For terrestrial plants, it is the rhizosphere and root system microbiota that fuel growth and health. Rhizosphere and root system microbiota provide available nutrients and synthesize phytohormones to induce plant growth, nutrition, and resilience to environmental stress. In addition, root exudates interact with rhizosphere microbiota, and it has been proven that beneficial microorganisms enhance plant growth and development (Liu et al. 2021; Qu et al. 2020).

In recent years, interest has increased in the manipulation of soil microbiota to reduce environmental stressors that impede plant growth and soil health. In this regard, a critical role is played by soil microorganisms in encouraging the growth of plants, including those grown under stressful conditions. However, a significant number of studies have provided evidence of adverse impacts on these communities when exposed to ZnO-NPs. A pH of 8.58 in treated soil samples, wherein ZnO-NPs was suspended at concentrations of 0.5, 1.25, and 2.5 mg/kg of soil with sizes at 65.8 nm, induced significant reductions in soil bacterial, actinomycete, and fungal populations, relative to the control. The higher application rate of 2.5 mg/kg resulted in more significant inhibition of enzyme activity as compared to the lower rate of 0.5 mg/kg, mainly because of increased specific surface area of NPs that enhanced adsorption. It showed considerable reductions in soil dehydrogenase activity and microbial biomass carbon and therefore exerted an adverse impact upon overall microbial activity (Verma et al. 2022). Dehydrogenases are considered key soil enzymes as their activities indicate general microbial activity, as they exist in all living microbial cells. They transfer hydrogen from organic substrates to inorganic acceptors, therefore allowing the biological oxidation of organic matter. The rate will indicate redox processes in microbiology and the oxidative capacity of microorganisms in the soil (Wolińska and Stępniewska 2012). The ratio of microbial biomass carbon to soil organic carbon is considered an indicator of standard quality for organic matter, status, diversity, and maturity of the microbial community (Zavyalova, Vasbieva, and Fomin 2020). On the other hand, another study reported that ZnO-NPs (hemispherical, 50 ± 10 nm) increased the richness of the soil microbial community at a pH of 7.24. The prevailing bacterial and fungal classes in those samples were *Proteobacteria* (38.7–44.6%) and *Ascomycota* (59.9–80.9%), respectively. Elevated ZnO-NPs concentrations drastically increased the enrichment of the order *Sphingomonadales* linked with the degradation of persistent organic pollutants, although relative abundance of *Solirubrobacterales, Catenulisporales*, and the phylum *Armatimonadetes* was diminished (Liu et al. 2021).

5.5.2 Effect of Titanium NPs

Among the five major types of NPs, which are applied extensively due to their highest consumption and stability in applications, for example, antibacterial treatments, water purification, and cosmetics with anticorrosive and photocatalytic properties, titanium dioxide NPs are the most common type. As indicated by Keller et al. (2013), about 88 000 tons of TiO_2 NPs are produced per year globally. A large scale of production and usage with inefficient disposal methods has resulted in the unwanted release of many of them into the environment, creating a risk for both specific organisms and entire ecosystems (Minetto, Ghirardini, and Libralato 2016; Khosravi-Katuli et al. 2017). The input of TiO_2 NPs into agricultural soils may be due to irrigation, the application of sewage sludge, or the direct use of nanofertilizers and nanopesticides, thus significantly raising their concentrations in the soil (Gogos, Knauer, and Bucheli 2012). While natural soils contain typically around 0.5% TiO_2, the soil microorganisms are well adjusted to this natural form. However, due to increased surface area and reactivity, as well as high solubility, TiO_2 NPs may alter ecosystem functions than their natural counterparts differently (Scheffer et al. 2002; Nowack and Bucheli 2007; Gardea-Torresdey, Rico, and White 2014). Because these NPs work at the nanoscale, they may penetrate the microbial cells and cause damage at the cellular and DNA levels as well (Auffan et al. 2012). Besides, NPs can also alter the microbial community in the soil rhizosphere, particularly when the NPs are used as antimicrobial agents, such as TiO_2 and AgO_2, which are used in antimicrobial soaps.

Studies have indicated that the toxicity of the NPs toward microbes is selective and differs among NPs as well as among one type of bacterial communities or functional groups (Emami-Karvani and Chehrazi 2011; Priester et al. 2012). Microbial populations in the rhizosphere play a significant role in stimulating plant growth, protecting plants, and helping them take up nutrients (Hodge, Campbell, and Fitter 2001; Smith and Read 2008). The selective inhibition of some of the microbial populations could adversely affect the plant's uptake of nutrients. However, concerning the integrated effect of NPs on the growth of plants and the populations in rhizosphere microbiology, there have been only a few studies done in this regard (Feng et al. 2013).

5.5.3 Effect of Ag-NPs

Ag-NPs have gained a lot of interest these days in applications as diverse as clothing, food package surfaces, cleaning agents, and also some form of medical uses, thanks to their antimicrobial properties (Haider and Kang 2015). The overall use of these applications is projected to increase its environmental concentration. One pathway is through the application of sewage sludge as a fertilizer, and its use is supposed to alter their form to silver sulfides (Kampe et al. 2018). Doolette et al. (2016) investigated the impact of Ag-NPs (44 nm), sulfide-coated silver nanoparticles (Ag_2S-NPs; 152 nm), and silver ions (Ag^+ from $AgNO_3$) on the composition of soil microbial communities by metagenomic sequencing. The scientists concluded that the toxic concentrations of both types of NPs for OTUs (Operational Taxonomic Units) were comparable to, or lower than, those of Ag^+. Surprisingly, when an 80% protection level for OTUs was considered, the Ag_2S-NPs exhibited far less toxicity.

Other studies have reported an inhibition of the activities of dehydrogenase and urease by soils contaminated with 50 mg/kg of Ag-NPs (20 nm). In addition, the bacterial community composition was drastically affected by reductions in *Acidobacteria* and *Verrucomicrobiota* and an increase of *Proteobacteria*. The structure and abundance of both bacterial and archaeal *amoA* genes were also affected at a high level (McGee et al. 2017). Once again, identical results have been found in soils from regions of the low Arctic, whereby exposure to 20 nm Ag-NPs at a concentration of 660 mg/kg lowered characteristic bacterial fatty acids and reduced richness and evenness of bacterial and fungal DNA sequences alongside 50% drop in the microbial respiration. The nitrogen-fixing *Rhizobiales* bacteria were most sensitive to the Ag-NPs (Kumar et al. 2014). Ag-NPs, on the other hand, showed greater significant impacts than the silver microparticles of 3 μm on microbial communities.

In forest soil, at concentrations of 10 and 100 mg/kg, a reduction in the biomass carbon content of microbes by 33–45% was registered with the presence of Ag-NPs sized between 1 and 10 nm. Moreover, a higher metabolic quotient revealed that there was stress caused by environmental alterations due to the NPs (Carbone et al. 2014). Biological activities in calcareous soils of various textures and salinity were further investigated at Ag-NP concentrations of 0–50 mg/kg (40 nm), and findings indicated that the impacts of the NPs were dependent upon the soil type and concentration. The greatest sensitivity was found in soils with low clay content, and no observable effects were detected at concentrations below 20 ppm (Rahmatpour et al. 2017).

5.5.4 Effect of Iron NPs

Iron is one of the most abundant elements in the Earth's crust and can be found in a wide variety of biogeochemical environments (Braunschweig, Bosch, and Meckenstock 2013). The most widely used ZVFe-NPs, FeO-NPs, and $Fe_3O_4^-$ NPs are effective in the cleanup process of soil pollution through reductive transformations and the detoxification of various contaminants (Tiwari, Hasan, and Pandey 2017). Despite this, the detailed impacts of NP on soil microbial communities are poorly understood. Recent research efforts have begun to explore these potential impacts. Research has reported that the reactivity and toxicity of NP depend on factors such as iron's crystal structure, NP size, charge, solubility, and organic matter (Braunschweig et al. 2013). Furthermore, the cytotoxicity of the bacterial iron-based NPs is associated with oxidative stress related to reactive oxygen species. This may be due to the reaction between the reduced forms of iron with oxygen or ionic transport disruption resultant cell membrane affinity of the NPs (Braunschweig, Bosch, and Meckenstock 2013). For instance, Shah et al. (2014) showed that ZVFe-NPs in soil at concentrations of 550 mg/kg did not affect the bacterial communities, and the deviations were rather due to environmental conditions than to the NPs themselves. Conversely, quasi-spherical FeO-NPs (10.0 ± 2.5 nm) at concentrations of 1 and 10 mg/kg positively affected nitrogen and carbon cycles due to the enhancement of microbial metabolic activity, with the potential of nitrification, unlike that of silver NPs (He et al. 2016). Iron has been reported to be pivotal in the metabolic and biochemical activities of microbes due to functional processes, such as synthesis of nitrate, dinitrogen fixation, and DNA synthesis.

5.5.5 Effect of Copper NPs

Lately, copper and copper oxide NPs (Cu-NPs and CuO-NPs) have attracted much attention for their beneficial properties, such as being excellent conductors of thermal and electric energy, capable of lubricating, catalyzing, and being relatively inexpensive. Due to the above-mentioned properties, they are applied in numerous consumer commodities. However, there is also research that reported their impacts on soil microorganisms. Frenk et al. (2013) investigated the effect of different concentrations of CuO-NPs on bacterial activity in two different soils: SL (sandy loam) and SCL (sandy clay loam) types. The authors established that the influence of CuO-NPs was stronger for SL soil than for the SCL soil type, where the most significant impact following the amendment with CuO-NPs was found in the reduction of bacterial hydrolytic activity, community structure, and oxidative potential. Although the total impact was less serious for SCL soil, some considerable decreases in its oxidative potential and changes in community composition were detected as well. By contrast, the density of *Bacilli* declined after exposure to 0.1 and 1% CuO-NPs in SL and SCL soils, respectively, as well as other groups, such as *Rhizobiales* and *Sphingobacteriaceae*, were negatively affected. The interaction between clay and organic matter could neutralize the toxicity of CuO-NPs. A related study examined the effects of CuO-NPs on the microbial communities that associate with the salt marsh plants *Phragmites australis* and *Halimione portulacoides* compared to ionic copper (Fernandes et al. 2017). CuO-NPs and ionic copper significantly altered the microbial community, suggesting that CuO-NPs may have had a disruptive effect on the ecosystem functions. Xu et al. (2015) studied the impact of commercial CuO-NPs at different concentrations (0, 100, 500, and 1000 mg/kg soil) on microbes in flooded paddy soil, along with TiO_2-NPs (20 nm). The outcomes showed that the alterations caused by CuO-NPs were more negative for soil microbes than those induced by TiO_2-NPs, based on a decrease in microbial biomass and significant reduction in enzymatic activities, such as phosphatase, urease, and dehydrogenase. An increased dissolution of CuO-NPs resulted in lower nutrient availability and increased oxidative stress among the microbes in the soil. More importantly, Zhai et al. (2017) studied the influence of various shapes and dimensions of metal nanoparticles (MNPs/MONPs) on their dissolution capacity towards the metabolic functions of soil bacteria. They showed that although size is an important factor for antimicrobial activity, shape also factors in. For example, spherical Cu-NPs at 500 nm are more cytotoxic to soil microbial communities compared with rod-like Cu-NPs, at 78 nm, which may be related to differences in shape and dissolution potential.

5.6 Assessment and Monitoring of NM Impacts

5.6.1 Long-term Effects on Soil Ecosystem Services

In the last few years, nanotechnology was applied to agricultural and food chain products with much interest. Uncontrolled application of NMs may be harmful to the soil ecosystem. NPs can change the physicochemical characteristics of the soil; thus, their effect may be seen directly and indirectly on the different organisms dwelling in the soil. This has created increased interest in investigating the ways in which NPs may be favorable to or deleterious to soil biota, including bacteria and fungi, respectively (Watson et al. 2015; Rabbani, Ahmed, and Park 2016). For instance, Ge, Schimel, and Holden (2011) reported that nano-ZnO negatively impacts the bacterial communities in the soil by altering enzyme activities as well as decreasing microbial biomass and diversity.

In fact, intensely conducted studies into NP effects on soil ecology have been seen over the last 20 years, especially the last 10 years. NPs are in constant interaction with biological processes in the soil, where they can penetrate plants and animals, affect populations of microbes, influence biogeochemical cycles, and at the same time provide potential toxicological risks. A measure of this balance within the soil-plant-microbe system is the level of organic carbon. From the research done, it has proven that TiO_2 NPs work to increase the carbon mineralization by a high extent in soils containing high organic matter soils, thus increasing the organic matter content in the soils (Simonin and Richaume 2015). Once more, high numbers of C_{60} NPs have been proved to cause ecological imbalances due to the increased content of organic matter in the soil (Chen et al. 2019).

Further research shows that another way through which the NPs affect the soil is by altering its nitrogen cycle. In this regard, it has been shown that metal-based NPs influence the nitrogen cycle in the soil. For example, Ag-NPs can inhibit key nitrogen-fixing, nitrifying, and denitrifying microorganisms in controlled settings, thus disrupting these important functions in soils. The population of denitrifying bacteria in the soil would probably be reduced by Ag_2S NPs, thereby changing the cycle of the nitrogen transformation, which can lead to the limitation of conversion of nitrates and nitrites to nitrogen gas (N_2; Wu et al. 2020a). Furthermore, carbon NPs have been shown to adversely impact nitrogen fixation in legumes, likely through an over 30% reduction in the infusion of nitrogen into the soil as a consequence of this mutualism between the legume and the rhizobium (Wang et al. 2017).

Another area of interest of MNPs is related to alterations in other biogeochemical cycles, such as the sulfur and micronutrient cycles. For example, the application of ZnO-NPs in soils could lead to decreases in *Sulfitobacter* population, thus impairing the sulfur cycle (Wu et al. 2020b). It may be that engineered nanoparticles enhance plant and animal nutrient uptake, indicative of loss of soil fertility and more significant nutrient loss (Conway and Keller 2016). The presence of titanium dioxide NPs introduces the possibility of depletion in essential metal and nonmetal elements, such as manganese, iron, and phosphorus, in soil. Changes in soil properties depend on characteristics adopted by such NPs, which include structure, molecular size, and hydrophobicity. Solid NPs show better protection against ion decomposition as compared to dissolved ones that can easily change the ionic balance of the soil. All these influences of NPs interfere with the nutrient absorption and utilization by plant and animals as well as affect the balanced nutrient cycling in the soil ecosystem.

5.6.2 Potential Risks and Benefits

So, without soil, human existence cannot sustain or be productive for agriculture; therefore, this has appeared as a significant threat to soil microorganisms, annelids, and arthropods due to soil contamination with the numerous NPs. So, soil microorganisms, annelids, and arthropods, are significantly threatened with the increasing presence of these contaminants. Although various studies have demonstrated the positive impacts of NPs (Khan et al. 2020; Mohammadi et al. 2020), other investigations have reported their harmful effects on soil microorganisms, plants, and the structure and composition of soil communities as a whole (Rajput et al. 2019; Meier et al. 2020). Released metal-based NPs in agricultural fields are important sink sites, and thus, the subsequent release and uptake of such metal-based NPs from agricultural environments in relation to plant growth in agricultural fields are required to assess the toxicity impact of NPs (Chen 2018).

Agricultural soils usually have high dissolved organic matter contents, which is highly significant for trace metals' biogeochemical cycling and colloid mobility in aquatic systems (Aiken, Hsu-Kim, and Ryan 2011). The study proposes that natural organic matter adsorption on the surfaces of NPs minimizes aggregation and stabilizes them under various environmental conditions (Keller et al. 2010). Up to recent times, a lot of research works have been carried out on the transport behavior of NPs in environmental settings and their interaction with vegetation, bioavailability, toxicity, and their fate in ecosystems (Bradford et al. 2002; Judy, Bertsch, and One 2014; Yadav et al. 2018). Crucially, the critical component in the analysis of agricultural ecosystems is soil and plant systems. However, more studies treat metal-based NPs as separate entities in the system rather than an integrated system. Alarmingly, no regulations exist for the use of NPs in the ecosystems of the agricultural sector, which may be a future threat to environmental stability and health. Therefore, extensive in vitro and field-based studies are warranted to mitigate the harm to the non-targeted soil organisms and key agricultural microbes. Research regarding the changes affected by NPs to the physicochemical properties of the soil is also warranted since such properties are of significant influence on soil biology and fertility. Another promising area for research can be undertaken to investigate the changes at the transcriptional level brought about in plants and microbes with respect to altering their very vital biological processes triggered by the exposure to NPs.

5.7 Mitigation Strategies and Future Directions

5.7.1 Approaches to Minimize Negative Impacts

The development and application of wide range of metallic NPs in the natural environment may pose threat to the beneficial microbes of soils, such as beneficial bacteria and fungi (Rajput et al. 2018). A scientific report indicated that application of titanium oxide (TiO_2) and copper oxide (CuO) containing NPs severely restricted soil microbial biomass and enzymatic activities in the flooded soils of rice rhizosphere regions (Xu et al. 2015). Several studies have highlighted the negative impacts of NMs on crop growth, including reduced root and shoot development, lower rate of photosynthesis, slower rate of germination, diminished chlorophyll pigmentation, and increased toxicity level (Moon et al. 2014; Hong et al. 2015; Tripathi et al. 2017). Since the excessive use or high concentration of NPs poses significant threat to soil, plant, and environment, it is essential to prevent their negative impacts through appropriate approaches.

Greener nanoscience is an innovative approach that utilizes green chemistry principles to synthesize safer eco-friendly NMs (Dahl, Maddux, and Hutchison 2007). This approach helps to develop synthetic methods for producing greener NMs consistently, with well-defined composition, structure, and purity (Hutchison 2008). Green nanoscience is based on 12 well-known principles of green chemistry that help in designing and production of safer NMs and thus eliminating hazards to human health and environment (McKenzie and Hutchison 2004). Table 5.1 outlines how these 12 principles of green chemistry can be applied to develop environmentally friendly NMs (Table 5.1). Green nanotechnology offers three key benefits: it facilitates the synthesis of new innovative NMs, enhances commercialization by reducing safety uncertainties, and provides advanced manufacturing process. Additionally, it helps protect financial investments in this technology by addressing consumers' concerns about potential risks (Hutchison 2008).

Table 5.1 Production and application of greener nanomaterials (NMs) based on 12 principles of green chemistry (Hutchison, 2008) / American Chemical Society.

Principle no.	Principles of green chemistry	Production of green NMs	Application in nanoscience
Principle 1 Principle 2	Prevent waste Atom economy	Design of safer NMs	Primarily focused on producing safer NMs with desired physical properties, avoiding the use of toxic substances in nanocomposites and evaluating the biological impacts of various nanoparticles with varying sizes and surface areas.
Principle 3 Principle 4	Minimally perilous chemical synthesis Formulating safer chemicals	Design for minimal environmental effects	Primarily utilizes nontoxic bio-NMs to eliminate hazardous substances in NM formulations. It also evaluates the degradation procedure and environmental fate of these NMs.
Principle 5 Principle 6	Green reaction media Design for energy optimization	Planning for minimizing waste	Utilize selective nano synthesis procedure to improve the purity of nanoformulations and thus eliminating solvent-intensive purification process. It also utilizes bottom-up approaches to increase the efficiency of NMs.
Principle 7 Principle 8	Renewable feedstock Reduce derivatives	Plan for risk reduction in process	Involves in development of advanced NM synthesis that utilize more natural and renewable benign feedstock and thereby eliminating the use of highly toxic reagents in the preparation phase.
Principle 9 Principle 10	Catalysis Design for decomposition	Plan for increasing NM efficacy	Develop new synthetic methods and improve the integration of raw materials, usage of alternative catalyst, and reaction medium to enhance reaction selectivity.
Principle 11 Principle 12	Real-time monitoring Built-in safety chemistry	Plan for improving energy use efficacy	Real-time monitoring system to control the process of NM synthesis. Synthetic pathway should run at an ambient temperature rather than at elevated temperature to optimize energy use efficacy as well as safety.

Hence, to minimize the negative impact of NMs on natural environment, it is crucial to formulate green NMs following the seven well distinctive principles of green chemistry. Furthermore, it is also essential to conduct pre-application testing, use controlled application procedures, and regularly monitor soil health to combat the negative impact of NMs on soil microbial communities. Additionally, developing less toxic formulations through green chemistry, integrating NMs in ways that reduce direct contact with microbes, employing remediation techniques, adhering to regulatory frameworks, and supporting ongoing research are essential strategies or approaches for safe and effective NM use in agriculture and environment.

5.8 Regulatory and Policy Considerations

Government initiatives are vigorously encouraging the improvement and approbation of smart-release nutrient NMs for sustainable industrialization. The objective of scheme is the development and adoption of innovative technologies. This approach was demonstrated by the National Nanotechnology Initiative (NNI), a combined effort involving more than 25 government body in the United States. The NNI's main goal is to promote the progression and employment of nanotechnology, with a special set on nanotechnology-based formulations that's smart-release fertilizers. Advancement of environmentally friendly technologies is conjointly explored by these initiatives and lead sustainable evolvement in the agricultural area. In addition, the United States launched the Small Business Innovation Research (SBIR) plan to boost technical innovation and expansion of pocket-size businesses (Roco 2005).

The SBIR plan boosts pocket-size businesses in their research and the attempts of product development. The SBIR initiative has been engaged by the USDA to reinforce the improvement of smart-release fertilizers based on NMs (Fuglie and Toole 2014). First, nanotechnology incorporating plan into agriculture was delineated by the USDA in 2002 (Manjunatha, Biradar, and Aladakatti 2016). Thereafter, academics and agriculture both have led the way in intermingling the techniques of nanotechnology into agricultural sector. The control in utilization of NPs as fertilizers is subject to sophisticated federal, state, and international laws in the United States. The US government executed many tax prompts, to provoke the capitalization of nanotechnology-related goods and operations. Advanced Energy Project Credit was initiated by the Congress, which supply tax credits for controlled-release fertilizers, including NMs (Berk 2010).

The IRS (Internal Revenue Service) launched the Orphan Drug Tax credit, which allows tax credits for the enhancement of medicine for uncommon diseases. Many scientists reported that these benefits could be applied to smart-release fertilizers based on NMs (Peabody, Ruby, and Cannon 1995). Numerous initiatives have been enforced by the UK government to facilitate the development and dissemination of nanotechnology-based applications, involving controlled-release formulation fertilizers using nanotechnology. In the United Kingdom, the Technology Strategy Board (TSB) was established to provide funding for research and development initiatives in many fields, along with nanotechnology (Gupta and Dantsiou 2013). The TSB is a nondepartmental public organization funded by the UK government. Besides, government of United Kingdom launched the SBRI to support small-type businesses in their technological advancement, and TSB oversees the SBRI.

United Kingdom implemented various tax plans, to aid with the capitalization of nanotechnology-based produce. According to Griffith, Miller, and O'Connell (2010), Patent Box plan acts as providers to minimum corporate taxes for companies involved in nanotechnology-based innovative research and development. The Patent Box could be accessed by the UK patent owners or exclusive licensees. The UK government established the Research and Development Tax Credit to encourage R&D spending (Fowkes, Sousa, and Duncan 2015). All enterprises are eligible for this tax benefit. Australia and Canada have also sanctioned laws to assist the commercialization of nanotechnology-based products, involving controlled-release fertilizers. Canada launched the Industrial Research Assistance Program (IRAP) based on nanotechnology, to encourage R&D initiatives (Niosi 2006).

IRAP is commanded by the National Research Council of Canada (NRC; Doern and Levesque 2002). Additionally, in Canada, for R&D investments, the Scientific Research and Experimental Development (SR&ED) plan allows a tax credit (Parsons and Phillips 2007). The organizations and governments enact laws that are essential and promote the prudent use of nanofertilizers. Minimum use of NFs in sensitive locations offers financial incentives as per imposed guidelines by the Government. Although the advantages of nanofertilizers could not be denying, their use might be restricted by a number of international restrictions. China, the United States, and the European Union could tighten their regulations due to concerns about potential hazards to the environment and public health related to the use of nanofertilizers (López-Moreno, Cassé, and Correa-Torres 2018).

As a result, it is important for both manufacturers and users of nanofertilizers to be aware of and abide by local laws. Research on nanotechnology has advanced; however, in many nations, there are no established laws for items that use nanotechnology, which has slowed down progress (Adholeya et al. 2017). To determine what stands in the way of effective legislative enactment, a thorough analysis of relevant topics must be conducted. Businesses that produce nanofertilizers must comply with complicated laws, which take time and money. Governments should also find out how the general population feels about nanotechnology in agriculture. Progress requires the establishment of standards, testing facilities, and certification of NMs in addition to legal frameworks.

5.9 Future Research Prospects and Knowledge Gaps

Nanotechnology offers a transformative potential for agriculture by mitigating farming hazards. Its application remains underexplored in agricultural sectors. Utilizing NMs could enhance understanding of plant–microorganism interactions and alleviate soil stressors.

Limited research exists on the impact of nanotechnology on the rhizospheric microbiome due to experimental constraints (Rajput et al. 2023). Clarifying the role of NMs in improving soil properties and managing stresses is essential for sustainable crop production. A comprehensive understanding of NM interactions with plants and soil microbes could enhance crop yield and reduce reliance on conventional fertilizers. A thorough evaluation of nanotechnology's advantages and disadvantages is necessary prior to its application for soil fitness and crop improvement. In agriculture ecosystem, NMs can endure transformations that could generate new potentially toxic pollutants through interactions with biological macromolecules.

To optimize the utilization of NMs, existing knowledge gaps and challenges must be addressed. Knowledge deficits hinder the full realization of NMs' potential and their widespread adoption globally. Specific gaps across various domains are outlined below.

5.9.1 Method of Application of Nanofertilizers

A significant knowledge gap in NFs' application pertains to the interaction of NMs with soil, plants, and environmental factors. Understanding plant absorption of NMs and their interaction with soil chemistry is crucial. Further investigation is requisite to explicate the usefulness of nanofertilizers and their interactions with different conventional fertilizers, soil types, and crop varieties, and the most effective application methods for NPs.

5.9.2 Formation of Successful Execution Mechanisms

Capitalization of nanofertilizers stances encounters in forming effectual delivery systems. Optimum distribution methods are compulsory to boost plant contact due to the size restrictions of NPs. Cost-effective delivery mechanisms are intended for wide range of formulations and commercialization of nanofertilizers. Despite their potential to enhance crop productivity and decrease fertilizer usage, delivery costs hinder their widespread implementation (Bratovcic et al. 2021).

5.9.3 Assessing the Financial Possibility of Extensive Production

Evaluating the financial possibilities of wide-range NF formulation requires investigating production costs. Various factors, such as raw material, labor, and equipment expenses, are crucial in decisive overall formulation costs (Jagadeesh et al. 2022).

5.9.4 Significance of Nanofertilizers on Environment

The ecological suggestions of extensively used nanofertilizers formulation permit careful attention. Although introducing nanotechnology in the field of agriculture could significantly minimize the environmental burdens of traditional fertilizers, its yield may cause adverse effects. A comprehensive evaluation of the ecological effect of large-scale NF manufacturing is crucial, along with strategies to mitigate negative impacts. The environmental effects of NF-based formulation use remain unexplored and need more study (Bhardwaj et al. 2022). Particularly, application of NF may induce enhanced plant yield and curtail reliance on conventional fertilizers, consequently alleviating their detrimental environmental effects.

5.10 Conclusion

Nanofertilizers furnish a hopeful key to strengthen agricultural sustainability. By enhancing nutrient uptake mechanism, they boost crop productivity although reducing environmental pollution from fertilizer runoff and leaching. Nanofertilizers also improve soil physical parameter, such as water retention capacity and plant resilience, resulting in better quality agricultural products. Furthermore, the adoption of nanofertilizers is greatly obstructed by many challenges. Nanofertilizers are highly expensive compared to traditional fertilizers, which could be a financial bother for small and marginal farmers. Moreover, the lack of standardization makes it difficult to ensure persistent quality and safety, hampering regulatory management. To find out the possible benefits of nanofertilizers, it is crucial to notify these challenges through policy enterprises that help research, development, and the adoption of sustainable agricultural practices.

References

Adholeya, A., Das, R., Dubey, M. et al. (2017). *Zero Draft Policy on Regulation of Nanoproducts in Agriculture*. Energy and Resources Institute (TERI).

Ahmed, F. and Rodrigues, D.F. (2013). Investigation of acute effects of graphene oxide on wastewater microbial community: a case study. *Journal of Hazardous Materials* 256: 33–39.

Aiken, G.R., Hsu-Kim, H., and Ryan, J.N. (2011). Influence of dissolved organic matter on the environmental fate of metals, nanoparticles, and colloids. *Environmental Science & Technology* 45 (8): 3196–3201.

Alivisatos, A.P., Gu, W., and Larabell, C. (2005). Quantum dots as cellular probes. *Annual Review of Biomedical Engineering* 7 (1): 55–76.

Alvarez, L.H. and Cervantes, F.J. (2012). Assessing the impact of alumina nanoparticles in an anaerobic consortium: methanogenic and humus reducing activity. *Applied Microbiology and Biotechnology* 95: 1323–1331.

Auffan, D.M., Santaella, D.C., Thiéry, P.A. et al. (2012). Ecotoxicity of inorganic nanoparticles: from unicellular organisms to invertebrates. In: *Encyclopedia of Nanotechnology* (ed. P.B. Bhushan), 623. Springer.

Bandala, E.R. and Berli, M. (2018). Nanomaterials: new agrotechnology tools to improve soil quality?. *Agricultural Nanobiotechnology: Modern Agriculture for a Sustainable Future*, 127–140.

BCC Research (2024). BCC research: global markets reports and industry analysis. https://www.bccresearch.com/?srsltid=AfmBOop9LxGwpeDjM6Nz-gSCXQSVOrh4df3f3wa5N2lAIKe53qOD8_2K (accessed 10 October 2024).

Berk, H.R. (2010). Tax credit transactions corner. *Journal of Passthrough Entities* 13: 45.

Bhardwaj, A.K., Arya, G., Kumar, R. et al. (2022). Switching to nanonutrients for sustaining agroecosystems and environment: the challenges and benefits in moving up from ionic to particle feeding. *Journal of Nanobiotechnology* 20 (1): 19.

Bradford, A., Handy, R.D., Readman, J.W. et al. (2009). Impact of silver nanoparticle contamination on the genetic diversity of natural bacterial assemblages in estuarine sediments. *Environmental Science & Technology* 43 (12): 4530–4536.

Bradford, S.A., Yates, S.R., Bettahar, M. et al. (2002). Physical factors affecting the transport and fate of colloids in saturated porous media. *Water Resources Research* 38 (12): 6301–6312.

Bratovcic, A., Hikal, W.M., Said-Al Ahl, H.A. et al. (2021). Nanopesticides and nanofertilizers and agricultural development: scopes, advances and applications. *Open Journal of Ecology* 11 (4): 301–316.

Braunschweig, J., Bosch, J., and Meckenstock, R.U. (2013). Iron oxide nanoparticles in geomicrobiology: from biogeochemistry to bioremediation. *New Biotechnology* 30 (6): 793–802.

Buzea, C., Pacheco, I.I., and Robbie, K. (2007). Nanomaterials and nanoparticles: sources and toxicity. *Biointerphases* 2 (4): MR17–MR71.

Cao, H., Qiao, Y., Liu, X. et al. (2013). Electron storage mediated dark antibacterial action of bound silver nanoparticles: smaller is not always better. *Acta Biomaterialia* 9 (2): 5100–5110.

Carbone, S., Antisari, L.V., Gaggia, F. et al. (2014). Bioavailability and biological effect of engineered silver nanoparticles in a forest soil. *Journal of Hazardous Materials* 280: 89–96.

Chai, H., Yao, J., Sun, J. et al. (2015). The effect of metal oxide nanoparticles on functional bacteria and metabolic profiles in agricultural soil. *Bulletin of Environmental Contamination and Toxicology* 94 (4): 490–495.

Chen, H. (2018). Metal-based nanoparticles in agricultural systems: behavior, transport, and interaction with plants. *Chemical Speciation and Bioavailability* 30 (1): 123–134.

Chen, M., Sun, Y., Liang, J. et al. (2019). Understanding the influence of carbon nanomaterials on microbial communities. *Environmental International* 126: 690–698.

Chen, Y., Wang, D., Zhu, X. et al. (2012). Long-term effects of copper nanoparticles on wastewater biological nutrient removal and N_2O generation in the activated sludge process. *Environmental Science & Technology* 46 (22): 12452–12458.

Coll, C., Notter, D., Gottschalk, F. et al. (2016). Probabilistic environmental risk assessment of five nanomaterials (nano-TiO_2, nano-Ag, nano-ZnO, CNT, and fullerenes). *Nanotoxicology* 10 (1): 4.

Concha-Guerrero, S.I., Brito, E.M.S., and Piñón-Castillo, H.A. (2014). Effect of CuO nanoparticles on isolated bacterial strains from agricultural soil. *Journal of Nanomaterials* 2014: 148743.

Conway, J.R. and Keller, A.A. (2016). Gravity-driven transport of three engineered nanomaterials in unsaturated soils and their effects on soil pH and nutrient release. *Water Research* 98: 250–260.

Cvjetko, P., Milošić, A., Domijan, A.M. et al. (2017). Toxicity of silver ions and differently coated silver nanoparticles in *Allium cepa* roots. *Ecotoxicology and Environmental Safety* 137: 18–28.

Dahl, J.A., Maddux, B.L., and Hutchison, J.E. (2007). Toward greener nanosynthesis. *Chemical Reviews* 107 (6): 2228–2269.

Das, P., Sarmah, K., Hussain, N. et al. (2016). Novel synthesis of an iron oxalate capped iron oxide nanomaterial: a unique soil conditioner and slow release eco-friendly source of iron sustenance in plants. *RSC Advances* 6 (105): 103012–103025.

Das, P., Xenopoulos, M.A., Williams, C.J. et al. (2012). Effects of silver nanoparticles on bacterial activity in natural waters. *Environmental Toxicology and Chemistry* 31 (1): 122–130.

Doern, G.B. and Levesque, R. (2002). *The National Research Council in the Innovation Policy Era: Changing Hierarchies, Networks and Markets*. University of Toronto Press.

Doolette, C.L., Gupta, V.V., Lu, Y. et al. (2016). Quantifying the sensitivity of soil microbial communities to silver sulfide nanoparticles using metagenome sequencing. *PLoS One* 11 (8): e0161979.

Dresselhaus, M.S., Dresselhaus, G.A.P., and Avouris, P. (2004). Carbon nanotubes. *MRS Bulletin* 33 (7): 237.

Drexler, K.E. (1986). *Engines of Creation: The Coming Era of Nanotechnology*. Anchor Books.

Ebbs, S.D., Bradfield, S.J., Kumar, P. et al. (2016). Accumulation of zinc, copper, or cerium in carrot (*Daucus carota*) exposed to metal oxide nanoparticles and metal ions. *Environmental Science: Nano* 3: 114–126.

Ebrahimi, A., Galavi, M., Ramroudi, M. et al. (2016). Effect of TiO_2 nanoparticles on antioxidant enzymes activity and biochemical biomarkers in Pinto bean (*Phaseolus vulgaris* L.). *Journal of Molecular Biology and Molecular Biology* 6 (1): 58–66.

Emami-Karvani, Z. and Chehrazi, P. (2011). Antibacterial activity of ZnO nanoparticle on gram-positive and gram-negative bacteria. *African Journal of Microbiology Research* 5: 1368–1373.

Feng, Y., Cui, X., He, S. et al. (2013). The role of metal nanoparticles in influencing arbuscular mycorrhizal fungi effects on plant growth. *Environmental Science & Technology* 47: 9496–9504.

Fernandes, J.P., Almeida, C.M.R., Andreotti, F. et al. (2017). Response of microbial communities colonizing salt marsh plants rhizosphere to copper oxide nanoparticles contamination and its implications for phytoremediation processes. *Science of the Total Environment* 581: 801–810.

Fowkes, R.K., Sousa, J., and Duncan, N. (2015). Evaluation of research and development tax credit. *HM Revenue and Customs*. HMRC Working Paper, 17.

Frenk, S., Ben-Moshe, T., Dror, I. et al. (2013). Effect of metal oxide nanoparticles on microbial community structure and function in two different soil types. *PLoS One* 8 (11): e84441.

Fuglie, K.O. and Toole, A.A. (2014). The evolving institutional structure of public and private agricultural research. *American Journal of Agricultural Economics* 96 (3): 862–883.

Gardea-Torresdey, J.L., Rico, C.M., and White, J.C. (2014). Trophic transfer, transformation, and impact of engineered nanomaterials in terrestrial environments. *Environmental Science & Technology* 48: 2526–2540.

Ge, Y.G., Schimel, J.P., and Holden, P.A. (2011). Evidence for negative effects of TiO_2 and ZnO nanoparticles on soil bacterial communities. *Environmental Science & Technology* 45 (4): 1659–1664.

Giovannetti, M., Salvioli di Fossalunga, A., Stringlis, I.A. et al. (2023). Unearthing soil-plant-microbiota crosstalk: looking back to move forward. *Frontiers in Plant Science* 13: 1082752.

Glick, B.R. (2012). Plant growth-promoting bacteria: mechanisms and applications. *Scientifica* 2012: 963401.

Gogos, A., Knauer, K., and Bucheli, T.D. (2012). Nanomaterials in plant protection and fertilization: current state, foreseen applications, and research priorities. *Journal of Agricultural and Food Chemistry* 60: 9781–9792.

Griffith, R., Miller, H., and O'Connell, M. (2010). *Corporate Taxes and Intellectual Property: Simulating the Effect of Patent Boxes*. Institute for Fiscal Studies.

Gupta, R. and Dantsiou, D. (2013). Understanding the gap between 'as designed' and 'as built' performance of a new low carbon housing development in the UK. In: *Sustainability in Energy and Buildings*, 567–580. Springer.

Haider, A. and Kang, I.K. (2015). Preparation of silver nanoparticles and their industrial and biomedical applications: a comprehensive review. *Advanced Materials Science and Engineering* 2015: 1–16.

Hansen, S.F., Heggelund, R.L., Besora, P.R. et al. (2016). Nanoproducts: what is actually available to European consumers? *Environmental Science: Nano* 3: 169–180.

He, S., Feng, Y., Ni, J. et al. (2016). Different responses of soil microbial metabolic activity to silver and iron oxide nanoparticles. *Chemosphere* 147: 195–202.

Hodge, A., Campbell, C.D., and Fitter, A.H. (2001). An arbuscular mycorrhizal fungus accelerates decomposition and acquires nitrogen directly from organic material. *Nature* 413: 297–299.

Hong, J., Rico, C.M., Zhao, L. et al. (2015). Toxic effects of copper-based nanoparticles or compounds to lettuce (*Lactuca sativa*) and alfalfa (*Medicago sativa*). *Environmental Science: Processes & Impacts* 17 (1): 177–185.

Hutchison, J.E. (2008). Greener nanoscience: a proactive approach to advancing applications and reducing implications of nanotechnology. *ACS Nano* 2 (3): 395–402.

Jagadeesh, P., Puttegowda, M., Boonyasopon, P. et al. (2022). Recent developments and challenges in natural fiber composites: a review. *Polymers and Composites* 43: 2545–2561.

Jiling, C., Youzhi, F., Xianggui, L. et al. (2016). Arbuscular mycorrhizal fungi alleviate the negative effects of iron oxide nanoparticles on bacterial community in rhizospheric soils. *Frontiers in Environmental Science* 4: 10.

Joudeh, N. and Linke, D. (2022). Nanoparticle classification, physicochemical properties, characterization, and applications: a comprehensive review for biologists. *Journal of Nanobiotechnology* 20 (1): 262.

Judy, J.D., Bertsch, P.M., and One, C. (2014). Bioavailability, toxicity, and fate of manufactured nanomaterials in terrestrial ecosystems. In: *Advances in Agronomy* (ed. D.L. Sparks), Volume 123: 1–64. Academic Press.

Kampe, S., Kaegi, R., Schlich, K. et al. (2018). Silver nanoparticles in sewage sludge: bioavailability of sulfidized silver to the terrestrial isopod *Porcellio scaber*. *Environmental Toxicology and Chemistry* 37 (6): 1606–1613.

Kang, S., Mauter, M.S., and Elimelech, M. (2009). Microbial cytotoxicity of carbon-based nanomaterials: implications for river water and wastewater effluent. *Environmental Science & Technology* 43 (7): 2648–2653.

Kasemets, K., Ivask, A., Dubourguier, H.C. et al. (2009). Toxicity of nanoparticles of ZnO, CuO, and TiO_2 to yeast *Saccharomyces cerevisiae*. *Toxicology in Vitro* 23 (6): 1116–1122.

Keller, A.A. and Lazareva, A. (2014). Predicted releases of engineered nanomaterials: from global to regional to local. *Environmental Science & Technology Letters* 1: 65–70.

Keller, A.A., McFerran, S., Lazareva, A. et al. (2013). Global life cycle releases of engineered nanomaterials. *Journal of Nanoparticle Research* 15: 1–17.

Keller, A.A., Wang, H., Zhou, D. et al. (2010). Stability and aggregation of metal oxide nanoparticles in natural aqueous matrices. *Environmental Science & Technology* 44 (6): 1962–1967.

Khan, Z.S., Rizwan, M., Hafeez, M. et al. (2020). Effects of silicon nanoparticles on growth and physiology of wheat in cadmium contaminated soil under different soil moisture levels. *Environmental Science and Pollution Research* 27 (5): 4958–4968.

Khosravi-Katuli, K., Prato, E., Lofrano, G. et al. (2017). Effects of nanoparticles in species of aquaculture interest. *Environmental Science and Pollution Research* 24: 17326–17346.

Klaine, S.J., Alvarez, P.J., Batley, G.E. et al. (2008). Nanomaterials in the environment: behavior, fate, bioavailability, and effects. *Environmental Toxicology and Chemistry* 27: 1825–1851.

Kumar, N., Palmer, G.R., Shah, V. et al. (2014). The effect of silver nanoparticles on seasonal change in Arctic tundra bacterial and fungal assemblages. *PLoS One* 9 (6): e99953.

Lakshmi, J.V., Sharath, R., Chandraprabha, M.N. et al. (2012). Synthesis, characterization and evaluation of antimicrobial activity of zinc oxide nanoparticles. *Journal of Biochemical Technology* 3: S151–S154.

Li, N., Ma, L., Wang, J. et al. (2010). Interaction between nano-anatase TiO_2 and liver DNA from mice in vivo. *Nanoscale Research Letters* 5: 108–115.

Liu, Y., Li, Y., Pan, B. et al. (2021). Application of low dosage of copper oxide and zinc oxide nanoparticles boosts bacterial and fungal communities in soil. *Science of the Total Environment* 757: 143807.

Lofts, S., Criel, P., Janssen, C.R. et al. (2013).Modelling the effects of copper on soil organisms and processes using the free ion approach: towards a multi-species toxicity model. *Environmental Pollution* 178: 244–253.

López-Moreno, M.L., Cassé, C., and Correa-Torres, S.N. (2018). Engineered nanomaterials interactions with living plants: benefits, hazards and regulatory policies. *Current Opinion in Environmental Science & Health* 6: 36–41.

Lugtenberg, B. and Kamilova, F. (2009). Plant-growth-promoting rhizobacteria. *Annual Review of Microbiology* 63: 541–556.

Maliszewska, I. (2016). Effects of biogenic gold nanoparticles on microbial community structure and activities. *Annals of Microbiology* 66: 785–794.

Manjunatha, S., Biradar, D., and Aladakatti, Y.R. (2016). Nanotechnology and its applications in agriculture: a review. *Journal of Farm Science* 29 (1): 1–13.

Masrahi, A., VandeVoort, A.R., and Arai, Y. (2014). Effects of silver nanoparticle on soil-nitrification processes. *Archives of Environmental Contamination and Toxicology* 66: 504–513.

Maurer-Jones, M.A., Ian, L., Gunsolus, I.L. et al. (2013). Toxicity of engineered nanoparticles in the environment. *Analytical Chemistry* 85 (6): 3036–3049.

McGee, C.F., Storey, S., Clipson, N. et al. (2017). Soil microbial community responses to contamination with silver, aluminum oxide, and silicon dioxide nanoparticles. *Ecotoxicology* 26: 449–458.

McKenzie, L.C. and Hutchison, J.E. (2004). Green nanoscience. *Chimica Oggi* 22 (9): 30–33.

Meier, M.J., Dodge, A.E., Samarajeewa, A.D. et al. (2020). Soil exposed to silver nanoparticles reveals significant changes in community structure and altered microbial transcriptional profiles. *Environmental Pollution* 258: 113816.

Minetto, D., Ghirardini, M.A.V., and Libralato, G. (2016). Saltwater ecotoxicology of Ag, Au, CuO, TiO_2, ZnO and C60 engineered nanoparticles: an overview. *Environmental International* 92: 189–201.

Mohammadi, H., Amani-Ghadim, A.R., Matin, A.A. et al. (2020). Fe^0 nanoparticles improve physiological and antioxidative attributes of sunflower (*Helianthus annuus*) plants grown in soil spiked with hexavalent chromium. *Biotech* 10 (1): 19.

Mohanty, A., Wu, Y., and Cao, B. (2014). Impacts of engineered nanomaterials on microbial community structure and function in natural and engineered ecosystems. *Applied Microbiology and Biotechnology* 98: 8457–8468.

Moon, Y.S., Park, E.S., Kim, T.O. et al. (2014). SELDI-TOF MS-based discovery of a biomarker in *Cucumis sativus* seeds exposed to CuO nanoparticles. *Environmental Toxicology and Pharmacology* 38 (3): 922–931.

Mubeen, B., Hasnain, A., Wang, J. et al. (2023). Current progress and open challenges for combined toxic effects of manufactured nano-sized objects (MNOs) on soil biota and microbial community. *Coatings* 13: 212.

Ng, L.Y., Mohammad, A.W., Leo, C.P. et al. (2013). Polymeric membranes incorporated with metal/metal oxide nanoparticles: a comprehensive review. *Desalination* 308: 15–33.

Niosi, J. (2006). Success factors in Canadian academic spin-offs. *Journal of Technology Transfer* 31 (4): 451–457.

Nowack, B. and Bucheli, T.D. (2007). Occurrence, behavior and effects of nanoparticles in the environment. *Environmental Pollution* 150 (1): 5–22.

Nyberg, L., Turco, R.F., and Nies, L. (2008). Assessing the impact of nanomaterials on anaerobic microbial communities. *Environmental Science & Technology* 42 (6): 1938–1943.

Olkhovych, O., Volkogon, M., Taran, N. et al. (2016). The effect of copper and zinc nanoparticles on the growth parameters, contents of ascorbic acid, and qualitative composition of amino acids and acylcarnitines in *Pistia stratiotes* L. (Araceae). *Nanoscale Research Letters* 11: 218.

Parsons, M. and Phillips, N. (2007). *An Evaluation of the Federal Tax Credit for Scientific Research and Experimental Development*. Department of Finance.

Pawlett, M., Ritz, K., Dorey, R.A. et al. (2013). The impact of zero-valent iron nanoparticles upon soil microbial communities is context dependent. *Environmental Science and Pollution Research* 20: 1041–1049.

Peabody, J.W., Ruby, A., and Cannon, P. (1995). The economics of orphan drug policy in the US. *Pharmacoeconomics* 8 (5): 374–384.

Pradhan, A., Seena, S., Pascoal, C. et al. (2011). Can metal nanoparticles be a threat to microbial decomposers of plant litter in streams? *Microbial Ecology* 62: 58–68.

Priester, J.H., Gea, Y., Mielke, R.E. et al. (2012). Soybean susceptibility to manufactured nanomaterials with evidence for food quality and soil fertility interruption. *Proceedings of the National Academy of Sciences of the United States of America* 109: E2451–E2456.

Priester, J.H., Moritz, S.C., Espinosa, K. et al. (2017). Damage assessment for soybean cultivated in soil with either CeO_2 or ZnO manufactured nanomaterials. *Science of the Total Environment* 579: 1756–1768.

Qu, Q., Zhang, Z., Peijnenburg, W. et al. (2020). Rhizosphere microbiome assembly and its impact on plant growth. *Journal of Agricultural and Food Chemistry* 68: 5024–5038.

Rabbani, M.M., Ahmed, I., and Park, S.J. (2016). Application of nanotechnology to remediate contaminated soils. In: *Environmental Remediation Technologies for Metal-contaminated Soils* (ed. H. Hasegawa, I.M. Rahman Mofizur, and A.R. Azizur Rahman), 219–229. Springer.

Rafique, M., Sadaf, I., Rafique, M.S. et al. (2017). A review on green synthesis of silver nanoparticles and their applications. *Artificial Cells, Nanomedicine, and Biotechnology* 45 (7): 1272–1291.

Rahmatpour, S., Shirvani, M., Mosaddeghi, M.R. et al. (2017). Dose–response effects of silver nanoparticles and silver nitrate on microbial and enzyme activities in calcareous soils. *Geoderma* 285: 313–322.

Rajput, V., Minkina, T., Sushkova, S. et al. (2019). ZnO and CuO nanoparticles: a threat to soil organisms, plants, and human health. *Environmental Geochemistry and Health* 42: 147–158.

Rajput, V.D., Kumari, A., Upadhyay, S.K. et al. (2023). Can nanomaterials improve the soil microbiome and crop productivity? *Agriculture* 13: 231.

Rajput, V.D., Minkina, T., Sushkova, S. et al. (2018). Effect of nanoparticles on crops and soil microbial communities. *Journal of Soils and Sediments* 18: 2179–2187.

Roco, M.C. (2005). Environmentally responsible development of nanotechnology. *Environmental Science & Technology* 39 (5): 106A–112A.

Salata, O.V. (2004). Applications of nanoparticles in biology and medicine. *Journal of Nanobiotechnology* 2: 3.

Scheffer, F., Schachtschabel, P., Blume, H.P. et al. (2002). *Scheffer/Schachtschabel: Lehrbuch der Bodenkunde*. Spektrum Akademischer Verlag GmbH.

Servin, A.D., De la Torre-Roche, R., Castillo-Michel, H. et al. (2017). Exposure of agricultural crops to nanoparticle CeO_2 in biochar-amended soil. *Plant Physiology and Biochemistry* 110: 147–157.

Shah, V., Collins, D., Walker, V.K. et al. (2014). The impact of engineered cobalt, iron, nickel and silver nanoparticles on soil bacterial diversity under field conditions. *Environmental Research Letters* 9 (2): 024001.

Shandilya, N., Le, B.O., Bressot, C. et al. (2015). Emission of titanium dioxide nanoparticles from building materials to the environment by wear and weather. *Environmental Science & Technology* 49 (4): 2163–2170.

Simonin, M. and Richaume, A. (2015). Impact of engineered nanoparticles on the activity, abundance, and diversity of soil microbial communities: a review. *Environmental Science and Pollution Research* 22: 13710–13723.

Simonin, M., Martins, J.M.F., Le Roux, X. et al. (2017). Toxicity of TiO_2 nanoparticles on soil nitrification at environmentally relevant concentrations: lack of classical dose-response relationships. *Nanotoxicology* 11 (2): 247–255.

Singh, D. and Kumar, A. (2016). Impact of irrigation using water containing CuO and ZnO nanoparticles on *Spinach oleracea* grown in soil media. *Bulletin of Environmental Contamination and Toxicology* 97: 548–553.

Smith, S.E. and Read, D.J. (2008). *Mycorrhizal Symbiosis*, 3e. Academic Press.

Solanki, A., John, D.K., and Ki-Bum, L. (2008). Nanotechnology for regenerative medicine: nanomaterials for stem cell imaging. *Nanomedicine* 3: 567–578.

Sun, Q., Li, Y., Tang, T. et al. (2013). Removal of silver nanoparticles by coagulation processes. *Journal of Hazardous Materials* 261: 414–420.

Suppan, S. (2013). *Nanomaterials in Soil*. Institute for Agriculture and Trade Policy.

Taniguchi, N. (1974). On the basic concept of nanotechnology. In: *Proceedings of the International Conference on Production Engineering*, Tokyo, 18–23.

Tiwari, S., Hasan, A., and Pandey, L.M. (2017). A novel bio-sorbent comprising encapsulated *Agrobacterium fabrum* (SLAJ731) and iron oxide nanoparticles for removal of crude oil co-contaminant, lead Pb (II). *Journal of Environmental Chemical Engineering* 5 (1): 442–452.

Tourinho, P.S., van Gestel, C.A., Lofts, S. et al. (2012). Metal-based nanoparticles in soil: fate, behavior, and effects on soil invertebrates. *Environmental Toxicology and Chemistry* 31 (8): 1679–1692.

Tripathi, D.K., Singh, S., Singh, S. et al. (2017). Nitric oxide alleviates silver nanoparticles (AgNPs)-induced phytotoxicity in *Pisum sativum* seedlings. *Plant Physiology and Biochemistry* 110: 167–177.

Tripathi, S., Champagne, D., and Tufenkji, N. (2012). Transport behavior of selected nanoparticles with different surface coatings in granular porous media coated with *Pseudomonas aeruginosa* biofilm. *Environmental Science & Technology* 46 (13): 6942–6949.

Van Der Heijden, M.G., Bardgett, R.D., and Van Straalen, N.M. (2008). The unseen majority: soil microbes as drivers of plant diversity and productivity in terrestrial ecosystems. *Ecology Letters* 11 (3): 296–310.

Van, N.L., Ma, C., Shang, J. et al. (2016a). Effects of CuO nanoparticles on insecticidal activity and phytotoxicity in conventional and transgenic cotton. *Chemosphere* 144: 661–670.

Van, N.L., Rui, Y., Cao, W. et al. (2016b). Toxicity and bio-effects of CuO nanoparticles on transgenic Ipt-cotton. *Journal of Plant Interactions* 11 (1): 108–116.

Vance, M.E., Kuiken, T., Vejerano, E.P. et al. (2015). Nanotechnology in the real world: redeveloping the nanomaterial consumer products inventory. *Beilstein Journal of Nanotechnology* 6: 1769–1780.

Verma, Y., Singh, S.K., Jatav, H.S. et al. (2022). Interaction of zinc oxide nanoparticles with soil: insights into the chemical and biological properties. *Environmental Geochemistry and Health* 44: 221–234.

Wang, Y., Chang, C.H., Ji, Z. et al. (2017). Agglomeration determines effects of carbonaceous nanomaterials on soybean nodulation, dinitrogen fixation potential, and growth in soil. *ACS Nano* 11 (6): 5753–5765.

Watson, J.L., Fang, T., Dimkpa, C.O. et al. (2015). The phytotoxicity of ZnO nanoparticles on wheat varies with soil properties. *Biometals* 28: 101–112.

Wolińska, A. and Stępniewska, Z. (2012). Dehydrogenase activity in the soil environment. *Dehydrogenases* 10: 183–210.

Wu, J., Bai, Y., Lu, B. et al. (2020). Silver sulfide nanoparticles reduce nitrous oxide emissions by inhibiting denitrification in the earthworm gut. *Environmental Science & Technology* 54 (18): 11146–11154.

Wu, X., Liu, Y., Shang, Y. et al. (2021). Peat-vermiculite alters microbiota composition towards increased soil fertility and crop productivity. *Plant and Soil* 470: 21–34.

Xu, C., Peng, C., Sun, L. et al. (2015). Distinctive effects of TiO_2 and CuO nanoparticles on soil microbes and their community structures in flooded paddy soil. *Soil Biology and Biochemistry* 86: 24–33.

Yadav, A.N., Verma, P., Singh, B. et al. (2017). Plant growth-promoting bacteria: biodiversity and multifunctional attributes for sustainable agriculture. *Advances in Biotechnology and Microbiology* 5: 555671.

Yadav, S.K., Patel, J.S., Kumar, G. et al. (2018). Factors affecting the fate, transport, bioavailability, and toxicity of nanoparticles in the agroecosystem. In: *Emerging Trends in AgriNanotechnology: Fundamental and Applied Aspects* (ed. H.B. Singh and A.K. Yadav), 118. CABI.

Yadav, T., Mungray, A.A., and Mungray, A.K. (2014). Fabricated nanoparticles: current status and potential phytotoxic threats. *Reviews of Environmental Contamination and Toxicology* 230: 83–110.

Yang, E.J., Kim, S., Kim, J.S. et al. (2012). Inflammasome formation and IL-1β release by human blood monocytes in response to silver nanoparticles. *Biomaterials* 33 (28): 6858–6867.

You, T., Liu, D., Chen, J. et al. (2018). Effects of metal oxide nanoparticles on soil enzyme activities and bacterial communities in two different soil types. *Journal of Soils and Sediments* 18: 211–221.

Zavyalova, N.E., Vasbieva, M.T., and Fomin, D.S. (2020). Microbial biomass, respiratory activity and nitrogen fixation in soddy-podzolic soils of the pre-Ural area under various agricultural uses. *Eurasian Soil Science* 53: 383–388.

Zhai, Y., Hunting, E.R., Wouterse, M. et al. (2017). Importance of exposure dynamics of metal-based nano-ZnO, -Cu and -Pb governing the metabolic potential of soil bacterial communities. *Ecotoxicology and Environmental Safety* 145: 349–358.

Zheng, D., Wang, N., Wang, X. et al. (2012). Effects of the interaction of TiO_2 nanoparticles with bisphenol A on their physicochemical properties and in vitro toxicity. *Journal of Hazardous Materials* 199: 426–432.

6

Silver Nanoparticles' Emerging Roles in Enhancing Crop Plant Growth and Yield

Anuradha Kumari[1], *Anumanchi Sree Manogna*[2], *Prabhat Kumar*[1], *and Ilora Ghosh*[1]

[1] *School of Environmental Sciences, Jawaharlal Nehru University, New Delhi, India*
[2] *SASTRA Deemed University, Thanjavur, Tamil Nadu, India*

6.1 Introduction

The influence of nanotechnology on an industry is the leap in science with a recently added influence on several industries, such as agriculture (Khot et al. 2012; Mariyam et al. 2023). Nanotechnology has shown the integration of nanotechnology in agriculture that could raise the superior agrarian harvest, stimulate plant vigor, and aid in sustainable cultivation operations (Kah et al. 2013; Yadav et al. 2023). The regulating materials are regulated on a mini scale of 100 nm to yield nanoparticles (NPs) with unique physical, chemical, and biological properties (Ahire et al. 2022). Within the existing scenario, past agricultural practices are ongoing sustainable development goals facing many challenges (Atapattu et al. 2024). For instance, the agrarian system is vulnerable to the persistent effects of inappropriate agrochemical use (Rashid et al. 2023), that is, insecticides, pesticides, fertilizers, manures (contaminated with pathogens, heavy metals, and pharmaceuticals), and other chemicals that protect the plants. These manures and the contaminated chemicals can threaten environmental pollution and also lead to bad conditions of soil and water systems (Nam et al. 2024). Nanotechnology advances the ability to address the abovementioned challenges through augmentation of the efficacies of fertilizers, insecticides, and pesticides; crop resilience to biological, chemical, and environmental stressors; and gene expression and functions in plants for improved overall plant well-being and development (Iavicoli et al. 2017; Tang et al. 2023). Discrete attributes of NPs, such as elevated surface area/volume ratio and higher reactivity, make them extremely effective as agriculture practices (Saritha, Anju, and Kumar 2022). It is an important feature in agriculture because of its ability to surge crop harvest, improve soil, and diminish farming techniques' environmental effects (Mousavi and Rezaei 2011).

Nanotechnology-based Sustainable Agriculture, First Edition. Edited by Pardeep Singh et al.

One of the foremost challenges contemporary agriculture faces today, obtaining sustainable methods, the shortage of arable land, and the growing food demand, is being addressed with nanotechnological methods and products (Mittal et al. 2020). New instruments and advanced methods of precision agriculture, the regulation of crop infection, and the effective distribution of nutrients all contribute to more robust and productive farming systems (Prasad, Kumar, and Prasad 2014).

Since they can boost crop growth and harvest, silver nanoparticles (AgNPs) have gained enormous appeal in agronomy (Mehmood et al. 2018; Tariq et al. 2022). Constituted of silver atoms in a nanostructure, these NPs are enhanced in unique properties that do not exist on larger forms (Altammar 2023). Among their many features, one of the most significant is their formidable antimicrobial abilities (Bruna et al. 2012). AgNPs have shown promise in effectively restraining the progress of numerous plant pathogens, including bacteria, fungi, and viruses, resulting in decreased plant disease incidence and improved plant growth (Sharon et al. 2010). This microbicidal effect results from the release of silver ions (Ag^+), which can reach the microbial cell membranes to cause damage and cell mortality (Fan et al. 2021). Besides their antimicrobic role, AgNPs also enhance plant physiology in myriad ways (Rai, Yadav, and Gade 2009). In seed treatment, AgNPs enhance seed germination and seedling vigor resulting in healthier crop development and standardization (Rahman et al. 2023). Advance research studies prove that the AgNPs can enter the seeds through which the water and vital essential nutrients for germination can be taken up inside the seeds, accelerating the germination process. In addition to that, AgNPs have been identified as essential in increasing and delivering nutrients in the plant. It can optimize the utilization of key nutrients and trace elements to encourage the growth and development of plants. However, this is only important when simply dealing with nutrient deficiencies or imbalances when common fertilization methods cannot perform optimally. Furthermore, AgNPs are also recorded to aid in situations of abiotic stress, such as drought, salinity, and toxicity, from heavy metals (Junedi, Mukhopadhyay, and Manjari 2023). Additionally, it can also develop plants' capacity to deal with harsh environmental surroundings by increasing an ability to stay in water and for ion balance and detoxification. Thus, it assists plants despite the stiff circumstances. Additionally, AgNPs foster the development of plants with raised photosynthesis and root growth that leads to higher biomass production and better-quality yields (Luomin Cui et al. 2023).

Sustainable farming requires plant growth enhancement and productivity. Sustainable agriculture aims to achieve existing agricultural requirements for nourishment, nutrients, and fiber from food while also allowing future generations to do the same. It is concerned with eco-friendly, economically possible, and socially accountable actions. The employment of AgNPs in agriculture is consistent with principles of sustainable development and introduces new paradigms that engage in crop yield augmentation and ecological effect reduction (Tilman and Clark 2014). These NPs can act as carriers for fertilizers and pesticides, enhancing targeted delivery and thus reducing the need for volumes at which effective pest and nutrient management can be achieved (Khan et al. 2023). It boosts the effectiveness of farming inputs; however, at the same time minimizing environmental contamination and the emergence of pest resistance (Tudi et al. 2021). In addition, the capacity of AgNPs to stimulate healthier plant growth and increase crop yield (Saha et al. 2024) can provide assurance to food security as it is used. The same piece of land will yield higher crop production, minimizing the agricultural encroachment on natural ecosystems. This approach increases the conservation of biodiversity and natural resources while reducing stress on vital habitats. On the contrary, AgNPs have been massively applied in agriculture, making AgNPs environmental effects and safety issues also worthy of apprehension (Kim et al. 2023). To better understand the fate and movement of AgNPs in soil and water, its potential impacts on nontarget organisms, and its potential human health consequences, systematic research is needed (Sharma et al. 2022). Establishment of a regulatory framework for the biosafety and reasonable use of nanotechnology in agriculture are indispensable. Finally, it has been demonstrated that AgNPs could lead to high potential in crop growth enhancement, increasing yields, and practicing sustainable agriculture (Manzoor et al. 2023). Its unique features, such as antimicrobial effect, greater nutrient absorption, and stress resistance, can provide great support for modern and sustainable farming systems (Gupta and Shah 2022). However, requiring a meticulous evaluation of environmental impacts and regulatory considerations of their long and sustainable use is vital in processing plans and methods. This chapter compiles all the information concerning the implementation of AgNPs to agricultural practices from all array of aspects. There is more research and innovation needed to discover the whole benefit potential of agricultural practices using nanotechnological innovations.

6.2 AgNPs: Synthesis and Characterization

Nanotechnology has been progressing and evolving in agricultural applications for quite some time. AgNPs are counted among the numberless nanomaterials that have attracted more than deserved attention due to their extraordinary physicochemical characteristics and distinct antimicrobial effect (Rafique et al. 2017). Therefore, it emerges that the AgNPs may be a potential tool to improve all aspects of crop yield and crop production by combating pathogens, stimulating plant growth development, and facilitating nutrient uptake (Gogos, Knauer, and Bucheli 2012). Consequently, it becomes crucial to find out the synthesis and characterizations of NPs for the best solutions for AgNPs' production in agriculture (Sharma, Yngard, and Lin 2009). In this section, characterization techniques, important in the determination of the physical, structural, optical, and functional properties of AgNPs, are discussed, which ensure that AgNPs can safely and effectively be applied in agricultural practices (Sondi and Salopek-Sondi 2004; Prasad, Kumar, and Prasad 2014).

6.2.1 Methods for Synthesizing AgNPs

Several approaches have been reported for the synthesis of AgNPs and can be broadly divided into chemical, physical, and biological methods. The decision of the synthesis method affects the properties of the NPs in so far as there are several benefits and limitations related to every method (Iravani et al. 2022).

6.2.1.1 Chemical Methods

The selection of the highest producing synthesis method of AgNPs is, therefore, chemical synthesis, mainly because of the above-highlighted advantages. Usually, these methods reduce the silver ions (Ag^+) in solution to metallic silver (Ag^0) NPs (Fahim et al. 2024). A detailed discussion of some of the chemical methods used is detailed below.

Chemical reduction: This is a method of silver salt, where silver nitrate ($AgNO_3$) is reduced by a reducing agent like sodium borohydride ($NaBH_4$), ascorbic acid, or hydrazine. The size and shape of the resulting NPs may depend on the choice of reducing agent and reaction conditions (Chaudhary et al. 2023).

Tollens' method: Using a reducing sugar, Ag^+ ions are reduced in this method by Tollens' reagent (ammoniacal silver nitrate). The AgNPs obtained by this method are highly uniform.

Polyol process: In this process, silver salts are reduced in a polyol medium (e.g. ethylene glycol) at elevated temperature. As a reducing agent and stabilizing agent, the polyol reacts to produce well-dispersed NPs.

Microemulsion technique: In this method, the controlled growth of AgNP's is done based on water in oil microemulsions. The confined reaction environment for producing NPs of uniform size and shape is provided by microemulsions (Malik, Wani, and Hashim 2012).

6.2.1.2 Physical Methods

Most of the time, physical methods for the synthesis of AgNPs include physical processes to condense silver atoms into NPs. Key physical methods include the following:

Evaporation–condensation: In this method, thermal evaporation of a silver source from a silver salt occurs, and particles condense into NPs in a controlled atmosphere. By adjusting the evaporation temperature and nature of the inert gas used, the size and distribution of NPs are controlled.

Laser ablation: The ablation of silver targets in a liquid medium produces AgNPs. Producing pure NPs without the requirement of any chemical reagents is one of the advantages of this method.

Arc discharge: In this technique, two silver electrodes submerged in solution within a dielectric liquid are used to produce an electric arc. However, the silver ions are vaporized and then condensed into NPs (Lung et al. 2007; Tien et al. 2008).

6.2.1.3 Biological Methods

Biological systems, such as plants, microorganisms, and enzymes, are the best choice for synthesizing AgNPs via biological methods. These methods are considered pro-environmental and sustainable. Some of the key biological methods include the following:

Plant extracts: The silver salt solution is reduced to AgNPs by different plant extracts having natural reducing agents (e.g. polyphenols). The plant extract components constitute both a reducing and stabilizing agent (Das et al. 2020).

Microbial synthesis: Biosynthesis of AgNPs is possible involving some bacteria, fungi, and algae that can reduce Ag^+ ions. Extracellular enzymes of these microorganisms mediate the reduction process (Mikhailova 2020).

Enzyme-mediated synthesis: Nitrate reductase can serve as enzymes that help reduce silver ions to AgNPs. The precision of this method allows the size and shape of the NPs to be controlled (Shankar et al. 2023; Sharma et al. 2021).

6.2.2 Factors Influencing the Synthesis Process and NP Properties

Several factors can influence the size, shape, and stability of AgNPs and the pathways in which they are synthesized. The externalization of these factors is essential in tailoring the properties of AgNPs for the applications.

6.2.2.1 Concentration of Silver Precursor

An essential factor in determining the synthesis process is the concentration of the silver precursor (e.g. $AgNO_3$). Usually, the larger the NPs, the greater the silver ion availability for the reduction (Li et al. 2019). Low concentrations, on the other hand, prefer the formation of smaller NPs.

6.2.2.2 Reducing Agent Type and Concentration

The reduction rate and the particle final size and morphology of the NPs are strongly influenced by the choice and concentration of a reducing agent. Not surprisingly, strong reducing agents like $NaBH_4$ enable fast reduction and small-sized NPs, while less strong reducing agents result in slow reduction and large-sized NPs (Smith et al. 2022).

6.2.2.3 Stabilizing Agents

To avoid the coagulation of the NPs and to have control over their size, stabilizing agents (capping agents) are used. They can include surfactants, polymers, and biomolecules among other types of stabilizers. According to Duman et al. (2024), the outer charge density that the NPs have, the stability, and biocompatibility of those structures rely on the form and concentration of these agents.

6.2.2.4 pH of the Reaction Medium

The pH of the reaction medium can affect the ionization status of the reducing and stabilizing agents, the reduction mechanism, and the stability of NPs. For instance, the reduction rate and smaller dimension particles may be achieved at a faster rate by increasing the pH (Hosseingholian et al. 2023; Szczyglewska, Feliczak-Guzik, and Nowak 2023).

6.2.2.5 Temperature

In addition, temperature is the second critical factor since it controls the kinetics of the reduction process and nucleation together with the growth of the NPs. In most cases, the process takes place with increased rates at elevated temperatures resulting in smaller particles, while at low temperatures the process is sluggish and results in large particle formation (Rafique et al. 2017).

6.2.3 Characterization Techniques for Evaluating the Size, Shape, and Stability of AgNPs

The characterization of AgNPs is useful for the evaluation of the properties and compatibility of AgNPs in various applications. The finding of the size, shape, and stability of AgNPs, as illustrated in Figure 6.1, is analyzed in terms of several analytical possibilities.

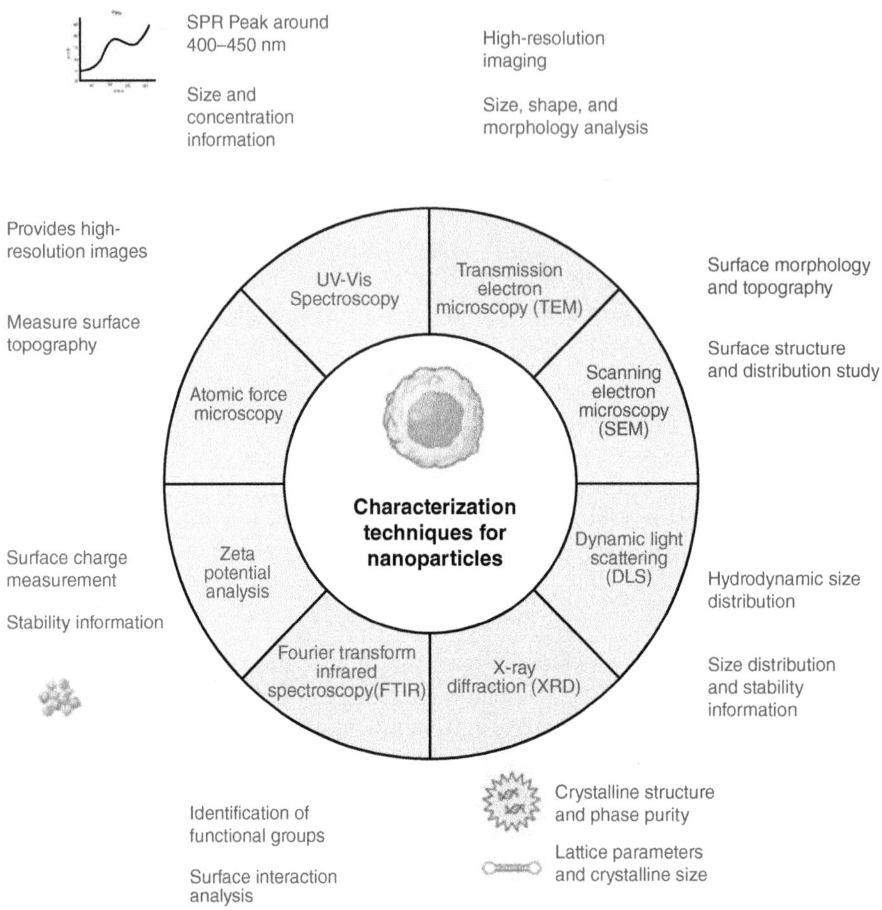

Figure 6.1 Silver nanoparticles' characterization techniques.

6.2.3.1 UV-vis Spectroscopy

UV-visible spectroscopy is first used to determine the size of the synthesized AgNPs. The synthesized AgNPs were characterized using UV-vis spectroscopy and displayed a characteristic surface plasmon resonance (SPR) band between 400 and 450 nm. This peak provides information on the size as well as the distribution density of the NPs (Dhaka et al. 2023).

6.2.3.2 Transmission Electron Microscopy

Transmission electron microscopy (TEM) is capable of providing high-resolution images of the cross-sectional structure of thin sections of fibers. The subsequent generation of high-resolution images by TEM in turn enables a systematic assessment of the size, shape, and contour of the AgNPs. Therefore, it is a very important tool for nanoscale material characterization (Malatesta 2021).

6.2.3.3 Scanning Electron Microscopy

The surface characteristics and roughness of AgNPs are analyzed using scanning electron microscopy (SEM). In comparison to TEM, SEM is utilized to examine the surface structure characteristics and the general distribution of NPs on the substrates (Sharma et al. 2022).

6.2.3.4 Dynamic Light Scattering

This technique is also known as photon correlation spectroscopy, which measures the size of any particles present in the solution through observation of the particles' fluctuations in time. Dynamic light scattering (DLS) is a technique used in the determination of the hydrodynamic size distribution profile of NPs in a colloidal suspension. The size distribution and stability of particles in solution are derived from variations in the intensity of scattered light, which provides us with insightful data regarding such qualities (Johnson and Gabrielson 2011).

6.2.3.5 X-ray Diffraction

The utility of X-ray diffraction (XRD) is to facilitate the identification of the crystalline structure and phase purity of synthesized AgNPs. XRD analysis provided information on the lattice parameters of the NP and the crystallite size from the diffraction pattern as explained by Chen et al. (2020).

6.2.3.6 Fourier Transform Infrared Spectroscopy

The identification of functional groups on the surface of AgNPs is done using the technique known as Fourier transform infrared (FTIR) spectroscopy. Possibly, with this technique, an understanding of the NPs and the way the surface modifiers and other surface modifications interact can be explored (Hosseingholian et al. 2023).

6.2.3.7 Zeta-potential Analysis

Zeta potentials are employed to assess the surface charge of NPs in a colloidal suspension system. From the zeta potential, the absolute values provide information about the stability of the NP suspension, raising the absolute value means that there will be better stability by electrostatic repulsion (Clogston and Patri 2011).

6.3 Antimicrobial Properties of AgNPs

AgNPs possess unique antibacterial properties, which extend their uses in various fields of biotechnology. The effectiveness of their use arises in a way that interferes with the bacterial cell walls and destroys their cellular functions and kills them. It has also been expressed that AgNPs are capable of producing toxic reactive oxygen species (ROS), which disrupt cellular entities encompassing DNA, lipids, and proteins (Bruna et al. 2021; Misirli, Sridharan, and Abrantes 2021). Apart from Gram-positive and Gram-negative bacteria, fungi, and viruses, they are also effective against many other pathogenic organisms. The literature review indicates that the use of AgNPs can improve the effectiveness of conventional antibiotics; hence, they might be useful for the treatment of antibiotic-resistant bacteria (Misirli, Sridharan, and Abrantes 2021). For more of their evidence of usefulness in the treatment and prevention of infections with negligible toxicity to human cells, further studies are required.

6.3.1 Mechanisms of Action of AgNPs Against Plant Pathogens

The antimicrobial ability of AgNPs has become of interest in global studies since it has established the ability of acting as a broad-spectrum agent against a wide variety of plant pathogens, such as bacteria, fungi, and viruses. AgNPs function through various manners of actions, which include direct adhesion on microbial cells and the formation of ROS.

6.3.1.1 Disruption of Cell Membrane Integrity

Adhesion of AgNPs to bacteria and fungi cell membranes can cause an increase in membrane permeability. This connection ends in cell death and causes cellular contents to flow out. Not only do these NPs penetrate the cell membrane, but also the internal elements are disrupted by the NPs, disrupting vital cellular processes. This interaction between the lipid bilayer and membrane proteins and the cells compromises the integrity of the membrane and the cell death occurs as a result of the loss of their essential internal content (Morones et al. 2005; Rai, Yadav, and Gade 2009).

6.3.1.2 Generation of ROS

The AgNPs can facilitate the reaction of hydroxyl radicals and the other free radicals, including the superoxide anion radical. The ROS can decompose lipids, proteins, and DNA, which in turn results in oxidative stress and cell death. Oxidative damage is one of the main mechanisms for the antimicrobial properties of AgNPs. ROS even though consumes a high amount of energy can injure irreversible infrastructures of a cell and kill microbial cells (Kim et al. 2007).

6.3.1.3 Interaction With Biomolecules

AgNPs can interact with a large number of biomolecules, as they are present in the cell, such as enzymes and DNA. These NPs can halt enzyme function by directly reacting to the enzyme via contacting thiols of enzymes. It hinders the necessary cell survival pathways by binding enzymes. This, in turn, may engender suppression of the two processes such as replication and transcription, thereby preventing microbial reproduction and contact with DNA. One of the leading mechanisms by which AgNPs exhibit their antimicrobial activity is this disruption of cell functions (Rai et al. 2012).

6.3.1.4 Inhibition of Signal Transduction

AgNPs' influence on microbial cells shows the ability to disrupt signal transduction pathways. It may also prevent the formation of biofilms that greatly contribute to the pathogenic capabilities of the majority of plant pathogens. These pathways include actions vital to the pathogen's ability to infect and survive; thus their inhibition by AgNPs, while AgNPs for their part do not have targets specific or unique to the pathogen (Lok et al. 2006).

6.3.1.5 Release of Silver Ions

Among the other main consequences of AgNPs' antibacterial activity, it is worth mentioning the mechanism associated with the emission of silver cations (Ag^+). Due to the high reactivity of these ions, they chelate with intracellular molecules, which may in turn enhance the mechanisms through which the NPs kill the cells. Antimicrobial action of the total of the silver ions formed through the interaction of the microbial cell membrane and the intracellular structures (Pal, Tak, and Song 2007).

6.3.2 Effects of AgNPs on Pathogen Growth Inhibition and Disease Suppression in Crops

AgNPs have been promisingly applied in the inhibition of the growth of various plant pathogens and in suppressing crop diseases. Many studies have demonstrated the ability of AgNPs to retard bacterial, fungal, and viral infections in agricultural settings.

6.3.2.1 Bacterial Pathogens

In particular, AgNPs are effective at killing bacterial pathogens, such as *Xanthomonas campestris, Pseudomonas syringae*, and *Erwinia amylovora*. These same pathogens are responsible for bacterial spot, bacterial speck, and fire blight diseases. AgNPs application can greatly decrease bacterial density leading to lower incidence and severity of disease. AgNP treatments, for example, have been shown to reduce the severity of bacterial spots on tomato plants so that crops are highly healthier and produce more (Ansari et al. 2023).

6.3.2.2 Fungal Pathogens

The importance of fungal pathogens, including *Botrytis cinerea, Fusarium oxysporum*, and *Phytophthora infestans*, as major yield loss pathogens in crops is demonstrated. Therefore, these fungi are inhibited from spore germination and mycelial growth by AgNPs, which is also helpful in controlling such diseases as gray mold, *Fusarium* wilt, and late blight. AgNPs' fungicidal properties arise from their capacity to impede on cell membranes and produce ROS that are fatal to fungal cells. Studies show that treatments with AgNP can significantly reduce the incidence of gray mold in strawberries and hence preferably improve the fruit quality and also prolong the shelf life (Rai et al. 2012).

6.3.2.3 Viral Pathogens

While the research on the anti-viral activity of AgNPs in plants is relatively limited, there is evidence that AgNPs would inhibit viral replication and attenuate the severity of viral diseases. As an example of such application, AgNPs have been shown to inhibit the replication of the tobacco mosaic virus (TMV) and protect plants against viral infections. According to Galdiero et al. (2011), AgNPs cause anti-viral activity due to their interaction with viral particles and inhibition of binding of viral particles to host cells.

6.3.3 Potential Applications of AgNPs as Antimicrobial Agents in Crop Protection

Having potent antimicrobial properties, AgNPs prospect as crop protection agents. With their application, it is possible to integrate them into multiple agricultural practices improving crop health and yield.

6.3.3.1 Seed Treatment

Seed treated with AgNPs can protect seedlings from soil-borne pathogens. Consequently, seeds can be coated with AgNPs to produce a protective barrier from diseases in seedlings and can facilitate healthy plant growth. Seeds treated with AgNPs were found to have better resistance in soil-borne pathogens, such as *Pythium* and *Rhizoctonia* (Krishnaraj et al. 2012), and led to better germination rates.

6.3.3.2 Foliar Sprays

AgNPs may be applied as foliar sprays to be effective in controlling foliar pathogens. AgNPs can be sprayed onto leaves and have been shown to prevent the establishment and development of bacterial, fungal, and viral pathogens on the leaves, thereby lowering foliar disease pressure. For example, foliar application of AgNPs has been shown to control powdery mildew in cucumbers and downy mildew in grapes producing more healthy plants and improved yield (Jo, Kim, and Jung 2009).

6.3.3.3 Soil Amendments

AgNPs incorporated into soil amendments can help manage soil-borne diseases. The AgNPs in the soil can suppress the growth of pathogenic microorganisms and increase the plant's resistance to the disease and increase soil microbiome health. It has been shown through research that AgNPs when added to soil reduce the incidence of root rot caused by *Phytophthora* on multiple crops (Min et al. 2009).

6.3.3.4 Postharvest Treatments

The AgNPs can be used in postharvest treatments to increase the shelf life of agricultural produce. The potential use of AgNPs to reduce the loss of postharvest and ensure the quality and safety of fresh produce by preventing microbial contamination during storage and transportation. An example is AgNPs applied on harvested fruits and vegetables that inhibit the growth of the spoilage organisms and pathogens, keeping fresh and maintaining the nutritional quality of produce (Lieu, Dang, and Nguyen 2024).

6.4 Seed Treatment With AgNPs

With the aid of novel strategies to improve the plants' capacity of nutrient uptake efficiently, NP implementation has the potential to revolutionize agriculture (Elemike et al. 2019). NPs are very capable of interacting with a live cell at a molecular level for a wide range of potential agricultural applications. The ability of NPs to penetrate live tissue allows them to travel to other parts of the plant as well (Salem and Fouda 2021). Many of

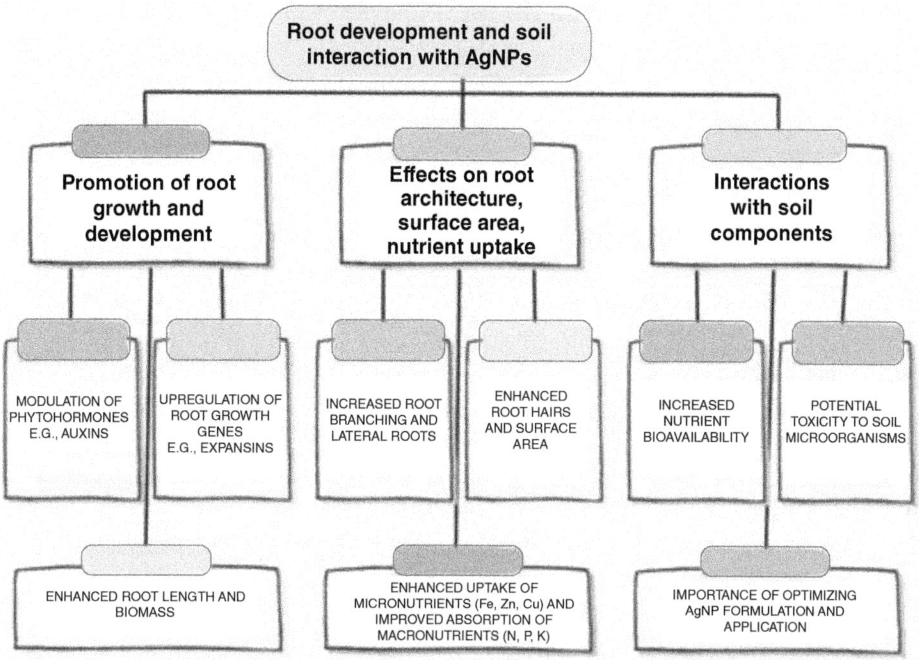

Figure 6.2 Application of silver nanoparticles in agricultural yield enhancement.

these technologies now make use of knowledge-intensive technology to improve crop yield and output. Recently, metal-based NPs, such as AgNPs, have become more frequently used as agents for seed pretreatment. Several crop plants have already been shown to benefit from these NPs in terms of improving seed germination, seedling growth, and stress tolerance (Mohamed et al. 2017). Only a few of these investigations used a seed priming technique in which seeds were re-dried to their original moisture levels prior to sowing. Furthermore, seed nanopriming does not have same mechanism as presowing seed treatment without drying seeds (Mahakham et al. 2017). Some of the AgNPs' utilization for crop yield enhancement are depicted in Figure 6.2.

6.4.1 Effects of AgNP Seed Treatment on Germination Rates and Seedling Vigor

Metal-based NPs, since they are more effective and more environmentally acceptable substitutes for conventional agriculture methods, e.g. nanopesticides, nanosensors, and nanofertilizers, should be integrated in agriculture without compromising on these requirements. Effect of AgNPs on germination of pea seeds (*Pisum sativum*) and subsequent germination performance. The study shows that the percentage of germination of the peas exposed to a concentrated solution of AgNPs with a concentration of 2.5 mM for a duration of three hours was much higher as compared to the control (100 $\pm$ 0.00%; Rahman et al. 2023). Additionally, biogenic AgNPs affect the growth of seedlings and the germination of seeds in *Solanum lycopersicum*. Treatment of seeds with AgNPs resulted in a significantly greater germination rate and seedling growth than untreated seeds. The bactericidal activity of the compound was also investigated in regard to *Clavibacter michiganensis* subsp. *michiganensis* (cmm) infection of *S. lycopersicum* (Noshad, Hetherington, and Iqbal 2019).

6.4.2 Influence of AgNPs on Seedling Establishment and Early Growth Stages

Bean growth and yield may be enhanced by AgNPs in extreme temperatures. This study implemented a controlled application of AgNPs to bean seeds that were aimed for the application of sowing, in the spirit of precision agriculture. Additionally, the studies reported are proposed to suggest the presowing treatment of AgNPs in order to promote AgNPs absorption and uptake in seed, which might induce immediate and delayed biological effects (Mahakham et al. 2017). Statistically significantly improved growth parameters, such as plant height (cm), fresh biomass (g), number of shoots/plants, root weight (g), tomato yield/plant (g), and dried biomass (g), were obtained in AgNPs treated plants (Prażak et al. 2020). We, therefore, recommend bio-based AgNPs as a means to attenuate deleterious effects of chemical-based bactericides.

6.4.3 Optimization of AgNPs' Application Methods for Seed Treatment

Acharya et al. (2020) used agroindustrial byproducts (AIB) to synthesize AgNPs and turmeric oil nanoemulsions (TNE) as nanopriming agents to treat simultaneously triploid (Maxima) and diploid (Riverside) watermelon seeds. Triploid seeds treated with AgNPs had a significantly greater seedling emergence rate at 14 days post sowing than seeds treated with other treatments. Acharya et al. (2020) found that soluble sugars (glucose and fructose) were higher during the germination process of seedlings treated with AgNPs when compared to their control at 96 hours. By using the green and environmentally friendly nanotechnological technique of seed priming with AgNPs, seed germination, growth, and yield can be enhanced and fruit quality maintained.

6.5 Nutrient Uptake and Transport Enhancement

However, AgNPs have become fairly successful agents for enhancing plant nutrient uptake and movement. Studies have indicated that the application of AgNPs can greatly enhance nutrient absorption, increasing in most crops the growth and productivity of the crop. On AgNPs, Guzmán-Báez et al. (2021) observe that stimulation of root growth occurs in particular response to 5 mg/L AgNPs. The AgNPs increase nutrient absorption and result in a corresponding increase in chlorophyll and key amino acid levels in the plant, which are critical for plant health, and productivity increased plant metabolic activities. Besides these, the studies have also indicated that AgNPs induce the expression of genes for the transportation products, such as nitrate reductase and ferredoxin. This is further supporting an important role of AgNPs for nutrient assimilation.

6.5.1 Role of AgNPs in Improving Nutrient Absorption by Crop Plants

At a molecular level, the study shows that AgNPs increase plant cell membrane permeability and fluidity, a fundamental property of cells regarding nutrient and water uptake. The size and surface properties of AgNPs and their characteristics promote optimal nutrient transport pathways, so they may penetrate plant tissues very efficiently. It also demonstrated that the use of AgNPs (25 mg/L) increased growth markers, such as biomass and root length and nutrient absorption efficiency (Jhanzab et al. 2015). Besides this, AgNPs make the plants absorb nutrients, and the NPs also boost the metabolic activities inside them, as evidenced in higher levels of vital amino acids and chlorophyll, the two things absolutely critical for plant health and productivity. Furthermore, AgNPs were shown to induce the expression of genes required for nutrient assimilation, e.g. ferredoxin and nitrate reductase.

6.5.2 Mechanisms of AgNP-mediated Nutrient Uptake and Transport Within Plants

AgNPs were evaluated as a means of improving plant nutrient absorption and transport. These processes are facilitated by AgNPs using very different pathways and interactions at the cellular level.

6.5.2.1 Direct Uptake by Roots

The cells of the root hair can directly absorb AgNPs. The AgNPs are so small that they enable them enter into plant tissues through cell wall and plasma membrane penetrating. There are many pathways by which this uptake can occur, including endocytosis, passive diffusion through cells' porous walls, and through cells' natural NP sieves (Sembada and Lenggoro 2024).

Endocytosis: AgNPs can enter plant cell through a process called endocytosis. The cell membrane unexpectedly surrounds the NPs, forming vesicles that then carry the nanoscale particles into the cytoplasm. Incorporation of both clathrin-dependent and independent routes permits efficiency in the absorption process (Awashra and Młynarz 2023).

Symplastic and apoplastic pathways: Two pathways are present for AgNPs to disseminate through plant tissues: the symplastic route going through linked plasmodesmata within the cytoplasm and the apoplastic route via cell walls and interspaces between cells. AgNPs cross these channels and can then disseminate throughout the plant by moving across these channels (Ali, Mehmood, and Khan 2021).

6.5.2.2 Translocation Through Xylem

The AgNPs are transported by the xylem, a vascular tissue that normally transports nutrients and minerals. AgNPs can easily be transported to the shoot system from the root to increase nutrient uptake of the plant. According to some studies, AgNPs can reach different parts of plants, such as leaves and fruits, affecting their nutrients (Yan and Chen 2019).

6.5.2.3 Influence on Cellular Mechanisms

AgNPs can help in the modification of cell membranes, enhancing their permeability by further improving their water absorption and nutrients (Mikhailova 2020). Normally, NPs have the ability to create new pores in the cell wall, which help large NPs to easily enter the cell. AgNPs have the potential to modify gene expression associated with nutrient transport and increase protein molecules involved in dietary uptake and absorption (Francis et al. 2024).

6.5.2.4 Oxidative Stress and Defense Mechanisms

AgNPs have the ability to improve the absorption of nutrients. However, they can increase oxidative stress, causing plants to activate their defense mechanisms, and secondary metabolites are formed, helping plants to grow (Chandrashekar et al. 2023).

6.5.2.5 Formation of New Pores

AgNPs have the ability to form new large pores in the cell wall, which allows larger molecules to be easily penetrated into the cell enhancing its nutrition (Mittal et al. 2020).

6.5.2.6 Influence of Ag^+ Ions

Plant roots can absorb AgNPs and silver ions (Ag^+). Ag^+ ions in the presence of AgNPs cause them to absorb at a higher rate than when they are dissolved (Zhang et al. 2023).

6.5.3 Effects of AgNPs on Nutrient Availability in Soil and Nutrient Utilization Efficiency by Plants

Recently, AgNPs have been applied in agriculture because they can increase plant nutrient utilization efficiency and increase the availability of nutrients in soil (Duhan et al. 2017). Alteration of soil properties, microbiology, and plant interaction impact on soil nutrient can occur through the influence of AgNPs.

6.5.3.1 AgNPs' Effects on Soil's Nutrient Availability

The AgNPs change the physical and chemical properties of the soil to increase its capacity of nutrient absorption. Adding AgNPs to the soil matrix could increase the availability of essential nutrients (N, P, and K) through reaction with soil organic matter and minerals (Guzmán Báez et al. 2021). The studies have shown that AgNPs can be utilized to improve soil structure that would in turn maximize water retention and aeration, which is important in the uptake of nutrients. Research done by Tripathi et al. (2017) has also demonstrated that some beneficial microbial populations can be stimulated, without increasing pathogen colonization. A switch in microbial diversity like this is important because beneficial bacteria are very necessary to break down organic matter, then release nutrients and such. In a study it was shown that actually AgNPs contained in soil helped to increase soil nutrient availability by decreasing pathogenic and increasing beneficial bacteria in soil (Hänsch and Emmerling 2010).

6.5.3.2 Plant Nutrient Utilization Efficiency

One possibility of the enhanced ability of AgNPs to increase nutrient uptake efficiency in plants is through the promotion of root growth and the surface area available for nutrient absorption in plants. Increased root length and biomass are closely related to improved nutrient uptake capabilities, and AgNPs can increase root length and biomass significantly. As an example, the increase in the concentrations of nitrogen (N), phosphorus (P), and potassium (K) in the leaves of AgNPs-treated plants indicated enhancement of the plant nutrient absorption. Furthermore, AgNPs could trigger metabolic physiological processes in plants required for the assimilation of nutrients. AgNPs may increase the expression of genes involved in nutrient transport and metabolism resulting in increased nutrient uptake efficiency (Zhang et al. 2023).

6.6 Stress Tolerance Improvement

Besides the drought, AgNPs improve plants' resilience to other abiotic stressors, such as salinity and toxicity from heavy metals. In the presence of oxidative stress, these NPs alleviate oxidative stress and increase nutrient uptake in support of plant growth and productivity (Abasi et al. 2022; Dilnawaz, Kalaji, and Misra 2023).

6.6.1 Mechanisms of Stress Tolerance Enhancement

6.6.1.1 ROS Management

The role of AgNPs in activating the antioxidant machinery of plants is very much important. Potential oxidative stress involves the production of ROS when plants are stressed. AgNPs have been applied to increase the production of the key antioxidant enzyme, superoxide dismutase (SOD), catalase (CAT), peroxidase (POD), and hydrogen peroxide (H_2O_2) as a product of SOD activity (Jadczak et al. 2020). These enzymes are critical for scavenging ROS and for mitigating oxidative damage. This leads to an increase in antioxidant activity, which helps the cell stay healthy. The use of AgNPs is shown to reduce oxidant stress and protect cells from oxidative stress damage.

6.6.1.2 Methylglyoxal Detoxification

During glycolysis, methylglyoxal is a harmful byproduct, so it accumulates during times of stress and poses a serious threat to cellular integrity. AgNPs were shown to aid the detoxification of methylglyoxal and improve protection from its detrimental effect on plant cells. It is important for this detoxification mechanism in order to maintain metabolic balance within stressful conditions (Abasi et al. 2022).

6.6.1.3 Enhanced Nutrient Uptake

AgNPs lead to greatly increased nutrient absorption via the promotion of root development and an increase in the amount of surface area for nutrient uptake. It is the increased availability of such basic nutrients, in particular nitrogen and phosphorus, that helps plant development and vigor and boosts their ability to resist stress factors (Rahman et al. 2023).

6.6.1.4 Gene Expression Regulation

AgNPs can change the expression of genes involved in stress response mechanisms. This regulation improves the plants' responsiveness to environmental adversities allowing the plant to survive and prosper in these less favorable conditions (Abasi et al. 2022).

6.6.1.5 Physiological Enhancements

The ability to increase chlorophyll levels and to improve photosynthetic efficiency has been shown with AgNPs. Such enhancement helps both plants to keep energy production even in the most challenging conditions and helps plant development and growth (Dilnawaz, Kalaji, and Misra 2023).

6.6.1.6 Hormonal Regulation

However, AgNPs affect plant hormone levels, especially abscisic acid (ABA), which is known to be responsible for stress responses. In drought-like conditions, high levels of ABA promote the closure of the stomata decreasing water loss. Moreover, AgNPs activate a stress response similar to stress, even in the absence of classic environmental challenges, increasing stress tolerance and ABA synthesis (Alabdallah and Hasan 2021). The ability of this hormonal modulation to seed physiological equilibrium and the bolstering of resistance to abiotic stressors is required.

6.6.2 Mitigation of Abiotic Stresses by AgNPs

6.6.2.1 Drought Stress Mitigation

AgNPs enhanced plants' resistance to drought, as well as antioxidant defenses. When plants go through drought, they build up ROS and that causes oxidative stress. The activity of antioxidant enzymes SOD and CAT is stimulated by AgNPs and reduces oxidative damage and scavenges for ROS. This research shows that AgNP could be applied to drought-stressed wheat and rice crops to improve water use efficiency (Jhanzab et al. 2015; Iqbal et al. 2019; Abasi et al. 2022).

6.6.2.2 Salinity Stress Alleviation

Plant development is adversely influenced by salinity, which is a critical abiotic stressor. AgNPs are found to enhance seed germination and root development under conditions of high salt concentration. AgNPs stimulate root development and increase the absorption surface area. It is being researched that low concentrations of AgNPs can counteract the effects of high salinity on plants by boosting osmotic adjustment and, hence, turgor pressure as well as cellular functions (Kumari et al. 2022). This example occurs in the species *Lathyrus sativus*, where seed germination and root elongation increase while high salt stress is present; however, the germination is improved and root elongation enhanced with AgNPs' application. The observation highlights the potential of AgNPs to be strong biostimulants (Thabet and Alqudah 2024).

6.6.2.3 Heavy Metal Stress Reduction

Additionally, AgNP has a vital role in helping heavy metals damage plants. To do this, they elevate the expression of the metallic reducing proteins, metallothioneins, which are proteins that bind heavy metals and sequester them away from their damaging effects (Nayak et al. 2022). In addition, AgNPs improve the overall health of plants subjected to heavy metal exposure through increasing growth and development and increasing antioxidant capacity in response to plant stress induced by metals.

6.7 Promotion of Photosynthesis and Biomass Accumulation

AgNPs have shown considerable effectiveness in improving photosynthesis and promoting biomass accumulation in plants. Their unique properties enable them to engage with plant systems in ways that enhance both photosynthetic performance and overall growth (Tripathi et al. 2024).

6.7.1 Mechanisms of AgNPs-mediated Photosynthesis Promotion

6.7.1.1 Improved Chlorophyll Content

The presence of AgNPs has been shown to enhance the concentration of photosynthetic pigments in plants, including chlorophyll and carotenoids. A foliar application of AgNPs at concentrations between 20 and 40 parts per million (ppm) significantly boosts chlorophyll levels in wheat and fenugreek plants when compared to untreated controls (Prasad 2014; Abasi et al. 2022).

6.7.1.2 Enhanced Photosynthetic Efficiency

The incorporation of AgNPs into photosynthetic systems found in plants can possibly increase the efficiency of plants. These NPs can influence the change in the activity of genes that have a bearing on photosynthesis and the improvement of enzymes that fix carbon. As a result, there is enhanced light absorption and enhanced power generation as compared to other methods (Ansari et al. 2023).

6.7.1.3 Increased Nutrient Uptake

AgNPs accelerate nutrient uptake at root through rootlet proliferation and eventual increase in the absorption area of roots. Increase in photosynthesis and biomass accumulation in the improved accessibility of essential nutrients, e.g. nitrogen and phosphorus (Mehmood 2018).

6.7.1.4 Impact on Stomatal Conductance

Stomatal activity is strongly regulated in plants to maintain water balance (via transpiration), and the possibility of AgNPs impacting gas exchange via stomata poses regulatory challenges for the whole of the internal environment. When the heat stress conditions were applied on wheat plants, it was observed from the research that treatment of AgNPs at a concentration of 50 mg/L enhanced the stomatal conductance considerably when compared to the control group. Due to this improvement in stomatal performance, the plants can still accomplish photosynthesis and growth even if the stress they are under (Jhanzab et al. 2015; Iqbal et al. 2019).

6.7.2 Promotion of Biomass Accumulation

6.7.2.1 Root Growth Promotion

By promoting root growth and development, AgNPs have been shown to enhance anchoring as well as absorption of nutrients. It also increases the biomass of the whole plant (Singh et al. 2024).

6.7.2.2 Shoot Growth Enhancement

AgNPs change the behavior of stomata, which plays an important role in gas exchange and water balance in plants. A study on the wheat plants exposed to the heat stress plots that AgNP application at the concentration of 50 mg/L increased the stomatal conductance significantly in comparison with the control group. Increase in this stomatal activity enables the plant to carry out photosynthesis and growth even under conditions of stress (Prasad 2014; Jhanzab et al. 2015).

6.7.2.3 Stress Tolerance Improvement

In addition to current challenges associated with drought, salinity, and toxicity triggered by heavy metals, AgNPs may also aid in enhancing plants' stress resistance to diverse abiotic stressors. Using these NPs, plant growth and productivity under stressful conditions are encouraged by reduced oxidative stress and elevated nutrient uptake (Guzmán-Báez et al. 2021).

6.7.2.4 Influence of AgNP Treatment on Crop Yield

AgNPs have a major impact on crop yield. Several studies have demonstrated that AgNPs significantly improve yield components as evidenced by seed weight and fruit sizes. For instance, research in mung beans demonstrated that AgNP treatment promoted development and better photosynthetic efficiency and yielded increased yields. Just as applications of AgNP concentrations equivalent to 25–50 mg/L to wheat increased chlorophyll levels and plant stature, thereby increasing the capacity for higher yields (Mehmood 2018; Wasaya et al. 2020). Secondly, the use of AgNPs has been associated with higher nutrient uptake, specifically potassium, phosphorus, and nitrogen, which are essential for good plant growth and yield achievement (Guzmán-Báez et al. 2021).

6.8 Root Development and Soil Interaction

Across a number of plant species, AgNPs have been shown to affect root development and soil interaction positively. Research reveals that minimal AgNP concentrations can promote root growth by enabling cell division in the root meristem. Also, AgNPs may increase nutrient absorption and soil quality to increase agricultural productivity (Syu et al. 2014). However, the influence of AgNPs on plant is still variable, and it is necessary to appropriately utilize their impact on the farming sector.

6.8.1 Promotion of Root Growth and Development by AgNPs

In recent years, much interest has been shown in the ability of AgNPs to promote plant growth and its development. One aspect that is so significant about this is that their root growth could be enhanced. Therefore, any enhancement of root development can allow the plant to grow better and have a higher yield. The results of research findings explain that AgNPs can stimulate root development through several mechanisms. Modulating the levels of phytohormones, especially auxins, is one primary way for regulating the differentiation and elongation of roots (Guzmán-Báez et al. 2021). AgNP exposure increases auxin activity and longer and stronger roots (Tripathi and Singh 2016). In the case of crops (maize, wheat), they have successfully shown significant increases in root length and biomass due to treatment of AgNPs indicating their great potential as effective growth enhancers (Sharma et al. 2012). Moreover, it was known that AgNPs can play a role in upregulating root growth and development-related genes. Our results demonstrate that AgNPs upregulate genes that are involved in cell wall loosening and expansion, including genes encoding expansins and xyloglucan endotransglycosylases. This gene modulation helps root growth, such as elongation and branching (Zhang et al. 2024), suggesting that AgNPs contribute to the improvement of root growth in a molecular way.

6.8.2 Effects of AgNPs on Root Architecture, Root Surface Area, and Nutrient Uptake

The plant root's ability to fastidiously search soil in search of water and nutrients is determined by the structural configuration of these plants. When AgNPs are present, root architecture is significantly modified, giving roots a morphology of a more extensive and branched root system. Advantages that they offer are they increase the root surface area that ultimately increases the plant's ability to take in the needed water and nutrients (Rizwan et al. 2017).

Besides, plants that were treated with AgNPs have a higher number of lateral roots and increase root surface area with the soil to absorb more water and more nutrients. According to Gogos et al. (2012), root hairs and lateral roots are used to allow a plant to explore more volume of the soil. In that situation, it is critical for plants to be able to take up what they need to live just as acutely, and extra lateral roots and root hairs are especially helpful.

Additionally, essential nutrients are taken up better with AgNPs and root morphology changed. The results of the studies were that the absorption by macronutrients (nitrogen [N], phosphorus [P], and potassium [K]) was higher with the effect of AgNP-treated plants. These macronutrients are extremely critical for function in a number of physiological processes (including biosynthesis, energy storage and transfer, and protein synthesis). Changes in AgNPs could increase nutrient availability via modification of the chemistry of the soil (Pérez-de-Luque 2017) and also presumably enlarge root surface area.

In addition, AgNPs are also able to alter the uptake of metals, such as iron (Fe), zinc (Zn), and copper (Cu), which are vital for enzymatic functions and plant health (Dimkpa and Bindraban 2018). This would lead to higher plant uptake of these micronutrients, more plant growth and biomass higher biomass production, and stress resilience.

6.8.3 Interactions Between AgNPs and Soil Components Affecting Plant Growth

The interaction between AgNPs and soil components is complex and leads to both positive and negative plant development impacts. Understanding these dynamics is critical to optimize the use of AgNPs in agricultural environments. The availability of micronutrients is greatly improved. Silver ions (Ag^{+}) released from AgNPs in soil interaction with organic matter and soil particles remove nutrients from the soil (Ma et al. 2015). This interaction can improve the bioavailability of nutrients by making the nutrients easier to reach the roots of the plants primarily. For instance, AgNPs could displace iron from soil colloids making it available for plant absorption, particularly useful in iron-deficient soils (Cruz et al. 2021). However, there is a major concern regarding the AgNPs toxicity to beneficial soil microorganisms. In general, these microorganisms are important in the cycling of nutrients, decomposition of organic matter, and maintaining soil structure (Przemieniecki et al. 2024). If AgNPs are present in elevated concentrations, these underlying soil functions could be disturbed as soil microbes are suppressed by the NPs. As shown by Grün et al. (2019), the research has indicated that the AgNPs can lower the numbers of beneficial microbes, e.g. nitrogen-fixing bacteria and mycorrhizal fungi, and can cause detrimental effects on plant growth. An additional consideration is the aggregation of AgNPs within the soil. NP aggregate may reduce the efficacy of enhancing root growth and nutrient uptake. The stability and interactions of AgNPs with soil components are determined mostly by their size and surface chemistry (Sharma and Kumar 2024). Maximizing positive implications while minimizing negative effects from AgNPs therefore requires refining the formulation and application technique of the particles.

6.9 Sustainable Agriculture Applications

AgNPs have recently gained special attention for their sustainable application in agriculture for raising agricultural yield and improving resistance to environmental stress. Their physicochemical character makes them highly efficient biostimulants able to manipulate plant development and to increase crop productivity.

6.9.1 Potential Benefits and Challenges of Integrating AgNPs Into Agricultural Practices

6.9.1.1 Potential Benefits

Enhanced crop growth: Several studies have indicated that AgNPs could enhance the growth parameters, such as chlorophyll content, root and shoot biomass, and seed germination (Khan et al. 2023). For example, we find that AgNPs can have a notable effect on increasing levels of nitrogen, phosphorus, and potassium in plants, thereby improving overall photosynthesis and growth efficiency (Guzmán-Báez et al. 2021).

Tolerance of abiotic stresses: AgNPs are shown to impart enhanced drought, salinity, and heavy metal toxicities to plants. In unsuitable circumstances, they counteract oxidative stress, and this strengthens the antioxidant defenses, so they promote cellular vitality. As a result, this enhanced resilience translates into better productivity under favorable growing conditions (Kumari et al. 2022).

Antimicrobial properties: AgNPs possess antimicrobial properties that could intensely address plant diseases controlled by bacterial and fungal pathogens. One of the characteristics it bears is the need to reduce dependence on chemical pesticides and advocate for more sustainable approaches to agriculture (Khan et al. 2023).

6.9.1.2 Challenges

Toxicity concerns: The different concentrations of AgNPs can inhibit plant growth and induce phytotoxic effects at elevated levels and in some instances provide beneficial effects at lower concentrations. The influence of the size of the NPs and application techniques is both highly dependent on the particular plant species and highly dependent on the application technique. Optimal concentrations for a variety of crops are still a hard task (Kale, Parishwad, and Patil 2021).

Environmental impact: Introducing AgNPs into the environment is also apprehensive and concerns are raised whether AgNPs could be accumulated in soil and in aquatic ecosystems. Accumulation of this type could harm non-target organisms and disturb ecological balance. It is important to be able to understand the long-term environmental implications involved in that use (Ferdous and Nemmar 2020).

Regulatory and economic barriers: Regulatory challenges with respect to environmental effects and safety evaluations of AgNPs in integrated agricultural practices exist. In addition, the production and use of these devices are quite costly and may be beyond even widespread acceptance by farmers (Mishra et al. 2017).

6.9.2 Considerations for the Safe and Responsible Use of AgNPs in Crop Production

Optimizing application methods: AgNPs' effects on plants are influenced by the method of application. Previous studies have shown that for certain crops (e.g. oil seed rape, sunflower, and rice), it is possible to significantly increase germination, growth, and yield through the use of low concentrations (5–60 mg/L) of AgNPs in seed priming (Nile et al. 2022). On the other hand, administration of higher amounts (100–1000 mg/L or application on soil) may induce phytotoxic effects and tissue accumulation of harmful substances in plant tissues (Mehmood 2018). Since the AgNPs are directly exposed to the environment, it is necessary to optimize the foliar spraying of AgNPs.

Determining safe concentrations: Typically, AgNPs are dose-dependent, where lower concentrations encourage growth and higher concentrations prevent it. For example, AgNPs increased nutrient uptake and root growth of tomato seedlings at 25 mg/L concentration, whereas increased concentrations caused damage (Ansari et al. 2023). However, safe concentration limits for several crops and cultivars are necessary to define optimal benefits and reduce potential risks.

Assessing environmental fate and impacts: Discharge of AgNPs into the ecosystem raises concerns about their persistence, bioaccumulation (exposure in sufficient amounts), and impact on non-target organisms. Other research found that AgNPs have been found in reclaimed water, sewage sludge, as well as in agricultural soils (Yan and Chen 2019). They need to be evaluated for their long-term ecological implications (i.e. risks) and should be incorporated into broader risk assessments that can help inform mitigation strategies.

Promoting sustainable and responsible use: The integration of AgNPs into sustainable agricultural practices needs to be done in a responsible manner. This includes the use of an eco-friendly synthesis method, the minimized environmental dispersion of the AgNPs, and formulation with the potential to undergo natural degradation (Mittal et al. 2020).

Research, policies, and guidelines for the safe application of AgNPs in agriculture need to be established by researchers, policymakers, and stakeholders for better results.

6.10 Conclusion

The incorporation of AgNPs into agricultural methodologies may offer a promising strategy for improving crop output, increasing nutrient uptake, as well as enhancing resilience against environmental stressors. The physicochemical properties of AgNPs can lead to beneficial interactions with plant systems that may enhance agricultural performance and long-term ecological sustainability. AgNPs offer a flexible and optimistic path forward for Agronomy 4.0, taking us toward improved plant production and the sustainability of the agricultural sector as a whole. Not only do they improve the absorption and reduce the stress on already stressed nutrients but also they contribute to benefitting the soil, which in turn benefits environmentally sustainable practices. They have proved themselves as invaluable tools in the struggle against the pressure that modern agriculture finds itself under, such as the effects of the environment and climate change. The safe integration of AgNPs into agricultural practices, though, hinges on how application methods can be refined, what concentration levels are safe, and how ecology is affected long term. These aspects are addressed by AgNPs that can be useful to promote sustainable agricultural development, to increase food security and increase the resilience of agriculture against environmental challenges. Therefore, AgNPs can be a promising alternative for sustainable agricultural practices promoting crop growth, enhancing yield, and bearing resilience to environmental adversities as well as eco-friendly practices.

References

Abasi, F., Raja, N.I., Mashwani, Z.U.R. et al. (2022). Biogenic silver nanoparticles as a stress alleviator in plants: a mechanistic overview. *Molecules* 27 (11): 3378.

Acharya, P., Jayaprakasha, G.K., Crosby, K.M. et al. (2020). Nanoparticle-mediated seed priming improves germination, growth, yield, and quality of watermelons (*Citrullus lanatus*) at multi-locations in Texas. *Scientific Reports* 10 (1): 5037.

Ahire, S.A., Bachhav, A.A., Pawar, T.B. et al. (2022). The augmentation of nanotechnology era: a concise review on fundamental concepts of nanotechnology and applications in material science and technology. *Results in Chemistry* 4: 100633.

Alabdallah, N.M. and Hasan, M.M. (2021). Plant-based green synthesis of silver nanoparticles and its effective role in abiotic stress tolerance in crop plants. *Saudi Journal of Biological Sciences* 28 (10): 5631–5639.

Ali, S., Mehmood, A., and Khan, N. (2021). Uptake, translocation, and consequences of nanomaterials on plant growth and stress adaptation. *Journal of Nanomaterials* 2021 (1): 6677616.

Altammar, K.A. (2023). A review on nanoparticles: characteristics, synthesis, applications, and challenges. *Frontiers in Microbiology* 14: 1155622.

Ansari, M., Ahmed, S., Abbasi, A. et al. (2023). Plant mediated fabrication of silver nanoparticles, process optimization, and impact on tomato plant. *Scientific Reports* 13 (1): 18048.

Atapattu, A.J., Ranasinghe, C.S., Nuwarapaksha, T.D. et al. (2024). Sustainable agriculture and sustainable development goals (SDGs). In: *Emerging Technologies and Marketing Strategies for Sustainable Agriculture*, 1–27. Hershey, PA: IGI Global.

Awashra, M. and Młynarz, P. (2023). The toxicity of nanoparticles and their interaction with cells: an in vitro metabolomic perspective. *Nanoscale Advances* 5 (10): 2674–2723.

Bruna, T., Maldonado-Bravo, F., Jara, P. et al. (2021). Silver nanoparticles and their antibacterial applications. *International Journal of Molecular Sciences* 22 (13): 7202.

Chandrashekar, H.K., Singh, G., Kaniyassery, A. et al. (2023). Nanoparticle-mediated amelioration of drought stress in plants: a systematic review. *3 Biotech* 13 (10): 336.

Chaudhary, A., Kumar, T., Kumar, A. et al. (2023). A review on general concept and preparation methods together with characterization techniques of silver nanoparticles. *Research Journal of Pharmacy and Technology* 16 (6): 2819–2824.

Clogston, J.D. and Patri, A.K. (2011). Zeta potential measurement. In: *Characterization of Nanoparticles Intended for Drug Delivery* (ed. S. McNeil), 63–70. Humana Press.

Cruz, N.C., Farto, M., Mourinha, C. et al. (2021). Dissolution of Ag nanoparticles in agricultural soils and effects on soil exoenzyme activities. *Environments* 8 (3): 22.

Dhaka, A., Mali, S.C., Sharma, S. et al. (2023). A review on biological synthesis of silver nanoparticles and their potential applications. *Results in Chemistry* 6: 101108.

Dilnawaz, F., Kalaji, M.H., and Misra, A.N. (2023). Nanotechnology in improving photosynthesis under adverse climatic conditions: cell to canopy action. *Plant Nano Biology* 4: 100035.

Dimkpa, C.O. and Bindraban, P.S. (2017). Nanofertilizers: new products for the industry? *Journal of Agricultural and Food Chemistry* 66 (26): 6462–6473.

Duhan, J.S., Kumar, R., Kumar, N. et al. (2017). Nanotechnology: the new perspective in precision agriculture. *Biotechnology Reports* 15: 11–23.

Duman, H., Eker, F., Akdaşçi, E. et al. (2024). Silver nanoparticles: a comprehensive review of synthesis methods and chemical and physical properties. *Nanomaterials* 14 (18): 1527.

Elemike, E.E., Uzoh, I.M., Onwudiwe, D.C. et al. (2019). The role of nanotechnology in the fortification of plant nutrients and improvement of crop production. *Applied Sciences* 9 (3): 499.

Fahim, M., Shahzaib, A., Nishat, N. et al. (2024). Green synthesis of silver nanoparticles: a comprehensive review of methods, influencing factors, and applications. *JCIS Open* 16: 100125.

Fan, X., Yahia, L.H., and Sacher, E. (2021). Antimicrobial properties of the Ag, Cu nanoparticle system. *Biology* 10 (2): 137.

Ferdous, Z. and Nemmar, A. (2020). Health impact of silver nanoparticles: a review of the biodistribution and toxicity following various routes of exposure. *International Journal of Molecular Sciences* 21 (7): 2375.

Francis, D.V., Abdalla, A.K., Mahakham, W. et al. (2024). Interaction of plants and metal nanoparticles: exploring its molecular mechanisms for sustainable agriculture and crop improvement. *Environment International* 190: 108859.

Galdiero, S., Falanga, A., Vitiello, M. et al. (2011). Silver nanoparticles as potential antiviral agents. *Molecules* 16 (10): 8894–8918.

Gogos, A., Knauer, K., and Bucheli, T.D. (2012). Agricultural applications of nanotechnology: a review. *Journal of Nanobiotechnology* 10: 9781–9792.

Grün, A.L., Manz, W., Kohl, Y.L. et al. (2019). Impact of silver nanoparticles (AgNP) on soil microbial community depending on functionalization, concentration, exposure time, and soil texture. *Environmental Sciences Europe* 31 (1): 1–22.

Guzmán-Báez, G.A., Trejo-Téllez, L.I., Ramírez-Olvera, S.M. et al. (2021). Silver nanoparticles increase nitrogen, phosphorus, and potassium concentrations in leaves and stimulate root length and number of roots in tomato seedlings in a hormetic manner. *Dose-Response* 19 (4): 15593258211044576.

Hänsch, M. and Emmerling, C. (2010). Effects of silver nanoparticles on the microbiota and enzyme activity in soil. *Journal of Plant Nutrition and Soil Science* 173 (4): 554–558.

Hosseingholian, A., Gohari, S.D., Feirahi, F. et al. (2023). Recent advances in green synthesized nanoparticles: from production to application. *Materials Today Sustainability* 24: 100500.

Iavicoli, I., Leso, V., Beezhold, D.H. et al. (2017). Nanotechnology in agriculture: opportunities, toxicological implications, and occupational risks. *Toxicology and Applied Pharmacology* 329: 96–111.

Iqbal, M., Raja, N.I., Mashwani, Z.U.R. et al. (2019). Assessment of AgNPs exposure on physiological and biochemical changes and antioxidative defence system in wheat (*Triticum aestivum* L) under heat stress. *IET Nanobiotechnology* 13 (2): 230–236.

Iravani, S., Korbekandi, H., Mirmohammadi, S.V. et al. (2014). Synthesis of silver nanoparticles: chemical, physical and biological methods. *Research in Pharmaceutical Sciences* 9 (6): 385–406.

Jadczak, P., Kulpa, D., Drozd, R. et al. (2020). Effect of AuNPs and AgNPs on the antioxidant system and antioxidant activity of lavender (*Lavandula angustifolia* Mill.) from in vitro cultures. *Molecules* 25 (23): 5511.

Jhanzab, H.M., Razzaq, A., Jilani, G. et al. (2015). Silver nano-particles enhance the growth, yield and nutrient use efficiency of wheat. *International Journal of Agronomy & Agricultural Research* 7 (1): 15–22.

Jo, Y.K., Kim, B.H., and Jung, G. (2009). Antifungal activity of silver ions and nanoparticles on phytopathogenic fungi. *Plant Disease* 93 (10): 1037–1043.

Junedi, M.A., Mukhopadhyay, R., and Manjari, K.S. (2023). Alleviating salinity stress in crop plants using new engineered nanoparticles (ENPs). *Plant Stress* 9: 100184.

Kah, M., Beulke, S., Tiede, K. et al. (2013). Nanopesticides: state of knowledge, environmental fate, and exposure modeling. *Critical Reviews in Environmental Science and Technology* 43 (16): 1823–1867. doi: 10.1080/10643389.2012.671750.

Kale, S.K., Parishwad, G.V., and Patil, A.S.H.A.S. (2021). Emerging agriculture applications of silver nanoparticles. *ES Food & Agroforestry* 3: 17–22.

Khan, S., Zahoor, M., Khan, R.S. et al. (2023). The impact of silver nanoparticles on the growth of plants: the agriculture applications. *Heliyon* 9 (6): e16928.

Khot, L.R., Sankaran, S., Maja, J.M. et al. (2012). Applications of nanomaterials in agricultural production and crop protection: a review. *Crop Protection* 35: 64–70. doi: 10.1016/j.cropro.2012.01.007.

Kim, D.Y., Patel, S.K., Rasool, K. et al. (2023). Bioinspired silver nanoparticle-based nanocomposites for effective control of plant pathogens: a review. *Science of the Total Environment* 908: 168318.

Kim, J.S., Kuk, E., Yu, K.N. et al. (2007). Antimicrobial effects of silver nanoparticles. *Nanomedicine: Nanotechnology, Biology and Medicine* 3 (1): 95–101.

Krishnaraj, C., Ramachandran, R., Mohan, K. et al. (2012). Optimization for rapid synthesis of silver nanoparticles and its effect on phytopathogenic fungi. *Spectrochimica Acta Part A: Molecular and Biomolecular Spectroscopy* 93: 95–99.

Kumari, S., Khanna, R.R., Nazir, F. et al. (2022). Bio-synthesized nanoparticles in developing plant abiotic stress resilience: a new boon for sustainable approach. *International Journal of Molecular Sciences* 23 (8): 4452.

Lieu, D.M., Dang, T.K.T., and Nguyen, T.H. (2024). Green synthesized silver nanoparticles, a sustainable approach for fruit and vegetable preservation: an overview. *Food Chemistry: X* 23 (9): 101664.

Lok, C.N., Ho, C.M., Chen, R. et al. (2006). Proteomic analysis of the mode of antibacterial action of silver nanoparticles. *Journal of Proteome Research* 5 (4): 916–924.

Lung, J.K., Huang, J.C., Tien, D.C. et al. (2007). Preparation of gold nanoparticles by arc discharge in water. *Journal of Alloys and Compounds* 434: 655–658.

Ma, C., White, J.C., Dhankher, O.P. et al. (2015). Metal-based nanotoxicity and detoxification pathways in higher plants. *Environmental Science & Technology* 49 (12): 7109–7122.

Mahakham, W., Sarmah, A.K., Maensiri, S. et al. (2017). Nanopriming technology for enhancing germination and starch metabolism of aged rice seeds using phytosynthesized silver nanoparticles. *Scientific Reports* 7 (1): 8263.

Malatesta, M. (2021). Transmission electron microscopy as a powerful tool to investigate the interaction of nanoparticles with subcellular structures. *International Journal of Molecular Sciences* 22 (23): 12789.

Malik, M.A., Wani, M.Y., and Hashim, M.A. (2012). Microemulsion method: a novel route to synthesize organic and inorganic nanomaterials: 1st nano update. *Arabian Journal of Chemistry* 5 (4): 397–417.

Manzoor, M.A., Xu, Y., Xu, J. et al. (2023). Nanotechnology-based approaches for promoting horticulture crop growth, antioxidant response and abiotic stresses tolerance: an overview. *Plant Stress* 11: 100337.

Mariyam, S., Upadhyay, S.K., Chakraborty, K. et al. (2023). Nanotechnology, a frontier in agricultural science, a novel approach in abiotic stress management and convergence with new age medicine: a review. *Science of the Total Environment* 912: 169097.

Mehmood, A. (2018). Brief overview of the application of silver nanoparticles to improve growth of crop plants. *IET Nanobiotechnology* 12 (6): 701–705.

Mikhailova, E.O. (2020). Silver nanoparticles: mechanism of action and probable bio-application. *Journal of Functional Biomaterials* 11 (4): 84.

Min, J.S., Kim, K.S., Kim, S.W. et al. (2009). Effects of colloidal silver nanoparticles on sclerotium-forming phytopathogenic fungi. *The Plant Pathology Journal* 25 (4): 376–380.

Mishra, S., Keswani, C., Abhilash, P.C. et al. (2017). Integrated approach of agri-nanotechnology: challenges and future trends. *Frontiers in Plant Science* 8: 471.

Misirli, G.M., Sridharan, K., and Abrantes, S.M. (2021). A review on nanostructured silver as a basic ingredient in medicine: physicochemical parameters and characterization. *Beilstein Journal of Nanotechnology* 12 (1): 440–461.

Mittal, D., Kaur, G., Singh, P. et al. (2020). Nanoparticle-based sustainable agriculture and food science: recent advances and future outlook. *Frontiers in Nanotechnology* 2: 579954.

Mohamed, A.K.S., Qayyum, M.F., Abdel-Hadi, A.M. et al. (2017). Interactive effect of salinity and silver nanoparticles on photosynthetic and biochemical parameters of wheat. *Archives of Agronomy and Soil Science* 63 (12): 1736–1747.

Mousavi, S.R. and Rezaei, M. (2011). Nanotechnology in agriculture and food production. *Journal of Applied Environmental and Biological Sciences* 1 (10): 414–419.

Nayak, S., Goveas, L.C., Kumar, P.S. et al. (2022). Plant-mediated gold and silver nanoparticles as detectors of heavy metal contamination. *Food and Chemical Toxicology* 167: 113271.

Nile, S.H., Thiruvengadam, M., Wang, Y. et al. (2022). Nano-priming as emerging seed priming technology for sustainable agriculture—recent developments and future perspectives. *Journal of Nanobiotechnology* 20 (1): 254.

Noshad, A., Hetherington, C., and Iqbal, M. (2019). Impact of AgNPs on seed germination and seedling growth: a focus study on its antibacterial potential against *Clavibacter michiganensis* subsp. *michiganensis* infection in *Solanum lycopersicum*. *Journal of Nanomaterials* 2019 (1): 6316094.

Pal, S., Tak, Y.K., and Song, J.M. (2007). Does the antibacterial activity of silver nanoparticles depend on the shape of the nanoparticle? A study of the gram-negative bacterium *Escherichia coli*. *Applied and Environmental Microbiology* 73 (6): 1712–1720.

Pérez-de-Luque, A. (2017). Interaction of nanomaterials with plants: what do we need for real applications in agriculture? *Frontiers in Environmental Science* 5: 12.

Prasad, R. (2014). Synthesis of silver nanoparticles in photosynthetic plants. *Journal of Nanoparticles* 2014 (1): 963961.

Prasad, R., Kumar, V., and Prasad, K.S. (2014). Nanotechnology in sustainable agriculture: present concerns and future aspects. *African Journal of Biotechnology* 13 (6): 705–713.

Prażak, R., Święciło, A., Krzepiłko, A. et al. (2020). Impact of Ag nanoparticles on seed germination and seedling growth of green beans in normal and chill temperatures. *Agriculture* 10 (8): 312.

Przemieniecki, S.W., Ruraż, K., Kosewska, O. et al. (2024). The impact of various forms of silver nanoparticles on the rhizosphere of wheat (*Triticum aestivum* L.)–shifts in microbiome structure and predicted microbial metabolic functions. *Science of the Total Environment* 914: 169824.

Rafique, M., Sadaf, I., Rafique, M.S. et al. (2017). Silver nanoparticles: synthesis, properties, toxicology, applications, and perspectives. *Nanotechnology, Science and Applications* 10: 1–16.

Rahman, M.S., Chakraborty, A., Kibria, A. et al. (2023). Effects of silver nanoparticles on seed germination and growth performance of pea (*Pisum sativum*). *Plant Nano Biology* 5: 100042.

Rai, M., Yadav, A., and Gade, A. (2009). Silver nanoparticles as a new generation of antimicrobials. *Biotechnology Advances* 27 (1): 76–83.

Rai, M.K., Deshmukh, S.D., Ingle, A.P. et al. (2012). Silver nanoparticles: the powerful nanoweapon against multidrug-resistant bacteria. *Journal of Applied Microbiology* 112 (5): 841–852.

Rashid, A., Schutte, B.J., Ulery, A. et al. (2023). Heavy metal contamination in agricultural soil: environmental pollutants affecting crop health. *Agronomy* 13 (6): 1521.

Rizwan, M., Ali, S., Qayyum, M.F. et al. (2017). Effect of metal and metal oxide nanoparticles on growth and physiology of globally important food crops: a critical review. *Journal of Hazardous Materials* 322: 2–16.

Saha, B., Biswas, S., Datta, S. et al. (2024). Sustainable nano solutions for global food security and biotic stress management. *Plant Nano Biology* 9: 100090.

Salem, S.S. and Fouda, A. (2021). Green synthesis of metallic nanoparticles and their prospective biotechnological applications: an overview. *Biological Trace Element Research* 199 (1): 344–370.

Saritha, G.N.G., Anju, T., and Kumar, A. (2022). Nanotechnology-big impact: how nanotechnology is changing the future of agriculture? *Journal of Agriculture and Food Research* 10: 100457.

Sembada, A.A. and Lenggoro, I.W. (2024). Transport of nanoparticles into plants and their detection methods. *Nanomaterials* 14 (2): 131.

Sharma, N.K., Vishwakarma, J., Rai, S. et al. (2022). Green route synthesis and characterization techniques of silver nanoparticles and their biological adeptness. *ACS Omega* 7 (31): 27004–27020.

Sharma, P., Bhatt, D., Zaidi, M.G.H. et al. (2012). Silver nanoparticle-mediated enhancement in growth and antioxidant status of *Brassica juncea*. *Applied Biochemistry and Biotechnology* 167: 2225–2233.

Sharma, R. and Kumar, V. (2024). Nano enabled agriculture for sustainable soil. *Waste Management Bulletin* 2: 152–161.

Sharma, V.K., Yngard, R.A., and Lin, Y. (2009). Synthesis and applications of silver nanoparticles. *Journal of Colloid and Interface Science* 1: 1–18.

Singh, A., Sharma, A., Singh, O. et al. (2024). In-depth exploration of nanoparticles for enhanced nutrient use efficiency and abiotic stresses management: present insights and future horizons. *Plant Stress* 14: 100576.

Sondi, I. and Salopek-Sondi, B. (2004). Silver nanoparticles as antimicrobial agent: a case study on *E. coli* as a model for gram-negative bacteria. *Journal of Colloid and Interface Science* 275 (1): 177–182.

Syu, Y.Y., Hung, J.H., Chen, J.C. et al. (2014). Impacts of size and shape of silver nanoparticles on *Arabidopsis* plant growth and gene expression. *Plant Physiology and Biochemistry* 83: 57–64.

Szczyglewska, P., Feliczak-Guzik, A., and Nowak, I. (2023). Nanotechnology–general aspects: a chemical reduction approach to the synthesis of nanoparticles. *Molecules* 28 (13): 4932.

Tang, Y., Zhao, W., Zhu, G. et al. (2023). Nano-pesticides and fertilizers: solutions for global food security. *Nanomaterials* 14 (1): 90.

Tariq, M., Mohammad, K.N., Ahmed, B. et al. (2022). Biological synthesis of silver nanoparticles and prospects in plant disease management. *Molecules* 27 (15): 4754.

Thabet, S.G. and Alqudah, A.M. (2024). Unraveling the role of nanoparticles in improving plant resilience under environmental stress condition. *Plant and Soil* 503: 313–330.

Tien, D.C., Liao, C.Y., Huang, J.C. et al. (2008). Novel technique for preparing a nano-silver water suspension by the arc-discharge method. *Reviews on Advanced Materials Science* 18 (8): 752–758.

Tripathi, D.K. and Singh, S. (2016). Silver nanoparticles alter the auxin signaling pathway in *Arabidopsis* roots. *Environmental Science & Technology Letters* 3 (8): 300–304.

Tripathi, D.K., Tripathi, A., Shweta et al. (2017). Uptake, accumulation and toxicity of silver nanoparticle in autotrophic plants, and heterotrophic microbes: a concentric review. *Frontiers in Microbiology* 8: 07.

Tripathi, S., Mahra, S., Sharma, S. et al. (2024). Interaction of silver nanoparticles with plants: a focus on the phytotoxicity, underlying mechanism, and alleviation strategies. *Plant Nano Biology* 9: 100082.

Tudi, M., Daniel Ruan, H., Wang, L. et al. (2021). Agriculture development, pesticide application and its impact on the environment. *International Journal of Environmental Research and Public Health* 18 (3): 1112.

Wasaya, A., Yasir, T.A., Sarwar, N. et al. (2020). Improving growth and yield of mungbean (*Vigna radiata* L.) through foliar application of silver and zinc nanoparticles. *Pure and Applied Biology* 9 (1): 790–797.

Yadav, A., Yadav, K., Ahmad, R. et al. (2023). Emerging frontiers in nanotechnology for precision agriculture: advancements, hurdles and prospects. *Agrochemicals* 2 (2): 220–256.

Yan, A. and Chen, Z. (2019). Impacts of silver nanoparticles on plants: a focus on the phytotoxicity and underlying mechanism. *International Journal of Molecular Sciences* 20 (5): 1003.

Zhang, H.Y. and Su, W.H. (2024). Classification, uptake, translocation, and detection methods of nanoparticles in crop plants: a review. *Environmental Science: Nano* 11 (5): 1847–1870.

Zhang, N., Sun, J., Yin, L. et al. (2023). Silver nanoparticles: from in vitro green synthesis to in vivo biological effects in plants. *Advanced Agrochem* 2 (4): 313–323.

7

Effect of Nanomaterials on the Physiological Status of Crop Plants

Akanksha Rout[1,#], *Komal Jalan*[2,#], *and Pradipta Banerjee*[3]

[1] *Department of Biological Sciences, BITS Pilani K K Birla Goa Campus, Goa, India*
[2] *Department of Botany, University of Delhi, Delhi, India*
[3] *Department of Surgery, University of Pittsburgh, Pittsburgh, PA, USA*
[#] *These authors contributed equally*

Nanotechnology in medicine is going to have a major impact on the survival of human race

– Bernard Marcus

7.1 Introduction

Recently, it has become evident from the literature studies that nanotechnology is a powerful tool that has a significant role in promoting sustainable agriculture. Nanoparticles (NPs) are those particles that have dimensions less than 100 nm and are used at the nanoscale level. They possess unique physical and chemical characteristics and a higher surface-to-volume ratio, which enhances their energy and electrical properties when compared to bulk materials. These properties allow NPs or more efficient delivery of fertilizers or pesticides into the plants, making them highly advantageous for agriculture applications (Al-Antary and Ghidan 2023). Various methods of synthesis of NPs such as physical, chemical, and biological methods offer accurate control over the composition, shape, and size of NPs, allowing their tailored application in the field of agriculture. Nanomaterials (NMs) have been characterized using advanced microscopic and spectroscopy techniques that provide detailed nanoscale imaging and measurements, ensuring that they meet the precise specifications required to optimize crop yield and maintain sustainability. The interaction between plants and NMs is relatively more complex. NPs can enhance plant growth and improve transpiration rate, chlorophyll content, photosynthetic efficiency, root and shoot growth, biomass production, nutrient availability, and seed germination. Plants can increase their resistance to biotic and abiotic stresses including drought and salinity by modifying multiple stress-related pathways (Verma, Patel, and Kushwah 2021; Yadav et al. 2024). Although fertilizers are essential for plant growth and development, they are often unavailable to the plants due to their

Nanotechnology-based Sustainable Agriculture, First Edition. Edited by Pardeep Singh et al.

leaching and photolysis. Nanotechnology addresses this problem by utilizing nanofertilizers that have controlled-release properties to deliver nutrients more effectively. This reduces chemical usage and their residues in food (Rajput et al. 2021; Tang et al. 2023). Engineered nanomaterials (ENM) interact with plants at the molecular level providing novel solutions to longstanding challenges in agriculture. These advancements open new avenues to improve crop productivity, stress resilience, and nutritional quality, which marks the beginning of a new era in precision agriculture (Kah, Tufenkji, and White 2019).

Thus, in this chapter, we explore how NMs affect the physiological welfare of crop plants in many faceted ways that can enhance food security and sustainability while addressing environmental and safety concerns. We examine the effects of NMs on genes and biochemical regulation. This shall also include gene expression and signaling pathways associated with stress response and stress-responsive enzymes. We also revisit these effects as beneficial and detrimental. Case studies done are based on studies related to growth promotion and pest control, phytotoxicity, and long-term impact on the environment. Potential future research areas are also presented taking into consideration real-world applications like nanofertilizers and precision farming. Finally, the environmental and safety concerns are discussed. The classification refers to the importance of ecotoxicology studies, the assessment of the risk, and the need for further development of the rather weak regulations regarding the use of NMs in agriculture. Figure 7.1 shows the advantageous effects observed by enhancing crop plants' physiological status.

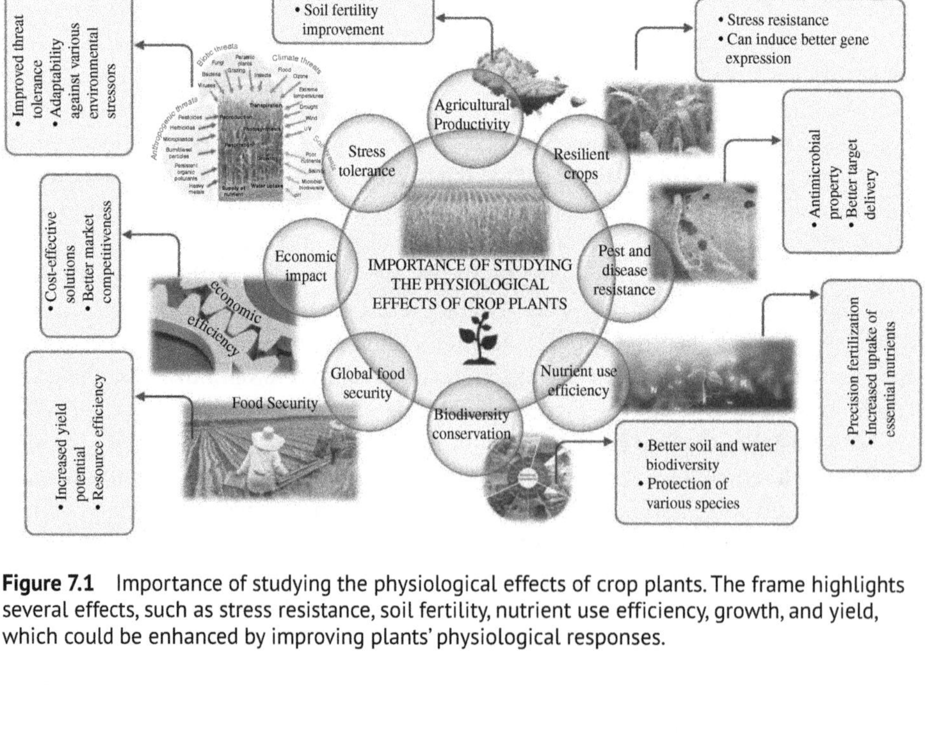

Figure 7.1 Importance of studying the physiological effects of crop plants. The frame highlights several effects, such as stress resistance, soil fertility, nutrient use efficiency, growth, and yield, which could be enhanced by improving plants' physiological responses.

7.2 Types of NPs

Based on the origin, there are three types of NPs: natural NPs, incidental NPs, and engineered NPs.

NPs are classified into three types based on their origin: natural, incidental, and engineered. Natural NPs have been present since earth's formation and are found in lunar and volcanic dust, mineral composites, and forest fire residues. Incidental NPs are developed from human activities like emission from vehicles and industrial processes. Engineered NPs (ENPs) are created for their customized application. They are further classified into several categories, which have been discussed in the following section.

Metal and metal oxide–based NPs include copper (Cu), gold (Au), iron (Fe), silver (Ag), Al_2O_3, FeO, TiO_2, CeO_2, and ZnO. AgNPs can increase the length of both roots and shoots in plants and exhibit antimicrobial activity against both Gram-positive and negative bacteria. AuNPs affect various plant processes, such as plant growth, water transport, photosynthesis, and antioxidant response. They have adaptogenic properties that enhance plants' resistance under adverse conditions. ZnO-NPs have the potential to improve soil fertility. They can do this by making essential nutrients available to the plants. TiO_2 NPs can increase nitrogen metabolism and seed germination rate. CuONPs can increase antioxidant enzyme levels, while CeO_2 NPs can increase total chlorophyll content. Al_2O_3 NPs can enhance the expression of stress-induced miRNA, and Fe_2O_3 NPs have also shown to affect photosynthesis. The significant roles of these NPs have been listed in detail in Table 7.1.

Fullerene (C_{70}), fullerol [$C_{60}(OH)_{20}$], single-walled and multi-walled carbon nanotubes (SWCNTs and MWCNTs, respectively), and single-walled carbon nanohorns (SWCNHs) are carbon-based NPs. Fullerenes are spherical and possess antioxidant properties. SWCNTs and MWCNTs can assist plants in water uptake, thereby promoting plant growth (Lahiani et al. 2013). CNTs show species-specific responses. They have increased water uptake and plant growth in wheat and maize but have inhibited root elongation in tomatoes (Mahor et al. 2021). These NPs can also show toxic effects. For instance, MWCNTs can accumulate reactive oxygen species (ROS), decrease cell proliferation, and cause apoptosis (Tan, Lin, and Fugetsu 2009; Zaytseva and Neumann 2016). Graphene has hexagonal lattice structure. Its distinctive electrical properties have the potential to promote plant growth.

Magnetic NPs are composed of metals such as cobalt (Co), iron (Fe), and nickel (Ni). They can be manipulated by magnetic fields making them suitable for precise delivery of agrochemicals due to their controlled-release capability. They influence key physiological processes, which have been discussed in detail in the following section (Ulbrich et al. 2016).

Nanopolymers also have controlled-release property, which is used for efficient delivery of nutrients and pesticides. Commonly used nanopolymers are liposomes, polyethylene glycol (PEG), and polyvinyl pyrrolidone (PVP; Yang et al. 2009). Chitosan nanopolymers have a branched structure and show high functionality. Dendrimers are nano-sized, highly branched polymers, usually symmetrical around a core, and form spherical three-dimensional (3D) structures. They can encapsulate fertilizers or pesticides to improve their effectiveness. This can reduce environmental damage, which makes them valuable in promoting sustainable agricultural practices (Mittal et al. 2020).

Quantum dots are semiconductors and have unique optical properties, such as strong fluorescence (Pang and Gong 2019). Their bright and stable fluorescence makes them ideal for

Table 7.1 Types of nanoparticles, physiological effects, and the plant species in which they are studied.

Types of NPs	Examples	Positive effects	Negative effects	References
Metal-based NPs	Ag	• Promotes growth in onion and tomato and increases PSII quantum yield in Indian mustard. • Increases length of roots and shoots and dry weight in maize, barley, and cucumber (*Cucumis sativus*). Decahedral Ag NPs showed highest root growth. • Enhances yield in wheat. • Increases antioxidant enzymes in response to ROS. • Antimicrobial activity and pathogen control. Prolong the vase life of flowers.	• At higher doses, it reduces shoot and root length in wheat, tomato, onion, mung bean, and rice. • Reduced germination in maize and barley, reduced transpiration, and total chlorophyll content in field pumpkin. • Damages cell wall in rice. • Affect plant hormones. Nitric oxide can regulate its toxic effects.	Nawaz and Bano (2020); Noori et al. (2020); Casillas-Figueroa et al. (2020); Tripathi et al. (2017); Matras et al. (2023)
	Au	• Optical properties such as surface plasmon resonance used for biosensing and imaging. • Enables accurate plant monitoring. • Increases germination rate in cucumber and lettuce. • Have positive effect on gibberellin and negative effect on ethylene. • Stimulates light absorption in soybean.	• Increase in accumulation on root surface and plant cells. • Reduces quantum yield in soybean. • Larger Au NPs reduces overall photosynthetic efficiency.	Panichikkal and Krishnankutty (2022); Barazzouk, Kamat, and Hotchandani (2005)
	ZnO	• Photocatalytic, antimicrobial property, and controls pathogens. • Enhances soil fertility. • Increases flowering and seed productivity. • Improves growth and yield in groundnut at lower doses. • Increases biomass in mung bean. • Increases micronutrients in cucumber. • Improves shoot and root growth, somatic embryogenesis, and shoot regeneration.	• Higher doses negatively regulate seed germination and root growth, biomass, photosynthetic efficiency in rice, wheat, soybean, rape seed, alfalfa, and onion. • DNA fragmentation in onion.	Sheteiwy et al. (2021); Laware and Raskar et al. (2014); Prasad et al. (2012); Rastogi et al. (2017)

TiO_2	• Increases RuBisco activase level, fresh and dry weight, nitrogen metabolism, germination rate, and PSII quantum yield in spinach. • Increases lycopene and chlorophyll content, transpiration rate, PSII quantum yield, and crop yield in tomato. Increases seed germination, root elongation, and chlorophyll level in wheat. • Increases nitrate reductase activity and seed germination in soybean. • Causes splitting of water, oxygen evolution, and photophosphorylation in mung bean.	• Reduced germination in wheat and maize. Inhibits leaf growth, affects transpiration, and decreases mitotic index. • Dose-dependent genotoxicity and ROS in onion.	Demir, Kaya, and Kaya (2014); Pakrashi et al. (2014); Raliya, Biswas, and Tarafdar (2015); Choi et al. (2021)
NiO	No significant effect.	• Decrease in the surface area of leaf. • Negative effect on photosynthetic pigments. • In tomato, promotes the release of caspase 3 proteases from mitochondria, cell death, ROS, and antioxidants formation. • Inhibition of root growth due to mitochondrial dysfunction and oxidative stress. • Translocation within root cells caused phytotoxicity.	Soares et al. (2016); Faisal et al. (2013)
CuO	• Dose-dependent increase in antioxidant enzymes and biomass in wheat. • Influences plant growth in mustard.	• Inhibition of seedling growth and reduced biomass in tomato and mung bean. • Causes phytotoxicity, for example, rice. • Affects auxin signaling. • Increases DNA damage at higher doses in radish and rye grass.	Song et al. (2016); Faraz et al. (2022)

(Continued)

Table 7.1 (Continued)

Types of NPs	Examples	Positive effects	Negative effects	References
	CeO_2	• When coated with citric acid, increases total chlorophyll level, stem length, and catalase activity. • Fruit maturation at lower doses in tomato. • Increases shoot and root length and biomass in *Coriandrum sativum*.	• Destruction of vascular bundles. • Decreased level of Zn, Mg, Fe, and P in xylem sap of Bt cotton. • Lower level of IAA and gibberellic acid in conventional cotton. • Compromises the quality of rice grain by decreasing the level of few essential components including chlorophyll level. • Decreases seed germination rate, root and shoot growth in tomato, soyabean, maize, cucumber, and alfalfa.	Nhan et al. (2015); Morales (2013); Rico et al. (2013); Wang et al. (2012)
	Al_2O_3	• Increases expression of stress-induced miRNA in tobacco. • Increases growth in lettuce. • Increases biomass, root length, number of fronds, and photosynthetic efficiency.	• Decrease in biomass, average root length and leaf number at higher doses in tobacco. Reduced plastid pigments. • Unmodified alumina NPs can retard root elongation in some species.	Burklew et al. (2012); Hayes et al. (2020)
	Fe_2O_3, Fe_3O_4	• Increases root length, photosynthesis, and seed germination index when applied in lower doses • Prevents oxidative damage in wheat, soyabean, pepper. • Increases expression of ferric reductase in Citrus sp. • Increases nitrogen fixation in soyabean. • Improves germination characteristics and biofortify wheat by increasing shoot length and iron content. • Improves germination and growth in alfalfa.	Decrease in germination index at higher doses.	Hu et al. (2017); Yang et al. (2023); Feng et al. (2022)

	CdO	Increases total amino acids at lower doses with no effect on photosynthesis.		Rastogi et al. (2017)
Carbon-based NPs	Fullerene	• Antioxidant property used for plant protection in rice and onion. Increases biomass and fruit yield in *field pumpkin.* • Reduces the accumulation of pesticides in soyabean and maize.	Reduced biomass at higher doses in soyabean.	Khodakovskaya et al. (2013); Mahor et al. (2021); Kole et al. (2013)
	SWCNTs	• Electrical conductivity, mechanical strength. • Increases germination rate, seedling growth, flowering, and fruiting in crops such as tobacco, tomato, rice, and soyabean. • Water absorption and retention in chickpea.		Lahiani et al. (2015); Singh et al. (2024)
	MWCNTs	• Electrical conductivity and mechanical strength. • Variable effects on root elongation. • Improves water uptake in rice. • Stimulates cell growth in tobacco. • Enhances seed germination in soyabean, barley, and maize and increases flower count and nutrient uptake in tomato. Enhances dehydrogenase activity and overall plant growth.	• ROS accumulation in protoplast. • Chromatin condensation in *Arabidopsis*. • Decreases the length of root in lettuce. • Possible inhibitory effects at higher concentrations, negative impact on rice plants leading to delay in flowering and reduced seed set.	Khodakovskaya et al. (2013); Demirer et al. (2018)
	Graphene	• Electrical properties. Enhances plant growth and development. • Helps in herbicide detection in wheat, maize, peanut, garlic, tomato, onion, soybean, and barley		Hu and Xianyu (2021)

(Continued)

Table 7.1 (Continued)

Types of NPs	Examples	Positive effects	Negative effects	References
Magnetic NPs	SPIO	• Superparamagnetic, biocompatible, chemically stable, and used for targeted delivery and controlled release of agrochemicals. • Promotes seed germination and pathogen resistance. Enables controlled release of agrochemicals. Increases chlorophyll content in soyabean. • Improves photosynthetic efficiency.		Ulbrich et al. (2016); Ahmad et al. (2022); Ghafariyan et al. (2013)
Nanopolymers and dendrimers	Chitosan nanopolymers and polyamidoamine (PAMAM) dendrimers	• Chitosan shows high functionality due to branched structure that allows multiple attachment sites for active molecules. • Increases efficiency of herbicide. • Can monitor plant and soil health. • Increases chlorophyll content, number of shoots, and spike length in wheat.		Yang et al 2009; Mittal et al. (2020); Ahmad et al. (2022)
Quantum Dots	Cadmium selenide/zinc sulfide (CdSe/ZnS)	• Size-dependent, fluorescent, used for imaging, nutrient uptake tracing, and monitoring nutrient management. • Helps in pathogen detection.	• Reduces root growth. • Increases ROS formation. • Decreases cell viability.	Pang and Gong (2019)
Composite NPs	Silica/inorganic (SiO_2 SiO_2/TiO_2), silica/polymer (SiO_2/PEG)	• Enhances surface area and high electrical conductivity, enhances solubility, enhances photocatalytic properties, and reduces toxicity. • Delivers genes and proteins in plants. • Used in plant genetic engineering. • Can alleviate Chromium (VI) toxicity in peas.	Toxic effect by decreasing pH.	Martin-Ortigosa et al. (2014)

studying physiological processes at the cellular level. In agriculture, quantum dots of cadmium, selenium, and zinc sulfide as well as conjugated amino acids can move through the plant vascular system (Schwab et al. 2016). They are used for imaging and tracing nutrients within plants. They can also monitor real-time interactions between plants and microorganisms.

Composite NPs consist of hybrids of different NPs or their combination with larger materials, such as core–shell NP composed of silica/inorganic, silica/polymer, or polymer/inorganic materials. They provide better solubility, high electrical conductivity, increased surface area, photocatalytic properties, simpler functionalization, and lower toxic effects in comparison to single-component nanoparticles (NPs). They can be used in plant genetic engineering mechanisms (Janczak and Aspinwall 2012; Martin-Ortigosa et al. 2014).

7.3 Synthesis and Characterization of NMs

7.3.1 Methods of Synthesis

NMs can be developed using several methods. This includes physical, chemical, and biological methods. Each method has its own beneficial and detrimental effects in relation to size control, shape, and purity. Physical methods of synthesis include laser ablation, physical vapor deposition, and laser pyrolysis (Krishnia, Thakur, and Thakur 2022). These techniques can break down bulk materials into NPs. They do this through physical or thermal processes. This method can form highly pure and crystalline NMs with minimal impurities. This makes them ideal for high-performance applications (Harish et al. 2022). Chemical methods are known for their adaptability and scalability. They include plasma-assisted chemical vapor deposition, sol–gel processes, and hydrothermal synthesis. In this method, chemical reactions convert precursor liquids or gases into NMs. This method allows accurate control of the NP morphology, structure, and surface characteristics (Baig, Kammakakam, and Falath 2021). Biological methods use microbes, plant extracts, and enzymes to develop NMs in an environment-friendly way. These NMs exhibit superior biocompatibility and reduced toxicity as compared to other methods (Kumari et al. 2023).

7.3.2 Techniques for Characterization

Characterizing NMs is important for understanding their properties. It helps to confirm their suitability for specific applications. Several sophisticated techniques are available for this purpose, including microscopic techniques, X-ray diffraction (XRD), dynamic light scattering (DLS), and spectroscopic techniques (Jagadeesh, Rangappa, and Siengchin 2024).

Scanning electron microscope (SEM) is particularly used to analyze the morphological and topographical features of the surface of NMs. Coupling SEM with energy-dispersive spectroscopy (EDS) can reveal composition of different elements in NMs. This is essential for their comprehensive characterization (Akhtar et al. 2018). Transmission electron microscope (TEM) provides atomic resolution images. These images reveal the internal structure of NMs. It is useful for visualizing the shape and size distribution of NMs. It also helps in studying the crystal nature of NMs (Kim and Hochella 2015). X-rays have been utilized to determine the crystalline arrangement and phase composition of NMs. It provides details on the arrangement of atoms within a NM. XRD is also used for monitoring the purity of phases. It can identify secondary phases or impurities in NMs to ensure the accuracy of NMs for its agricultural use (Kundu et al. 2019). DLS is used for calculating size distribution of NP in suspension using a light-scattering technique. It is utilized in measuring the hydrodynamic diameter of NP in terms of stability and the aggregation tendency (Shukla et al. 2023).

Other microscopic techniques (secondary ion mass microscopy, atomic force microscopy, scanning probe, high-resolution optical microscopy) and spectroscopic techniques and the combination of micro-spectroscopy techniques (such as vibrational microspectroscopy, Raman spectroscopy and Raman confocal microspectroscopy, Fourier transform infrared spectroscopy [FTIR] and FTIR microspectroscopy, laser-induced breakdown spectroscopy, atomic spectroscopy, photothermal/photoacoustic, μ-particle-induced X-ray emission, and synchrotron radiation) are helpful to detecting and quantifying the uptake, translocation, and accumulation of synthesized NMs. Recent literature suggests that combining multiple techniques, it is possible to have a better understanding of the interaction of NMs with crop plants (de Paiva Pinheiro et al. 2024).

7.3.3 NM–crop Plant Interaction

NM–crop plant interactions include the method of using NMs as well as their uptake by the plants, translocation to different tissues, accumulation, and distribution within plant systems.

7.3.3.1 Method of Use of NMs

Application of NMs on plants involves several critical considerations. These include determining the minimum effective dose, scalability, and accessibility of nanoformulations to reach target tissues. Plant cell walls and cell membranes can act as barriers to the movement of the NMs. To facilitate easier access, researchers have used preprocessing techniques, such as the isolation of protoplasts, which can internalize NPs like silicon NPs and polystyrene nanospheres. However, the practical use of protoplast is limited due to the excessive costs and scalability issues. The biolistic approach, which involves the physical delivery of NPs into plant tissues, also faces limitations like those of protoplast techniques, particularly in terms of cost and large-scale applicability.

NPs can be applied to the plant's aerial parts by foliar application in powdered form or as aqueous suspension using spray techniques. Foliar application of NMs involves crossing the cuticle barrier via lipophilic pathways (through cuticular waxes) or hydrophilic pathways (through polar aqueous pores and/or stomata). The most feasible pathway for NM access is through stomata having a size restriction of more than 10 nm (Eichert et al. 2008). Once applied to leaf surfaces, NPs take entry through cuts, wounds, stomata, or through the bases of epicuticular structures, such as trichomes, which can increase NP deposition on shoot surfaces. NPs then translocate to various other tissues through phloem. ENPs can adhere to plant surfaces and are absorbed through micrometer- and nanometer-sized openings. The shape, size, concentration, and surface charge of NPs all influence their uptake rate (Tarafdar et al. 2012). Repeated moistening and drying of foliar-applied solutions can improve absorption of NPs. Lipophilic NPs are taken up through a polar fluid area of the cuticle, while larger NPs are taken up through stomata and hydathodes, which are cuticle-free structures. The variation in stomatal permeability makes this pathway unpredictable. Despite their benefits, the potential risks and challenges associated with foliar application include foliar heating that may result in impaired gaseous exchange due to stomatal obstruction. NPs become prone to aggregation in aqueous media because of their small size and high surface energy, which reduces their bioavailability (Nair 2016).

Research has also shown NP entry through roots. However, more study needs to be done to clearly understand the mechanism. It has been observed that root exudates, especially mucilage secreted by root tips and hairs, can also cause NP to aggregate and clog pore channels, which hinders its entry into the plant. Alterations in zeta potentials of NPs suggest that the exudates from root can modify NP characteristics and behavior (Siddiqi and Husen 2022). The size of the NPs is especially important for their uptake through roots.

Hydroponic systems offer a promising avenue for NP application. The controlled condition of hydroponics makes it possible for constant monitoring of NP exposure. It enhances the effectiveness of experiment by minimizing extraneous interactions.

7.3.3.2 Uptake, Translocation, Accumulation, and Distribution

Uptake of NMs is affected by numerous factors, including plant species, type of the NM, and their application method. NMs can enter plants through roots or foliar application on leaves. SWCNTs have been found to penetrate tobacco cell walls and membranes and increase water channel activity, resulting in significant physiological improvements.

Research has shown that fullerene uptake is minimal in onion cells due to structural barriers, such as cell-wall pores. However, it accumulates in rice roots through mechanisms such as osmotic pressure, capillary forces, cell-wall pores, and plasmodesmata. Following uptake, NMs can be translocated through two primary routes: the apoplastic and symplastic routes. Apoplastic route involves the movement of NMs in the extracellular space, eventually traveling through the xylem vessels. Symplastic transport system involves the movement of NMs from the cytoplasm of one cell to the next cell through plasmodesmatal connections (González-Melendi et al. 2008). NMs can also accumulate in different plant tissues. In a study on pumpkins (*Cucurbita maxima*), Fe_3O_4 NPs were absorbed from the roots and translocated through the plant tissues with approximately 45.5% of the administered NPs accumulated in the roots, while only 0.6% in leaves. When the same NPs were applied to lima bean (*Phaseolus limensis*), no uptake or transport was observed (Zhu et al. 2008). Fullerene and fullerols have also been observed to be absorbed and accumulated in plants. For example, in rice (*Oryza sativa*), fullerene was absorbed by roots and transported to shoots, while MWCNTs were not (Lin et al. 2009). Research indicates that different NPs exhibit distinct uptake and translocation efficiencies. For instance, water-soluble SWCNTs were found to enter the cell wall and the cell membrane via fluidic endocytosis in Bright Yellow (BY-2) cells, while another study reported no uptake of SWCNTs by cucumber roots after 48 hours (about 4 days) of exposure (Cañas et al. 2008; Liu et al. 2009), while another study have showed that γ-Fe_2O_3NP migrates through cell wall space from the root epidermis to the endodermis and accumulates in vacuoles in case of maize, but are inefficient in translocating iron from roots to shoots (Li et al. 2016). In contrast, research on CeO_2 NP in *Phaseolus vulgaris* showed that when they are applied directly to leaves, they translocate to floral organs and accumulate in seeds (Salehi et al. 2021).

These findings highlight the complexity and variability in NP behavior within plant systems and suggest the need for further research to understand their useful impact on plant growth and development.

7.4 Physiological Effects on Crop Plants

7.4.1 Impact on Photosynthesis

Recent literature indicates that NPs significantly impact photosynthesis. They can influence chlorophyll content and improve photosynthetic efficiency.

7.4.1.1 Changes in Chlorophyll Content

Changes in chlorophyll content of plants directly affect photosynthesis. There have been studies that indicate some NPs, for example, superparamagnetic iron oxide NPs (SPIONs), TiO_2, ZnO, CuO, SiO_2, and Fe_2O_3, can increase chlorophyll levels. For example, SPIONs increase chlorophyll content in soybeans. SiO_2 and TiO_2 NPs can improve carbonic anhydrase activity, chlorophyll formation, and RuBisCo activase levels, which can increase photosynthetic rate and efficiency (Ghafariyan et al. 2013; Xie et al. 2012; Linglan et al. 2008). Giraldo et al. (2014) demonstrated that SWCNTs can serve as artificial antennae. They can capture light wavelengths far beyond their normal spectrum, such as UV rays, green light rays, and near-infrared rays, and incorporating them into isolated chloroplasts can triple photosynthetic activity and increase electron transport rates. Govorov and Carmeli (2007) proposed that metal NPs could influence energy conversion efficiency in photosynthetic systems. For instance, Au and AgNPs can interact with chlorophyll and enhance photosynthesis by developing a novel hybrid system that is expected to produce 10 times more excited electrons. Plasmon resonance and fast electron-hole separation. Small AuNPs in soybeans can increase chlorophyll fluorescence quenching, promoting efficient energy transfer and photosynthetic activity. However, larger AuNPs may absorb more chlorophyll molecules, potentially reducing overall photosynthetic efficiency if not perfectly balanced (Falco et al. 2011; Barazzouk, Kamat, and Hotchandani 2005). This protective effect indirectly supports photosynthesis under stress conditions. AgNPs have been shown to increase PSII activity in plants like *Brassica juncea* (Sharma et al. 2012). In contrast, some NPs like CeO_2, Fe_3O_4, and $CoFe_2O_4$ have been observed to reduce chlorophyll content of crops such as rice and sunflower, leading to reduction in photosynthesis. Studies have demonstrated that CeO_2 NPs can effectively decrease the chlorophyll content in rice. However, in isolated chloroplast, CeO_2 NPs have enhanced photosynthetic activity by scavenging ROS. This helps in protecting chloroplasts from ROS damage (Giraldo et al. 2014; Rico et al. 2013). High concentrations of TiO_2 can damage microalgae by reducing its chlorophyll content and affecting protein and lipid levels (Juhel et al. 2011). CuO NPs have been observed to decrease biomass and photosynthesis in duckweed (*Lemna gibba*) causing significant changes in PSII electron transport (Perreault, Samadani, and Dewez 2014). The effect of NPs on photosynthesis is highly concentration dependent. For example, smaller amounts of ferrofluid increased chlorophyll levels in maize plants, while higher concentrations led to inhibited chlorophyll a level due to localized heating of magnetic NPs in plant tissues. This results from the absorption of electromagnetic energy by magnetic NPs, which affects the redox reactions related to photosynthesis (Răcuciu, Miclăuş and Creangă 2009). Similarly, the concentration-dependent effects of AuNPs on PSII chlorophyll fluorescence quenching, for example, vary significantly with the size and concentration of the NPs; smaller AuNPs enhance photosynthetic efficiency, while larger AuNPs may lead to reduced photosynthetic efficiency (Falco et al. 2011; Barazzouk, Kamat, and Hotchandani 2005).

7.4.1.2 Effects on Photosynthetic Rate and Efficiency

NPs significantly affect the photosynthetic rate and efficiency in plants, influencing key physiological processes. NPs, such as aluminum oxide (alumina and Al_2O_3 NPs), SiO_2, and TiO_2, have been shown to enhance photosynthetic efficiency in various plants. SiO_2 NPs enhance photosynthetic performance by enhancing carbonic anhydrase activity, facilitating CO_2 supply to RuBisCo and promoting the synthesis of photosynthetic pigments, while TiO_2 NPs protect chloroplasts from senescence, stimulate chlorophyll formation, activate RuBisCo, and upregulate RuBisCo activase gene expression and protein levels. Additionally, they boost water conductance, transpiration, electron transport chain activity, PSII photoreduction, oxygen evolution, and chlorophyll photophosphorylation under both visible and ultraviolet light and the net photosynthetic rate (Sharma et al. 2012; Siddiqui et al. 2015). Magnetic NPs coated with perchloric acid slightly inhibited photosynthesis in germinated maize seedlings. At higher concentrations, NP triggered oxidative stress in leaf cells, resulting in brown spots on leaves, reduced the rate of photosynthesis, and lowered metabolic activity (Răcuciu, Miclăuş and Creangă 2009).

7.4.2 Growth and Development

7.4.2.1 Seed Germination and Root Development

NPs can have a significant effect on plant growth and development, including processes such as seed germination, root and shoot growth, and biomass production (Das et al. 2020). NP application can have both positive and negative impacts on seed germination as well and that depends on various phases of plant growth and maturation.

Metal NPs have been widely researched for their impact on seed germination and initial phases of plant development. AuNPs have been observed to improve seed germination and plant growth in lettuce, cucumber, and Indian mustard (*Brassica juncea*). AgNPs had a greater effect on root growth than leaf growth. While AgNPs enhanced germination in trumpet weed (*Eutrochium fistulosum*), they had minimal effect on other wetland species (Siddiqui et al. 2015).

SiO_2 NP has a positive effect on seed germination. They can enhance seed germination in tomatoes (*Solanum lycopersicum*) at lower concentrations and increase nutrient availability to maize seeds, leading to optimal growth conditions and better germination rates. These NPs are effective in improving seed germination in stressed conditions, such as saline stress in tomato and zucchini. Amorphous silica NPs (noncrystalline) have been safely used as nanobiopesticides and are designated harmless for humans by the World Health Organization and the US Department of Agriculture (Khan et al. 2022). Scientists have discovered that silica NP can increase lettuce (*Lactuca sativa*) germination. TiO_2 NPs can enhance nitrate reductase activity and improve nutrient uptake in crops such as soybean and spinach. In mature spinach seeds, TiO_2 has demonstrated a positive effect on seed germination and vigor when applied at proper concentration. They also promote plumule and radicle growth in the seedling of *Brassica napus*. Compared to the control, the use of nanosized TiO_2 (2–10 ppm) particles on bulk wheat plants improved germ-ination time, shoot length, and seedling growth. However, higher amounts exhibited inhibitory effects. ZnONPs have exhibited beneficial effects on seed germination in zucchini, peanuts (*Arachis hypogaea*), wheat (*Triticum aestivum*), cucumber (*Cucumis sativus*), and onions (*Allium cepa*), whereas an inhibitory effect was observed in alfalfa (*Medicago sativa*), tomato, ryegrass (*Lolium perenne*), and maize. Additionally, reduced germination index and inhibition of root growth were noted in rice and pea, as well as phytotoxic effects at higher doses in peas (Siddiqui et al. 2015; Cyriac et al. 2020). For enhancing root and shoot length in chickpea, 75% of ZnONPs have been found optimal (Dulta et al. 2021). Hydroxyapatite (HAP) nanorods at a concentration of 1 mg/mL improved chickpea germination and growth (Bala et al. 2014). It has been reported that using Fe_2O_3 NPs for seed priming improved seed germination, shoot length, and iron content in wheat (Sundaria et al. 2019). Similarly, MgONPs, prepared using marine brown algae, have been shown to enhance seed germination in green gram (Anand et al. 2020). In tobacco plants, Al_2O_3 NPs inhibit root elongation, biomass production, and leaf count. Their effects became more apparent with the increase in their concentration (Burklew et al. 2012). The genotoxic effects of CeO_2 NPs in soybeans and CuONPs in radish and ryegrass cause lesions on DNA and subsequent DNA damage (Siddiqui et al. 2015). Phytotoxicity from metal NPs is complex, often due to ion dissolution and associated toxicity.

Seed priming with NPs has recently gained attention for enhancing seed vigor and synchronization, leading to improved growth under stressful conditions. Seed nanopriming with turmeric oil nanoemulsions and AgNPs has been proven to enhance seed germination and yield. It also helped in maintaining quality of the fruit under both field and greenhouse conditions, as observed in watermelon (Acharya 2020).

MWCNTs enhance the absorption of water, calcium, and iron in barley, soybean, maize, tomato, Bt cotton (*Gossypium hirsutum*), urdbean (*Vigna mungo*), Indian mustard, and rice (*Oryza sativa*). This enhancement has a positive effect on seed germination and plant growth, and this may be attributed to upregulated gene expression related to cell division, cell wall formation, and water transport. Carbon nanohorns also help in seed germination and promote organ growth. However, little is known about it so far (Villagarcia et al. 2012; Tiwari et al. 2014; Lahiani et al. 2015). CNTs can have variable effects. For example, SWCNTs reduce biomass in zucchini, whereas MWCNTs have no such effect. MWCNTs can trigger tomato seed germination by penetrating the seed coat (Siddiqui et al. 2015). Tomato plants grown in soil, supplemented with MWCNT, produced double the flowers and fruits when compared to untreated plants (Khodakovskaya et al. 2013). MWCNTs can enhance germination and growth in crops such as corn, barley, and soybean. Its higher doses can have detrimental effects, as observed in crops such as maize, peanut, garlic, and wheat (Lahiani et al. 2013; Srivastava and Rao 2014).

7.4.2.2 Root and Shoot Growth and Biomass

Several NPs including SiO_2 and TiO_2 have demonstrated the ability to increase biomass production in plants, such as spinach and soybean, by increasing nutrient uptake and nitrate reductase activity. ZnONPs have shown significant advancements in biomass of the plant, root, and shoot growth, particularly in the rhizosphere of cluster beans (*Cyamopsis tetragonoloba*), chickpea, and mung bean. Additionally, ZnONPs have also been found to stimulate somatic embryogenesis and shoot regeneration in plantlets. However, ZnONPs, at elevated doses, have shown to retard root development in maize, cucumber, soybean, ryegrass, carrot, and cabbage (Siddiqui et al. 2015). Al_2O_3 NPs have been observed to enhance the growth of duckweed (*Lemna gibba*). This effect is due to morphological adjustments like longer roots and higher photosynthetic efficiency (Juhel et al. 2011). NiONPs suppressed root growth in tomato seedlings primarily due to oxidative stress and mitochondrial dysfunction (Faisal et al. 2013). The intracellular translocation of NiONPs within root cells has been linked with phytotoxicity, alteration in organelle structure, and elevated activity of enzymes related to oxidative stress.

MWCNTs have been demonstrated to enhance root cell elongation, dehydrogenase activity, increased biomass, flowering, and fruit yield in plants (Siddiqui et al. 2015). They can even enhance therapeutic properties in plants (Khodakovskaya et al. 2013). Studies have documented that MWCNTs inhibit growth in rice plants with specific effect on flowering and seed set when suspended in natural organic matter (Tan, Lin, and Fugetsu 2009). Fullerol NPs have shown significant increase in biomass, yield, water content, and fruit yield in bitter melon (*Momordica charantia*) by up to 128% (Kole et al. 2013).

AuNPs have been reported to increase leaf number, leaf area, plant height, and chlorophyll content, leading to greater crop yield (Gopinath et al. 2014). AgNPs increased shoot/root length and leaf area in plants like *Brassica juncea*, common bean, and maize. In barley, AgNPs promote root elongation, while in lettuce it has inhibitory effects. Decahedral AgNPs promote root growth more effectively than spherical ones. At lower concentrations, AgNPs have been toxic to seedlings of *Arabidopsis*. It did not show any significant negative effect on zucchini plants when grown hydroponically apart from reduced biomass and transpiration rate. The medium used for NP application can also influence toxicity as it has shown different effects in agar and soil cultures. CuNPs had a greater negative effect on mung bean seedlings than wheat seedlings. It also inhibited root growth in zucchini plants. Hydroxyapatite (HAP) NPs were also reported to damage mung bean cells and trigger apoptosis (Jiang et al. 2014).

7.4.3 Nutrient Uptake and Assimilation

NPs can influence both nutrient uptake and assimilation in plants. SeNPs can reduce cadmium absorption while increasing the concentration of essential minerals such as Mg, Fe, and Zn. They can also increase photosynthetic pigments in rice. NPs like AgNPs and Fe_3O_4 NPs can modify primary metabolites essential for the fundamental functions in plants. TiO_2 and AgNPs can modify secondary metabolites such as phenols, terpenes, alkaloids, and flavonoids. These changes can improve the plants' resistance to biotic and abiotic stresses, thereby playing a key role in defense mechanisms. NPs can also induce toxicity in plants through mechanisms, such as oxidative stress, metal ion release, and NP aggregation. This toxicity can interfere with metabolic processes and cause damage to cellular components like lipids and DNA. NPs by interacting with plant hormones and by stimulating the synthesis of nitric oxide can decrease oxidative damage by modulating phytohormone homeostasis and activates plant defense mechanisms as well as lowers the mobilization of harmful substances. For example, gibberellic acid has been found to mitigate the toxic effects of TiO_2 NPs, thereby improving the nutrient status growth of the plant.

NPs also impact other physiological parameters in plants. For instance, nanoscale zero-valent iron (nZVI) triggers plasma membrane H+-ATPase activity, reducing apoplastic pH, increasing leaf area, and widening stomatal pore in *Arabidopsis*. Studies have shown that the increase in stomatal aperture is driven by the upregulation of the *AHA2* gene that could potentially enhance CO_2 intake in plants. TiO_2 NPs have shown no significant effects on key physiological processes even when the NPs size varied from 14 to 655 nm in wheat, rapeseed, and *Arabidopsis*. SiO_2 NPs affect plant height, shoot, and root biomass and mineral content showing increase in Cu and Mg in shoots and Na in the roots of Bt cotton. CeO_2 NPs reduce Ca content while increasing other nutrients in soybean pods.

The use of NMs can significantly increase crop productivity by increasing seed germination, nutrient assimilation, and enzymatic activities and improving stress tolerance. These findings demonstrate the complex and wide range of effects of NPs on nutrient dynamics and physiological processes. Further research is required to fully understand optimal NP application and to develop guidelines for the safe and beneficial use of NP in agriculture.

The types of NPs and their positive and negative physiological effects along with the plant species in which these NPs have been studied are shown in Table 7.1.

7.5 Molecular and Biochemical Responses

NMs with their unique properties cause significant changes in crop plants at the physiological, biochemical, and molecular levels. Research indicates that NMs can alter gene expression profiles and influence critical signaling pathways, leading to changes in the biochemical and physiological status of crop plants.

7.5.1 Gene Expression and Signaling Pathways

Molecular alterations affect various physiological and biochemical processes in crops. Gene expression studies show that NMs can induce molecular changes distinct from those caused by standard environmental stimuli.

7.5.1.1 Changes in Gene Expression Profiles

Research has documented alterations in gene expression related to oxidative stress, including superoxide dismutase (SOD), by the exposure to NPs like Ag NPs and TiO_2 NPs. For instance, exposure to Ag NPs alters gene expression in *Arabidopsis thaliana*. More genes were found to be upregulated rather than downregulated, particularly those involved in metal and oxidative stress responses, plant defense, and hormonal regulation. Furthermore, microarray-based analyses in *A. thaliana* have revealed NP-specific changes in gene expression upon exposure to ZnO, TiO_2, and fullerene soot. For example, ZnO exposure led to the downregulation of genes involved in cell organization, such as microtubule processes, and upregulated genes are mainly associated with stress response pathways, while TiO_2 affected genes responsive to abiotic stimuli. Scientists have reported that the tomato plants treated with MWCNTs showed upregulation of stress-responsive genes and water channel genes such as *LeAqp2*. These effects were detected using sophisticated imaging techniques like photothermal and photoacoustic methods in live samples (Khodakovskaya et al. 2011).

Exposure to high concentrations of NPs, such as 1% Al_2O_3, leads to significant upregulation of microRNA (miRNA) genes like *miR395, miR397, miR398*, and *miR399* in plants. These miRNAs are linked to various stress responses, with *miR398* showing a strong correlation with SOD expression. The impact of NPs also negatively impacts plant defense mechanisms by weakening the plants' immunity resulting in higher bacterial loads. The negative impact of NPs can be lowered by using salicylic acid, a plant hormone known for its role in defense response (García-Sánchez, Bernales, and Cristobal 2015).

7.5.1.2 Key Signaling Pathways Affected

NPs can induce oxidative stress via the overproduction of ROS, triggering activation of mitogen-activated protein kinase (MAPK) signaling pathways, leading to transcriptional changes, and altered secondary metabolism. This ROS-induced signaling cascade plays an important part as it is seen to affect various metabolic pathways. The MAPK cascade is of great interest. It is seen to play a crucial role in the plant's response to both biotic and abiotic stimuli. As it is the cycle that is affecting the stimuli due to the NM exposure. Recent studies have shown that Y_2O_3 NPs at concentrations above 2 mM particularly activate the MAPK signaling pathway. It is seen to reduce the lignin synthesis-related gene expression. Also, it is seen to enhance abscisic acid (ABA) and ethylene signaling pathways in *Arabidopsis* seedlings. Studies indicate that various interconnected signaling pathways are influenced by NPs, with the MAPK cascade being central to these responses. The other notable signaling pathways affected by NM exposure is the calcium-dependent protein kinase (CDPK) cascade. Salicylic acid has been known to regulate various signaling networks involved in plant responses to NMs. Studies emphasize salicylic acid's role in generation of systemic acquired resistance (SAR) through nano-enabled applications (Yang, Zhou, and Chu 2023). Research on soybean plants, when exposed to copper hydroxide nanowires (CNWs), showed that alteration in bioaccumulation of micronutrients, such as Co, Mn, Zn, and Fe, affects nutrient signaling circuits. This indicates that CNWs can significantly alter nutrient signaling pathways in plants (Majumdar et al. 2021). Another study revealed that carbon-based NMs affect whole-transcriptome responses in crops. They show significant changes in major stress-response genes and transcription factors (TFs), such as aquaporins, dehydrins, heat shock proteins, NAC, and WRKY, which were either upregulated or downregulated depending on the CBN treatment. Some of which are ABA signaling, InsP3 signaling, and MAPK signaling (Rezaei Cherati et al. 2021). Studies on Mo-based nanofertilizers and Cu-based nanopesticides demonstrated changes in protein profiles related to various metabolic pathways in wheat plants. These changes highlight the tissue-specific downregulation in response to Cu exposure, suggesting that NMs can have differential effects on plant metabolism (Li and Keller 2023).

The major signaling pathways seem to have been affected by NPs, including the modulation of phytohormone signaling. It is this signaling pathway that plays a vital role in stress tolerance. Various investigational effects of metallic NPs on signaling pathways have shown significant changes in phytohormone regulation. For instance, the exposure of plants to ZnO, SiO_2, and ZnO/SiO_2 composite NPs has been shown to induce the expression of ethylene response genes. This later on leads to the downregulation of the expression of cytokinin-response genes. All these are due to alterations in the signaling pathways involving these hormones. Additionally, the auxin signaling pathway is seen to be affected due to CuO NPs. All this happens due to a decrease in the expression of auxin signaling F-box proteins. These proteins play a role in the negative feedback regulation of auxin signaling.

These findings suggest that NPs can profoundly influence gene expression and signaling pathways in plants. And all these pathways are to be studied as it shall help know the various impacts at the nanoscale level.

- **TFs in ZnO-NP Response**
  The importance of understanding the molecular mechanisms of NM-plant interactions is highly crucial. It is observed that the effect of ZnO-NPs on the plant's response to salt stress involves different TFs, including AP2/EREBP, WRKYs, NACs, and bZIPs. Besides, these TFs are well documented to participate in osmotic stress and hormone response, which implies a cross talk between NMs, signal transduction, and stress tolerance in crops (Ahmed, Tóth, and Decsi 2024).
- **Nitrogen-Doped Carbon Dots and Growth Effects**
  Jing et al. studied plant growth promotion and salt stress alleviation in *A. thaliana*. The study shows the differential signaling pathways in detail that are associated with nitrogen-doped carbon dots' (N-CDs) growth boosting and tolerance properties. The results imply that the growth promoting and stress tolerance parameters in plants are regulated via diverse mechanisms accounting for the intricate nature of N-CDs' action. Furthermore, it makes it obligatory to choose the correct signaling pathways influenced by various types of NMs and the conditions of their actions (Jing et al. 2023).
- **Dose-Dependent Effects and Multifaceted Interactions**
  Different plants of Gramineae were influenced by the several types of NPs. The types of NMs included iron-based, titanium-containing, zinc, and copper NPs, and the effects were in the dose-dependent manner. The researchers established that the effects of such NPs on photosynthesis, chlorophyll content, and antioxidant enzyme activity depend on the chemical nature, size, concentration of NPs, and the plant species, implying the involvement of multiple layers of the interactive relationship between NMs and these signaling mechanisms (Li et al. 2024). Consequently, the reviews on the impacts of NMs, as well as on interactions between nanoadditives and crops, allow to indicate several key signaling pathways. It showed how well plant reactions are to such innovative materials. These include ROS-mediated signaling, MAPK and CDPK cascades, SIR, nutrient signaling, regulation of stress-response genes, and hormones. These signaling pathways are intricate and frequently depend on the dosage level. It means that the effects of NMs on agricultural systems in the future should be studied in collaboration with various specialties. Further studies in this area will be beneficial in finding an optimal approach to crop production in times of new unfavorable conditions.

7.5.2 Enzymatic Activity and Stress Responses

7.5.2.1 Alterations in Enzymatic Activities

Metal-based NPs are recognized as biosafe for the organisms, as they are put forward with an ability to stimulate seed germination and plant growth, suppress diseases, and protect the plants because of their antimicrobial features. For instance, reduction in oxidative stress and osmotic stress is seen. As well as NMs are shown to assist in photosynthesis, in the uptake of water and nutrients, and in decreasing water deficit. Also, NMs can stimulate the development of some genes that are associated with adaptations to drought stress, including LEA proteins and the aquaporins, which enhance the performance of crops in drought-stressed environments. Hence, the use of carbon-based NMs, metallic/metallic oxides, metalloids, and nonmetallic NPs can, in the long run, enhance the growth, development, and production of drought-stressed crops (Maswada et al. 2020).

NMs are finally seen to improve the effectiveness and reduce waste of agro inputs with the intention of increasing plant growth, development, and yield. It can be used as carriers to enhance the transport and storage of nutrients in crop plants. In addition, since nanotechnology enables management of nanoscale processes, instruments like nanosensors and nanofilters can be created for crop observation and regulation. Thus, offering a holistic concept of agricultural sustainability (Awasthi et al. 2020).

The effect of NMs on the physiological status of crop plants with special attention to the enzymatic activity had become a subject of concern in the recent past. A number of papers described the interaction of plants with different types of NPs and the influence of the obtained NMs on the adaptive reactions and stress tolerance of plants under different stress conditions.

- **Biogenic Metal Colloidal Solutions**
  In one experiment, researchers aimed at studying doses of biogenic metal colloidal (BMC) solutions to study enzyme activity. The BMC solution is one that includes Ag, copper, iron, zinc, and manganese NPs on the activities of antioxidant enzymes and stress of soya bean plants. Two different concentrations of 120 and 240 mg/L NP solutions were used by the researchers to conduct experiments. The authors discovered that a lower concentration of copper (120 mg/L) affected ROS generation in the photosynthesizing tissues by increasing its content of thiobarbituric acid reactive substances (TBARS) by 12%. And a higher concentration of iron (240 mg/L) reduced the amount of oxidative processes by 19% indicating decrease in possible oxidative damage. The study also evidenced that the 120 mg/L of pre-sowing treatment with vegetative treatment significantly influenced a decrease in TBARS levels by 10% compared to the control (Taran et al. 2016).
- **SOD and Catalase Activity**
  The researchers noted that in relation to the TBARS content, SOD activity rose. As such, the catalase activity declined with the increase in the given parameter across all the treatment variants. These results suggested that the NP treatments affected the changes in these related enzyme activities on stress responses, and the higher concentration of NPs could enhance the antioxidant defense system (Taran et al. 2016).
- **Review on Antioxidant Enzymes**
  In a review article, a study about the effect of NPs on enzymatic activities was done. And mostly, the activity of plant stress responses and antioxidants was studied including ascorbate peroxidase (APX) and CAT. The review drew attention to the fact that they are capable of stimulating/altering the processes of oxidative stress and

abiotic stress responses between NPs and plants. This implies that NMs can affect the enzymatic activities that plants employ as a defense mechanism (Yang, Cao, and Rui 2017).

- **Carbon Nanomaterials**
  A study of the carbon nanomaterials (CNMs) for the antioxidant system of tomato seedlings was done. It revealed that the use of CNMs was done in the form of CNTs and graphene. The results showed that various enzymes were stimulated. An increase in phenolic compounds, flavonoids, ascorbic acid, and glutathione content in seedling was seen. Moreover, CNM treatment stimulated the activity of the effective antioxidant enzymes such as APX and catalase. Therefore, these NMs possess the ability to increase stress tolerance in tomato crops (González-García et al. 2019).
- **Chitosan NPs**
  There is work pertaining to the application of chitosan NPs (CHT-NPs) for alleviating oxidative stress in durum wheat. This study demonstrated that *N*-acetylcysteine (NAC) when incorporated into the CHT-NPs enhanced the levels of ascorbic acid (ASA), which is one of the significant antioxidants with the critical role for the defense of oxidative stress in plants. It can be concluded that there is a potential for using NMs as a tool to increase the crop tolerance to abiotic stress (Picchi et al. 2021).
- **Silicon NPs**
  A study was done to determine changes in enzymatic and nonenzymatic antioxidants in muscadine grape plants under hypoxia stress with the help of silicon NP (SiNP). Inhibition of locomotive activity under hypoxia stress stimulated an increase in enzymatic antioxidants, such as SOD, POD, and CAT, and other nonenzymatic antioxidants, such as ASA and glutathione, after the application of SiNP treatment (Iqbal et al. 2021).
- **AuNPs**
  Production of AuNPs by biochemical methods with plant extract from *Nicotiana tabacum* and the influence of Xanthi leaf protein and their effectiveness on the apoplastic and symplastic water relations, growth indices, and organic solutes, with special reference to ROS scavenging activity in tobacco plants exposed to stress conditions, have been reviewed. The experiment proved that moderate AuNPs, at 200–300 ppm in the present study, improved the photosynthetic characteristics. While excess, at 400–500 ppm, deteriorated the energy production in plants (Jalil et al. 2019).
- **Aspartic-Acid-Coated Magnetite NPs**
  An experimental study was carried out to study the impact of magnetite NPs that were coated with aspartic-acid-coated magnetite NPs (A-MNPs). It was studied specifically with respect to the growth of maize, in addition to the expression of antioxidant enzymes as well as chlorophyll content. The results of the study revealed that treatment of maize plants with A-MNPs promoted plant growth and elicited the enzymatic activities of CAT and POX. Moreover, there was an improvement in the antioxidant mechanisms of the maize plants due to treatment with the NPs (Răcuciu, Tecucianu, and Oancea 2022).

- **Metal-Based NPs**
  NMs, such as Fe, Cu, Co, and ZnO based, were used to investigate drought tolerance promotion in soybean plants. The authors also proved that the use of these NPs enhanced the expression of drought-responsive genes to stress in the plants, indicating a positive relationship between NMs and plant stress response mechanisms (Linh et al. 2020).
- **Cerium Oxide NPs**
  A study based on the protective role of cerium oxide (CeO_2) NPs on the salt stress tolerance in *Brassica napus L* was conducted. The study revealed that CeO_2 NPs triggered a change in root anatomical characteristics. This change therefore enhanced the plant's salt stress tolerance. It was observed that NMs altered the ion movement across the root cells. This led to differing plant physiological attributes under saline conditions (Rossi, Zhang, and Ma 2017).
- **AuNPs**
  When a study was conducted to investigate the growth stimulation of Au NPs in *Arabidopsis* seedlings, it reported a decrease in stress in the plants. The differential gene expression profiling and protein variant assessment was done in response to Au NPs treatment. Its results showed that stress-related genes were downregulated and growth-related genes were upregulated. So, it can be concluded that plants' physiological status could be enhanced by Au NPs through stress-response pathway modulation (Ferrari et al. 2021).
- **Graphene Oxide and Amine-functionalized GO**
  An investigation was done on the physiological and biochemical changes in wheat (*Triticum aestivum*) under the effect of unfunctionalized graphene oxide (GO) and amine-functionalized GO (G-NH). The results of the analysis pointed to the fact that concentrations of GO were more than 1000 μg/mL. It thus showed that GO produced strong cytotoxic and oxidative effects on seedlings, whereas G-NH stimulated growth parameters. This demonstrates how NMs have a differential effect on plant physiology (Chen et al. 2018).
- **Tin NPs**
  A study on the tin-nanoparticle (Sn NP)-treated maize seedlings was done to observe the various biochemical responses. The exposure showed that it affected the oxidative stress in the maize seedling. This led to an enhanced activity of antioxidant enzymes that include guaiacol peroxidase, polyphenol oxidase, and CAT (Pudake et al. 2019).

7.5.2.2 Responses to Oxidative and Abiotic Stress

1. **Endophytic Microbes and NMs**
   The interaction between endophytic microbes and NMs was studied in reducing the abiotic stress in plants. The particularity of the study was to stress on the fact that NMs can stimulate the activity of antioxidative enzymes that contribute to the reduction of the oxidase stress in plants. Some of the eco-friendly technologies that are suggested to minimize the stress impact in crops include the applications of nano-growth enhancers from helpful microbes (Adeleke et al. 2022).

2. **Arabidopsis and NP Exposure**
   A focused study was done to check on the first sign of interaction between *Arabidopsis* and three various NPs including TiO_2, Ag, and CNTs in both biotic as well as abiotic stresses. It was then studied that exposure to NPs hinders the transcriptional responses to the microbial pathogens. Hence, changing the ability of a plant to respond to infection. It further concluded that NP exposure might be implicated with the physiological alterations connected with the fundamental enzymatic activities in plants responsible for stress stimulus (García-Sánchez, Bernales, and Cristobal 2015).

- **Synergistic Relationship with Beneficial Microorganisms**
  A study was done to see how the occurrence of the NM-plant interactions can affect gene expression for stress-responsive genes and signaling, which in turn leads to changes in enzymatic functions and the physiological well-being of the crop. These molecular events are important in charting out appropriate approaches to exploit the benefits of NMs in boosting up the stressed yield in crops through abiotic and oxidative stresses (García-Sánchez, Bernales, and Cristobal 2015; Ferrari et al. 2021; Nawaz et al. 2023).
- **Fullerenol NPs and Drought Stress**
  The efficiency of fullerenol NPs in mitigating drought stress in sugar beet plants was studied. The work concluded that FNPs prevented the build-up of the proper osmolyte proline and upregulated the antioxidant enzymes catalase, APX, and glutathione peroxidase in plants affected by drought stress, thereby demonstrating the significance of FNPs in stress management in plants (Borišev et al. 2016).
- **Se-NPs and Temperature Stress**
  A discussion on the effects of Se-NPs on the elevated temperature stress in grain sorghum was conducted. It revealed that the foliar spray of Se-NPs strengthened the antioxidant defense system, which in turn lowered oxidative damage, thereby enhancing pollen germination and seed yield under stressful conditions (Djanaguiraman et al. 2018).
- **ZnO-NPs and Salinity Stress**
  A study showed the management of salinity effects on crops due to the role of zinc ZnO NPs. It showed alleviating salinity stress in plants. The review also showed how ZnO NPs can contribute positively to the crops' physiological condition by modulating the enzyme activity and stress signaling networks (Ahmed, Tóth, and Decsi 2024).
- **NMs and Stress Response Pathways**
  A paper surveyed the contribution of NMs to enhance the abiotic stress tolerance in plants and their significant impact on enzymatic-based systems, particularly showing that the NMs affect the antioxidant system and help reduce the effects of oxidative stress in crops (Aguirre-Becerra et al. 2022). From the literature review above, it can be stated that the incorporation of different forms of NMs has an enormous impact on enzymatic functions and stress responses in crops. Many researchers have pointed out that NPs increase the activity of antioxidant enzymes, including SOD, catalase, APX, and glutathione peroxidase, which are important in combating oxidative stress. Also, the addition of NMs helped increase the content of nonenzymatic antioxidants, such as components of ascorbic acid, glutathione, and phenolic

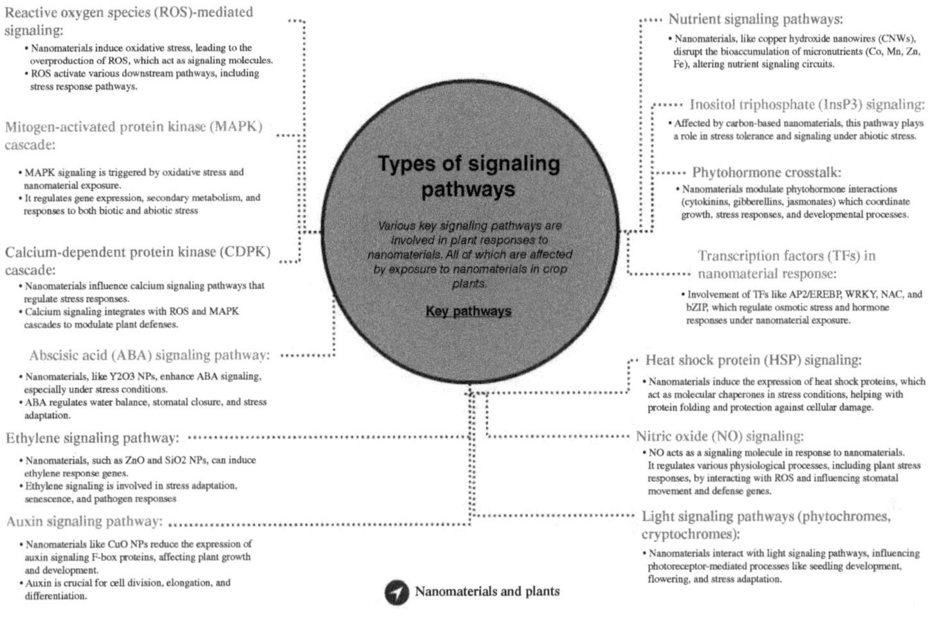

Figure 7.2 Categories of signaling pathways associated with plant's interaction with nanomaterials.

compounds in the plant, which support the plant's defense against abiotic stresses. It can be presumed that this scheme of application of NMs and biostimulants can be successfully utilized to enhance the physiological parameters of the crop plants under stress. Thus, the various studies mentioned above confirm the enhanced enzymatic activities and stress responses in crop plants due to NMs.

In Figure 7.2, we have outlined several other signaling pathways that are affected by NM toxicity in plants. Such pathways are the ROS signaling, MAPK pathways, CDPK signaling, and phytohormones, which coordinate the stress, growth, and development signals. NMs influence these pathways based on oxidative stress, nutrient assimilation, and defense in such a way which may help or hinder plant adaptation to different environmental conditions.

7.6 Case Studies and Experimental Findings

7.6.1 Positive Effects

7.6.1.1 Enhanced Growth and Yield

1. **Nutrient Delivery and Utilization:** Nutrient delivery and utilization seem to be one of the ways that NMs can be used to increase crop growth and production. Some attributes peculiar to NMs include a remarkably high surface area-to-volume ratio and nutrient delivery capabilities. Therefore, the resulting effects include improved uptake

and efficiency of nutrient use in crop plants (Aguirre-Becerra et al. 2022). For example, the use of NMs as nanofertilizers enhances nutrient uptake, as seen in the case of zinc, nitrogen, and phosphorus uptake and/or assimilation in plants, leading to a better increase in the plant yield (Mgadi et al. 2024).

2. **Alleviation of Abiotic Stresses:** Besides, the use of NMs can also be useful in combating such abiotic stresses as water dormancy, salinity, and severe temperatures that threaten agricultural yields. It has been documented that the use of various kinds of NPs, such as ZnO, TiO_2, and Ag NPs, has a positive impact on the physiological condition of the crop plants by enforcing the antioxidant activity (Wohlmuth et al. 2022; Ijaz et al. 2023; Ehsan et al. 2022), further advancing the process of photosynthesis and efficient nutrient use. Perennials especially can benefit from the following effects under these stresses. They can indeed experience enhanced growth rates, biomass production, and yields.
3. **Crop-specific Effects:** Studies show that NMs can enhance the growth and yield of crops across the crop species. For instance, phytogenically synthesized NMs, namely Ag zinc oxide (Ag/ZnO), enhanced the biochemical and yield performance of wheat varieties, including protein content, grain numbers per spike, spike length, and grain yield, according to the study (Ehsan et al. 2022). Likewise, investigations on common sunflower plants have also revealed that ZnO and TiO_2 NPs can be applied physically and biochemically to increase photosynthetic efficiency, stomatal conductance, and grain yields (Kolenčík et al. 2020).
4. **Nano-priming:** Nano-priming using seed treatment with NMs has been reported to improve seed germination, growth characteristics, and yield-contributing factors in different crop plants (Lee and Kasote 2024; Jaskulski et al. 2022). By reducing all these forms of abiotic stress, the different physiological aspects of the plant are put in a position to thrive and enhance their rates of growth, thereby improving overall productivity and growth.
5. **Post-harvest Preservation:** Additionally, the incorporation of nanoelements in post-harvest technologies such as nanoliposome and nanoemulsion also prevents crop spoilage and increases shelf life, thereby promoting sustainable crop production (Jiang et al. 2021).

7.6.1.2 Improved Resistance to Pests and Diseases

- **Disease and Pest Resistance**

  The advantages associated with NMs in crop plants are not contradictory to aspects such as enhanced growth and production. NMs have also been shown to help increase plants' resistance to diseases and pests, which can indirectly lead to better crop yields. For example, metal-based NMs, including copper oxide (CuO) and iron oxide (Fe_3O_4), in an experimental study confirmed control of diseases and promotion of plant growth in bean varieties.

- **NMs in Enhancing Crop Resistance**

1. NMs can enhance some of the physiological and chemical performances of crop plants. The authors note that the exposure of NPs of zinc, Ag, and selenium to the surfaces of the crop can intensify the interaction of that surface area with pests. Therefore, leading

to better resistance to these pests may be obtained. They also mention that nano-based approaches can provoke a plant's response to pests and diseases and control microbial processes to increase plants' defense (Tipu et al. 2021).

2. Researchers showed that NMs can enhance the plants' resistance to both biotic factors like pests and diseases and abiotic factors like drought and salinity. The authors also note that basic details of the molecular processes, which are involved in NP regulation of plant response, have to be known in order to use NPs in agriculture efficiently (Ali, Mehmood, and Khan 2021).
3. A study showed how NMs augment plant defense mechanisms. Some of them describe how the processes involved in the utilization of NPs to increase the resistance of plants against diseases are still under research (Singh et al. 2024).

- **Integrated Use of NMs with Biostimulants**
  Biostimulants seem to function as natural biofertilizers. Thus, incorporating NMs and tracking how this may enhance the biostimulant activity can improve nutrient acquisition, growth, and physiological health of plants, with the possible outcome of increasing crop resistance to pests and diseases (Singh et al. 2024).
- **Nano-priming and Seed Germination**
  Nano-priming triggers the generation of ROS and antioxidant defenses, which appears fundamental to stress tolerance and pest and disease resistance. It influences the status of biochemical signaling networks and directly affects the secretion of hormones, while decreasing the content of reactive oxygen species (ROS) to enhance disease tolerance in crops (Magnabosco et al. 2023).
- **Specific Crop-pathogen Systems**
  The application of effectual metal NPs is highly studied in controlling bacterial diseases in cereal staples. Researchers also stressed on the high antimicrobial effectiveness of these NMs that can substitute poisonous pesticide and make the crops more resistant to bacterial threats in an environmentally friendly way. The most beneficial application of NPs is seen in gene delivery for the creation of insect-resistant plant varieties. Furthermore, identification of diseases with the help of nanosensors also helps to develop the approach to their management and crops resistance. Next comes the concept of nanopesticides, which helps in developing the withstanding ability of crop plants to pests and diseases as well as protect the environment from the dangerous influence of pesticide, which negatively affects the nontargeted organisms (Shekhar et al. 2021).

7.6.2 Negative Effects

7.6.2.1 Phytotoxicity and Growth Inhibition

1. **Phytotoxicity of NMs**
   NMs can promote plant growth and nutrition and protect the plants from different stress conditions. But the side effects like phytotoxicity and growth hindering impacts are also to be seen along with it. Physiological responses of crop plants are dependent on the type of NMs, concentration of NMs, the pathways through which the NMs affect the plants, and the time that the plants are exposed to the NMs.

So, the possibility of phytotoxicity of NMs can be explained in several ways, particularly, the behavior of free ions originating from the nanomatrix. They can cause tissue damage and membrane defects in plant cells due to the generation of ROS. Features such as the ability of the plants to absorb, transport, and internalize NMs are considered critical in determining the phytotoxicity of NMs (Hossain, Yasmeen, and Komatsu 2020).

2. **Tunneling Formation in Roots**
   Some previous research stated that NMs can alter plant root cell structure, as studies show effects including cell abrasion, metaxylem reduction, and "tunneling-like formation" in the root apical meristem. Such changes can negatively affect the plant's vascular tissue and nutrient transport, thereby causing detrimental effects on growth and biomass yields (Li and Keller 2023).

3. **Negative Oxidative Stress Responses**
   NMs have also been reported to have phytotoxic effects, thereby affecting the plants' oxidative stress, which in turn affects DNA fragmentation, lipid peroxidation, and changes in the antioxidant enzyme system. This oxidative stress response can affect lot of other physiological activities in the plant, for instance, photosynthetic activities, nutrient assimilation, and even the overall performance of the plant (Sarraf et al. 2022; Baskar et al. 2018).

4. **Growth Inhibition**
   NMs have been shown to inhibit crop growth under the influence of some types of ENMs. For example, silver, copper, molybdenum, and zinc oxide negatively affect the germinability of seeds, length of roots, and growth of sprouts of crops such as maize, cabbage, eggplant, and wheat. The extent of growth retardation is dose-dependent. So, the cell density of the NMs is also affected. A higher concentration tends to have a more severe impact on the development of various vital aspects of plant growth. Furthermore, the solubility and coagulation of the NMs in soil or growth media may also determine the amount of utilizable concentration, thereby influencing plant productivity (Li and Keller 2023).

7.6.2.2 Long-term Environmental Impact

1. **Disruption of Essential Physiological Processes**
   Different physiological processes, such as antioxidant system function, growth, yield, photosynthesis, mineral nutrient uptake, oxidative stress, cytotoxicity, and genotoxicity, are influenced in crop plants due to NM exposure. These disruptions could result in changes in plant morphology, biochemistry, physiology and molecular genetics, oxidative stress, cellular damage, and inhibition of physiological processes. These NMs also cause formation of ROS and affect plant biomass and nutrient uptake and even cause lethality in some plant species (Thiruvengadam, Chi, and Kim 2024). The above-mentioned study of relationships between NMs and plants shows that there is still much about these processes to learn to reduce stress and promote plant production. Finally, these physiological alterations depict the need to find an understanding of NMs' impacts to promote growth and development of resistance and stress endurance to reduce the impacts of unfavorable conditions.

2. **Impact on Soil Health and Ecosystem Stability**
   Metal-based NPs such as ZnO, Cu, and Zn are very much capable of exerting adverse interference in nutrient uptake and water relations in crop plants, further affecting soil health and the sustainability of the ecosystem. Literature suggests that these NPs when present in the soil can be absorbed by plants and alter the microbial content of the soil, leading to an unstable ecosystem. For instance, studies have shown that increased concentrations of ZnO NPs in soil could alter nutrient uptake and water status, which in turn affects soil microbial population and overall ecosystem stability (Strekalovskaya, Perfileva, and Krutovsky 2024).

   In the same way, metal NPs can lead to oxidative stress and disrupt cellular homeostasis in plants, which leads to alteration in transport of nutrients and effects on the soil. This causes toxic effects on the elements of the terrestrial ecosystem and possible impacts on human health through the food chain efficacy. Metal NPs when released into the environment in an uncontrolled manner pose toxicity risks to all forms of life and their functions. All this is further dependent on the size of the particles, their surface chemistry, and the rate at which they are applied (Shah et al. 2021). However, new methods include use of nano-biochar which has proven to reduce some of these toxic impacts while improving microbial life, nutrient status of the soil, and crop yield. But further studies need to be conducted to determine other effects and use of NMs for agriculture safely.

3. **Trophic Transfer and Food Safety Concerns**
   The incorporation of NMs in crop plants can be hazardous toward trophic transfer and food safety concerns. Studies showed that these substances affect the bodily functions of plants and the surrounding ecosystems. In a similar way, the presence of NPs in the environment causes what may be called nanopollution, which influences the soil and water ecosystems (Gowtham et al. 2024). While exposed plants can store or accumulate massive amounts of NMs if taken in large quantities, they become hazardous at the next higher trophic level and cause possible future impacts on the food chain. They also act toxicologically in plants through the formation of ROS that leads to oxidative stress, genotoxicity, and alteration in plant nutrient uptake, thus causing long-term effects on plant growth and development. As much as NMs seem to be advantageous such as through increase in crop yield, their potential harm emphasizes the need to understand how they interact with plants for effective and safe agriculture and food production (Francis et al. 2024). Even though NMs improve the availability of nutrients and protect plants from different stresses, the phytotoxicity and the effect on plant growth must be considered and investigated in further detail. Parameters like the particular type of NM, its dosage in the fertilizer and spray, ways of exposure, and its time length can affect the crop plant's physiological reactions greatly. Thus, the knowledge of all the parameters, concerning the levels of NMs' exposure and the presence of their byproducts, is important for evaluation of the organism's environmental risk. Some questions are still unanswered regarding the transportation of NPs in soil. Some of which include the adsorption and desorption of particles, dissolution, volatility, and leaching to groundwater. The various dissipation processes NPs experience in the soil environment includes the physiochemical process of hydrolysis and

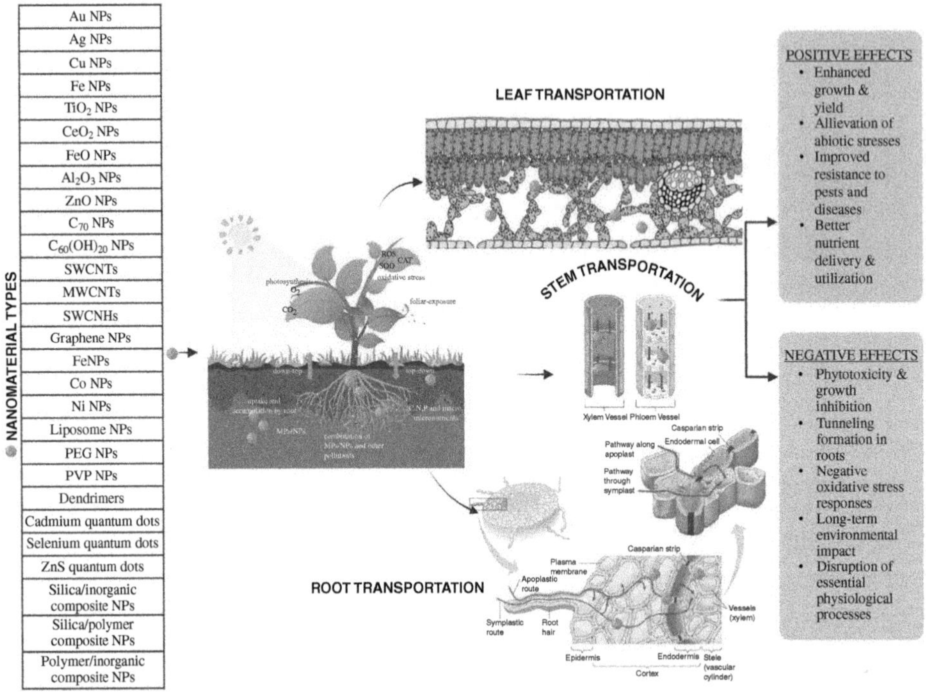

Figure 7.3 Translocation and bioremediation potential of different nanomaterials in agri-crops.

photodegradation, as well as biodegradation through microbial and other living organisms such as fauna and plants. Figure 7.2 shows an overview of NMs that affect the various physiological processes and their positive and negative impacts of them on the plant and its environment.

Figure 7.3 demonstrates the movement of several types of NM through the plant stem from roots and up to the leaves. It discloses possible benefits as well as drawbacks of the ingestion of NM. These are increased growth and yield, abiotic stress management, resistance to pests and diseases, and efficient nutrient uptake. The negative impacts are the growth inhibitions, phytotoxicity, oxidative stress induction, root tunneling formation, and disorders of important vital activities. NMs, including metal NPs, quantum dots, and polymer/inorganic hybrids, are included.

7.7 Practical Applications and Future Prospects

The effects of NMs on the growth and development and the future potential of crop plants have been well studied. Various literature studies have revealed that nanotechnology can be used in agriculture to increase crop yield, reduce stress on crops, and promote environmentally friendly practices, with certain issues and risks.

7.7.1 Current Applications in Agriculture: Implementation and Realization

7.7.1.1 Nanofertilizers and Nutrient Delivery

Among all the uses of NMs in agriculture, the most common one is the creation of nanofertilizers. Nanotechnology experiments can enhance the delivery of nutrients to crop plants through the use of engineered NMs. By increasing the surface area and specificity to release properties, the plants have higher efficiency of absorption of nutrients, causing less loss of fertile materials through leaching or runoff (Gade et al. 2023). Review on the use of nanofertilizers has shown that its use results in better plant growth, increased yield, and enhanced nutrient level in various crops. Micronutrient delivery like zinc, iron, and boron by NM has also been applied to deliver the specified nutrients to deal with nutritional deficits and enhance the physiological status of the crop plants (Shang et al. 2019).

7.7.1.2 Nanopesticides and Crop Protection

NMs applicability as nanopesticides and nanoherbicides seem to enhance the performance and selectivity of the agrochemicals. Due to the specific characteristics of NMs, including small size and high surface area, the bioavailability and penetration of active ingredients, as well as opportunity for pest and weed control, are improved (Kumar et al. 2021). Moreover, the incorporation of organic pesticides and herbicides with nanocarrier systems decreases their vulnerability to leaching, runoff, and exposure of nontarget species. Such an approach is referred to as "smart delivery systems," with the goal to enhance the effectiveness and responsible use of agrochemicals.

7.7.1.3 Nanosensors and Precision Agriculture

NMs have also been used in manufacturing high-performance nanosensors with applications in precision agriculture. These can be made to check factors ranging from stress in the plant, nutrient deficiencies, and disease incidences, leading to the need to check various control measures to be applied immediately. The combination of nanosensors with endemic precision farm practices also results in improved resource management, consequently reducing the impacts of agrochemicals on crops.

7.7.1.4 Stress Mitigation and Crop Resilience

One of the most attractive fields that involve NMs is agriculture. And it deals with increasing the crop resistance to abiotic and biotic stresses. NMs have been demonstrated to decrease the dangerous effects of stresses, including drought, salinity, heavy metal toxicity, and heat shock by regulating the plant's physiological and biochemical processes. It is also seen that NMs can elevate the activities of antioxidant enzymes, the efficiency of photosynthesis, and the stress involved in the synthesis of biomolecules in crop plants. This in turn increases the stress tolerance of plants. Many major applications are thus seen in enhancing the efficiency of soil whose use is with regard to climatic change and other environmental factors in the production of food supply (El-Saadony et al. 2022).

7.8 Environmental and Safety Considerations

7.8.1 Ecotoxicology of NMs

7.8.1.1 Impact on Soil, Water, and Nontarget Organisms

Despite the great interest in NPs due to their general and specific usage in industries and in the agricultural sector, their higher production in mass seems to be detrimental. The usage seems to be detrimental causing food chain disruption, thus affecting humans in the same way. Not only living beings, the NPs have the potential to negatively impact soil and water ecosystems. As water is an essential part of plant growth and physiology, the effects seem to be visible in several ways. Few of which are listed in the following.

- **Soil Contamination**
  It should be noted that accumulation of NPs in the soil is possible, hence causing contamination. They can alter their physical and chemical status, which results in changes in their solubility and toxicity. It is also observed that NPs lead to oxidatively stressed plants, thereby posing a potential threat to humans through ingestion of plant products (Khalid et al. 2022). They also influence microbial species and can decline the soil enzymes, thereby affecting the overall health of the soil.
- **Plant Toxicity**
  NPs can be toxic to plants through the formation of ROS resulting in cytotoxic and genotoxicity impacts. This can lead to programmed cell death in different species of plants. Such as copper oxide NPs can reduce biomass in rice seedling by formulating a huge amount of biomass accumulation, leading to stomatal and ultrastructural changes. Likewise, rare earth oxide NPs can negatively reduce the photosynthetic activity in the radish and affect hormonal content synthesis.
- **Impact on Plant Development**
  NPs can affect plant morphological and physiological changes, both at the initial stages or at the mature stage. They can accumulate in plants and, consequently, affect the human body in a negative manner. If the plant is consumed, it can lead to ailments such as heart diseases, lung diseases, and gastrointestinal diseases, among others.
- **Nutrient Assimilation**
  It should be noted that the incorporation of NPs can influence the absorption and transport of plant nutrients, and the study of these phenomena is vast and important for understanding the prospects of sustainable agriculture and food security.
- **Water Pollution**
  Generally, the NPs in aquatic environments reduce the growth rate and photo-physiological productivity and increase the level of oxidative stress in marine phytoplankton. Among NPs, AgNPs, TiO_2 NPs, and ZnO NPs are known to freely and continuously contaminate water systems. It is this that may affect water organisms and possibly bioaccumulate in food chains.

Thus, understanding the behavior and effects of NPs is crucial for developing safe usage guidelines and mitigating their negative impacts.

- **Comparative Studies Among Models Belonging to Different Environments**
  Ten studies searching for multiple categories were identified. They exposed the ecosystem to several NMs and assessed the ecotoxicity differences in the various organisms grouped according to the targets, for example, bacterial, invertebrate, and vertebrate. Among all the investigated NPs, ZnO and AgNPs were the most commonly used particles, with a total of 60% of all the studies, whereas the rest were for TiO_2, FeNP, porphyrin NMs, and PAMAM or polyamidoamine-coated Au NP. Variation in sensitivity among the model species that target the different categories of the environment was noticed when exposed to (i) PAMAM-coated AuNPs in mammals, FW algae, and marine bacteria; (ii) ZnO NPs in plants, crustaceans, and insects; and (iii) Ag NPs in FW crustaceans, worms, and plants. For instance, the Ag NPs are genotoxic to different organisms, including mammals, plants, and aquatic life forms. The influence of magnetite NPs is not as damaging to the environment as presented by a study that compared the effects of both in various models including human red blood cells-MACs, onion root tip cells, fish embryo, crustaceans, and mammals. These studies suggest drawing conclusions on the toxicity of NMs. It is not only important to test for its toxicity level but also to assess the ability of NMs to react with other prevailing pollutants in the environment (Gambardella and Pinsino 2022).

7.8.2 Risk Assessment and Management

Ethical analyses show a possibility of environmental effects of NMs applied in farming. The dangers of regulatory compliance and various other issues indicate that efficient and responsible use of nano-properties in farming is achievable. Hazard characterization of NPs is done with an aim of determining the endangerment and routes of human contact. In addition, it is also important to focus on the potential beneficial applications of nanotechnology, such as energy production, environmental clean-up, and increase in product utility, while assessing and mitigating the risk factors along with these advantages.

7.8.2.1 Preliminary Activity in Risk Ranking

The first steps pertain to the critical stages of defining and describing the sources of NPs in the environment. Next, assessing the risks involves defining the dose–response curve for various endpoints and probability of exposure. After both dose–response and exposure have been measured, the two values are then compared to identify the level of risk.

7.8.2.2 Hazard Assessment

Risk assessment considers the capability of NPs to cause harm and provides anthropological benefits and the sort of impact that may occur. It shows short-term or long-term or even physiological or genetic effects. There are also other possible adverse effects that have to be taken into account be it interference with the performance of the atmosphere and/or the stratosphere. It also checks the changes in stability of soil and the behavior of mineral nutrients in the presence of the compound.

7.8.2.3 Dose–Response Assessment

After the hazard identification process, the next step is dose–response assessment in the risk assessment process. This entails the making of dose–effect relations whether through a series of experiments with animals or a mathematical model. With regard to NPs, there are indications that the exposure measures normally used in terms of mass concentration may not be as applicable as those expressed in terms of surface area, due to differences in the reactivity of the surfaces of particles, which can affect toxicity.

7.8.2.4 Exposure Assessment

Exposure assessment is the possibility of exposure to NMs right from their manufacturing process. Knowledge of quantitative elements of production, separation, alteration, treatment, packaging, transportation, and attached losses, and wastes is significant. Concerning environmental exposure, it is necessary to possess empirical information to have a better idea of the dumping and transport of NPs in the air, soil, and aqueous environment.

7.8.2.5 Risk Characterization

The last element of the risk assessment procedure is risk characterization. It integrates the findings from hazard identification, dose–response, and exposure assessment and evaluates and communicates risk as it exists to the exposed populations. This is usually done by recognizing the exposure concentration and comparing it with a level believed to have no impact.

7.9 Conclusion

Nanotechnology has a major impact on crop physiology, yield, and disease prevention. However, for successful implementation of nanotechnology-based approaches, proper risk assessment related to NMs is needed. This chapter sheds light on the effective implementation of NMs in crop improvement techniques for better future agriculture prospects and sustainable agriculture. NMs have the potential to modulate the nutrient uptake, and anatomical as well as genome-level changes in plants that may lead to better yield and productivity. Further research is needed to understand how the NPs regulate the molecular and cellular processes in plants. Moreover, more research is warranted to evaluate whether the use of NPs is environmentally friendly. A general observation from the previously published work is that at low concentrations NPs can enhance nutrient uptake, promote root and shoot growth, and improve overall plant biomass, without causing any significant damage to soil. For instance, copper oxide NP has been shown to increase certain metabolites and enhance growth, but excessive levels can lead to decreased biomass and reduced photosynthetic pigment levels. There are studies indicating that NP-like Ag/ZnO can enhance photosynthesis and growth rates in wheat seedlings, leading to increased biomass. However, high concentrations of NP can be toxic, reducing biomass production. The use of NP as nanofertilizers or in combination with plant growth regulators can lead to increased biomass and crop yields. The effects of NP on plants vary depending on factors such as type, concentration, method of application, plant species, and environmental conditions. Thus, it can be stated that the outcome of nanotechnology-based approaches depends on various factors including the type, concentration, and method of application of the NPs, as well as the plant species and environmental conditions. NPs offer promising potential for enhancing nutrient uptake and improving plant health, and their application requires careful consideration to balance their beneficial effects with potential risks of toxicity and metabolic disruption. Understanding the specific interactions between NPs, plant species, and environmental conditions is crucial for optimizing their use in agriculture.

NMs are increasingly being explored in agriculture to enhance crop growth and productivity. These materials, typically ranging in size from 1 to 100 nm, offer unique physicochemical properties that make them ideal for improving nutrient uptake, stress tolerance, pest resistance, and overall plant health. However, their effects on crop physiology can be both beneficial and detrimental, depending on the material, concentration, and environmental conditions. NPs like zinc oxide (ZnO) and titanium dioxide (TiO_2) can promote seed germination and seedling growth by enhancing water uptake and enzyme activation. Carbon nanotubes (CNTs) improve water absorption by creating channels in the seed coat, leading to faster germination. NMs such as iron oxide (Fe_3O_4) NPs and nanofertilizers facilitate the absorption of essential nutrients, improving plant nutrition. Silicon NPs boost photosynthetic efficiency by increasing chlorophyll content, photochemical activity, and stomatal regulation. Silica (SiO_2) NPs and silver (Ag) NPs help mitigate the effects of abiotic stress, such as drought, salinity, and heat, by reducing oxidative damage. Chitosan NPs activate the plant's defense mechanisms, making them more resistant to pathogens and pests. NMs such as gold (Au) NPs have been shown to influence the production of hormones like gibberellins and auxins, promoting cell division and elongation. However, there are some negative impacts of NMs on plant physiology as well. At higher concentrations, NMs such as silver (Ag) NPs and copper oxide (CuO) NPs can generate excessive ROS, causing lipid peroxidation, DNA damage, and enzyme inhibition. Overaccumulation of NMs in plant tissues can interfere with nutrient uptake, leading to deficiencies and reduced photosynthetic activity. Some NMs may block root pores or interfere with aquaporins, disrupting water transport and nutrient absorption. Persistent use of NMs may lead to bioaccumulation in soil and plant tissues, potentially affecting soil microbiota and long-term crop productivity. The factors influencing the impact of NMs include the following:

- Type and Size of NM: Smaller NPs generally have more significant physiological effects due to better penetration and surface reactivity.
- Concentration and Dosage: Low concentrations tend to be beneficial, whereas excessive levels can induce toxicity.
- Crop Species: The response to NMs can vary among different plant species due to differences in uptake and metabolism.
- Environmental Conditions: Soil type, pH, temperature, and water availability affect the uptake and efficacy of NMs.

NMs present immense potential for sustainable agriculture by improving crop productivity, stress resilience, and nutrient use efficiency. However, careful optimization of their dosage and environmental impact assessments are crucial to prevent toxicity and ensure long-term agricultural sustainability. Integrating nanotechnology with conventional agricultural practices offers a promising way forward, but further research is needed to understand the complex interactions between NMs, plants, and the environment.

7.10 Future Prospects

The future prospects of NMs in agriculture are vast and promising. As research continues to advance, the development of more sophisticated NMs and their integration into agricultural practices will likely lead to even greater efficiencies and sustainability. Potential future applications include the use of nanotechnology for genetic modification, the development of NM-based plant growth regulators, and the creation of advanced nanocomposites for improved agricultural tools and infrastructure.

The convergence of nanotechnology with other emerging technologies, such as biotechnology, information technology, and artificial intelligence, could further enhance the capabilities and applications of NMs in agriculture. This interdisciplinary approach could pave the way for the development of next-generation agricultural systems that are more resilient, sustainable, and capable of meeting the demands of a growing global population.

References

Acharya, P. (2020). *Seed Priming with Green Synthesized Nanoparticles Improve Germination, Growth, and Yield while Maintaining Quality of Onion and Watermelon* (Doctoral dissertation).

Adeleke, B.S., Akinola, S.A., Adedayo, A.A. et al. (2022). Synergistic relationship of endophyte-nanomaterials to alleviate abiotic stress in plants. *Frontiers in Environmental Science* 10: 1015897.

Aguirre-Becerra, H., Feregrino-Perez, A.A., Esquivel, K. et al. (2022). Nanomaterials as an alternative to increase plant resistance to abiotic stresses. *Frontiers in Plant Science* 13: 1023636.

Ahmad, A., Hashmi, S.S., Palma, J.M. et al. (2022). Influence of metallic, metallic oxide, and organic nanoparticles on plant physiology. *Chemosphere* 290: 133329.

Ahmed, M., Tóth, Z., and Decsi, K. (2024). The impact of salinity on crop yields and the confrontational behavior of transcriptional regulators, nanoparticles, and antioxidant defensive mechanisms under stressful conditions: A review. *International Journal of Molecular Sciences* 25 (5): 2654.

Akhtar, K., Khan, S.A., Khan, S.B. et al. (2018). Scanning electron microscopy: Principle and applications in nanomaterials characterization. In *Characterization of Nanomaterials*, 113–145. Cham: Springer International Publishing.

Al-Antary, T.M. and Ghidan, A.Y. (2023). Strengths and weaknesses of metal oxide nanoparticles in agriculture. In: *Nanometal Oxides in Horticulture and Agronomy*, 353–376. Cham: Springer.

Ali, S., Mehmood, A., and Khan, N. (2021). Uptake, translocation, and consequences of nanomaterials on plant growth and stress adaptation. *Journal of Nanomaterials* 2021 (1): 6677616.

Anand, K.V., Anugraga, A.R., Kannan, M. et al. (2020). Bio-engineered magnesium oxide nanoparticles as nano-priming agent for enhancing seed germination and seedling vigour of green gram (Vigna radiata L.). *Materials Letters* 271: 127792.

Awasthi, G., Singh, T., Tiwari, Y. et al. (2020). A review on nanotechnological interventions for plant growth and production. *Materials Today: Proceedings* 31: 685–693.
Baig, N., Kammakakam, I., and Falath, W. (2021). Nanomaterials: A review of synthesis methods, properties, recent progress, and challenges. *Materials Advances* 2 (6): 1821–1871.
Bala, N., Dey, A., Das, S. et al. (2014). Effect of hydroxyapatite nanorod on chickpea (Cicer arietinum) plant growth and its possible use as nano-fertilizer. *Iranian Journal of Plant Physiology* 4 (3): 1061–1069.
Barazzouk, S., Kamat, P.V., and Hotchandani, S. (2005). Photoinduced electron transfer between chlorophyll a and gold nanoparticles. *The Journal of Physical Chemistry B* 109 (2): 716–723.
Baskar, V., Nayeem, S., Kuppuraj, S.P. et al. (2018). Assessment of the effects of metal oxide nanoparticles on the growth, physiology and metabolic responses in in vitro grown eggplant (Solanum melongena). *3 Biotech* 8: 1–12.
Borišev, M., Borišev, I., Župunski, M. et al. (2016). Drought impact is alleviated in sugar beets (Beta vulgaris L.) by foliar application of fullerenol nanoparticles. *PLoS One* 11 (11): e0166248.
Burklew, C.E., Ashlock, J., Winfrey, W.B. et al. (2012). Effects of aluminum oxide nanoparticles on the growth, development, and microRNA expression of tobacco (Nicotiana tabacum). *PloS One* 7 (5): e34783.
Cañas, J.E., Long, M., Nations, S. et al. (2008). Effects of functionalized and nonfunctionalized single-walled carbon nanotubes on root elongation of select crop species. *Environmental Toxicology and Chemistry: An International Journal* 27 (9): 1922–1931.
Casillas-Figueroa, F., Arellano-García, M.E., Leyva-Aguilera, C. et al. (2020). Argovit™ silver nanoparticles effects on allium cepa: Plant growth promotion without cyto genotoxic damage. *Nanomaterials* 10 (7): 1386.
Chen, J., Yang, L., Li, S. et al. (2018). Various physiological response to graphene oxide and amine-functionalized graphene oxide in wheat (Triticum aestivum). *Molecules* 23 (5): 1104.
Choi, H.G. (2021). Effect of TiO_2 nanoparticles on the yield and photophysiological responses of cherry tomatoes during the rainy season. *Horticulturae* 7 (12): 563.
Cyriac, J., Melethil, K., Thomas, B. et al. (2020). Synthesis of biogenic ZnO nanoparticles and its impact on seed germination and root growth of Oryza sativa L. and Vigna unguiculata L. *Materials Today: Proceedings* 25: 224–229.
Das, S., Mukherjee, A., Sengupta, G. et al. (2020). Overview of nanomaterials synthesis methods, characterization techniques and effect on seed germination. In: *Nano-materials as Photocatalysts for Degradation of Environmental Pollutants*, 371–401. Amsterdam: Elsevier.
de Paiva Pinheiro, S.K., da Silva Pontes, M., Miguel, T.B.A.R. et al. (2024). Nanoparticles and plants: A focus on analytical characterization techniques. *Plant Science* 348: 112225.
Demir, E., Kaya, N., and Kaya, B. (2014). Genotoxic effects of zinc oxide and titanium dioxide nanoparticles on root meristem cells of Allium cepa by comet assay. *Turkish Journal of Biology* 38 (1): 31–39.
Demirer, G.S., Zhang, H., Matos, J. et al. (2018). High aspect ratio nanomaterials enable biomolecule delivery and transgene expression or silencing in mature plants. *bioRxiv*. doi: 10.1101/179549.
Djanaguiraman, M., Belliraj, N., Bossmann, S.H. et al. (2018). High-temperature stress alleviation by selenium nanoparticle treatment in grain sorghum. *ACS Omega* 3 (3): 2479–2491.

Dulta, K., Koşarsoy Ağçeli, G., Chauhan, P. et al. (2021). A novel approach of synthesis zinc oxide nanoparticles by Bergenia ciliata rhizome extract: Antibacterial and anticancer potential. *Journal of Inorganic and Organometallic Polymers and Materials* 31: 180–190.

Ehsan, M., Raja, N.I., Mashwani, Z.U.R. et al. (2022). Effects of phytogenically synthesized bimetallic Ag/ZnO nanomaterials and nitrogen-based fertilizers on biochemical and yield attributes of two wheat varieties. *Nanomaterials* 12 (17): 2894.

Eichert, T., Kurtz, A., Steiner, U. et al. (2008). Size exclusion limits and lateral heterogeneity of the stomatal foliar uptake pathway for aqueous solutes and water-suspended nanoparticles. *Physiologia Plantarum* 134 (1): 151–160.

El-Saadony, M.T., Saad, A.M., Soliman, S.M. et al. (2022). Role of nanoparticles in enhancing crop tolerance to abiotic stress: A comprehensive review. *Frontiers in Plant Science* 13: 946717.

Faisal, M., Saquib, Q., Alatar, A.A. et al. (2013). Phytotoxic hazards of NiO-nanoparticles in tomato: A study on mechanism of cell death. *Journal of Hazardous Materials* 250: 318–332.

Falco, W.F., Botero, E.R., Falcão, E.A. et al. (2011). In vivo observation of chlorophyll fluorescence quenching induced by gold nanoparticles. *Journal of Photochemistry and Photobiology A: Chemistry* 225 (1): 65–71.

Faraz, A., Faizan, M., Hayat, S. et al. (2022). Foliar application of copper oxide nanoparticles increases the photosynthetic efficiency and antioxidant activity in Brassica juncea. *Journal of Food Quality* 2022 (1): 5535100.

Feng, Y., Kreslavski, V.D., Shmarev, A.N. et al. (2022). Effects of iron oxide nanoparticles (Fe_3O_4) on growth, photosynthesis, antioxidant activity and distribution of mineral elements in wheat (Triticum aestivum) Plants. *Plants* 11 (14): 1894.

Ferrari, E., Barbero, F., Busquets-Fité, M. et al. (2021). Growth-promoting gold nanoparticles decrease stress responses in Arabidopsis seedlings. *Nanomaterials* 11 (12): 3161.

Francis, D.V., Abdalla, A.K., Mahakham, W. et al. (2024). Interaction of plants and metal nanoparticles: Exploring its molecular mechanisms for sustainable agriculture and crop improvement. *Environment International* 190: 108859.

Gade, A., Ingle, P., Nimbalkar, U. et al. (2023). Nanofertilizers: The next generation of agrochemicals for long-term impact on sustainability in farming systems. *Agrochemicals* 2 (2): 257–278.

Gambardella, C. and Pinsino, A. (2022). Nanomaterial ecotoxicology in the terrestrial and aquatic environment: A systematic review. *Toxics* 10 (7): 393.

García-Sánchez, S., Bernales, I., and Cristobal, S. (2015). Early response to nanoparticles in the Arabidopsis transcriptome compromises plant defence and root-hair development through salicylic acid signalling. *BMC Genomics* 16: 1–17.

Ghafariyan, M.H., Malakouti, M.J., Dadpour, M.R. et al. (2013). Effects of magnetite nanoparticles on soybean chlorophyll. *Environmental Science & Technology* 47 (18): 10645–10652.

Giraldo, J.P., Landry, M.P., Faltermeier, S.M. et al. (2014). Plant nanobionics approach to augment photosynthesis and biochemical sensing. *Nature Materials* 13 (4): 400–408.

González-García, Y., López-Vargas, E.R., Cadenas-Pliego, G. et al. (2019). Impact of carbon nanomaterials on the antioxidant system of tomato seedlings. *International Journal of Molecular Sciences* 20 (23): 5858.

González-Melendi, P., Fernández-Pacheco, R., Coronado, M.J. et al. (2008). Nanoparticles as smart treatment-delivery systems in plants: Assessment of different techniques of microscopy for their visualization in plant tissues. *Annals of Botany* 101 (1): 187–195.

Gopinath, K., Gowri, S., Karthika, V. et al. (2014). Green synthesis of gold nanoparticles from fruit extract of Terminalia arjuna, for the enhanced seed germination activity of Gloriosa superba. *Journal of Nanostructure in Chemistry* 4: 1–11.

Govorov, A.O. and Carmeli, I. (2007). Hybrid structures composed of photosynthetic system and metal nanoparticles: Plasmon enhancement effect. *Nano Letters* 7 (3): 620–625.

Gowtham, H.G., Shilpa, N., Singh, S.B. et al. (2024). Toxicological effects of nanoparticles in plants: Mechanisms involved at morphological, physiological, biochemical and molecular levels. *Plant Physiology and Biochemistry* 210: 108604.

Harish, V., Ansari, M.M., Tewari, D. et al. (2022). Nanoparticle and nanostructure synthesis and controlled growth methods. *Nanomaterials* 12 (18): 3226.

Hayes, K.L., Mui, J., Song, B. et al. (2020). Effects, uptake, and translocation of aluminum oxide nanoparticles in lettuce: A comparison study to phytotoxic aluminum ions. *Science of the Total Environment* 719: 137393.

Hossain, Z., Yasmeen, F., and Komatsu, S. (2020). Nanoparticles: Synthesis, morphophysiological effects, and proteomic responses of crop plants. *International Journal of Molecular Sciences* 21 (9): 3056.

Hu J., Guo H., Li J. et al. (2017). Comparative impacts of iron oxide nanoparticles and ferric ions on the growth of Citrus maxima. *Environmental Pollution* 221: 199–208.

Hu, J. and Xianyu, Y. (2021). When nano meets plants: A review on the interplay between nanoparticles and plants. *Nano Today* 38: 101143.

Ijaz, M., Khan, F., Ahmed, T. et al. (2023). Nanobiotechnology to advance stress resilience in plants: Current opportunities and challenges. *Materials Today Bio* 22: 100759.

Iqbal, Z., Sarkhosh, A., Balal, R.M. et al. (2021). Silicon alleviate hypoxia stress by improving enzymatic and non-enzymatic antioxidants and regulating nutrient uptake in muscadine grape (Muscadinia rotundifolia Michx.). *Frontiers in Plant Science* 11: 618873.

Jagadeesh, P., Rangappa, S.M., and Siengchin, S. (2024). Advanced characterization techniques for nanostructured materials in biomedical applications. *Advanced Industrial and Engineering Polymer Research* 7 (1): 122–143.

Jalil, S.U., Zahera, M., Khan, M.S. et al. (2019). Biochemical synthesis of gold nanoparticles from leaf protein of Nicotiana tabacum L. cv. xanthi and their physiological, developmental, and ROS scavenging responses on tobacco plant under stress conditions. *IET Nanobiotechnology* 13 (1): 23–29.

Janczak, C.M. and Aspinwall, C.A. (2012). Composite nanoparticles: The best of two worlds. *Analytical and Bioanalytical Chemistry* 402: 83–89.

Jaskulski, D., Jaskulska, I., Majewska, J. et al. (2022). Silver nanoparticles (AgNPs) in urea solution in laboratory tests and field experiments with crops and vegetables. *Materials* 15 (3): 870.

Jiang, H., Liu, J.K., Wang, J.D. et al. (2014). The biotoxicity of hydroxyapatite nanoparticles to the plant growth. *Journal of Hazardous Materials* 270: 71–81.

Jiang, M., Song, Y., Kanwar, M.K. et al. (2021). Phytonanotechnology applications in modern agriculture. *Journal of Nanobiotechnology* 19: 1–20.

Jing, X., Liu, Y., Liu, X. et al. (2023). Nitrogen-doped carbon dots enhanced seedling growth and salt tolerance with distinct requirements of excitation light. *RSC Advances* 13 (18): 12114–12122.

Juhel, G., Batisse, E., Hugues, Q. et al. (2011). Alumina nanoparticles enhance growth of Lemna minor. *Aquatic Toxicology* 105 (3–4): 328–336.

Kah, M., Tufenkji, N., and White, J.C. (2019). Nano-enabled strategies to enhance crop nutrition and protection. *Nature Nanotechnology* 14 (6): 532–540.

Khalid, M.F., Iqbal Khan, R., Jawaid, M.Z. et al. (2022). Nanoparticles: The plant saviour under abiotic stresses. *Nanomaterials* 12 (21): 3915.

Khan, S., Mansoor, S., Rafi, Z. et al. (2022). A review on nanotechnology: Properties, applications, and mechanistic insights of cellular uptake mechanisms. *Journal of Molecular Liquids* 348: 118008.

Khodakovskaya, M.V., de Silva, K., Nedosekin, D.A. et al. (2011). Complex genetic, photothermal, and photoacoustic analysis of nanoparticle-plant interactions. *Proceedings of the National Academy of Sciences* 108 (3): 1028–1033.

Khodakovskaya, M.V., Kim, B.S., Kim, J.N. et al. (2013). Carbon nanotubes as plant growth regulators: Effects on tomato growth, reproductive system, and soil microbial community. *Small* 9 (1): 115–123.

Kim, B. and Hochella Jr, M.F. (2015). Analytical transmission electron microscopy and scanning transmission electron microscopy techniques for the characterization of nanomaterial composition, phase and crystallinity. *Frontiers of Nanoscience* 8: 123–152.

Kole, C., Kole, P., Randunu, K.M. et al. (2013). Nanobiotechnology can boost crop production and quality: First evidence from increased plant biomass, fruit yield and phytomedicine content in bitter melon (*Momordica charantia*). *BMC Biotechnology* 13: 1–10.

Kolenčík, M., Ernst, D., Urík, M. et al. (2020). Foliar application of low concentrations of titanium dioxide and zinc oxide nanoparticles to the common sunflower under field conditions. *Nanomaterials* 10 (8): 1619.

Krishnia, L., Thakur, P., and Thakur, A. (2022). Synthesis of nanoparticles by physical route. In: *Synthesis and Applications of Nanoparticles*, 45–59. Singapore: Springer Nature Singapore.

Kumar, A., Choudhary, A., Kaur, H. et al. (2021). Smart nanomaterial and nanocomposite with advanced agrochemical activities. *Nanoscale Research Letters* 16: 1–26.

Kumari, S., Raturi, S., Kulshrestha, S. et al. (2023). A comprehensive review on various techniques used for synthesizing nanoparticles. *Journal of Materials Research and Technology* 27: 1739–1763.

Kundu, M., Pramanik, P., Maity, A. et al. (2019). Engineered nanomaterials: Classification and strategies for physicochemical characterization and advanced analytical techniques for the measurement of nanomaterials in plant samples. In: *Advances in Phytonanotechnology: From Synthesis to Application*, 17–43. Amsterdam: Elsevier. doi: 10.1016/B978-0-12-815322-2.00002-X.

Lahiani, M.H., Chen, J., Irin, F. et al. (2015). Interaction of carbon nanohorns with plants: uptake and biological effects. *Carbon* 81: 607–619.

Lahiani, M.H., Dervishi, E., Chen, J. et al. (2013). Impact of carbon nanotube exposure to seeds of valuable crops. *ACS Applied Materials & Interfaces* 5 (16): 7965–7973.

Laware, S. L. and Raskar, S. (2014). Influence of zinc oxide nanoparticles on growth, flowering and seed productivity in onion. *International Journal of Current Microbiology Science* 3 (7): 874–881.

Lee, J.H. and Kasote, D.M. (2024). Nano-priming for inducing salinity tolerance, disease resistance, yield attributes, and alleviating heavy metal toxicity in plants. *Plants* 13 (3): 446.

Li, J., Hu, J., Ma, C. et al. (2016). Uptake, translocation and physiological effects of magnetic iron oxide (γ-Fe2O3) nanoparticles in corn (Zea mays L.). *Chemosphere* 159: 326–334.

Li, P., Xia, Y., Song, K. et al. (2024). The impact of nanomaterials on photosynthesis and antioxidant mechanisms in Gramineae plants: Research progress and future prospects. *Plants* 13 (7): 984.

Li, W. and Keller, A.A. (2023). Assessing the impacts of Cu and Mo engineered nanomaterials on crop plant growth using a targeted proteomics approach. *ACS Agricultural Science & Technology* 4 (1): 103–117.

Lin, S., Reppert, J., Hu, Q. et al. (2009). Uptake, translocation, and transmission of carbon nanomaterials in rice plants. *Small* 5 (10): 1128–1132.

Linglan, M., Chao, L., Chunxiang, Q. et al. (2008). Rubisco activase mRNA expression in spinach: Modulation by nanoanatase treatment. *Biological Trace Element Research* 122: 168–178.

Linh, T.M., Mai, N.C., Hoe, P.T. et al. (2020). Metal-based nanoparticles enhance drought tolerance in soybean. *Journal of Nanomaterials* 2020 (1): 4056563.

Liu, Q., Chen, B., Wang, Q. et al. (2009). Carbon nanotubes as molecular transporters for walled plant cells. *Nano Letters* 9 (3): 1007–1010.

Magnabosco, P., Masi, A., Shukla, R. et al. (2023). Advancing the impact of plant biostimulants to sustainable agriculture through nanotechnologies. *Chemical and Biological Technologies in Agriculture* 10 (1): 117.

Mahor, A., Singh, P.P., Bharadwaj, P. et al. (2021). Carbon-based nanomaterials for delivery of biologicals and therapeutics: A cutting-edge technology. *C* 7 (1): 19.

Majumdar, S., Long, R.W., Kirkwood, J.S. et al. (2021). Unraveling metabolic and proteomic features in soybean plants in response to copper hydroxide nanowires compared to a commercial fertilizer. *Environmental Science & Technology* 55 (20): 13477–13489.

Maswada, H.F., Mazrou, Y.S., Elzaawely, A.A. et al. (2020). Nanomaterials. Effective tools for field and horticultural crops to cope with drought stress: A review. *Spanish Journal of Agricultural Research* 18 (2): 15.

Matras, E., Gorczyca, A., Pociecha, E. et al. (2023). Silver nanoparticles affect wheat (Triticum aestivum L.) germination, seedling blight and yield. *Functional Plant Biology* 50 (5): 390–406.

Mgadi, K., Ndaba, B., Roopnarain, A. et al. (2024). Nanoparticle applications in agriculture: Overview and response of plant-associated microorganisms. *Frontiers in Microbiology* 15: 1354440.

Mittal, D., Kaur, G., Singh, P. et al. (2020). Nanoparticle-based sustainable agriculture and food science: Recent advances and future outlook. *Frontiers in Nanotechnology* 2: 579954.

Morales, M.I. (2013). *Impact of Cerium Oxide Nanoparticles on Cilantro (Coriandrum sativum)*. The University of Texas at El Paso, El Paso, TX, USA. https://www.proquest.com/openview/bfa696265fbe82d97b3fa4b39192bb82/1?cbl=18750&pq-origsite=gscholar

Nair, R. (2016). Effects of nanoparticles on plant growth and development. In: *Plant Nanotechnology: Principles and Practices* 95–118. Cham: Springer. doi: 10.1007/978-3-319-42154-4_5.

Nawaz, A., Rehman, H.U., Usman, M. et al. (2023). Nanobiotechnology in crop stress management: An overview of novel applications. *Discover Nano* 18 (1): 74.

Nawaz, S. and Bano, A. (2020). Effects of PGPR (Pseudomonas sp.) and Ag-nanoparticles on enzymatic activity and physiology of cucumber. *Recent Patents on Food, Nutrition & Agriculture* 11 (2): 124–136.

Nhan, L.V., Ma, C., Rui, Y. et al. (2015). Phytotoxic mechanism of nanoparticles: Destruction of chloroplasts and vascular bundles and alteration of nutrient absorption. *Scientific Reports* 5 (1): 11618.

Noori, A., Donnelly, T., Colbert, J. et al. (2020). Exposure of tomato (Lycopersicon esculentum) to silver nanoparticles and silver nitrate: Physiological and molecular response. *International Journal of Phytoremediation* 22 (1): 40–51.

Pakrashi, S., Jain, N., Dalai, S. et al. (2014). In vivo genotoxicity assessment of titanium dioxide nanoparticles by Allium cepa root tip assay at high exposure concentrations. *PloS One* 9 (2): e87789

Pang, C. and Gong, Y. (2019). Current status and future prospects of semiconductor quantum dots in botany. *Journal of Agricultural and Food Chemistry* 67 (27): 7561–7568.

Panichikkal, J. and Krishnankutty, R.E. (2022). Chitosan and gold nanoparticles supplementation for augmentation of indole-3-acetic acid production by rhizospheric Pseudomonas aeruginosa and plant growth enhancement. *Current Microbiology* 79 (6): 185.

Perreault, F., Samadani, M., and Dewez, D. (2014). Effect of soluble copper released from copper oxide nanoparticles solubilisation on growth and photosynthetic processes of Lemna gibba L. *Nanotoxicology* 8 (4): 374–382.

Picchi, V., Gobbi, S., Fattizzo, M. et al. (2021). Chitosan nanoparticles loaded with N-acetyl cysteine to mitigate ozone and other possible oxidative stresses in durum wheat. *Plants* 10: 691.

Prasad, T.N.V.K.V., Sudhakar, P., Sreenivasulu, Y. et al. (2012). Effect of nanoscale zinc oxide particles on the germination, growth and yield of peanut. *Journal of Plant Nutrition* 35 (6): 905–927.

Pudake, R.N., Mittal, J., Tripathi, R.M. et al. (2019). Biochemical responses of maize seedlings exposed to SnNPs. *Micro & Nano Letters* 14 (6): 645–649.

Răcuciu, M., Miclăuş, S., and Creangă, D.E. (2009). The response of plant tissues to magnetic fluid and electromagnetic exposure. *Romanian Journal of Biophysics* 19 (1): 73–83.

Răcuciu, M., Tecucianu, A., and Oancea, S. (2022). Impact of magnetite nanoparticles coated with aspartic acid on the growth, antioxidant enzymes activity and chlorophyll content of maize. *Antioxidants* 11 (6): 1193.

Rajput, V.D., Singh, A., Minkina, T.M. et al. (2021). Potential applications of nanobiotechnology in plant nutrition and protection for sustainable agriculture. In: *Nanotechnology in Plant Growth Promotion and Protection: Recent Advances and Impacts*, 79–92. Hoboken, NJ: Wiley & Sons Ltd. doi: 10.1002/9781119745884.ch5.

Raliya, R., Biswas, P., and Tarafdar, J.C. (2015). TiO_2 nanoparticle biosynthesis and its physiological effect on mung bean (Vigna radiata L.). *Biotechnology Reports* 5: 22–26.

Rastogi, A., Zivcak, M., Sytar, O. et al. (2017). Impact of metal and metal oxide nanoparticles on plant: A critical review. *Frontiers in Chemistry* 5: 78.

Rezaei Cherati, S., Shanmugam, S., Pandey, K. et al. (2021). Whole-transcriptome responses to environmental stresses in agricultural crops treated with carbon-based nanomaterials. *ACS Applied Bio Materials* 4 (5): 4292–4301.

Rico, C.M., Morales, M.I., Barrios, A.C. et al. (2013). Effect of cerium oxide nanoparticles on the quality of rice (Oryza sativa L.) grains. *Journal of Agricultural and Food Chemistry* 61 (47): 11278–11285.

Rossi, L., Zhang, W., and Ma, X. (2017). Cerium oxide nanoparticles alter the salt stress tolerance of Brassica napus L. by modifying the formation of root apoplastic barriers. *Environmental Pollution* 229: 132–138.

Salehi, H., Chehregani Rad, A., Raza, A. et al. (2021). Foliar application of CeO2 nanoparticles alters generative components fitness and seed productivity in bean crop (Phaseolus vulgaris L.). *Nanomaterials* 11 (4): 862.

Sarraf, M., Vishwakarma, K., Kumar, V. et al. (2022). Metal/metalloid-based nanomaterials for plant abiotic stress tolerance: An overview of the mechanisms. *Plants* 11 (3): 316.

Schwab, F., Zhai, G., Kern, M. et al. (2016). Barriers, pathways and processes for uptake, translocation and accumulation of nanomaterials in plants–Critical review. *Nanotoxicology* 10 (3): 257–278.

Shah, G.A., Ahmed, J., Iqbal, Z. et al. (2021). Toxicity of NiO nanoparticles to soil nutrient availability and herbage N uptake from poultry manure. *Scientific Reports* 11 (1): 11540.

Shang, Y., Hasan, M.K., Ahammed, G.J. et al. (2019). Applications of nanotechnology in plant growth and crop protection: A review. *Molecules* 24 (14): 2558.

Sharma, P., Bhatt, D., Zaidi, M.G.H. et al. (2012). Silver nanoparticle-mediated enhancement in growth and antioxidant status of Brassica juncea. *Applied Biochemistry and Biotechnology* 167: 2225–2233.

Shekhar, S., Sharma, S., Kumar, A. et al. (2021). The framework of nanopesticides: A paradigm in biodiversity. *Materials Advances* 2 (20): 6569–6588.

Sheteiwy, M.S., Shaghaleh, H., Hamoud, Y.A. et al. (2021). Zinc oxide nanoparticles: Potential effects on soil properties, crop production, food processing, and food quality. *Environmental Science and Pollution Research* 28 (28): 36942–36966.

Shukla, V., Niveria, K., Shashidhar, P. et al. (2023). Dynamic light scattering (DLS) particle size analysis for biomedical nanotechnology. In: *Analytical Techniques for Biomedical Nanotechnology*, 16-1. Bristol, UK: IOP Publishing.

Siddiqi, K.S. and Husen, A. (2022). Plant response to silver nanoparticles: A critical review. *Critical Reviews in Biotechnology* 42 (7): 973–990.

Siddiqui, M.H., Al-Whaibi, M.H., Firoz, M. et al. (2015). Role of nanoparticles in plants. In: *Nanotechnology and Plant Sciences: Nanoparticles and Their Impact on Plants* 19–35. Cham: Springer. doi: 10.1007/978-3-319-14502-0_2.

Singh, A., Rajput, V.D., Varshney, A. et al. (2024). Revolutionizing crop production: Nanoscale wonders-current applications, advances, and future frontiers. *Egyptian Journal of Soil Science* 64 (1): 221–258.

Singh, D., Sharma, A., Verma, S.K. et al. (2024). Impact of nanoparticles on plant physiology, nutrition, and toxicity: A short review. *Next Nanotechnology* 6: 100081.

Soares, C., Branco-Neves, S., de Sousa, A. et al. (2016). Ecotoxicological relevance of nano-NiO and acetaminophen to Hordeum vulgare L.: Combining standardized procedures and physiological endpoints. *Chemosphere* 165: 442–452.

Song, G., Hou, W., Gao, Y. et al. (2016). Effects of CuO nanoparticles on Lemna minor. *Botanical Studies* 57: 1–8.

Srivastava, A. and Rao, D.P. (2014). Enhancement of seed germination and plant growth of wheat, maize, peanut and garlic using multiwalled carbon nanotubes. *European Chemical Bulletin* 3 (5): 502–504.

Strekalovskaya, E.I., Perfileva, A.I., and Krutovsky, K.V. (2024). Zinc oxide nanoparticles in the "soil–bacterial community–plant" system: Impact on the stability of soil ecosystems. *Agronomy* 14 (7): 1588.

Sundaria, N., Singh, M., Upreti, P. et al. (2019). Seed priming with iron oxide nanoparticles triggers iron acquisition and biofortification in wheat (Triticum aestivum L.) grains. *Journal of Plant Growth Regulation* 38: 122–131.

Tan, X.M., Lin, C., and Fugetsu, B. (2009). Studies on toxicity of multi-walled carbon nanotubes on suspension rice cells. *Carbon* 47 (15): 3479–3487.

Tang, Y., Zhao, W., Zhu, G. et al. (2023). Nano-pesticides and fertilizers: solutions for global food security. *Nanomaterials* 14 (1): 90.

Tarafdar, J.C., Xiong, Y., Wang, W.N. et al. (2012). Standardization of size, shape and concentration of nanoparticle for plant application. *Applied Biological Research* 14 (2): 138–144.

Taran, N., Batsmanova, L., Kovalenko, M. et al. (2016). Impact of metal nanoform colloidal solution on the adaptive potential of plants. *Nanoscale Research Letters* 11: 1–6.

Thiruvengadam, M., Chi, H.Y., and Kim, S.H. (2024). Impact of nanopollution on plant growth, photosynthesis, toxicity, and metabolism in the agricultural sector: An updated review. *Plant Physiology and Biochemistry* 207: 108370.

Tipu, M.M.H., Baroi, A., Rana, J. et al. (2021). Potential applications of nanotechnology in agriculture: A smart tool for sustainable agriculture. In: *Agricultural Development in Asia-potential Use of Nano-materials and Nano-technology*. London: IntechOpen.

Tiwari, D.K., Dasgupta-Schubert, N., Villaseñor Cendejas, L.M. et al. (2014). Interfacing carbon nanotubes (CNT) with plants: Enhancement of growth, water and ionic nutrient uptake in maize (Zea mays) and implications for nanoagriculture. *Applied Nanoscience* 4: 577–591.

Tripathi, D.K., Singh, S., Singh, S. et al. (2017). Nitric oxide alleviates silver nanoparticles (AgNps)-induced phytotoxicity in Pisum sativum seedlings. *Plant Physiology and Biochemistry* 110: 167–177.

Ulbrich, K., Hola, K., Subr, V. et al. (2016). Targeted drug delivery with polymers and magnetic nanoparticles: Covalent and noncovalent approaches, release control, and clinical studies. *Chemical Reviews* 116 (9): 5338–5431.

Verma, D.K., Patel, S., and Kushwah, K.S. (2021). Effects of nanoparticles on seed germination, growth, phytotoxicity and crop improvement. *Agricultural Reviews* 42 (1): 1–11.

Villagarcia, H., Dervishi, E., de Silva, K. et al. (2012). Surface chemistry of carbon nanotubes impacts the growth and expression of water channel protein in tomato plants. *Small (Weinheim an der Bergstrasse, Germany)* 8 (15): 2328–2334.

Wang, Q., Ma, X., Zhang, W. et al. (2012). The impact of cerium oxide nanoparticles on tomato (Solanum lycopersicum L.) and its implications for food safety. *Metallomics* 4 (10): 1105–1112.

Wohlmuth, J., Tekielska, D., Čechová, J. et al. (2022). Interaction of the nanoparticles and plants in selective growth stages—usual effects and resulting impact on usage perspectives. *Plants* 11 (18): 2405.

Xie, Y., Li, B., Tao, G. et al. (2012). Effects of nano-silicon dioxide on photosynthetic fluorescence characteristics of Indocalamus barbatus McClure. *Journal of Nanjing Forestry University* 55 (02): 59.

Yadav, N., Bora, S., Devi, B. et al. (2024). Nanoparticle-mediated defense priming: A review of strategies for enhancing plant resilience against biotic and abiotic stresses. *Plant Physiology and Biochemistry* 213: 108796.

Yang, J., Cao, W., and Rui, Y. (2017). Interactions between nanoparticles and plants: Phytotoxicity and defense mechanisms. *Journal of Plant Interactions* 12 (1): 158–169.

Yang, L., Wu, X., Liu, F. et al. (2009). Novel biodegradable polylactide/poly (ethylene glycol) micelles prepared by direct dissolution method for controlled delivery of anticancer drugs. *Pharmaceutical Research* 26: 2332–2342.

Yang, W., Zhou, Z., and Chu, Z. (2023). Emerging roles of salicylic acid in plant saline stress tolerance. *International Journal of Molecular Sciences* 24 (4): 3388.

Yang, Y., Li, R., Wang, L et al. (2023). Enhanced germination and growth of alfalfa with seed presoaking and hydroponic culture in Fe2O3 magnetic nanoparticles. *Journal of Nanomaterials* 2023 (1): 9783977.

Zaytseva, O. and Neumann, G. (2016). Carbon nanomaterials: Production, impact on plant development, agricultural and environmental applications. *Chemical and Biological Technologies in Agriculture* 3: 1–26.

Zhu, H., Han, J., Xiao, J.Q. et al. (2008). Uptake, translocation, and accumulation of manufactured iron oxide nanoparticles by pumpkin plants. *Journal of Environmental Monitoring* 10 (6): 713–717.

8

Chitosan Nanoparticles as Nanosorbent for Potential Removal of Pollutant from the Soil

Abirami Geetha Natarajan[1,2], *Kripa V*[1], *Jothi Ganesan M*[1], *and Philip Bernstein Saynik*[1]

[1] *Department of Chemical Engineering, Sri Venkateswara College of Engineering, Sriperumbudur, India*
[2] *Department of Chemical and Petroleum Engineering, Schulich School of Engineering, University of Calgary, Calgary, Canada*

8.1 Introduction

8.1.1 Background on Soil Remediation

Soil plays a vital role in human activity and balances the ecosystem. It supports crop yield in agriculture by providing all required nutrients and minerals for their healthy growth. Additionally, it acts as water storage, which is essential for photosynthesis and other biological processes (Lal 2009; Weil and Brady 2022). Acting as a protective barrier, it hinders the penetration of pollutants into the groundwater, thereby cleaning the water. Furthermore, it controls water flow by directing it to water bodies like rivers and streams and manages water storage (McBratney, Field, and Koch 2014; Pimentel and Burgess 2013). As an essential component in the carbon cycle, soil helps to absorb carbon dioxide, functioning as a primary carbon sink. Effective land management practices can lower greenhouse gas emissions (Lehmann and Kleber 2015; Smith et al. 2008). Soil diversity is home to a wide range of organisms that are vital to ecosystem health and nutrient cycling. It also offers a solid basis for building and infrastructure (Amundson et al. 2015; Bardgett and Van 2014). The land holds cultural and historical significance as it preserves archaeological artifacts and embodies traditional knowledge related to soil management practices (Homburg and Sandor 2011; Sandor and Eash 1991). Healthy soil produces nutrient-rich food, vital for human health, and helps suppress soil-borne diseases, lowering the need for chemical interventions. Its significance is further shown by how it reduces pollution, prevents erosion, and supports the agricultural sector (Bünemann et al. 2018). Despite its numerous benefits, soil can become compromised when exposed to harmful substances. This degradation is commonly known as soil pollution.

Soil pollution refers to the presence of chemical substances in soil through anthropogenic or natural activities that have harmful effects on the environment. Excessive levels

Nanotechnology-based Sustainable Agriculture, First Edition. Edited by Pardeep Singh et al.

of chemicals like pesticides, petroleum hydrocarbons, alkalis, lead, mercury, and naphthalene have a negative impact on soil quality (Neethu et al. 2021). Soil pollution results from the introduction of harmful substances into the soil through diverse sources and activities, upsetting its natural equilibrium and overall health.

8.1.1.1 Primary Causes of Soil Pollution

Industrial Activities: Industries like refineries produce a considerable number of pollutants and are known to be significant contributors to soil pollution. Chemical industries release heavy metals, chemicals, and other toxic substances, which can pose long-term environmental and health risks. These toxins result in chronic diseases in humans and affect the growth of microorganisms and plants (Dehkordi et al. 2024).

Agricultural Practices: Pesticides, fertilizers, and herbicides are consumed significantly in agricultural activities. Pesticides and herbicides contain toxic compounds that can persist in the soil and affect microbial activity and soil fertility. Overusing pesticides, herbicides, and chemical fertilizers in agriculture introduces harmful chemicals into the soil. These substances can accumulate over time, resulting in nutrient imbalances and the accumulation of salts in the soil, eventually degrading its quality and causing soil pollution (Pimentel 2005).

Improper Waste Disposal: The improper handling of waste materials, illegal dumping, landfills, and hazardous waste sites can all contribute to the buildup of pollutants in the soil. Hazardous waste sites can potentially leak hazardous chemicals into the surrounding soil and groundwater (Pimentel and Burgess 2013). The improper disposal of electronic waste, or "e-waste," can also lead to soil contamination through the release of heavy metals and other toxic substances (Inglezakis and Moustakas 2015).

Accidental Spills: Major oilfields are often located within significant river deltas and wetlands posing risks to the structure, functioning, and ecosystem service of these areas during exploration activities. The contamination of soil with crude oil could negatively impact soil fertility and physical properties, and contribute to the degradation of the marshlands (Wang et al. 2013).

Urbanization and Construction: An increase in the demand for individual living spaces has led to increased environmental damage from the procurement, manufacture, and processing of building materials. In addition, the construction industry releases a significant amount of hazardous and nonhazardous waste from construction and demolition, which contaminates soil, air, and water sources. When urban agriculture is practiced on contaminated soil, the concentration of toxic elements in crops increases, and consuming the produce of these crops may be harmful to human health (Sandil and Kumar 2022).

Sewage and Wastewater: Improper trash disposal can break down into the soil, depositing various chemicals and pollutants. Heavy metals can enter the soil through the use of mineral fertilizers or polluted water for crop irrigation. Improper drainage and increased runoff can contaminate the surrounding land areas or streams (Mishra, Mohammad, and Roychoudhury 2016).

Mining Activities: The extraction of minerals and metals often involves the removal of large amounts of soil and rock, leading to soil erosion and degradation. Furthermore, the disposal of mining waste, such as tailings, can introduce heavy metals and other pollutants into the soil. Acid mine drainage, which happens when sulfide minerals in exposed rock surfaces react with air and water to produce sulfuric acid, can also contaminate nearby soils (Lottermoser and Lottermoser 2010).

Atmospheric Deposition: Pollutants such as heavy metals, sulfur dioxide, and nitrogen oxides can be deposited onto the soil from the atmosphere. These pollutants can originate from industrial emissions, vehicle exhaust, and the burning of fossil fuels. Acid rain, which is caused by the deposition of sulfur and nitrogen compounds, can further degrade soil quality by severely altering its pH and nutrient availability (Likens, Driscoll, and Buso 1996).

8.1.2 Need for Effective Remediation Techniques

Soil pollution is a serious ongoing issue. Every year, 9 million premature deaths could be attributed to a combination of factors, including exposure to contaminated air, soil, and water. Soil contamination is one of the major factors that trigger disability adjusted life years (DALYs) at higher ages (Münzel et al. 2023). Soil contamination poses a significant threat to human and environmental health. The problem arises when toxic substances, such as pesticides, heavy metals, and industrial chemicals, are introduced to the soil and disrupt its natural equilibrium (Alloway 2012). The toxins present in the polluted soil get absorbed into the crops through their xylem and phloem. The chemicals accumulate inside the crops and are consumed by humans and animals. This affects food safety and poses a severe risk to human health (Lu et al. 2015). Heavy metals (like cadmium and lead) and metalloids (arsenic) are two significant contributors to cardiovascular diseases, as evidenced by epidemiological and experimental data. Research has demonstrated that heavy metals and metalloids have a detrimental effect on the epigenetic control of gene expression (Münzel et al. 2023). Improper landfilling techniques can pave the way for leachates, to intrude into the soil. The leachate contains heavy metals and other toxic substances that lead to soil contamination. Researchers have found that heavy metals, generally detected in leachates are classified as biohazards since they are carcinogenic and mutagenic.

Although the soil acts as a natural filter, it can also become a medium for groundwater pollution. The contaminated soil introduces toxins and other pathogens into the groundwater, resulting in microbial infections, gastrointestinal illness, and other health problems.

Soil pollution has a series of negative impacts on agriculture by affecting soil fertility, nitrogen fixation, and crop yield. It creates an imbalance in soil fauna and flora. Contaminants cause toxicity and hinder the growth of agricultural plants, thereby affecting all the dependent organisms. It destroys the soil ecosystem. Excessive application of fertilizers and pesticides in agricultural regions can result in soil degradation by losing its properties. It ultimately leads to the futility of the soil. The degraded soil loses its properties and could lead to increased soil erosion, which can lead to natural disasters like landslides. This has also been found to affect construction activities as well.

Therefore, in order to protect the food chain, prevent harmful exposure, restore ecological balance, mitigate adverse effects on the economy, maintain ecosystems, protect public health, and comply with regulations, remediation of soil is necessary.

There are various physical and biological methods available to treat contaminated soil. Bioremediation, phytoremediation, soil washing, bioaugmentation, chemical oxidation/reduction, thermal desorption, encapsulation, electrokinetic remediation, and air sparging are a few techniques for treating contaminated soil. Each method has its advantages and disadvantages. For instance, bioremediation is an effective and environmentally friendly method, but it is very time-consuming and unsuitable for complex mixtures. There is a potential for bioaccumulation of toxins in the phytoremediation method. In the case of soil washing, it causes secondary pollution. Chemical oxidation/reduction, electrokinetic remediation, and thermal desorption are pretty expensive. Air sparging may cause air pollution. Hence, the choice of remediation method for each contaminated area is crucial. A combination of physical and biological methods can help us to overcome the limitations involved in soil remediation. Biopolymers, such as chitosan, offer a promising solution due to their natural origin and versatility in addressing soil contamination. Chitosan has drawn notice for its capacity to adsorb contaminants and improve soil recovery methods.

8.2 Chitosan

8.2.1 Chemical Structure and Properties

Chitosan is a biopolymer derived from a natural polysaccharide known as chitin. Chitin is found in the form of ordered crystalline microfibrils, structural elements found in the cell walls of fungi and yeast, as well as the exoskeleton of arthropods.

It is a linear polymer of mainly β-(1→4)-2-acetamido-2-deoxy-D-glucopyranose units and β-(1→4)-2-amino-2-deoxy-D-glucopyranose (Teixeira-Costa and Andrade 2021). Chitin exists in three crystalline polymorphs as anti-parallel *N*-acetylglucosamine chains (α-chitin), parallel chains of *N*-acetylglucosamine (β-chitin), and two β-chitin alternated with an α-chitin forming γ-chitin (Teixeira-Costa and Andrade 2021). Chitin is insoluble in most solvents, which complicates its utilization in other processes (Kim 2013).

Chitosan consists of D-glucosamine and *N*-acetyl-D-glucosamine. Based on the degree of deacetylation, the number of D-glucosamine and *N*-acetyl-D-glucosamine varies. Chitosan depicts structure similarity with cellulose. Instead of the hydroxyl group at position C-2, it contains an amino group (Lizardi-Mendoza, Monal, and Valencia 2016). It has three major functional groups: primary, secondary hydroxyl, and amine. Chitosan is chemically active due to the presence of the primary and secondary hydroxyl groups in each unit and an amino group in the deacylated unit. They help perform modifications to improve the mechanical properties (Islam, Bhuiyan, and Islam 2017). Chitosan attains polycationic characteristics while pronation under acidic conditions due to the presence of amine groups (Wahba 2020). Unlike other polymers, Chitosan is positively charged. The role of each of the functional groups of chitosan in adsorption is given in Table 8.1.

The temperature, acidity, degree of acetylation, polymer crystallinity, and polymer molecular weight are among the variables that affect chitosan solubility. The chitosan contains $-NH_2$ groups. The chitosan becomes soluble in acid media when it contains a large amount of pronated $-NH_2$ groups.

Table 8.1 Role of functional groups of chitosan.

Functional group	Structure	Role in adsorption
Amino (NH_2)	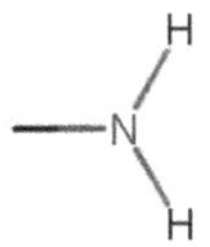	Provides a positive charge to the chitosan which attracts negative ions for adsorption
Hydroxyl (OH)		Promotes interaction with organic compounds
Acetyl ($COCH_3$)	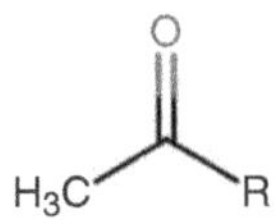	It is removed to produce highly deacetylated chitosan for better adsorption
N-acetylglucosamine	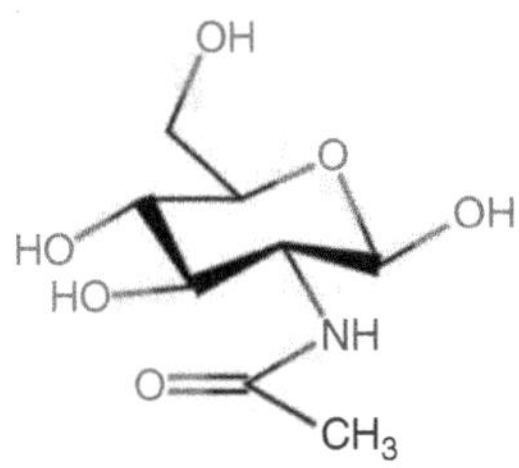	Provides structural integrity

When the degree of deacetylation of chitin attains approximately 50%, it becomes soluble in an acidic medium, which is then termed chitosan (Pal et al. 2021). The degree of deacetylation (DD) is a key parameter in evaluating the properties of chitosan. Chitosan shows unique characteristics like biocompatibility, biodegradability and nontoxic. It makes it suitable for various applications. The physical properties depend on the molecular weight, which can range from 10 000 to 1 million Daltons (Zargar, Asghari, and Dashti 2015). The following section contains various properties of chitosan.

8.2.1.1 Solubility

The solubility of the chitosan depends on the extent of its deacetylation. Higher solubility can be achieved by a higher degree of deacetylation. Chitosan is soluble only in an acidic medium due to the presence of amino groups that protonate and make it a polyelectrolyte. The key factors affecting the solubility are ionic concentration, degree of deacetylation, pH, and the nature of the acid used for dissolution. The pKa value will be approximately 6.5. The limited solubility of chitosan restricts its utilization in neutral pH conditions. Chitosan has better solubility at low molecular weight due to intra-chain hydrogen bonds and the distribution of acetyl groups along the chain. Recent research has shown chitosan to be soluble at neutral pH with the application of glycerol 2-phosphate (Zargar, Asghari, and Dashti 2015).

8.2.1.2 Viscosity

The viscosity of chitosan solution, which very much depends on the molecular weight of chitosan, is a critical property that influences its processing, formulation, and performance in various applications. Higher viscosity is achieved by high molecular weight, whereas lower viscosity is attained by low molecular weight. This is due to the increased interaction among the polymer chains. Depending upon the source and method of preparation, the molecular weight of chitosan varies. The average molecular weight is determined using Mark-Houwink Parameters (Teixeira-Costa and Andrade 2021). It relates intrinsic viscosity with the molecular weight of chitosan. The viscosity of chitosan plays an important role in soil remediation since it affects its efficiency in soil treatment. The viscosity of chitosan solution increases with an increase in its concentration and vice versa (Kim 2013).

8.2.1.3 Thermal Properties

Chitosan shows good thermal stability due to its polymeric nature and hydrogen bonding. The thermal degradation of chitosan depends on its degree of deacetylation. Chitosan was found to undergo thermal degradation only at temperatures greater than 200 °C, revealed through thermogravimetric analysis (Hong et al. 2007). Although chitosan has poor thermal conductivity, it is relatively higher than other polymers.

8.2.1.4 Biological Properties

Due to the physicochemical characteristics of chitosan, the following properties can be observed: Biocompatibility, antimicrobial activity, biodegradability, antioxidant activity, antitumoral, nontoxicity, bioadhesivity, hypolipidemic activity (Jiménez-Gómez and Cecilia 2020). Chitosan oligomers exhibit prebiotic properties, which help in the growth of gut bacteria. They also have physiological effects like macrophage activation, angiogenesis and anti-inflammatory effects.

8.2.1.5 Mechanical Properties

Chitosan has weak mechanical properties, which are limitations in its applications. So, some critical parameters can be enhanced for its practical usage and application. They are

- Elastic modulus and tensile strength – Improvised by chemical crosslinking.
- Ductility – Incorporating plasticizers or blending with other polymers enhances the flexibility of chitosan.
- Surface area – Improved through the use of porogens.

Chitosan, with enhanced mechanical properties, finds application in Bone tissue regeneration, drug delivery, and food packaging. They have film-forming ability, which can be used to produce coatings and membranes. They also have good adsorption properties, which can be used as a medium for filtration and separation.

8.2.2 Synthesis and Modification

Chitosan is obtained from the deacetylation of chitin. Sources of chitin include shrimp, crayfish, oysters, and lobsters (commonly known as crustacean shells; Madera-Santana, Herrera-Méndez, and Rodríguez-Núñez 2018). The Crustacean shells contain about

20–50% $CaCO_3$, 15–40% chitin, and 20–40% proteins. Chitosan can be prepared via chemical or biological methods. The primary chemical reaction of deacetylation is as follows,

$$\underset{\text{Chitin}}{(C_8H_{13}O_5)_n} + \text{nNaOH} \rightarrow \underset{\text{Chitosan}}{(C_6H_{11}O_4)_n} + \underset{\text{Acetate}}{nCH_3COONa}$$

8.2.2.1 Chemical Method

The chitin is washed, dried, and ground into fine powder. The steps involved here are demineralization (removal of minerals), deproteinization (removal of protein), and deacetylation (removal of acetyl group). In the demineralization step, hydrochloric acid (HCl) is added to the prepared powder to remove the $CaCO_3$ present in the exoskeleton. Then, the demineralized sample is added with NaOH at 60–100 °C to remove the proteins present. The absence of color indicates complete protein removal. Then, the obtained demineralized and deproteinated residue is cleaned using water and ethanol. Then, the purified chitin is dried and added to 50% NaOH at 600 °C for eight hours to obtain pure chitosan. Here, removal of the acetyl group is observed. This method has several limitations. It is a multistep process and not environmentally friendly. Waste generation is relatively higher, and calcium carbonate is not recovered. However, it produces a high deacetylation percentage with a shorter processing time and is widely used in industries.

8.2.2.2 Biological Method

8.2.2.2.1 Enzymatic Method

In this process, demineralization takes place with the help of organic acid HCl. The enzymatic route involves two processes: deproteinization and deacetylation, which occur with the help of enzymes. Microbes or fish entrails like grey triggerfish and Sardinella intestines act as proteinases and *Bacillus licheniformis* as deacetylases. It uses mild operating conditions like a temperature range between 25 and 50 °C. However, their reaction mechanisms are quite sophisticated and complex. This method has limitations as the enzymes are quite expensive and inefficient compared to the chemical method. The processing time is also quite a long time.

8.2.2.2.2 Fermentation Method

This method can be classified into two methods based on the microbial strains used for preparation. During demineralization, if the strain employed involves the production of lactic acid, then the method is termed lactic acid fermentation. If not, it is a non-lactic acid fermentation method. Generally, Lactic acid fermentation is preferred. In this process, a suitable strain (*Lactobacillus*) is made to secrete lactic acid for demineralization to take place. Fermentation takes place for more than seven days under controlled temperatures. The produced Lactic acid removes $CaCO_3$ from the shells. Then, the secreted enzymes from the microbes degrade the proteins present in the shell. After that, deacetylation takes place, producing chitosan (Mourya and Inamdar 2008).

8.2.3 Applications of Chitosan

Chitosan nanoparticles have diverse applications in various fields. Some of them are drug delivery, antimicrobial activity, tissue engineering, wastewater treatment, food industry, and agriculture.

Antimicrobial Activity: Chitosan has distinctive antimicrobial properties, which result in its effective growth-inhibiting effect (Aranaz et al. 2021). They can inhibit bacteria like *Staphylococcus aureus, Lactobacillus plantarum, Bacillus megaterium, Lactobacillus bulgaricus, Salmonella typhimurium, Escherichia coli, Pseudomonas fluorescens* and *Vibrio parahaemolyticus*. Though the exact mechanism is still unknown, one widely accepted mechanism is that the chitosan's amino group reacts with the microbial cell membrane, causing protein leakage inside the microbes. This causes effective inhibition of bacteria. Antifungal properties were also observed in *Botrytis cinerea, Pyricularia oryzae, Fusarium oxysporum*, and *Trichophyton equinum*. They follow different mechanisms for inhibiting protein synthesis in fungi (Reshad, Jishan, and Chowdhury 2021). Generally, antimicrobial activity depends upon the pH, acetylation degree, molecular weight, and concentration.

Drug Delivery: Chitosan has huge applications in the pharmaceutical industry due to its nontoxicity, biocompatibility, and mucoadhesive properties. The primary amino group (positively charged) in chitosan interacts with the biological surfaces to deliver a drug to a specific area. They are used in oral, ocular, nasal, buccal, pulmonary, and anticancer drug delivery. They also have remarkable properties in in situ gelation and efflux pump inhibition (Morin-Crini et al. 2019). Glycol chitosan-based nanoparticles have been used for siRNA targeting and codelivery of drugs, leading to improved anticancer efficacy (Lee et al. 2014). MEL-containing chitosan microparticles have been suggested for nasal drug delivery to reduce acute pain and enhance analgesia (Bartos et al. 2021).

They are also used for site-specific antibiotic delivery and gene and vaccine delivery. In gene delivery, chitosan binds with DNA molecules forming to protect from degradation. It facilitates DNA to enter the cell structure in nonviral gene therapy.

Tissue Engineering: Chitosan derivatives have novel properties that can be applied for regeneration and wound healing. They can be used in skin, bone, cartilage, ligament, and neural and vascular tissue engineering. They act as supporting scaffolds to restore the damaged organs. In bone engineering, chitosan nanoparticles (CSNPs) have been used to mimic the natural bone extracellular matrix, which can enhance osteogenic differentiation and bone healing (Singh, Mittal, and Benjakul 2021). The neural regeneration process involves autografts and allografts. CSNPs have a lesser chance of immune rejection since the tissues are taken from the patient. One such compelling example is the use of chitosan polyglycolic acid, which showed effective results in nerve regeneration. In corneal regenerative technology, artificial cornea tissues are generated, which can be used for transplantation.

Wastewater Treatment: CSNPs are used to treat wastewater from industries as they can bind with different species. They have demonstrated effective results in the removal of dyes and purification of water. They can be used to remove Ni, Cu, As, Cd, Pb, and Zn using different structures of chitosan (Domingues et al. 2012). Chitosan – hydrogel-based water systems are implemented to produce safe drinking water economically (Gassara et al. 2015). They have shown significant results in the adsorption of methylene blue dye and methylene violate. They are also used for filtration, flocculation, and adsorption-related processes.

Food Industry: Chitosan is used as food additives, antioxidant and antimicrobial agents, and packing and coating material. Due to their antimicrobial properties, they are employed as coatings to prolong the shelf life of food items. They are also employed in eco-friendly food packing as they can prevent microbial deterioration. Since chitosan can lower cholesterol levels by blocking the absorption of fats, it can be used as dietary fiber and can also be used for body fat loss. They are also used as animal feed additives due to the presence of nutrients.

Beverage Industry: Chitosan stabilizes and clarifies fruit juices and wines by removing suspended particles. It enhances the clarity of the beverages. It is used in the coagulation and flocculation processes of passion fruit and beer production, respectively (Bayda et al. 2019; Jimtaisong and Saewan 2014). It acts as a natural preservative by inhibiting the growth of undesired microorganisms.

Cosmetics: The cationic nature of the chitosan helps in the interaction of the product with hair and skin. It acts as bacteriostatic, antistatic and fungistatic. Chitosan derivatives are versatile compounds that are compatible with any formulation. The controlled release of bioproducts increases product efficacy. It can retain moisture and provide hydration to the skin and hair. It is also used as a deodorizing and dental care product. It enhances the formulation of cosmetic products and provides better performance (Tsaramirsis et al. 2022).

Agriculture: Chitosan is used to control plant diseases since it has antifungal and antibacterial properties. It acts as a protective barrier and enhances the resistance to plant diseases. It is employed as a soil amendment to reduce pathogen attacks and promote the overall growth of plants. It improves soil structure, water retention, and nutrient availability. Chitosan is a bio-stimulant that stimulates beneficial bacterial growth, promoting plant growth. Chitosan is very versatile, as it is effective in all soil types.

8.3 Nanotechnology and Soil Remediation

8.3.1 Introduction to Nanotechnology

Nanotechnology is a branch of science that deals with the manipulation and control of matter at dimensions between 1 and 100 nm, where the matter exhibits unique characteristics. The word "Nano" is derived from the Greek word meaning "dwarf" (Bayda et al. 2019). Nanoparticles have enhanced properties compared to bulk materials due to quantum effects and electrical, optical, and other phenomena. Nanoparticles possess a large surface area, higher reactivity, and smaller dimensions than their bulk counterparts. From research labs to large-scale production units, nanotechnology has found numerous applications. It is efficient, clean, and more affordable than existing materials. Nanotechnology is one of the revolutionary technologies of the twenty-first century. Nanomaterials have the potential to contribute to a sustainable society through CO_2 emission control, heavy metal remediation, energy generation, and more.

Nanotechnology was proposed by Richard Feynman, an American physicist in 1959. Japanese scientist Norio Taniguchi coined the term "Nanotechnology." He researched the fabrication of nanomaterials. The concept of nanotechnology grabbed the attention of many scientists, leading to various research efforts in this field. Nanomaterials can be produced in two ways: top-down and bottom-up methods. The top-down method involves fragmenting bulk materials into nanomaterials using various precision engineering techniques. The bottom-up method involves building up structures atom by atom using various physical and chemical methods, such as self-assembly. Due to their exceptional characteristics, these materials have garnered significant attention in many fields.

The advancements in nanotechnology have brought about revolutionary inventions in the fields of electronics and medicine. In medicine, targeted drug delivery is a promising way to improve the treatment of various diseases, including incurable diseases, using nanomaterials. It holds the potential to develop compact electronic gadgets with exemplary performance. Smart sensors, nano storage chips, and optoelectronics are a few nanotechnology inventions in electronics (Tsaramirsis et al. 2022). Nanopolymers and nanocomposites are used in packaging and material fabrication due to their superior strength, durability, and functionality (Huang et al. 2022). With high surface area and unique characteristics, nanocatalysts facilitate improved conversion rates, enhancing the efficiency of the process, leading to higher yield, less energy usage, and lower emission (Saleh 2018). Adsorption is among the most effective techniques in various pollutant removal where the usage of nano adsorbent further enhances the removal efficiency by its high surface area, pore size and distribution, chemical reactivity, and more. Regeneration is possible in this process and has high selectivity. Since the removal process is based on the adsorption of pollutants with weak forces, no secondary pollution occurs. Hence, this process is efficient and sustainable.

8.3.2 Types of Nano Adsorbents

Nanosorbents are adsorbent materials at nano sizes with very high surface area, pore sizes with increased pore distribution, and enhanced electrical and thermal conductivity with high affinity. These nanosorbents find their application in the removal of various pollutants contributing to air, water, and soil pollution. Nanosorbent must possess the following characteristics to be used in pollution removal:

1. It should be nontoxic.
2. It must have a high adsorption capability even at low concentrations of pollutants.
3. Regeneration must be easy.

Based on the surface property and functionalization, nanosorbents are divided into multiple groups. Nanoparticles like metallic oxide nanoparticles, nanostructured mixed oxides, and magnetic nanoparticles are widely used as nanosorbents. Carbon nanotubes, carbon nanoparticles, and carbon nanosheets are a few carbonaceous nanomaterials based on sorbent properties. Correspondingly, silicon nanotubes, silicon nanoparticles, silicon nanosheets, nano clays, xerogels, and aerogels are used for adsorption methods.

8.3.2.1 Metallic Oxide Nanoparticles

Metallic oxide nanoparticles (e.g. aluminum trioxide nanoparticles) can potentially remove heavy metals and dyes because of their high affinity and size quantization effect. Metallic oxides can be prepared in various shapes with various functional characteristics (Kumar et al. 2013).

8.3.2.2 Metallic Nanoparticles

Silver and gold nanoparticles are widely used in water treatment methods due to their antimicrobial properties and ability to degrade organic pollutants. Metallic nanoparticles exhibit high surface area, enhanced reactivity, and tunable surface chemistry (Natarajan et al. 2023).

8.3.2.3 Carbonaceous Nanoparticles

These nanoparticles have unique chemical, mechanical, and electrical properties because of their sp^2 hybridization. Among all carbonaceous nanoparticles, activated carbon is widely employed for the remediation of air, soil, and water contamination. Fullerenes, graphene-based nanomaterial, graphene nanocomposites, and carbon nanotubes are other nanoparticles used in adsorption (Shan et al. 2017).

8.3.2.4 Other Nanoparticles

Various silicon-based nanoparticles like silicon nanotubes, silicon nanoparticles, and silicon nanosheets are used in air pollution control. Nanofibers, nanoclays, polymer-based nanomaterials, xerogels, and aerogels are also used as nanosorbents in various adsorption processes (Khajeh, Laurent, and Dastafkan 2013).

8.3.3 Benefits of Using Nano Adsorbent

Nanosorbents find application in various adsorption processes because of their unique characteristics. Various research studies have been conducted to assess their potential to replace conventional adsorbents currently available on the market. Here are some of the key benefits of nanosorbents:

8.3.3.1 High Surface Area

The size reduction of the bulk material into nanomaterial has increased its surface area-to-volume ratio. This offers more active sites for adsorption compared to the conventional adsorbents. The specific cost of the material is also reduced due to the overall low mass usage.

8.3.3.2 Enhanced Reactivity

Nanosorbents show enhanced reactivity attributed to their superior surface energy and increased active sites. It enhances the adsorption rate of the pollutants onto the surface of the nanosorbent. It results in a high removal rate over a short reaction time. Further surface modification of nanosorbent improves its adsorption capacity, efficiency, stability, and inertness. It is a tool used to overcome the constraints present in the nanosorbent and improve its adsorption performance. It is also used to prevent particle aggregation (Manyangadze et al. 2020).

8.3.3.3 Selectivity and Functionalization

The nanosorbents can be functionalized with various compounds and functional groups, which enhances the selectivity of the nanosorbent towards the desired adsorbate. It also improves stability and adsorption capacity.

8.3.3.4 Small Intraparticle Diffusion Distance

Nanosorbents are characterized by their small particle size, which means the pollutants travel very short distances to reach the active sites. The distance traveled by the molecules/ions through the pores of the adsorbent to reach the active site is called intraparticle diffusion distance. Since the intraparticle diffusion distance is comparatively small, the rate of adsorption is relatively higher. It also reduces the reaction time and facilitates higher adsorption capacity.

8.3.3.5 Versatility

Nanoparticles can be fabricated from the majority of the bulk materials. The properties of the nanosorbent can be customized with surface modifications and external functionalization. Any specific property of a nanosorbent can be modified through available techniques to improve its selectivity towards a particular pollutant, increasing its overall performance (Anoob et al. 2024).

8.3.3.6 Reduced Secondary Pollution

Adsorption involves a weak van der waals force; hence, no secondary pollution occurs. The recovery of the adsorbed pollutants is also feasible. Most of the nanosorbents can be easily regenerated and reused without appreciable loss in their performance, thereby improving their economic feasibility and making them sustainable.

8.4 Chitosan Nano Adsorbent for Soil Remediation

8.4.1 Synthesis of Chitosan Nanoparticles

Chitosan can be synthesized at the nanoscale as nanoparticles, nanocomposites, and nanofibers. They have applications in different fields. Some well-known syntheses are ionic gelation, reverse micellization, emulsion solvent diffusion, microemulsion, emulsification cross-linking, and synthesis from bio-composites of living and nonliving materials.

8.4.1.1 Ionic Gelation

In this method, the positively charged amino group of chitosan reacts with anions to form hydrogels by inter and intramolecular cross-linkages. One of the common cross-linking agents is polyanion tripolyphosphate (TPP). The instantaneous production of nanoparticles can be observed when TPP is added to the chitosan solution. Usually, this process is preferred due to its simple nature and mild requirement conditions. This technique was used in the formulation of letrozole – an anticancer drug for breast cancer, using sodium tripolyphosphate (STPP) as the cross-linking agent (Gomathi et al. 2017). This method is also modified by combining ionic gelation with radical polymerization. The polymerization of acrylic and methacrylic monomers takes place with ionic gelation at higher temperatures with nitrogen steam. This is widely used in the formulation of compounds for drug delivery (Grenha 2012).

8.4.1.2 Emulsification and Cross-linking

This was the first method proposed for preparing chitosan nanoparticles. In this method, the amino group of chitosan interacts with the aldehyde group of glutaraldehyde to form a covalent cross-link by adding a cross-linking agent that can harden forming droplets. This leads to emulsion formation and the formation of nanoparticles. Nanoparticles produced by this method have shown antimicrobial properties in strains like *Staphylococcus aureus, Escherichia coli* and *Candida albicans* (Pan et al. 2019).

8.4.1.3 Emulsion Solvent Diffusion

In this process, an organic phase consisting of a hydrophobic drug is added to an aqueous chitosan solution along with a stabilizer. Upon mechanically stirred and high-pressure homogenization, an emulsion is formed. The prepared emulsion is diluted with water. Diffusion of organic solvent in water leads to polymer precipitation, producing the nanoparticles. They have shown promising results for the encapsulation of drugs. One of the major disadvantages is that higher shear forces during the precipitation of nanoparticles are observed.

8.4.1.4 Microemulsion

In this process, two solutions, a surfactant in N-hexane and chitosan in acetic acid, are prepared. Glutaraldehyde was added to the mixture of the two solutions, and nanoparticles were produced upon continuous stirring of the mixtures at room temperature. Here, cross-linking occurred between the amine group of chitosan and the aldehyde group of glutaraldehyde (Divya and Jisha 2018).

8.4.1.5 Reverse Micellization

In the process, w/o (water in oil) – based microemulsion is formulated by dissolving an organic solvent in a lipophilic surfactant. An aqueous solution containing chitosan and glutaraldehyde is added to the mixture and stirred continuously. The formation of reverse micelles is seen in a w/o system. In the conventional micelle preparation method, o/w – based systems are used, whereas for reverse micelles, vice-versa is observed. After continuous stirring, the formation of nanoparticles is observed. They are separated using the solvent evaporation method. It has the notable advantage of producing ultrafine nanoparticles with an average size of around 100 nm.

8.4.1.6 Synthesis from Biocomposites

Nanoparticles can be synthesized using living and nonliving materials. For example, they can be synthesized from mixtures of bacterial species and methacrylic acid and pseudomonas & gelatin.

The chitosan nanoparticles can be characterized using different techniques. A few of the commonly utilized techniques are discussed in Table 8.2.

Table 8.2 Characterization techniques for chitosan nanoparticles.

Characterization technique	Purpose
Fourier transform infrared spectroscopy (FTIR)	Used to obtain structural and spectral measurements and identifies functional groups and chemical bonds
X-ray diffraction (XRD)	To analyze crystalline structure
Scanning electron microscopy (SEM)	To study surface morphology showing their size, shape, and surface texture
Transmission electron microscopy (TEM)	To visualize nanostructures and chemical characteristics
Atomic force microscopy (AFM)	Provides topographical data like surface roughness and particle size
Zeta-potential analysis	To measure the surface charge
Brunauer–Emmett–Teller (BET) analysis	To determine the specific surface area and its porosity
Energy-dispersive X-ray spectroscopy (EDX or EDS)	Provides elemental composition analysis
Ultraviolet-visible (UV-vis) spectroscopy	To analyze the optical properties

8.4.2 Functionalization of Chitosan Nanoparticles

Functionalization is a widely used technique in surface chemistry, biological treatments, and material science. It is a process to improve or modify the material's properties by attaching a new functional molecule onto the surface of the material. Surface modification of the nanoparticles affects their size, surface chemistry, and surface charge, which in turn affects their properties like reactivity, selectivity, and so on (Thiruppathi et al. 2016).

Conjugation of molecules on the surface of the particles can done via various methods as follows:

8.4.2.1 Cross-linking

Cross-linking involves the use of bifunctional or multifunctional reagents and results in covalent bonding between chitosan chains. Some commonly used cross-linking agents are glutaraldehyde, epichlorohydrin, and genipin. The amino group of chitosan reacts with the agents, producing a network structure that improves mechanical strength and stability.

Example: Bifunctional molecules like epichlorohydrin are chemically cross-linked with chitosan through its amino group. The latter molecule possesses varied pore sizes and enhances mechanical properties (Ibrahim, Nada, and Eid 2018).

8.4.2.2 Grafting

Polymer chains or small molecules are also covalently attached to the surface of chitosan nanoparticles through grafting. This is done using free radical polymerization, atom transfer radical polymerization (ATRP), or any other chemical reaction.

Example: To improve the adsorption affinity and selectivity of biochar–chitosan composite towards heavy metals, carboxyl monomers are grafted on the composite by radical chain reaction. The resultant composite showed strong intermolecular forces and selectivity towards heavy metals (Zhang et al. 2019).

8.4.2.3 Functional Group Addition

Chemical reactions may be carried out to covalently link specific functional groups like carboxyl, thiol, and hydroxyl on chitosan nanoparticles for improved interaction with target molecules.

Example: Chitosan was chemically modified by reacting with succinic anhydride to introduce carboxyl groups. The resulting product presented better adsorption of copper cations (Lima and Airoldi 2003).

8.4.2.4 Electrostatic Interactions

Electrostatic interactions are attractive forces between oppositely charged molecules. Chitosan is positively charged under acidic conditions and can interact with negatively charged molecules or ions.

Example: The chitosan nanoparticles are functionalized with negatively charged nucleic acids like DNA and mRNA through electrostatic interactions to form nanocarrier for gene delivery applications (Karayianni et al. 2023).

8.4.2.5 Hydrogen Bonding

Hydrogen bonds are formed between a hydrogen atom attached to some electronegative (like oxygen or nitrogen) molecule and electrons. Even though this is a weak interaction, chitosan nanoparticles make qualitative differences.

Example: Numerous hydroxyl groups present in the polyvinyl alcohol are attached on the amino groups of chitosan nanoparticle through hydrogen bonds (Chetouani et al. 2017).

8.4.2.6 Hybrid Functionalization

The functionalization of nanoparticles by both covalent and non-covalent interactions helps modify complex properties and achieve multifunctional surfaces. This approach establishes stability and functionality.

Functionalization of chitosan nanoparticles offers versatile methods to enhance their properties. The addition of functional groups enhances the adsorption capacity and improves selectivity by increasing its affinity for specific pollutants. Improved mechanical and chemical stability are achieved by cross-linking and other modifications. In addition, biological functionalization ensures its biocompatibility for medical applications. The choice of functionalization method depends on the intended application and desired properties of the modified nanoparticles.

8.5 Key Research Studies

Over the last few years, numerous studies have analyzed the potential of chitosan as a nanosorbent to remediate soil contamination. Here, we have presented the major studies done on the potential of chitosan nanosorbents in soil remediation.

8.5.1 Removal of Herbicides

The surplus use of herbicides in agriculture to gain high yield has resulted in several short- and long-term effects on the environment and human health. Hormonal imbalances, cancer, and reproductive abnormalities are a few of the effects caused by herbicides (Singh et al. 2010). The excessive and improper use of herbicides causes contamination of herbicides in runoff water, which results in water and soil pollution. Microfiltration and oxidation processes are a few methods used to remove herbicides from water. The development of efficient and inexpensive methods was needed to remediate contaminated zones. Using a natural polymer membrane is an effective method to adsorb herbicides from contaminated zones (Montgomery 2007).

8.5.1.1 Removal of Diquat

Diquat, an herbicide used to control the growth of weeds and grasses, is widely used in agriculture. Chitosan and alginate are used to develop a bilayer membrane to adsorb the herbicide efficiently (Gunasekara et al. 2009). The chitosan solution was prepared by dissolving it in 2% vol. acetic acid solution, and the alginate solution was prepared by dissolving it in 0.1 M sodium hydroxide solution. The chitosan solution was poured into a polystyrene dish and dried. The alginate solution was then coated on the chitosan film to form a bilayer membrane. The membrane was neutralized using sulfuric acid followed by ethanol. The prepared membrane was immersed in diquat aqueous solutions having concentrations ranging from 5 μM to 200 μM. The adsorption of diquat by the membrane gave good results, as evidenced by 90% adsorption within 120 minutes. Results indicated the potential of biopolymer-based membranes in the removal of herbicides (de Moraes et al. 2013).

8.5.1.2 Removal of Atrazine

Atrazine is a chlorinated triazine systemic herbicide used to control grass and broadleaf weed growth. The use of cross-linked chitosan–silver nanoparticles in the removal of atrazine showed significant results (Saifuddin et al. 2011). The synthesis of cross-linked chitosan–silver nanoparticles involved mixing of $AgNO_3$ solution with chitosan solution until it is homogeneous. The mixture was subjected to a series of microwave irradiation and cooling. A brown solution was obtained, indicating the reduction of Ag^+ ions. The obtained brown solution was slowly dropped into sodium hydroxide and formed as beads. The beads were kept in sodium hydroxide solution for 24 hour, washed, and dried. The dried chitosan–silver nanoparticle beads were mixed with epichlorohydrin solution and subjected to a series of microwave irradiation and cooling to produce cross-linked chitosan–silver nanoparticle beads. The irradiated beads were washed and dried. The percentage removal of atrazine was determined after 60 minutes with a range of varying adsorbent dosages.

The formed cross-linked chitosan–silver nanoparticle beads were well dispersed, and non-aggregation confirms its stability. For the adsorption studies, 1 L of atrazine solution along with cross-linked chitosan–silver nanoparticle beads was kept in a mechanical shaker. The maximum removal percentage of 99% was observed using cross-linked chitosan–silver nanoparticle beads, whereas only 73% removal of atrazine was shown by pure chitosan beads. This study shows that chitosan can be amalgamated with metal nanoparticles. It also proves that chitosan has the potential for herbicide reclamation (Escudero-Oñate and Martínez-Francés 2018).

8.5.2 Pesticide Removal

Pesticides are extensively used to control pests in agricultural land. Pesticides can leach into ground water through soil causing soil and water pollution. It harms soil health, air purity, and water quality, leading to environmental impacts. It is also known to cause various acute and chronic health issues in humans.

Permethrin is a widely used pesticide belonging to the pyrethroid family (Hassaan and Nemr 2020). Chitosan beads modified with ZnO nanoparticles (CS-ZnONP) are employed as an adsorbent for the removal of permethrin.

For the preparation of the composite, ZnO powder was dissolved in acetic acid and nitric acid. Then, chitosan powder was added to the ZnO solution and kept under sonication. NaOH was added dropwise into the prepared solution until it reached pH 10. The mixture was then filtered, washed, and dried. The results of the solubility studies showed better swelling behavior of the CS-ZnONPs with 13.7%.

The adsorption study involved a range of adsorbent dosages from 0.01 to 1.5 g. The adsorption of the permethrin was conducted at neutral pH. The study showed a maximum permethrin removal efficiency of 99% demonstrating its excellent adsorption performance. These results indicated that CS-ZnONPs beads can potentially remove permethrin (Dehaghi et al. 2014).

8.5.3 Organic Pollution Degradation

8.5.3.1 Degradation of Trichloroethene

Aromatic compounds like phenol and trichloroethylene (TCE) are commonly used as solvents in chemical manufacturing. These aromatic compounds possess adverse effects on soil health. TCE is used as a degreasing and cleaning agent. It breaks down at the site of production or usage, readily passes through the soil, and accumulates in groundwater. It breaks down slowly and remains in the environment for a long time. It is known to cause kidney cancer and other medical ailments. Biodegradation of TCE is possible by bacteria encapsulation in chitosan beads by a cometabolic mechanism. The chitosan beads were prepared by dissolving chitosan powder in water and lactic acid. It was then mixed with resuspended *Pseudomonas putida* cells. The prepared mixture was slowly dropped into a sodium tripolyphosphate solution to form a bead-like structure. The formed beads were washed and placed into a growing medium. Compared to suspended systems, this immobilized system depicted a slower degradation rate with a similar transfer yield. The immobilized system demonstrated better tolerance to TCE toxicity, suggesting that cells within chitosan beads were protected from harsh conditions. Degradation kinetics showed slower rates in immobilized systems, but these had higher inhibition constants, indicating better sustainability at higher TCE concentrations compared to suspended cultures (Chen et al. 2007).

8.5.3.2 Oil Spill Remediation

Oil spills have adverse effects on aquatic ecosystems and incur huge losses. Aquatic organisms cannot ingest oil and can cause internal damage. It coats the feathers of birds and leads to hypothermia or drowning. It contaminates the nearby soil, thereby affecting plant growth and soil integrity. The spilled oil can accumulate in marine organisms and affect human health. The degradation of oil is very challenging and persists in the environment.

Chitosan–silica hybrid nanoparticles are used as an adsorbent for oil spill removal. The synthesis of nonmagnetic chitosan silica hybrid nanoparticles involves mixing si-chitosan (silane-containing chitosan), tetraethyl orthosilicate, deionized water, ethanol, and ammonia solution in a stirrer. The obtained material was washed and dried.

The magnetic chitosan silica hybrid nanoparticles were prepared by alkali hydrolysis and encapsulation. The adsorption study involves the addition of an oil-in-water sample and the magnetic chitosan–silica hybrid nanoparticles in a beaker. The magnetic chitosan–silica hybrid nanoparticles were removed using a magnet and weighed. The gain of weight indicates the adsorption of oil. The magnetic chitosan silica hybrid nanoparticles showed better removal performance than the nonmagnetic ones in all organic solvents.

8.5.4 Immobilization and Removal of Heavy Metals

8.5.4.1 Stabilization of Chromium in Soil

Heavy metal contamination affects both terrestrial and aquatic life. Chromium is one of the widely used heavy metals in tanneries, mining, and chemical industry. Chromium can exist in various oxidation states. In natural soil, chromium exists as trivalent chromium [Cr(III)] and hexavalent chromium [Cr(VI)]. The negative impact of Cr(VI) is over 100 times greater than that of Cr(III). Cr(VI) exposure reduces plant growth and affects agricultural produce yield (Liu et al. 2022). Lowering the mobility of Cr(VI) and its bioavailability in the soil can be achieved by the stabilization method, which changes the occurrence of Cr by adding passivating agents to the soil (Chen et al. 2015; Liu et al. 2022). Research has shown strong adsorption of chromium in aqueous solution using chitosan/bentonite composites (CSBT). The composites can also be used to stabilize the chromium-spiked soil.

The chromium-contaminated soil was prepared by mixing dried soil with potassium dichromate solution and kept for 14 days (Liu et al. 2022). Later, it was baked at 105 °C. Iron-pillared bentonite was prepared by treating bentonite with hydroxy iron. Chitosan solution was prepared by dissolving chitosan powder in acetic acid. Pillared bentonite was added to the chitosan solution and stirred with NaOH to form a gel. The prepared composite was washed, dried, and grounded. The CSBT composite was added in a centrifuge tube along with the chromium-contaminated soil of moisture 60–75%. It was then sealed and treated for 56 days. The result showed high adsorption of chromium ions from the soil. The residual form increased from 0.24 mg/kg to 49.82 mg/kg, which reduces the bioavailability of chromium (Liu et al. 2022).

8.5.4.2 Decontamination of Cu^{2+} from Soil

Copper is an essential element for metabolism and is extensively used in electrical wires and chemical processes. Excessive copper can accumulate in soil and restrict plant growth, thereby affecting total yield (Wyszkowski 2019). The carboxylated graphene oxide/chitosan/cellulose (GCCSC) was prepared by mixing cellulose and chitosan together and continuously stirring. The mixture was centrifuged, and the supernatant was separated. The obtained supernatant was mixed with carboxylated graphene oxide (GO-COOH) solution and continuously stirred. The mixture was slowly dropped into the coagulating bath to form GCCSC beads. The beads were washed, frozen, and lyophilized. For the Cu^{2+} adsorption study, soil was taken along with different concentrations of Cu^{2+} and GCCSC beads were introduced into it. The mixture was kept in a thermostatic shaker for 24 hours. It was then centrifuged, and the concentration of Cu^{2+} was determined. The GCCSC beads showed high immobilization of 99.6% at a low concentration of 60 mg/L and 91.5% at a higher concentration of 200 mg/L. The GCCSC beads retained their porous structure throughout the adsorption process. This porous structure prevented the direct contact between graphene and the soil while still allowing the diffusion of Cu^{2+} into the inner surface of the bead. These characteristics of GCCSC contributed to the high adsorption capacity (Zhao et al. 2019).

8.5.4.3 Removal of Cd(II)

Cadmium is a naturally occurring element which can cause severe health impacts. It is known to cause kidney damage and diarrhea and is a potent carcinogen. The presence of cadmium in the soil can affect plant growth by interfering with its physiological functions and causing chlorosis. Hence, it is necessary to remove cadmium from the soil using an efficient method.

$MgFe_2O_4$/chitosan carbon aerogel beads show promising results in the containment of cadmium in soil. For its preparation, hexahydrate salt of magnesium nitrate [$Mg(NO_3)_2.6H_2O$] along with the nonahydrate salt of iron (III) nitrate [$Fe(NO_3)_3.9H_2O$] were added to the chitosan solution and stirred. The obtained mixture was molded into microspheres and freeze-dried. The frozen microspheres were carbonized to produce $MgFe_2O_4$/chitosan carbon aerogel beads. The carbon aerogels were mixed with Cd-contaminated soil with 60% moisture content. After soil remediation, the carbon aerogels were removed from the soil. The regeneration study was conducted by reacting carbon aerogels with HCL to remove the adsorbed Cd(II) and rinse it with distilled water. The obtained carbon aerogels showed low density and high mechanical strength, which contributed to their separability and abrasion resistance. A maximum removal percentage of 75.7% was obtained in this process. Interestingly, a 96.43% regeneration percentage was obtained by this carbon aerogel, which supports its adsorption and regeneration capacity (Li et al. 2024).

8.5.4.4 Uranium (VI) Sorption

Uranium gets exposed to the environment via mining, reclamation and waste handling. The uranium can possibly enter into living organisms through skin or gamma radiation. Among all processes, mining leads to a large quantity of uranium in the environment. Uranium waste has 80% radioactivity, and this radiological toxicity imposes health risks on humans. It can cause hematopoietic, reproductive, and pulmonary effects.

Chitosan nano-based magnetic particles can be synthesized by co-synthesis followed by co-deposition. After adding the chitosan solution to the precursors and allowing the composite material to form, it was heated to 90 °C and then washed. The produced material was combined with epichlorohydrin to stabilize the chitosan. After this preparation, the sorbent was washed and dried. After graphing amino acids onto the conditioned substrate, it was freeze-dried.

The absorption experiments were conducted by mixing sorbent in a uranium (VI) solution. The solution was uniformly stirred at room temperature. After adsorption, the solvent particles were recovered by magnetic separation. The residual concentration of uranium (VI) was measured. The adsorption and desorption isotherms were evaluated to study the performance of the sorbent. The experiments were conducted with pH above 5.8 to avoid hydrolysis. The sorbents did not lose performance up to 5 adsorption cycles, indicating reusability. So, this study revealed the possibility of using chitosan nano-based magnetic particles as a removal method for uranium (VI) ions (Galhoum et al. 2017).

8.6 Advantages and Limitations

Chitosan nanoparticles have very high adsorption compared to chitosan in bulk form. Due to the presence of amino groups, it enhances the interaction with a range of substrates. Chitosan is biodegradable, less toxic, and bio-compatible. Chitosan can adsorb a wide range of chemical compounds due to the presence of various functional groups; it can also be modified with different materials to enhance its properties. The chitosan nanoparticles also exhibit large surface area, customizable pore size, and pore distribution. It has a very high affinity towards metal ions due to its chelating properties. The chitosan nanoparticles can be synthesized or combined with various materials, making them more versatile and effective. Chitosan nanosorbents were also found to have very high regenerative capacity.

A few limitations were, however, observed with chitosan. The formation of agglomerates is very challenging to control while synthesizing chitosan nanoparticles. It negatively affects the surface area and pore distribution. The insolubility of chitosan in neutral and alkaline solutions restricts its utilization in various experimental conditions. The usage of the desired degree of deacetylation is essential in plays a crucial role in determining properties. Other soil contaminants might affect the removal efficiency of desired contaminants. The other contaminants may be adsorbed on the binding sites, which reduces the overall removal efficiency. It requires frequent regeneration or replacement of nanosorbent to maintain the effectiveness of remediation. In some cases, the acidic medium can adversely affect the microorganisms present in the soil.

8.7 Future Perspective and Research Directions

One of the critical issues with the synthesis of chitosan nanoparticles is agglomeration. Developing more precise methods to synthesize chitosan nanoparticles without any formation of agglomerates would represent a significant advancement in production techniques. Additional focus on developing techniques to control the size, shape, and other properties of chitosan nanoparticles is essential to achieve desired properties for soil remediation. Functionalizing with specific functional groups or ligands would improve its affinity towards positively charged pollutants. Various molecular imprinting techniques can be employed to create selective binding sites to increase the selectivity of specific pollutants. Enhancing chitosan nanoparticles' mechanical strength and abrasion resistance would help improve their durability and reusability. Incorporating various other nanoparticles to form hybrid composites can enhance adsorption capacity and diverse adsorption sites to target a wide array of pollutants and regeneration capacity. Employing smart carriers that can direct through the soil matrix and release chitosan nanoparticles at specific sites or in response to environmental triggers could help in soil remediation applications. Other possible areas of research would be designing a more straightforward approach for studying regeneration cycles to ensure affordability and time efficiency and incorporating sensing technologies in the remediation process to monitor the real-time adsorption activity of nanosorbent. This will provide more insights about the performance of the nanosorbent and its mechanism. Combining chitosan nanoparticles with a biological remediation process will boost the adsorption rate and contribute towards a more sustainable removal process.

8.8 Conclusion

This chapter explores the utilization of chitosan nanoparticles in soil remediation. It has shown that chitosan nanosorbent is a potential adsorbent for a wide array of pollutants in contaminated soil. Various studies underscored the versatile properties of chitosan nanoparticles in adsorption, like enhanced adsorption capacity and increased surface area with abundant active sites. Chitosan has demonstrated its versatility in recreating various herbicides, pesticides, organic pollutants, and heavy metals. Chitosan-based bilayer membranes and cross-linked chitosan–silver nanoparticles demonstrated high removal efficiency and effectiveness towards herbicide removal. Functionalized chitosan with metal nanoparticles showed excellent removal rates for permethrin. It highlights its application in agricultural contamination. Integrating chitosan nanosorbent with bioremediation exhibited compatibility with other methods and enhanced adsorption performance.

The employment of magnetic chitosan nanoparticles improved the separation and regeneration process. Various studies highlighted the potential of chitosan to improve soil health and lower the bioavailability of toxic metals like Cr(VI), Cu^{2+}, Cd(II), and uranyl. It showed a very high affinity towards metal ions and ease of recovery. The chemical structure of chitosan supported the adsorption of various metal ions. In all the remediation studies, chitosan demonstrated a high adsorption rate, increased surface area, larger binding sites, functionalization versatility, biocompatibility, and biodegradability. In general, chitosan nanosorbent is a viable method for addressing soil contamination. Subsequent investigations should focus on enhancing these materials for more extensive uses, testing their durability, and analyzing their efficacy in various soil types and pollution situations. In accordance with the ongoing progress in nanotechnology, Chitosan nanoparticles are expected to play a key role in the development of novel and environmentally sustainable remediation solutions for soil pollution.

References

Alloway, B.J. (ed.). (2012). *Heavy Metals in Soils: Trace Metals and Metalloids in Soils and Their Bioavailability*, 22. Springer Science & Business Media. doi: 10.1007/978-94-007-4470-7.

Amundson, R., Berhe, A.A., Hopmans, J.W. et al. (2015). Soil and human security in the 21st century. *Science* 348 (6235): 1261071. doi: 10.1126/science.1261071.

Anoob, F., Arachchi, S., Azamathulla, H.M. et al. (2024). Nanoadsorbents as an effective wastewater treatment candidate for pharmaceutical contaminants; towards sustainable policy development. *Case Studies in Chemical and Environmental Engineering* 9: 100639. doi: 10.1016/j.cscee.2024.100639.

Aranaz, I., Alcántara, A.R., Civera, M.C. et al. (2021). Chitosan: an overview of its properties and applications. *Polymers* 13 (19): 3256. doi: 10.3390/polym13193256.

Bardgett, R.D. and Van Der Putten, W.H. (2014). Belowground biodiversity and ecosystem functioning. *Nature* 515 (7528): 505–511. doi: 10.1038/nature13855.

Bartos, C., Varga, P., Szabó-Révész, P. et al. (2021). Physico-chemical and in vitro characterization of chitosan-based microspheres intended for nasal administration. *Pharmaceutics* 13 (5): 608. doi: 10.3390/pharmaceutics13050608.

Bayda, S., Adeel, M., Tuccinardi, T. et al. (2019). The history of nanoscience and nanotechnology: from chemical–physical applications to nanomedicine. *Molecules* 25 (1): 112. doi: 10.3390/molecules25010112.

Bünemann, E.K., Bongiorno, G., Bai, Z. et al. (2018). Soil quality–a critical review. *Soil Biology and Biochemistry* 120: 105–125. doi: 10.1016/j.soilbio.2018.01.030.

Chen, T., Zhou, Z., Xu, S. et al. (2015). Adsorption behavior comparison of trivalent and hexavalent chromium on biochar derived from municipal sludge. *Bioresource Technology* 190: 388–394. doi: 10.1016/j.biortech.2015.04.115.

Chen, Y.M., Lin, T.F., Huang, C. et al. (2007). Degradation of phenol and TCE using suspended and chitosan-bead immobilized *Pseudomonas putida*. *Journal of Hazardous Materials* 148 (3): 660–670. doi: 10.1016/j.jhazmat.2007.03.030.

Chetouani, A., Elkolli, M., Bounekhel, M. et al. (2017). Chitosan/oxidized pectin/PVA blend film: mechanical and biological properties. *Polymer Bulletin* 74: 4297–4310. doi: 10.1007/s00289-017-1953-y.

Dehaghi, S.M., Rahmanifar, B., Moradi, A.M. et al. (2014). Removal of permethrin pesticide from water by chitosan–zinc oxide nanoparticles composite as an adsorbent. *Journal of Saudi Chemical Society* 18 (4): 348–355. doi: 10.1016/j.jscs.2014.01.004.

Dehkordi, M.M., Nodeh, Z.P., Dehkordi, K.S. et al. (2024). Soil, air, and water pollution from mining and industrial activities: sources of pollution, environmental impacts, and prevention and control methods. *Results in Engineering* 23: 102729. doi: 10.1016/j.rineng.2024.102729.

Divya, K. and Jisha, M.S. (2018). Chitosan nanoparticles preparation and applications. *Environmental Chemistry Letters* 16: 101–112. doi: 10.1007/s10311-017-0670-y.

Domingues, R.C.C., Junior, S.B.F., Silva, R.B. et al. (2012). Clarification of passion fruit juice with chitosan: effects of coagulation process variables and comparison with centrifugation and enzymatic treatments. *Process Biochemistry* 47 (3): 467–471. doi: 10.1016/j.procbio.2011.12.002.

Escudero-Oñate, C. and Martínez-Francés, E. (2018). A review of chitosan-based materials for the removal of organic pollution from water and bioaugmentation. In: *Chitin-Chitosan-Myriad Functionalities in Science and Technology* 1. London: IntechOpen doi: 10.5772/intechopen.76540.

Galhoum, A.A., Mahfouz, M.G., Gomaa, N.M. et al. (2017). Chemical modifications of chitosan nano-based magnetic particles for enhanced uranyl sorption. *Hydrometallurgy* 168: 127–134. doi: 10.1016/j.hydromet.2016.08.011.

Gassara, F., Antzak, C., Ajila, C.M. et al. (2015). Chitin and chitosan as natural flocculants for beer clarification. *Journal of Food Engineering* 166: 80–85. doi: 10.1016/j.jfoodeng.2015.05.028.

Gomathi, T., Sudha, P.N., Florence, J.A.K. et al. (2017). Fabrication of letrozole formulation using chitosan nanoparticles through ionic gelation method. *International Journal of Biological Macromolecules* 104: 1820–1832. doi: 10.1016/j.ijbiomac.2017.01.147.

Grenha, A. (2012). Chitosan nanoparticles: a survey of preparation methods. *Journal of Drug Targeting* 20 (4): 291–300. doi: 10.3109/1061186x.2011.654121.

Gunasekara, A.S., dela Cruz, I.D.P., Curtis, M.J. et al. (2009). The behavior of clomazone in the soil environment. *Pest Management Science* 65 (6): 711–716. doi: 10.1002/ps.1733.

Hassaan, M.A. and El Nemr, A. (2020). Pesticides pollution: classifications, human health impact, extraction and treatment techniques. *The Egyptian Journal of Aquatic Research* 46 (3): 207–220. doi: 10.1016/j.ejar.2020.08.007.

Homburg, J.A. and Sandor, J.A. (2011). Anthropogenic effects on soil quality of ancient agricultural systems of the American Southwest. *Catena* 85 (2): 144–154. Available from: doi: 10.1016/j.catena.2010.08.005.

Hong, P.Z., Li, S.D., Ou, C.Y. et al. (2007). Thermogravimetric analysis of chitosan. *Journal of Applied Polymer Science* 105 (2): 547–551. doi: 10.1002/app.25920.

Huang, H., Wang, D., Zhu, J. et al. (2022). Sand-based economical micro/nanocomposite materials for diverse applications. *ACS Applied Materials & Interfaces* 14 (38): 43656–43665. doi: 10.1021/acsami.2c08778.

Ibrahim, N.A., Nada, A.A., and Eid, B.M. (2018). Polysaccharide-based polymer gels and their potential applications. *Polymer Gels: Synthesis and Characterization*, 97–126. doi: 10.1007/978-981-10-6083-0_4.

Inglezakis, V.J. and Moustakas, K. (2015). Household hazardous waste management: a review. *Journal of Environmental Management* 150: 310–321. doi: 10.1016/j.jenvman.2014.11.021.

Islam, S., Bhuiyan, M.R., and Islam, M.N. (2017). Chitin and chitosan: structure, properties, and applications in biomedical engineering. *Journal of Polymers and the Environment* 25: 854–866. doi: 10.1007/s10924-016-0865-5.

Jiménez-Gómez, C.P. and Cecilia, J.A. (2020). Chitosan: a natural biopolymer with a wide and varied range of applications. *Molecules* 25 (17): 3981. doi: 10.3390/molecules25173981.

Jimtaisong, A. and Saewan, N. (2014). Utilization of carboxymethyl chitosan in cosmetics. *International Journal of Cosmetic Science* 36 (1): 12–21. doi: 10.1111/ics.12102.

Karayianni, M., Sentoukas, T., Skandalis, A. et al. (2023). Chitosan-based nanoparticles for nucleic acid delivery: technological aspects, applications, and future perspectives. *Pharmaceutics* 15 (7): 1849. doi: 10.3390/pharmaceutics15071849.

Khajeh, M., Laurent, S., and Dastafkan, K. (2013). Nanoadsorbents: classification, preparation, and applications (with emphasis on aqueous media). *Chemical Reviews* 113 (10): 7728–7768. doi: 10.1021/cr400086v.

Kim, S. (2013). Chitin and chitosan derivatives. In: *Advances in Drug Discovery and Developments*. Boca Raton: CRC Press eBooks. doi: 10.1201/b15636.

Kumar, K.Y., Muralidhara, H.B., Nayaka, Y.A. et al. (2013). Low-cost synthesis of metal oxide nanoparticles and their application in adsorption of commercial dye and heavy metal ion in aqueous solution. *Powder Technology* 246: 125–136. doi: 10.1016/j.powtec.2013.05.017.

Lal, R. (2009). Soil degradation as a reason for inadequate human nutrition. *Food Security* 1: 45–57. doi: 10.1007/s12571-009-0009-z.

Lee, S.J., Min, H.S., Ku, S.H. et al. (2014). Tumor-targeting glycol chitosan nanoparticles as a platform delivery carrier in cancer diagnosis and therapy. *Nanomedicine* 9 (11): 1697–1713. doi: 10.2217/nnm.14.99.

Lehmann, J. and Kleber, M. (2015). The contentious nature of soil organic matter. *Nature* 528 (7580): 60–68. doi: 10.1038/nature16069.

Li, Y., Ma, X., Du, Y., Chen, S., Wu, P., and Zhang, S. et al. (2024). Cd-contaminated soil remediation using $MgFe_2O_4$/chitosan carbon aerogel beads. *Journal of Environmental Chemical Engineering*, 12 (1): 111918. doi: 10.1016/j.jece.2024.111918.

Likens, G.E., Driscoll, C.T., and Buso, D.C. (1996). Long-term effects of acid rain: response and recovery of a forest ecosystem. *Science* 272 (5259): 244–246. doi: 10.1126/science.272.5259.244.

Lima, I. S. and Airoldi, C. (2003). Interaction of copper with chitosan and succinic anhydride derivative – a factorial design evaluation of the chemisorption process. *Colloids and Surfaces A: Physicochemical and Engineering Aspects* 229 (1–3): 129–136. doi: 10.1016/s0927-7757(03)00310-8.

Liu, Y., Jia, J., Zhang, H. et al. (2022). Enhanced Cr (VI) stabilization in soil by chitosan/bentonite composites. *Ecotoxicology and Environmental Safety* 238: 113573. doi: 10.1016/j.ecoenv.2022.113573.

Lizardi-Mendoza, J., Monal, W.M.A., and Valencia, F.M.G. (2016). Chemical characteristics and functional properties of chitosan. In: *Chitosan in the Preservation of Agricultural Commodities*, 3–31. Academic Press. doi: 10.1016/B978-0-12-802735-6.00001-X.

Lottermoser, B. and Lottermoser, B. (2010). *Mine Water*, 83–141. Springer Berlin Heidelberg. doi: 10.1007/978-3-642-12419-8.

Lu, Y., Song, S., Wang, R. et al. (2015). Impacts of soil and water pollution on food safety and health risks in China. *Environment International* 77: 5–15. doi: 10.1016/j.envint.2014.12.010.

Madera-Santana, T.J., Herrera-Méndez, C.H., and Rodríguez-Núñez, J.R. (2018). An overview of the chemical modifications of chitosan and their advantages. *Green Materials* 6 (4): 131–142. doi: 10.1680/jgrma.18.00053.

Manyangadze, M., Chikuruwo, N.H.M., Chakra, C.S. et al. (2020). Enhancing adsorption capacity of nano-adsorbents via surface modification: a review. *South African Journal of Chemical Engineering* 31 (1): 25–32. doi: 10.1016/j.sajce.2019.11.003.

McBratney, A., Field, D.J., and Koch, A. (2014). The dimensions of soil security. *Geoderma* 213: 203–213. doi: 10.1016/j.geoderma.2013.08.013.

Mishra, R.K., Mohammad, N., and Roychoudhury, N. (2016). Soil pollution: causes, effects and control. *Van Sangyan* 3 (1): 1–14.

Montgomery, J.H. (2007). *Groundwater Chemicals Desk Reference*. CRC Press. doi: 10.1201/9781420009132.

de Moraes, M.A., Cocenza, D.S., da Cruz Vasconcellos, F. et al. (2013). Chitosan and alginate biopolymer membranes for remediation of contaminated water with herbicides. *Journal of Environmental Management* 131: 222–227. doi: 10.1016/j.jenvman.2013.09.028.

Morin-Crini, N., Lichtfouse, E., Torri, G. et al. (2019). Applications of chitosan in food, pharmaceuticals, medicine, cosmetics, agriculture, textiles, pulp and paper, biotechnology, and environmental chemistry. *Environmental Chemistry Letters* 17 (4): 1667–1692. doi: 10.1007/s10311-019-00904-x.

Mourya, V.K. and Inamdar, N.N. (2008). Chitosan-modifications and applications: opportunities galore. *Reactive and Functional Polymers* 68 (6): 1013–1051. doi: 10.1016/j.reactfunctpolym.2008.03.002.

Münzel, T., Hahad, O., Daiber, A. et al. (2023). Soil and water pollution and human health: what should cardiologists worry about? *Cardiovascular Research* 119 (2): 440–449. doi: 10.1093/cvr/cvac082.

Natarajan, B., Kannan, P., Rather, J.A. et al. (2023). Recent developments in metal nanoparticles functionalized nanocomposite adsorbents for heavy metals removal from wastewaters. *Journal of the Taiwan Institute of Chemical Engineers* 147: 104942. doi: 10.1016/j.jtice.2023.104942.

Neethu, S., Midhun, S.J., Radhakrishnan, E.K. et al. (2021). Microbially synthesized nanomaterials for remediation of contaminated soil and water environment. In: *Microbe Mediated Remediation of Environmental Contaminants*, 157–176. Woodhead Publishing. doi: 10.1016/b978-0-12-821199-1.00015-8.

Pal, P., Pal, A., Nakashima, K. et al. (2021). Applications of chitosan in environmental remediation: a review. *Chemosphere* 266: 128934. doi: 10.1016/j.chemosphere.2020.128934.

Pan, C., Qian, J., Fan, J. et al. (2019). Preparation nanoparticle by ionic cross-linked emulsified chitosan and its antibacterial activity. *Colloids and Surfaces A: Physicochemical and Engineering Aspects* 568: 362–370. doi: 10.1016/j.colsurfa.2019.02.039.

Pimentel, D. (2005). Environmental and economic costs of the application of pesticides primarily in the United States. *Environment, Development and Sustainability* 7: 229–252. doi: 10.1007/s10668-005-7314-2.

Pimentel, D. and Burgess, M. (2013). Soil erosion threatens food production. *Agriculture* 3 (3), 443–463. doi: 10.3390/agriculture3030443.

Reshad, R.A.I., Jishan, T.A., and Chowdhury, N.N. (2021). Chitosan and its broad applications: a brief review. *Journal of Clinical and Experimental Investigations* 12 (4): em00779. doi: 10.2139/ssrn.3842055.

Saifuddin, N., Nian, C.Y., Zhan, L.W. et al. (2011). Chitosan-silver nanoparticles composite as point-of-use drinking water filtration system for household to remove pesticides in water. *Asian Journal of Biochemistry* 6 (2): 142–159. doi: 10.3923/ajb.2011.142.159.

Saleh, T.A. (2018). *Nanotechnology in Oil and Gas Industries*. Cham, Switzerland: Springer. doi: 10.1007/978-3-319-60630-9.

Sandil, S. and Kumar, R. (2022). Soil contamination from construction projects. In: *Ecological and Health Effects of Building Materials*, 205–244. Cham: Springer. doi: 10.1007/978-3-030-76073-1_12.

Sandor, J.A. and Eash, N.S. (1991). Significance of ancient agricultural soils for long-term agronomic studies and sustainable agriculture research. *Agronomy Journal* 83 (1): 29–37. doi: 10.2134/agronj1991.00021962008300010011x.

Shan, S.J., Zhao, Y., Tang, H. et al. (2017, May). A mini-review of carbonaceous nanomaterials for removal of contaminants from wastewater. In *IOP Conference Series: Earth and Environmental Science* 68 (1): 012003. IOP Publishing. doi: 10.1088/1755-1315/68/1/012003.

Singh, A., Mittal, A., and Benjakul, S. (2021). Chitosan nanoparticles: preparation, food applications and health benefits. *Science Asia* 47 (2021): 1–10. doi: 10.2306/scienceasia1513-1874.2021.020.

Singh, B., Sharma, D.K., Kumar, R. et al. (2010). Development of a new controlled pesticide delivery system based on neem leaf powder. *Journal of Hazardous Materials* 177 (1–3): 290–299. doi: 10.1016/j.jhazmat.2009.12.031.

Smith, P., Martino, D., Cai, Z. et al. (2008). Greenhouse gas mitigation in agriculture. *Philosophical Transactions of the Royal Society B: Biological Sciences* 363 (1492): 789–813. doi: 10.1098/rstb.2007.2184.

Teixeira-Costa, B.E. and Andrade, C.T. (2021). Chitosan as a valuable biomolecule from seafood industry waste in the design of green food packaging. *Biomolecules* 11 (11): 1599. doi: 10.3390/biom11111599.

Thiruppathi, R., Mishra, S., Ganapathy, M. et al. (2016). Nanoparticle functionalization and its potentials for molecular imaging. *Advanced Science* 4 (3): 1600279. doi: 10.1002/advs.201600279.

Tsaramirsis, G., Kantaros, A., Al-Darraji, I. et al. (2022). A modern approach towards an industry 4.0 model: from driving technologies to management. *Journal of Sensors* 2022 (1): 5023011. doi: 10.1155/2022/5023011.

Wahba, M.I. (2020). Enhancement of the mechanical properties of chitosan. *Journal of Biomaterials Science, Polymer Edition* 31 (3): 350–375. doi: 10.1080/09205063.2019.1692641.

Wang, Y., Feng, J., Lin, Q. et al. (2013). Effects of crude oil contamination on soil physical and chemical properties in Momoge wetland of China. *Chinese Geographical Science* 23: 708–715. doi: 10.1007/s11769-013-0641-6.

Weil, R.R. and Brady, N.C. (2022). *The Nature and Properties of Soils*, 15e. Pearson Education.

Wyszkowski, M. (2019). Soil contamination with copper and its effect on selected soil properties after applying neutralizing substances. *Polish Journal of Environmental Studies* 28 (4): 2465–2471. doi: 10.15244/pjoes/90357.

Zargar, V., Asghari, M., and Dashti, A. (2015). A review on chitin and chitosan polymers: structure, chemistry, solubility, derivatives, and applications. *ChemBioEng Reviews* 2 (3): 204–226. doi: 10.1002/cben.201400025.

Zhang, L., Tang, S., He, F. et al. (2019). Highly efficient and selective capture of heavy metals by poly (acrylic acid) grafted chitosan and biochar composite for wastewater treatment. *Chemical Engineering Journal* 378: 122215. doi: 10.1016/j.cej.2019.122215.

Zhao, L., Yang, S., Yilihamu, A. et al. (2019). Adsorptive decontamination of $Cu2^{+}$-contaminated water and soil by carboxylated graphene oxide/chitosan/cellulose composite beads. *Environmental Research* 179: 108779. doi: 10.1016/j.envres.2019.108779.

9

Plant-based Nanomaterials for Remediation of Heavy Metal Pollution in Soil

Swagata Lakshmi Dhali[1] *and Moumita Pal*[2]

[1] *Department of Genetics and Microbiology, Faculty of Science, Charles University, Prague, Czech Republic*
[2] *Department of Biosciences and Biomedical Engineering, Indian Institute of Technology Indore, Indore, India*

9.1 Introduction

Soil is a complex, dynamic, and indispensable part of our ecosystem. It is an essential commodity for the livelihood of all mankind, their survival, development, and ultimately human health. Soil health is fundamental in regulating ecosystems and crop health. However, in recent decades, its overall quality has progressively declined due to accelerated anthropogenic activities to meet the demands of the growing population. The degradation of soil quality is directly related to rapid urbanization and industrialization (Taghavi et al. 2024). For instance, mining, municipal trash dumping, industrial effluent discharge, and overuse of fertilizers have altogether increased the concentration of heavy metals (HMs) and metalloids such as lead (Pb), cadmium (Cd), mercury (Hg), nickel (Ni), arsenic (As), chromium, (Cr), and zinc (Zn) which cause severe soil pollution (Table 9.1). Moreover, the persistent nature of HMs leads to bioaccumulation and toxicity in both plants and the associated soil microbes. Hence, the contamination of soil with HMs is a global environmental issue (Mathur, Singh, and Ranjan 2022).

HMs are naturally occurring substances in the Earth's crust with high atomic weights (>20) and specific gravities (>5 g/cm^3). They can be classified further according to their utility: essential HMs (such as Fe, Cu, Cr, and Zn) and nonessential HMs (Pb, Li, Ba, etc.). Essential HMs at lower concentrations promote plant growth, however, both essential and nonessential HMs are toxic at higher concentrations for all organisms (Prakash 2023). The United Nations has set forth the sustainable developmental goal of improving land quality to achieve food security, improve nutrition, and promote sustainable agriculture. Additionally, the European Union is also targeting to make 75% of soils 'healthy' by 2030 (Lal et al. 2021). To meet these criteria, monitoring the generation of HMs and the remediation of polluted soil is essential.

HM removal is challenging, and there are several drawbacks to using traditional remediation techniques. For example, physical remediation techniques such as soil excavation, and immobilization can cause secondary pollution, and require high economic costs.

Nanotechnology-based Sustainable Agriculture, First Edition. Edited by Pardeep Singh et al.

Table 9.1 List of some heavy metals (HMs), their permissible limits, utility, effects on plants, and potential remediation strategies.

Heavy metal	Permissible limits (ppm)	Essential/ nonessential	Sources	Biotoxicity in plants	Remediation strategies	References
Arsenic (As)	20	Nonessential	Smelting, burning fossil fuel and waste, pesticides, mining, and wood preservatives	Oxidative damage, inhibition of protein synthesis, and decrease in chlorophyll content	Microbial remediation, phytoremediation, and nano-remediation	Patel et al. (2023); Chauhan et al. (2024); Rahman et al. (2023)
Lead (Pb)	0.1	Nonessential	Smelting, metal plating, mining, fertilizers, and pesticides	Delayed seed germination, enzymatic degradation, and root and shoot stunting	Microbial remediation and phytoremediation	Kinuthia et al. (2020)
Mercury (Hg)	0.05	Nonessential	Mining, smelting, and coal-fired power plant	Reduction of photosynthesis and transpiration, inactivation of enzymes and antioxidant defense systems, cell organelle, and membrane disruption	Immobilization, phytoremediation, microbial remediation, and electrokinetic remediation	Kinuthia et al. (2020); Francy et al. (2020); Goyal et al. (2020)
Cadmium (Cd)	0.003	Nonessential	Mining, zinc refining, smelting, battery manufacturing, phosphate fertilizers, and solar panels	Inhibition of seed germination, increased oxidative stress, reduced enzyme activity, and stunted growth of root and shoot	Phytoremediation, microbial remediation, and nanomaterial-mediated remediation	Kinuthia et al. (2020); Zhang et al. (2021); Gupta et al. (2021); Suhani et al. (2021)

Chromium (Cr)	0.1	Essential (micronutrient)	Metalworks, mining, paint manufacturing, paper and wood processing, dyeing, fertilizer production, and incineration of coal or municipal wastes for energy generation	Retarded seed germination, enzyme activity inhibition, photosynthesis inhibition, reduced nutrient uptake, oxidative damage, mutagenesis, and overall yield and biomass reduction	Phytoremediation, microbial reduction, electromigration, chemical oxidation, adsorption and precipitation, and nanoparticle-mediated remediation	Kinuthia et al. (2020); Prasad et al. (2021); Tumolo et al. (2020)
Nickel (Ni)	35	Essential (micronutrient)	Tannery, pigment manufacturing, industrial wastes, batteries, mining activities, and landfills	Reduced germination, photosynthesis, growth, tissue necrosis, chlorosis, low biomass production, and yield	Biochar, soil washing, adsorption, electrokinetic remediation, phytoremediation, microbial remediation, and chemical immobilization	Mustafa et al. (2023); El-Naggar et al. (2021)
Cobalt (Co)	15.2	Essential (trace element)	Cement industries, e-waste, paint industries, mining activities, batteries, and televisions	Inhibition of active transport, premature leaf closure, increased production of ROS, decrease in photosynthetic pigments, and nitrogen metabolism	Phytoremediation, immobilization, soil washing, and electrokinetic remediation	Mahey et al. (2020); Jiang et al. (2022)
Zinc (Zn)	50	Essential (micronutrient)	Smelting, mining, fertilizers, and manures	Reduced plant growth and yield	Biochar, phytoremediation, and immobilization	Osobamiro et al. (2019); Kumar et al. (2023); Xu et al. (2023)

Standard chemical techniques including precipitation, washing, etc. can be highly invasive to the soil and the environment. Hence, sustainability and cost-effectiveness are compromised by traditional methods. Therefore, implementing more effective measures is needed to address these problems (Sánchez-Castro et al. 2023); one such upcoming and promising field in HM soil bioremediation is nanotechnology.

Nano-remediation, specifically plant-based nanomaterials/nanoparticles (NPs) employing "green" technologies is a promising and cost-effective approach to clearing up polluted and contaminated regions. Generally, the bioremediation approach in HM remediation entails utilizing plants and bacteria to accumulate or segregate HM ions from the soil, increasing soil functionality and reducing pollution. Nanotechnology improves this method, bringing a new dimension to successful pollutant removal (Mathur, Singh, and Ranjan 2022).

A thorough analysis of current developments in the use of NPs to improve the effectiveness of plant-based nanoparticle (PBNP)-mediated remediation in soil contaminated with HMs is provided in this composition. This study highlights the pressing need for efficient remediation methods by examining the detrimental effects of HMs on plant productivity and soil microbiota. The subsequent discussion focuses on the efficacy of various PBNPs in remediating different HMs from contaminated soils.

9.2 Sources and Effects of HM Pollution in Soil

HM pollutants are biologically resistant, toxic substances that cause severe soil pollution. They are natural components of the soil present due to the natural weathering of soil parent material (Zhang and Wang 2020). The earth's crust is composed of 95% ingenious rocks and 5% sedimentary rocks. Generally, basaltic ingenious rocks are rich in HMs including Cu, Ni, Co, and Cd, whereas shale ingenious rocks are rich in Cu, Zn, Pb, Mn, and Cd. Natural activities such as erosion, leaching, volcanic eruptions, etc. cause HMs to enter the soil (Li et al. 2019). However, rapid urbanization and industrial activities over the past few decades have made HM pollution more rampant. Among natural causes and anthropogenic activities, the latter is the greater factor responsible for the creation of HM

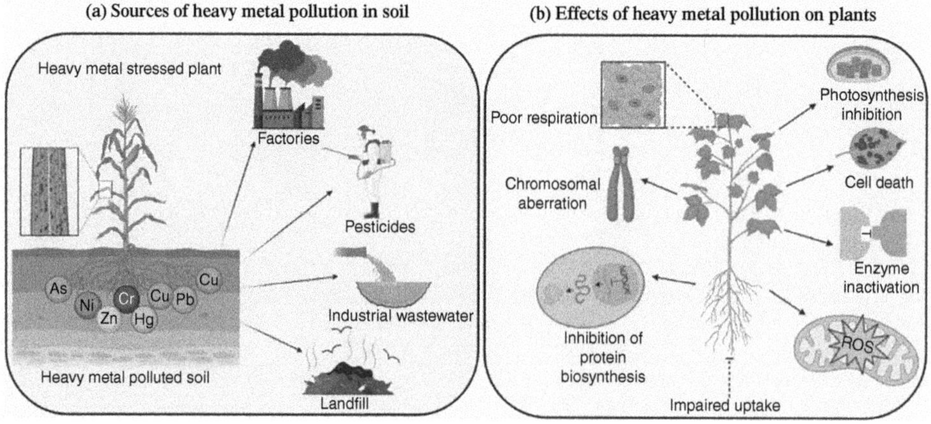

Figure 9.1 Overview of (a) sources of heavy metal (HM) pollutants in soil and (b) effects of HM pollution on plants.

pollution as portrayed in Figure 9.1. Advancements in agricultural practices, application of chemical pesticides, municipal waste disposal, and mining directly impact the severity of HM pollution (Zhang and Wang 2020).

9.2.1 Arsenic (As) Pollution

As is a severely hazardous substance that occurs in various chemical and oxidation states. In several countries including India, Spain, Mexico Argentina, and Chile, it is a natural contaminant in groundwater above permissible concentrations (10 μg/L). Anthropogenic sources of As in soil include smelting, burning fossil fuel and waste, pesticides, and mining activities (Patel et al. 2023). The inorganic form of As is extremely toxic and irrigation of food crops in As contaminated soil is one of the greatest threats to public health (WHO 2022). Alkaline soil enhances the mobility and phytoavailability of As (Patel et al. 2023). Crops absorb and accumulate As, which impacts their yield and enters the food chain. This metal reduces crop yield due to physiological and morphological effects such as stomatal conductance, oxidative stress, transpiration, protein synthesis, and chlorophyll content. Biotechnological strategies are preferred to physical and chemical techniques for As remediation. Some of the most eco-friendly methods of removing this HM contaminant from soil are phytoremediation, microbial remediation, mycoremediation, and nano-remediation (Beniwal, Yadav, and Ramakrishna 2023; Chauhan et al. 2024). These techniques have been shown to remediate As-contaminated soil while maintaining the soil microenvironment (Chauhan et al. 2024).

9.2.2 Lead (Pb) Pollution

Lead is another lethal HM with no known biological functions. However, it is extensively used in industries as a raw material leading to high levels of Pb accumulation. Due to poor management of these wastes, it leaches into the soil and can cause severe soil pollution. Other anthropogenic sources of Pb pollution include battery industrial waste, smelting, metal plating, mining, fertilizers, and pesticides. Pb pollution damages the microbial community in the soil and damages enzyme activity (Sun et al. 2023). Plants readily absorb Pb in the form of complexes and can be biomagnified through the food chain. Various physical, chemical, and biological methods are utilized to deal with Pb pollution in soil. Generally, microbial and phytoremediation techniques are utilized to extract or immobilize Pb ions in soil (Sun et al. 2023).

9.2.3 Cadmium (Cd) Pollution

Cadmium is a nonessential noxious metal with no known biological functions. It is highly carcinogenic even at low concentrations and France, Belgium, and China have the highest levels of Cd-contaminated soil. This HM can enter the soil through both natural and anthropogenic sources. Smelting, municipal sludge from sewage, fossil fuel combustion, fertilizers, and metallurgical practices are all major sources of Cd pollution. Plants uptake more Cd in acidic soil when it forms complexes (Mahmood et al. 2019). Plants display various stress responses to Cd pollution. Physiological responses include reduced seed germination and plant biomass. It also leads to changes in photosynthesis, electrolyte leakage, and altered transpiration rate. Besides these, many genetic abnormalities have also been noticed including gene mutation, DNA damage, and chromosomal aberrations. Overall, Cd accumulation leads to multifaceted effects in plants including biochemical, physiological, and molecular (El Rasafi et al. 2022).

9.2.4 Mercury (Hg) Pollution

Mercury is a highly toxic HM that exists in various forms: elemental, inorganic, and organic. All of these states pose varying levels of toxicity to plants and animals. Anthropogenic activities are the major source of Hg pollution in industrial areas. Soil is the largest reservoir of Hg globally, and crops grown in Hg-contaminated zones may contain elevated levels of Hg (Song et al. 2021). Due to Hg's high mobility and bioaccumulation, it gets easily absorbed into crops and vegetables and enters the food chain (Zhang et al. 2024). Since conventional techniques of Hg remediation are costly, bioremediation strategies are often used to overcome Hg pollution in soil. Several Hg-resistant bacteria and yeasts have been studied for their role in bioremediation. Phytoremediation and nanotechnology are also effective in removing this HM (Kumari et al. 2020). For example, in a recent study, carboxymethyl cellulose stabilized iron sulfide NPs were found to efficiently immobilize Hg from the soil via chemical precipitation, complexation, and ion exchange while following the principles of green chemistry (Lin et al. 2024). This demonstrates the need for more research in the field of green remediation using nanotechnology for HM remediation at a large scale.

9.2.5 Chromium (Cr) Pollution

Chromium is the 21st most abundant element in the Earth's crust. It is obtained majorly by the reduction of chromium oxide to sodium chromate which can leach out with water. South Africa, India, and Turkey are among the leading producers of this HM's ore. It is widely used in industries to produce anti-corrosive coating and alloys (Paul et al. 2024). Cr is a trace element for plants and promotes plant growth at low concentrations. However, the presence of this metal in high concentrations reduces yield and induces phytotoxicity. Excessive levels of this pollutant in soil damage plant roots reduce enzymatic activities and induce necrosis and chlorosis (Hassan et al. 2024). Additionally, Cr has a dose-dependent inhibitory effect on soil bacterial population. Bacterial population and community richness benefits agriculture; however, Cr pollution alters the dynamics. Cr pollution suppresses microbial diversity levels and composition in soil along with enzyme levels significantly (Zhu et al. 2023). Bioremediation techniques such as microbial remediation and phytoremediation are widely utilized for the eradication of Cr pollution from soil or water. Plants easily uptake Cr compounds due to their high solubility and accumulate them in roots. Bamboo plants are vital in this regard since they can accumulate over 95% of this HM from contaminated sites (Paul et al. 2024).

9.2.6 Zinc (Zn) Pollution

Zinc is an essential plant micronutrient and is abundantly present in chemical fertilizers, sewage sludge, pesticides, livestock manure, and compost (Rashid et al. 2023). It boosts photosynthesis, pigment biosynthesis, respiration, and enzyme activities at low concentrations.

However, it can adversely affect crops when present over optimal conditions and cause plant toxicity. Concentrations of Zn above the threshold hinder plant growth, enzyme activity, and plant mineral nutrition among others. Some species of bacteria are resistant to toxic levels of Zn in agricultural soils and can enhance plant growth parameters. For example, Jain et al. discovered four strains of zinc-tolerant bacteria which promoted plant growth even in high concentrations of Zn (Jain et al. 2020).

9.3 Effects of HMs on Plants

HMs are natural components; some are essential for plants and animals. These can only be uptake if present as soluble components in soil. Essential HMs including Cu, Zn, Fe, Ni, Co, Mo, and Mn are required in certain ranges, above which they can be toxic for plants. Plants accumulate both essential and nonessential metals equally and excessive accumulation of HM causes adverse effects such as oxidative stress, enzyme inhibition, and cell damage. These hinder plant metabolism and lead to a decrease in plant growth, which ultimately results in its death (Goyal et al. 2020). Hence excessive HM accumulation interferes with crop productivity and yields directly or indirectly.

Studies indicate that crop productivity is negatively impacted by HM pollution, which disrupts various essential physiological functions. As indicated in Figure 9.1, critical plant metabolic processes including photosynthesis and seed germination are hindered by the presence of reactive oxygen species (ROS) such as hydroxyl radicals, superoxide radicals, and hydrogen peroxide. The enzymes that normally neutralize these ROS are inhibited by the accumulation of HMs (Hafeez et al. 2023). HM cations can inhibit enzyme activity by attaching to cysteine, glutamine, or aspartate residues which form the active sites of numerous enzymes (Lebrazi and Fikri-Benbrahim 2018).

There is sufficient data to show that plants cope with the effect of HM toxicity using the help of certain microorganisms. For instance, arbuscular mycorrhizal fungi and plants form a symbiotic relationship to overcome HM toxicity. They help prevent metals from entering the plant roots by reducing bioavailability or removing them (Liu et al. 2022). Another important example is the rhizobium–plant symbiotic relationship which also serves as a coping mechanism for managing HM stress. The mechanisms behind these are of immense interest and are heavily researched. Rhizobium preferentially forms a symbiotic relationship with leguminous plants and helps them to persist in HM-polluted soil. In general, they are resistant to HM stress by immobilization of HMs, active transport of HMs by pumps, precipitation, and chelation by proteins (Lebrazi and Fikri-Benbrahim 2018). These mechanisms are also utilized for phytoremediation of HMs from soil which is discussed later.

Apart from plants, HM compounds also affect the indigenous microbiota of the soil. Elevated HM concentrations in soil decrease microbial diversity, biomass, and activity. Microbes consume more energy to survive in such polluted environments and put more pressure on sensitive microbes, thus causing changes in the diversity of the native species. Nevertheless, some microbes can remediate these environments using physiological mechanisms. For example, some soil microbes can utilize HMs as electron donors or acceptors during metabolism without causing any toxicity (Abdu, Abdullahi, and Abdulkadir 2016).

9.4 Conventional Remediation Techniques of Heavy Metal Soil Pollution

HMs are indestructible; they can persist indefinitely depending on the nature of the soil and the contaminating metals/metalloids. They can, however, transform from one oxidation state or organic complex state to another, to a less bioavailable, reactive, or toxic form. HM-contaminated soil remediation can be broadly classified into physical, chemical, and biological methods. The simultaneous use of multiple techniques is beneficial for the optimum remediation of HMs (Schommer et al. 2023).

9.4.1 Physical Methods

Physical methods of HM remediation are selected for quick results and prompt hazard management (Ashkanani et al. 2024). Physical techniques include soil washing, extraction, solidification, and thermal desorption. These techniques can be quite expensive since they might require the physical removal of the contaminated soil for treatment or ex-situ treatment. Besides, many of these treatments are only economically feasible on a small scale (Dhaliwal et al. 2020).

9.4.1.1 Soil Replacement

Soil substitution is the process of replacing polluted soil with clean soil to lower pollutant concentrations and improve soil functionality. HMs are removed from the soil generally by disposal. This can be accomplished by soil spading, which mixes contaminated soil deeper into clean soil, or by importing clean soil to dilute the metals and metalloids. This is not sustainable as it can lead to secondary pollution (Li et al. 2019).

9.4.1.2 Immobilization/Solidification

HM immobilization utilizes organic matter (e.g. leaves, manure, sewage sludge) or inorganic matter (e.g. lime, fly ash, cement) to stabilize HMs in soil. It also refers to solidifying polluted soil to segregate it from clean soil. Immobilization techniques engage various types of cement such as ordinary Portland cement (OPC), chemically bonded phosphate cement (CBPC), and alkali-activated cement (AAC), which are beneficial for reducing carbon footprints (Tyagi and Annachhatre 2023).

9.4.1.3 Thermal Desorption

This technique is based on the notion of HM volatility caused by the application of heat. Heating techniques such as steam, microwave, and infrared radiation are employed to heat the soil and convert the pollutant to a volatile form. Volatile metals are then isolated from the soil with a carrier gas or vacuum-negative pressure. Based on the temperature required for the technique, thermal desorption can be classified into high-temperature (320–560 °C) and low-temperature (90–320 °C) desorption (Dhaliwal et al. 2020). However, it is often done ex-situ making it energy-intensive and costly. Besides, the natural properties of the soil can be altered instigating further complications (Ashkanani et al. 2024).

9.4.2 Chemical Methods

Here, chemical agents are employed for rapid decontamination of HM-polluted soil. These techniques are preferred for rapid contaminant removal in case of severe environmental pollution. It is best suited for severe cases where physical and biological methods may not succeed (Li et al. 2019). Some of these techniques are described below.

9.4.2.1 Washing

Soil washing engages agents to chelate HMs some of which include oxalic acid, citric acid, and ethylenediaminetetraacetic acid (EDTA; Song and Nam 2023). Selection of specific agents is critical for success. However, overuse of washing agents can destroy native soil microorganisms. Therefore, the selection of alternative bioremediation techniques is ultimately beneficial.

9.4.2.2 Biochar

Due to its high porosity, and multiple functional groups, biochar is a solid biofuel that functions as an absorbent and catalyst. Produced from plant biomass, agricultural and industrial wastes, and manure, it is inexpensive and eco-friendly. Biochar is used in agriculture to retain water, improve soil fertility, and boost crop yield, and in bioremediation to remove organic pollutants and HMs. Its adsorption efficiency depends on soil type, feedstock, and pyrolysis conditions (Li et al. 2024; Srivastav et al. 2022).

9.4.2.3 Electrokinetic Method

The electrokinetic remediation technique establishes an electric field at polluted sites. The generated electric field mobilizes HMs and the ions are accumulated on the electrodes. The collected HM ions can be then treated. This rapid technique is effective even for low-permeability soil remediation (Kumar et al. 2023). However, the major drawbacks of this technique are that it is energy-intensive, less inefficient, and time-consuming (Sánchez-Castro et al. 2023).

9.4.3 Nanotechnology-assisted HM Remediation

This groundbreaking approach employs NPs to selectively adsorb and remove HM ions. Particles with a diameter ranging from 1 to 100 nm are known as nanomaterials and can be widely grouped according to their dimensions, and composition. When treated with specific chemicals, these minuscule particles display high adsorption capacities, rapid removal kinetics, and efficiency. Their massive surface area, numerous surface contact sites, excellent absorptivity, cost-effective, and rapid production are the fundamental reasons for their efficacy (Inobeme et al. 2023). Further details regarding NP-assisted HM remediation have been discussed in detail in subsequent sections.

9.4.4 Biological Methods

Confronted with the menace of increasing HM pollutants in the soil, bioremediation stands out as a cost-effective and environmentally friendly option. In these techniques, the natural metal-accumulating capability of microbes, and plants or other biological components extracted from microbes/ plants are utilized to remove HMs from affected soil (Gulzar and Mazumder 2022).

9.4.4.1 Phytoremediation

Phytoremediation uses plants to remove HMs from soil. Certain plants can extract, accumulate, and store HMs in their roots and shoots. Phytoextraction (absorption and transfer of HMs), phytovolatilization (release of HMs into the atmosphere), phytostabilization (limit HM mobility by binding within the plant rhizosphere), and phytoaccumulation (concentration and accumulation of HMs in plant parts) are all categories of the same (Bhat et al. 2022). These techniques are cost-effective, eco-friendly, and have great potential to extract HMs using their roots while adapting to the HM stress (Chengatt et al. 2023; Nnaji et al. 2023).

However, the major limitations of phytoremediation are (i) the plant should be able to grow in the contaminated site, (ii) plants can only remediate the soil area within the reach of its roots and the pollutants have to be soluble for uptake, (iii) it is harmful to the biodiversity of the area to introduce non-native plant species for remediation, (iv) the enriched phytoremediation plant biomass needs to be treated separately to avoid secondary pollution and (v) it is time-consuming. To overcome these limitations, phytoremediation is often combined with advanced techniques such as genetic engineering, and nanotechnology (Thakur et al. 2016; Zhang et al. 2023).

9.4.4.2 Microbial Remediation

Microbial remediation of HMs from soil is a green, low-cost, and sustainable alternative to physical and chemical methods. Although microbes cannot degrade HMs, they can convert them into less toxic forms or reduce their mobility (Chen et al. 2023). Microbes interact with HMs through biosorption, bioleaching, biotransformation, methylation, and enzymatic transformation, utilizing HMs as a nutrient source while mitigating their toxicity (Access et al. 2023). Phosphate-solubilizing microbes, known for promoting plant growth by converting insoluble phosphorus, can also stabilize, accumulate, and transform HMs in soil (Chen et al. 2023). Additionally, microbial surfactants offer a promising approach to HM remediation in soil.

9.4.4.3 Biosurfactants

Biosurfactants, such as glycolipids, phospholipids, lipoproteins, and polymers, are microbially produced compounds that facilitate soil leaching and solubilization, desorption, and dispersal of pollutants. These natural substitutes for chemical leaching agents aid in the bioremediation of HM-contaminated soil without causing secondary pollution. Unlike synthetic surfactants, biosurfactants are environmentally friendly and cost-effective, avoiding long-term environmental concerns (Sun et al. 2021; Yu et al. 2023). Rhamnolipids, produced by species like *Pseudomonas aeruginosa*, stabilize and flush HMs from soil, forming complexes for uptake by hyperaccumulator plants and mobilizing HMs for soil washing (Parus et al. 2023). These biosurfactants demonstrate significant potential for HM remediation, especially when combined with techniques like phytoremediation and filtration, warranting further exploration and application.

9.5 Role of NPs in Soil Remediation

Since there is a possibility of collateral contamination and the associated costs, the modern world needs substitutes for conventional HM-remediation techniques (Araújo, Castro, and Fiúza 2015). A recently developed promising and effective alternative solution to the problem is the use of nanotechnology for soil pollution. 'Nanotechnology' is the term given to the production and study of NPs, which are widely used in different scientific and technological fields. The demand for, and the creation of new materials highlight the positive impacts of nanotechnology in material sciences (Cai et al. 2019). Nanotechnology breakthroughs provide worldwide opportunities for effective remediation or restoration of contaminated soil. There have been claims that nanotechnology has enormous potential as a greener technology for the environment, especially in terms of eliminating the toxicities of certain metals and metalloids (Rajput et al. 2022). NPs provide environment-friendly, efficient, and economical ways of dealing with pollution thereby showing their potential for remediation. The increased demand for nanomaterials in pollutant remediation is driven by their extensive specific surface area, high chemical reactivity, and unique nanoscale properties (Araújo, Castro, and Fiúza 2015). Some of the advantages of employing NPs for soil bioremediation are as follows:

a. Reduced time and cost: The cost of remediation efforts in polluted soil can be decreased with the use of NPs. Their contribution can aid in reducing the duration of clean-up needed to restore contaminated areas (Medina-Pérez et al. 2019).
b. High surface area: High surface is the main reason behind the high reactivity of the NPs which in turn helps in the efficient HM remediation (Navya and Daima 2016).
c. High adsorption capacity and selectivity: NPs have a high surface area and can act as better adsorbents.
d. Efficiency: NPs have a high surface area and a high adsorption capacity, which increases the effectiveness of HM removal (Yang et al. 2019).

Despite these undeniable advantages of NPs, they can be potentially harmful. There are several drawbacks associated with the introduction of NPs to the soil since they can sometimes harm the soil ecosystem and terrestrial organisms. In one reported case, the high concentration of carbon nanomaterials such as fullerene harmed the plant. Therefore, there is a possibility of NPs being harmful to plant cells under certain circumstances (Kristanti et al. 2021). One of the most crucial factors influencing general soil health, microbial dynamics, soil nutrient availability, and plant growth and development is soil pH, which is known to be altered by the presence of some NPs in soil. The intensity of the negative impacts of NPs is influenced by the kind and NPs concentration in soil and the type of soil (Dhanapal et al. 2024). Therefore, it is important to determine the concentration, composition, and toxicity of NPs for the remediation of HMs.

9.6 Plant-based Nanomaterials (PBNPs) for Soil Remediation

Currently, there is an increasing demand for research focusing on the fabrication of NPs. The NPs fabrication process involves three methods, namely, chemical, physical, and green synthesis. The physical method of NPs fabrication involves thermal degradation, aerosol, and ultraviolet (UV) radiation all of which consist of high pressure and temperature. On the other hand, chemical synthesis employs sodium borohydride, a costly and hazardous chemical that utilizes dispersion stabilizers and organic solvents. To overcome the concerns about high energy consumption, the emission of hazardous and poisonous substances, the need for complicated equipment, and synthesis conditions, green synthesis techniques are gradually replacing physical and chemical processes (Ying et al. 2022).

Green chemistry techniques, which addressed longevity, performance maximization, conservation, and the avoidance and prevention of hazardous compounds, first appeared in the mid-1990s. A cutting-edge and emerging platform of the twenty-first century, biological manufacturing holds the power to fundamentally alter the course of nanomaterial applications across a wide range of fields. The term "phytonanotechnology" describes the fusion of nanotechnology with plants and their by-products. Due to the current emphasis on sustainability, phytonanotechnology is becoming very popular. It can synthesize NPs by extracting various plant parts, such as leaves, stems, roots, gum, seeds, and shoots, and by-products from fruit and vegetable waste (Omran, Whitehead, and Baek 2021). Several solvents, reactants, and capping agents can be used to fabricate metallic NPs in an environmentally friendly manner as explained in Figure 9.2. Plant biomolecules, such as polysaccharides, proteins, amino acids, tannins, phenolics, ketones, alkaloids, aldehydes,

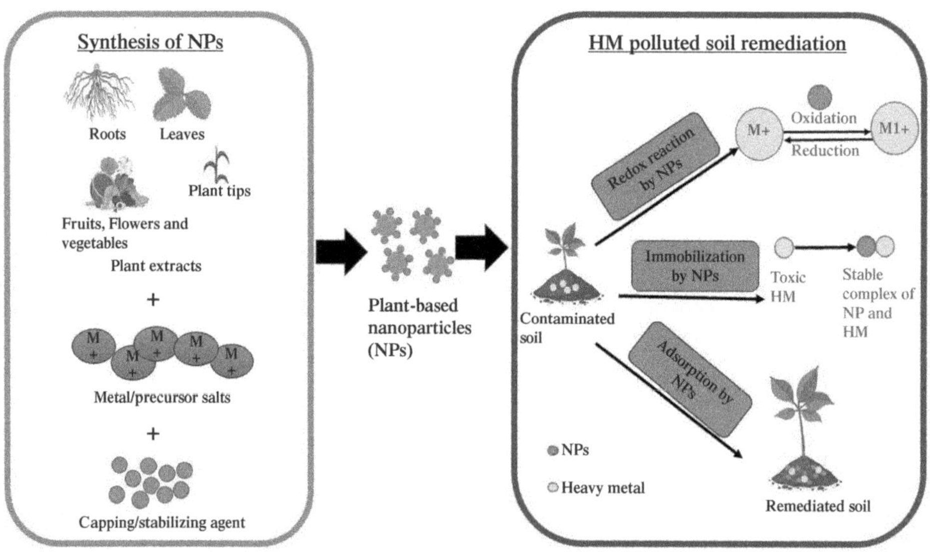

Figure 9.2 Schematic representation of the synthesis of plant-based nanoparticles (PBNPs) and their utilization in the remediation of HM-polluted soil.

terpenoids, saponins, vitamins, and flavones play a major role in the reduction of metals (Nakum and Bhattacharya 2022). When compared to conventional physical and chemical synthesis techniques, green synthesis has numerous advantages: it is less expensive, more sustainable, nontoxic, and pollutant-free (Ying et al. 2022).

A thorough analysis of current developments in the use of NPs to improve the effectiveness of plant-based NP-mediated remediation in soil contaminated with HMs is provided in this composition. In the subsequent sections, the types of PBNPs, their mechanisms of action, and some examples where PBNPs have succeeded in remediating contaminated soil with a range of HMs have been discussed in detail.

9.6.1 Overview of Different Types of Plant-Based Nanomaterials

9.6.1.1 Zero-valent Iron NPs

Due to its distinct structural and electronic characteristics, a promising material for eliminating harmful contaminants is the nanoscale zero-valent iron (nZVI) NPs (Abdelfatah et al. 2021). These particles were conventionally synthesized using co-precipitation, electrochemical synthesis, thermal decomposition, and microemulsion (Wu, He, and Jiang 2008; Cabrera et al. 2008). They have several unique traits, such as a large number of reactive surface sites, a high surface area and reactivity, and a strong ability to remove pollutants (Cai et al. 2019). Iron oxide surfaces exhibit strong adsorption forces of many types, including complex formations, ion exchange, electrostatic interactions, etc., all of which are essential for increasing the pollutants' and the adsorbent's affinity (Biftu, Ravindhranath, and Ramamoorty 2020).

9.6.1.2 Carbon-based Nanomaterials

A potential nanomaterial for soil remediation is carbon nanotubes (CNTs), especially for immobilizing HMs in contaminated soils (Matos, Correia, and Rasteiro 2017). These were first manufactured in the 1960s using the electric arc discharge technique (Szabó et al. 2010). There are two types of CNTs such as single-walled carbon nanotubes (SWCNTs) and multi-walled carbon nanotubes (MWCNTs; Chen et al. 2021). Since CNTs have a huge surface area, they can retain substantial quantities of HMs through their ability to attract and keep them on their surface (adsorption). An effective way to boost the soil's adsorption capacity and, as a result, enhance the immobilization of HMs in the soil matrix is to incorporate a modest quantity of scattered CNTs (Matos, Correia, and Rasteiro 2017). CNTs have demonstrated significant promise in the remediation of soil, especially in the treatment of hydrocarbons and the immobilization of HMs.

9.6.1.3 Copper NPs

Copper and its oxide form of NPs (CuO-NPs/Cu-NPs) have gained popularity as one of the most significant metal NPs due to their advantageous characteristics for plant growth, defense, and nutrition. The antiviral, antifungal, and antibacterial properties of these NPs, as well as their influence on the treatment of cancer, have also been utilized in nanomedicine. They can also be employed as catalysts to remove contaminants in the environment (Eid et al. 2023). The practical uses of copper-based NPs have garnered considerable attention owing to their wide availability, high activity, selectivity, affordability, and thermal stability (Bonthula et al. 2023). The most typical methods for its synthesis include photochemical reduction, electrochemical technique, chemical reduction, and thermal decomposition (Drbohlavova et al. 2009).

9.6.1.4 Quantum Dots

Over the last ten years, there has been a significant increase in the global scientific interest in quantum dots (QDs), which are promising nanoscale materials with diameters ranging from 1 to 10 nm. They display quantum confinement effects, dimensionless structures, and size-tuneable optical characteristics. The environmentally conscious, cost-effective, and energy- and time-saving green synthesis of QDs eliminates the need for costly, dangerous chemical solvents that can cause pollution (Omran, Whitehead, and Baek 2021). Over the past ten years, two main approaches have been taken to generate QDs: producing semiconductor particles at the nanoscale by colloidal chemistry and epitaxial growth and nanoscale patterning (Drbohlavova et al. 2009).

9.6.1.5 Polymer NPs

Waterways and contaminated soil can be remediated using polymer nanomaterials. Polymeric NPs are mostly used to remove PAHs from soil, which is a laborious process (Sathish et al. 2024). One of the best examples of an organic, environmentally friendly, and effective (nano)material is a microparticle/NP made of cellulosic material obtained from plants, specifically nanocellulose. Acid hydrolysis of cellulose fibers yields highly crystallite, rod-like cellulose nanocrystals (CNC) that are 10–20 nm in width and hundreds of nm in length. This process removes the amorphous parts of the fibers while leaving the crystalline component intact (Liu et al. 2024). Other conventional techniques of polymer NP synthesis are nanoprecipitation, evaporation, double emulsion, and emulsion solvent evaporation (Mandal, Banerjee, and Mandal 2022; Bohrey, Chourasiya, and Pandey 2016).

9.6.2 Mechanism of Action of PBNPs

a. *Adsorption:* A material can accumulate between the two-phase interfaces, such as the gas–solid, liquid–solid, or liquid–liquid interfaces, as part of the mass transfer process known as adsorption. The adsorbent is the name for the phase in which the substance/material gets collected, and adsorbate is the phrase for the accumulated substance/material. Pollutants (adsorbate) adhere to the surface of the adsorbent in this surface phenomenon. Depending on how the adsorbate and the adsorbent interact with one another, the adsorption is divided into two categories: physisorption (physical adsorption) and chemisorption (chemical adsorption; Gupta et al. 2021). Nanomaterials work well as sorbents when it comes to eliminating HM ions from wastewater. For this use, nanosorbents need to be completely recyclable, nontoxic, and possess excellent sorption capabilities and selectivity even at low concentrations of pollutants. They should simplify the process of removal of the adsorbed pollutants. Numerous nanomaterials, such as graphene, CNTs, carbon-based composites, nanometals or metal oxides, and polymeric sorbents, have strong adsorption capabilities in these applications (Wang 2012).

b. *Oxidation and Reduction:* There are two general ways by which HMs are reduced: NPs either directly reduce HMs or HMs are first adsorbed onto NPs, then reduced from high valent to low valent (Latif et al. 2020). When it comes to eliminating inorganic pollutants such as HMs, chemical oxidation is less successful than it is at eliminating organic pollutants like volatile organic molecules. Several different types of oxidations can occur. The most popular and widely used method is wet oxidation, which uses oxygen or air as the oxidizing agent (Ali et al. 2023).

c. *Immobilization:* Chemical immobilization reduces metal mobility, bioavailability, leaching, and translocation into plants by incorporating substances into contaminated soil. This effectively addresses HM removal. Complexation, sequestration, stabilization, adsorption, precipitation, and co-precipitation are all associated with this process. Due to their adsorption sites, porosity, surface areas, and retention capacities, immobilizing agents such as biochar, manure, compost, hydroxyapatite, bone, activated carbon, zeolite, chitosan, and clay serve as soil conditioners that improve soil physicochemical properties and plant tolerance to HM stress. Each of these agents targets a particular metal selectively. NPs are more efficient and effective at remediating HMs than bulk and macro-aggregated immobilizing agents because of their distinct morphological features, ease of delivery, quick soil dispersal, enhanced sorption capacity, and high affinity (Azeez et al. 2019).

d. *Nano-Phytoremediation:* Phytoremediation, derived from the Greek "phyton" (plant) and "remediare" (to remediate), involves plants and soil bacteria converting toxins into safe, often beneficial forms. This method utilizes hyperaccumulator plants, which absorb high levels of HMs, to clean contaminated soils. Nanotechnology enhances phytoremediation efficacy; NPs can remediate soils and water polluted with organic, inorganic, and HM contaminants. Integrating plants with nanomaterials can significantly boost plant growth and metal uptake, thereby improving phytoremediation efficiency in HM-contaminated soils (Prakash 2023). Nanomaterials can improve phytoremediation by effectively eliminating pollutants through surface processes, ion exchange, redox reactions, adsorption, surface complexation, and electrostatic interaction, applying plant growth regulators, or by improving plant growth and development, such as plant-growth-promoting rhizobacteria (PGPR), and using transgenic plants, and enhancing phytoavailability (Romeh 2022).

9.6.3 Examples of Successful Applications of Plant-based Nanomaterials in HM Remediation

Recent advances in environmental remediation have highlighted the potential of PBNPs for removing HMs from contaminated soils. These NPs, derived from various plant sources, have demonstrated significant efficacy in HM remediation, providing an eco-friendly and sustainable alternative to traditional methods. Several studies have reported successful applications of these NPs, which are discussed in Table 9.2.

Table 9.2 Successful examples of PBNPs in HM remediation from soil.

Nanoparticle	Plant extract	Target metal	Removal strategy	Efficiency (%)	References
Iron oxide nanoparticles	*Pittosporum undulatum, Melia azedarach*	Chromium	Reduction	84.1 and 96.2, respectively	Truskewycz, Shukla, and Ball (2018)
Iron oxide nanoparticles	*Schinus molle, Syzygium paniculatum* (var. *australe*)	Chromium	Reduction	43.7 and 58.7, respectively	Truskewycz, Shukla, and Ball (2018)
Iron oxide nanoparticles	*Eucalyptus globulus*	Chromium	Reduction	100	Andrade-Zavaleta et al. (2022)
Iron oxide nanoparticles	*Eucalyptus globulus*	Cadmium	Reduction	100	Andrade-Zavaleta et al. (2022)
Iron oxide nanoparticles	*Eucalyptus globulus*	Lead	Reduction	11.8	Andrade-Zavaleta et al. (2022)
Iron oxide nanoparticles	*Azadirachta indica* (neem), *Mentha longifolia* (mint)	Lead	Reduction	26.9 and 62.3, respectively	Francy et al. (2020)
Zero-valent extract of silver nanoparticles	*Theobroma cacao* (cocoa pod)	Lead and Cadmium	Immobilization	51.36 and 66.67, respectively	Azeez et al. (2019)
Silver nanoparticles	*Catharanthus roseus*	Chromium and Cadmium	Adsorption	47.84 and 5.68, respectively	Verma and Bharadvaja (2022)
Copper nanoparticles	*Catharanthus roseus*	Chromium and Cadmium	Adsorption	2.11 and 2.91, respectively	Verma and Bharadvaja (2022)
Iron nanoparticles	*Prangos ferulacea, Teucrium polium*	Arsenic	Adsorption	93.8	Karimi et al. (2019)
Silica nanoparticles	Corn cob ash	Arsenic	Adsorption	96	Balasubramaniam et al. (2020)

9.6.3.1 Removal of Cr

A study conducted by Truskewycz et al. reported a successful application of plant-based iron NPs for the remediation of the toxicogenic hexavalent Cr. They fabricated the iron oxide NPs using dried leaf extracts of four novel plant species: *Schinus molle, Melia azedarach, Pittosporum undulatum*, and *Syzygium paniculatum* (var. *australe*). The characterization study revealed that the NPs synthesized from the different plant species varied in their shape, size, and agglomeration, but they retained the basic elemental framework and amorphous nature. Furthermore, it has been observed that the hexavalent ion removal capacity was correlated to the particle size and dispersity. The aggregated NP produced from *S. paniculatum* and *S. molle* were between 30 and 100 nm with chromium removal efficacies of 58.7 and 43.7%. In contrast, the NPs produced from *M. azedarach* and *P. undulatum* extracts with sizes between 5 and 15 nm had 84.1 and 96.2% efficacies respectively after nine hours. Even with the maximum antioxidant concentration and highest reducing power capability, *S. paniculatum* was only able to remove 58.7% of the Cr (VI) after nine hours. These differences in the results suggest that factors beyond antioxidant content, such as particle size and dispersity, play a more crucial role in the efficiency of Cr (VI) remediation using plant-based iron oxide NPs (Truskewycz, Shukla, and Ball 2018).

9.6.3.2 Removal of Cr, Cd, and Pb

Another case study by Andrade-Zavaleta et al. reported the potential of green synthesized iron oxide NPs for HM remediation of contaminated soil. In this research, they synthesized iron oxide NPs employing *Eucalyptus globulus* leaves. These NPs were further characterized extensively using various techniques such as Fourier transform infrared spectroscopy (FTIR), scanning electron microscopy (SEM), energy-dispersive X-ray spectroscopy (EDX), and X-ray diffraction (XRD). FTIR studies demonstrated that the eucalyptus extract contains derivatives of phenolic compounds and simple phenols, which may be in charge of the reduction of metal ions and the subsequent creation of FeO NPs. These NPs have efficiently removed hexavalent chromium, cadmium, and to a lower extent lead. After 30 minutes, 100% of the chromium was reduced with the introduction of the FeO NPs colloids. When NPs were applied, electrons were transferred, and Cr (VI) was reduced to Cr (III) by oxidizing FeO to Fe (II) and Fe (III), resulting in the reduction of this metal. 100% of the total Cd was removed by applying FeO NPs colloids at various concentrations for 30 minutes. In the case of lead removal, the efficacy was too low i.e. 11.8% since the contact time was only 30 minutes, therefore, contact time is a very crucial factor in determining the remediation efficacy hence, this PBNP was found to have diverse activity (Andrade-Zavaleta et al. 2022).

9.6.3.3 Removal of Pb

A successful nano remediation of Pb was conducted by Francy et al., where they fabricated zero-valent iron NPs utilizing *Mentha longifolia* (mint) and *Azadirachta indica* (neem) leaves. These NPs were further characterized using Brunauer–Emmett–Teller isotherm (BET), transmission electron microscopy (TEM), SEM, etc. These NPs were used for Pb remediation from HM-polluted soil. The soil sample subjected to remediation was contaminated at an initial Pb concentration of approximately 250 mg/kg. According to the data, after 30 days, the Pb removal efficiency was 26.9% for particles made with neem leaves and 62.3% for particles made with mint leaves. Therefore these findings highlight the possibility of using plant precursors in the synthesis of nZVI NPs for effective HM remediation of soils (Francy et al. 2020).

9.6.3.4 Mitigation of Cd Toxicity

As mentioned earlier, crop productivity is seriously threatened by Cd intoxication in agricultural soil. To mitigate Cd toxicity, a study conducted by Ahmed et al. reported that biogenic silicon NPs synthesized utilizing rice leaf extract and sodium metasilicate

($Na_2SiO_30.5\ H_2O$) enhanced the activity of antioxidant enzymes, photosynthesis, and plant development. Biogenic silicon NPs improved plant silicon concentrations, which in turn decreased the Cd translocation. In plants treated with these NPs, the expression of genes associated with defense responses increased. Results from a greenhouse experiment showed that rapeseed plants treated with biogenic silicon NPs at a dose of 250 mg/kg had considerably better growth characteristics than the Cd-treated control group, including dry weight (32.6%) and fresh weight (33.3%). Furthermore, after treatment, the activity of antioxidant enzymes crucial for eliminating Cd-induced oxidative damage such as superoxide dismutase (19.1%), catalase (14.4%), ascorbate peroxidase (33.8%), and peroxidase (33.4%) enhanced remarkably. Therefore, these enzymes are crucial for eliminating Cd-induced oxidative damage and scavenging ROS. To address Cd toxicity in rapeseed plants, biogenic silicon NPs have demonstrated great promise in case of Cd toxicity due to their nontoxic, and sustainable nature (Ahmed et al. 2023).

9.6.3.5 Mitigating Arsenic Stress

A study conducted by Chatterjee et al. reported a successful application of plant-based iron NPs for the remediation of toxicogenic As contamination. These NPs were produced from the leaf extract of *Adiantum lunulatum*. These NPs were characterized by techniques such as UV–visible spectroscopy, zeta potential, EDX, FTIR, XRD, and TEM. The mean particle size of the quasi-spherical NPs was approximately 5 nm. The impact of FeO-NPs on rice seedling growth in response to As stress was assessed by applying an additional 100 mg/L dosage of FeO-NPs in the rice growing medium. When compared to seeds that germinated under merely 50 μm As (III) and 50 μm As (V), FeO-NPs significantly boosted germination of the seed by 9.8 and 15.4%, respectively, of seeds grown under 50 μm As (III) and 50 μm As (V) stress. As (III) had a more severe phytotoxic effect than As (V) on seedling growth, seed germination, and chlorophyll content. Nonetheless, IONP fertigation aids rice plants in overcoming the negative impact of As exposure. When seedlings were under As stress, FeO-NPs improved their development and vigor. Under conditions of As stress, IONP fertigation reduced the absorption and translocation of As by seedlings. Significant reductions in antioxidant activity and oxidative stress were brought about by the decreased As accumulation in seedlings treated with FeO-NPs. These results demonstrated that producing harmless and environmentally friendly FeO-NPs using *Adiantum lunulatum* leaf extract is a viable option since they can effectively enhance the growth of rice seedlings in stressful circumstances (Chatterjee et al. 2021).

9.7 Limitations of PBNPs

The green synthesis of NPs is an innovative and sustainable approach, but it has several drawbacks. A significant obstacle in green synthesis is the absence of exact control over the size and form of synthesized NPs. Gold NPs produced by green synthesis frequently have different sizes and shapes, which can be a major disadvantage in industrial settings where precise sizes and shapes are essential. A further drawback of green synthesis is the low yield of NPs compared to traditional synthesis methods. Agglomerate formation, inadequate metal ion reduction, and the reducing agent's instability can all lead to insufficient yields (Malik et al. 2023).

Furthermore, the utilization of various industrial chemical compounds, prolonged reaction times, and enormous energy consumption make it impractical at times. Certain green synthesis procedures demand very high temperatures and lengthy synthesis times, which might have an adverse environmental impact. Therefore, to make green-produced nanoscale materials practical, their efficiency needs to be increased (Ying et al. 2022).

9.8 Conclusion

For decades, environmental health has been compromised due to the HM pollution caused by anthropogenic activities. This article provides insights into the detrimental effects of multiple toxic HMs on plant growth and development, crop productivity, and soil microbiota. Due to several drawbacks associated with conventional remediation techniques, there is a need for better alternatives in an environmentally friendly and cost-effective way. Over the past decade, green synthesis of NPs has been a highly significant field of study. Several types of natural extracts (plant leaves, fruits, stems, roots, etc.) have been used to provide efficient sources for NP synthesis. Among these, plant extract is highly effective as a reducing and stabilizing agent for the fabrication of NPs which is the basis for green nanotechnology.

Numerous investigations have confirmed that PBNPs are capable of remediating HMs, not only from polluted soil but also in wastewater treatment. However, little is known about the potential risks associated with using these substances. Furthermore, it has been observed that many studies focus on zero-valent iron oxide NPs due to their potential and the high efficacy rates (~100%) of PBNPs in several cases. This review has also demonstrated the role of plant-based NPs in eradicating the toxicity of several HMs which further helped in seed germination and plant growth and development. Apart from this, PBNPs are also engaged in reducing HM stress in plants as opposed to remediation of the soil itself. However, much is still left unexplored; for example, the biosynthesis of NPs utilizing marine algae and marine plants is an unexplored area with much potential. The discovery of new plant extracts for antioxidant properties, and NP fabrication techniques may also lead to ground-breaking results. Therefore, there are still many opportunities to investigate novel environmentally friendly fabrication techniques based on biogenic synthesis.

References

Abdelfatah, A.M., Fawzy, M., Eltaweil, A.S. et al. (2021). Green synthesis of nano-zero-valent iron using *Ricinus communis* seeds extract: characterization and application in the treatment of methylene blue-polluted water. *ACS Omega* 6: 25397–25411.

Abdu, N., Abdullahi, A.A., and Abdulkadir, A. (2016). Heavy metals and soil microbes. *Environmental Chemistry Letters* 15 (1): 65–84.

Access, O., Abo-Alkasem, M.I., Hassan mat, H. et al. (2023). Microbial bioremediation as a tool for the removal of heavy metals. *Bulletin of the National Research Centre* 47: 1–15.

Ahmed, T., Masood, H.A., Noman, M. et al. (2023). Biogenic silicon nanoparticles mitigate cadmium (Cd) toxicity in rapeseed (*Brassica napus* L.) by modulating the cellular oxidative stress metabolism and reducing Cd translocation. *Journal of Hazardous Materials*. doi: 10.1016/J.JHAZMAT.2023.132070.

Ali, Q., Zia, M.A., Kamran, M. et al. (2023). Nanoremediation for heavy metal contamination: a review. *Hybrid Advances* 4: 100091.

Andrade-Zavaleta, K., Chacon-Laiza, Y., Asmat-Campos, D. et al. (2022). Green synthesis of superparamagnetic iron oxide nanoparticles with eucalyptus globulus extract and their application in the removal of heavy metals from agricultural soil. *Molecules* 27: 1367.

Araújo, R., Castro, A.C.M., and Fiúza, A. (2015). The use of nanoparticles in soil and water remediation processes. *Materials Today: Proceedings* 2: 315–320.

Ashkanani, Z., Mohtar, R., Al-Enezi, S. et al. (2024). AI-assisted systematic review on remediation of contaminated soils with PAHs and heavy metals. *Journal of Hazardous Materials* 468: 133813.

Azeez, L., Adejumo, A.L., Lateef, A. et al. (2019). Zero-valent silver nanoparticles attenuate Cd and Pb toxicities on *Moringa oleifera* via immobilization and induction of phytochemicals. *Plant Physiology and Biochemistry* 139: 283–292.

Balasubramaniam, T., Bakar, A.H.A., Uda, M.N.A. et al. (2020). Potential of synthesized silica nanoparticles (Si-NPs) using corn cob for arsenic heavy metal removal. *IOP Conference Series: Materials Science and Engineering* 864: 012187.

Beniwal, R., Yadav, R., and Ramakrishna, W. (2023). Multifarious effects of arsenic on plants and strategies for mitigation. *Agriculture* 13: 401.

Bhat, S.A., Bashir, O., Ul Haq, S.A. et al. (2022). Phytoremediation of heavy metals in soil and water: an eco-friendly, sustainable and multidisciplinary approach. *Chemosphere* 303: 134788.

Biftu, W.K., Ravindhranath, K., and Ramamoorty, M. (2020). New research trends in the processing and applications of iron-based nanoparticles as adsorbents in water remediation methods. *Nanotechnology for Environmental Engineering* 5: 12. doi: 10.1007/S41204-020-00076-Y.

Bohrey, S., Chourasiya, V., and Pandey, A. (2016). Polymeric nanoparticles containing diazepam: preparation, optimization, characterization, in-vitro drug release and release kinetic study. *Nano Convergence* 3: 1–7.

Bonthula, S., Bonthula, S.R., Pothu, R. et al. (2023). Recent advances in copper-based materials for sustainable environmental applications. *Sustainable Chemistry* 4: 246-271.

Cabrera, L., Gutierrez, S., Menendez, N. et al. (2008). Magnetite nanoparticles: electrochemical synthesis and characterization. *Electrochimica Acta* 53: 3436–3441.

Cai, C., Zhao, M., Yu, Z. et al. (2019). Utilization of nanomaterials for in-situ remediation of heavy metal(loid) contaminated sediments: a review. *Science of the Total Environment* 662: 205–217.

Chatterjee, A., Mridha, D., Banerjee, J. et al. (2021). Green synthesis of iron oxide nanoparticles and their ameliorative effect on arsenic stress relief in *Oryza sativa* seedlings. *Biocatalysis and Agricultural Biotechnology* 38: 102207.

Chauhan, R., Awasthi, S., Tiwari, P. et al. (2024). Biotechnological strategies for remediation of arsenic-contaminated soils to improve soil health and sustainable agriculture. *Soil and Environmental Health* 2: 100061.

Chen, H., Jiang, H., Nazhafati, M. et al. (2023). Biochar: an effective measure to strengthen phosphorus solubilizing microorganisms for remediation of heavy metal pollution in soil. *Frontiers in Bioengineering and Biotechnology* 11: 1127166.

Chen, X., Wang, J., Hayat, K. et al. (2021) Small structures with big impact: multi-walled carbon nanotubes enhanced remediation efficiency in hyperaccumulator *Solanum nigrum* L. under cadmium and arsenic stress. *Chemosphere* 276: 130130. doi: 10.1016/j.chemosphere.2021.130130.

Chengatt, A.P., Sarath, N.G., Sebastian, D.P. et al. (2023). Chelate assisted phytoextraction for effective rehabilitation of heavy metal(loid)s contaminated lands. *International Journal of Phytoremediation* 25: 981–996.

Dhaliwal, S.S., Singh, J., Taneja, P.K. et al. (2020). Remediation techniques for removal of heavy metals from the soil contaminated through different sources: a review. *Environmental Science and Pollution Research* 27: 1319–1333.

Dhanapal, A.R., Thiruvengadam, M., Vairavanathan, J. et al. (2024). Nanotechnology approaches for the remediation of agricultural polluted soils. *ACS Omega* 9: 13522–13533.

Drbohlavova, J., Adam, V., Kizek, R. et al. (2009). Quantum dots – characterization, preparation and usage in biological systems. *International Journal of Molecular Sciences* 10: 656.

Eid, A.M., Fouda, A., Hassan, S.E.D. et al. (2023). Plant-based copper oxide nanoparticles; biosynthesis, characterization, antibacterial activity, tanning wastewater treatment, and heavy metals sorption. *Catalysts* 13: 348.

El-Naggar, A., Ahmed, N., Mosa, A. et al. (2021). Nickel in soil and water: sources, biogeochemistry, and remediation using biochar. *Journal of Hazardous Materials* 419: 126421.

El Rasafi, T., Oukarroum, A., Haddioui, A. et al. (2022). Cadmium stress in plants: a critical review of the effects, mechanisms, and tolerance strategies. *Critical Reviews in Environmental Science and Technology* 52: 675–726.

Francy, N., Shanthakumar, S., Chiampo, F. et al. (2020). Remediation of lead and nickel contaminated soil using nanoscale zero-valent iron (nZVI) particles synthesized using green leaves: first results. *Processes* 8: 1453.

Goyal, D., Yadav, A., Prasad, M. et al. (2020). Effect of heavy metals on plant growth: an overview. In: *Contaminants in Agriculture: Sources, Impacts and Management*, 79–101. Cham: Springer.

Goyal, D., Yadav, A., Singh, T.B. et al. (2020). *Effect of Heavy Metals on Plant Growth: An Overview*. doi: 10.1007/978-3-030-41552-5_4.

Gulzar, A.B.M. and Mazumder, P.B. (2022). Helping plants to deal with heavy metal stress: the role of nanotechnology and plant growth promoting rhizobacteria in the process of phytoremediation. *Environmental Science and Pollution Research* 29 (27): 40319–40341.

Gupta, K., Joshi, P., Gusain, R. et al. (2021). Recent advances in adsorptive removal of heavy metal and metalloid ions by metal oxide-based nanomaterials. *Coordination Chemistry Reviews* 445: 214100.

Hafeez, A., Rasheed, R., Ashraf, M.A. Qureshi et al. (2023). Effect of heavy metals on growth, physiological and biochemical responses of plants. In: *Plants and their Interaction to Environmental Pollution: Damage Detection, Adaptation, Tolerance, Physiological and Molecular Responses*, 139–159. Amsterdam: Elsevier.

Hassan, M.U., Lihong, W., Nawaz, M. et al. (2024). Silicon a key player to mitigate chromium toxicity in plants: mechanisms and future prospective. *Plant Physiology and Biochemistry* 208: 108529.

Inobeme, A., Mathew, J.T., Adetunji, C.O. et al. (2023). Recent advances in nanotechnology for remediation of heavy metals. *Environmental Monitoring and Assessment* 195: 1–24.

Jain, D., Kour, R., Bhojiya, A.A. et al. (2020). Zinc tolerant plant growth promoting bacteria alleviates phytotoxic effects of zinc on maize through zinc immobilization. *Scientific Reports* 10: 1–13.

Jiang, M., Wang, K., Wang, Y. et al. (2022). Technologies for the cobalt-contaminated soil remediation: a review. *Science of the Total Environment* 813: 151908.

Karimi, P., Javanshir, S., Sayadi, M.H. et al. (2019). Arsenic removal from mining effluents using plant-mediated, green-synthesized iron nanoparticles. *Processes* 7: 759.

Kinuthia, G.K., Ngure, V., Beti, D. et al. (2020). Levels of heavy metals in wastewater and soil samples from open drainage channels in Nairobi, Kenya: community health implication. *Scientific Reports* 10: 1–13.

Kristanti, R.A., Mei, R., Liong, Y. et al. (2021). Soil remediation applications of nanotechnology. *Tropical Aquatic and Soil Pollution* 1: 35–45.

Kumar, V., Rout, C., Singh, J. et al. (2023). A review on the clean-up technologies for heavy metal ions contaminated soil samples. *Heliyon* 9: e15472.

Kumari, S., Amit, Jamwal, R.et al. (2020). Recent developments in environmental mercury bioremediation and its toxicity: a review. *Environmental Nanotechnology, Monitoring and Management* 13: 100283.

Lal, R., Bouma, J., Brevik, E. et al. (2021). Soils and sustainable development goals of the United Nations: an international union of soil sciences perspective. *Geoderma Regional* 25: e00398.

Latif, A., Sheng, D., Sun, K. et al. (2020). Remediation of heavy metals polluted environment using Fe-based nanoparticles: mechanisms, influencing factors, and environmental implications. *Environmental Pollution* 264: 114728. doi: 10.1016/J.ENVPOL.2020.114728.

Lebrazi, S. and Fikri-Benbrahim, K. (2018). Rhizobium-legume symbioses: heavy metal effects and principal approaches for bioremediation of contaminated soil. In: *Legumes for Soil Health and Sustainable Management*, 205–233. Singapore: Springer.

Li, C., Zhou, K., Qin, W. et al. (2019). A review on heavy metals contamination in soil: effects, sources, and remediation techniques. *Soil and Sediment Contamination: An International Journal* 28: 380–394.

Li, X., Li, R., Zhan, M. et al. (2024). Combined magnetic biochar and ryegrass enhanced the remediation effect of soils contaminated with multiple heavy metals. *Environment International* 185: 108498.

Lin, D., Hu, G., Li, H. et al. (2024). Green remediation of mercury-contaminated soil using iron sulfide nanoparticles: immobilization performance and mechanisms, effects on soil properties, and life cycle assessment. *Science of the Total Environment* 944: 173928.

Liu, W., Pang, B., Zhang, M. et al. (2024). Pickering multiphase materials using plant-based cellulosic micro/nanoparticles. *Aggregate* 5: e486.

Liu, Y., He, G., He, T. et al. (2022). Signaling and detoxification strategies in plant-microbes symbiosis under heavy metal stress: a mechanistic understanding. *Microorganisms* 11: 69.

Mahey, S., Kumar, R., Sharma, M. et al. (2020) A critical review on toxicity of cobalt and its bioremediation strategies. *SN Applied Science* 2: 1–12.

Mahmood, Q., Asif, M., Shaheen, S. et al. (2019). Cadmium contamination in water and soil. *Cadmium Toxicity and Tolerance in Plants: From Physiology to Remediation* 141–161.

Malik, A.Q., Mir, T.U.G., Kumar, D. et al. (2023). A review on the green synthesis of nanoparticles, their biological applications, and photocatalytic efficiency against environmental toxins. *Environmental Science and Pollution Research* 30 (27): 69796–69823.

Mandal, M., Banerjee, I., and Mandal, M. (2022). Nanoparticle-mediated gene therapy as a novel strategy for the treatment of retinoblastoma. *Colloids and Surfaces B: Biointerfaces* 220: 112899. doi: 10.1016/j.colsurfb.2022.112899.

Mathur, S., Singh, D., and Ranjan, R. (2022). Remediation of heavy metal(loid) contaminated soil through green nanotechnology. *Frontiers in Sustainable Food Systems* 6: 932424.

Matos, M.P.S.R., Correia, A.A.S., and Rasteiro, M.G. (2017). Application of carbon nanotubes to immobilize heavy metals in contaminated soils. *Journal of Nanoparticle Research* 19: 1–11.

Medina-Pérez, G., Fernández-Luqueño, F., Vazquez-Nuñez, E. et al. (2019). Remediating polluted soils using nanotechnologies: environmental benefits and risks. *Polish Journal of Environmental Studies* 28: 1013–1030.

Mustafa, A., Zulfiqar, U., Mumtaz, M.Z. et al. (2023). Nickel (Ni) phytotoxicity and detoxification mechanisms: a review. *Chemosphere* 328: 138574.

Nakum, J. and Bhattacharya, D. (2022). Various green nanomaterials used for wastewater and soil treatment: a mini-review. *Frontiers in Environmental Science* 9: 724814.

Navya, P.N. and Daima, H.K. (2016). Rational engineering of physicochemical properties of nanomaterials for biomedical applications with nanotoxicological perspectives. *Nano Convergence* 3: 1–14.

Nnaji, N.D., Onyeaka, H., Miri, T. et al. (2023). Bioaccumulation for heavy metal removal: a review. *SN Applied Sciences* 5: 1–12.

Omran, B.A., Whitehead, K.A., and Baek, K.H. (2021). One-pot bioinspired synthesis of fluorescent metal chalcogenide and carbon quantum dots: applications and potential biotoxicity. *Colloids Surf B Biointerfaces* 200: 111578. doi: 10.1016/J.COLSURFB.2021.111578.

Osobamiro, T.M., Awolesi, O., Alabi, O.M. et al. (2019) Heavy metal levels of soil samples collected from a major industrial area in Abeokuta, Southwestern Nigeria. *International Journal of Scientific and Research Publications (IJSRP)* 9: 92125.

Parus, A., Ciesielski, T., Woźniak-Karczewska, M. et al. (2023). Basic principles for biosurfactant-assisted (bio)remediation of soils contaminated by heavy metals and petroleum hydrocarbons – a critical evaluation of the performance of rhamnolipids. *Journal of Hazardous Materials* 443: 130171.

Patel, K.S., Pandey, P.K., Martín-Ramos, P. et al. (2023). A review on arsenic in the environment: contamination, mobility, sources, and exposure. *RSC Advances* 13: 8803–8821.

Paul, A., Dey, S., Ram, D.K. et al. (2024). Hexavalent chromium pollution and its sustainable management through bioremediation. *Geomicrobiology Journal* 41: 324–334.

Prakash, P. (2023). Nano-phytoremediation of heavy metals from soil: a critical review. *Pollutants* 3: 360–380.

Prasad, S., Yadav, K.K., Kumar, S. et al. (2021). Chromium contamination and effect on environmental health and its remediation: a sustainable approaches. *Journal of Environmental Management* 285: 112174.

Rahman, M.S., Reza, A.H.M.S., Siddique, M.A.B. et al. (2023). Accumulation of arsenic and other metals in soil and human consumable foods of Meherpur district, southwestern Bangladesh, and associated health risk assessment. *Environmental Sciences Europe* 35: 1–22.

Rajput, V.D., Minkina, T., Upadhyay, S.K. et al. (2022). Nanotechnology in the restoration of polluted soil. *Nanomaterials* 12: 769.

Rashid, A., Schutte, B.J., Ulery, A. et al. (2023). Heavy metal contamination in agricultural soil: environmental pollutants affecting crop health. *Agronomy* 13: 1521.

Romeh, A.A. (2022). Nano-phytoremediation: using plants and nanomaterials to environmental pollution remediation. In: *Nanotechnology for Environmental Pollution Decontamination*, 445–468. Florida: Apple Academic Press.

Sánchez-Castro, I., Molina, L., Prieto-Fernández, M.Á. et al. (2023). Past, present and future trends in the remediation of heavy-metal contaminated soil – remediation techniques applied in real soil-contamination events. *Heliyon* 9: e16692.

Sathish, T., Ahalya, N., Thirunavukkarasu, M. et al. (2024). A comprehensive review on the novel approaches using nanomaterials for the remediation of soil and water pollution. *Alexandria Engineering Journal* 86: 373–385.

Schommer, V.A., Vanin, A.P., Nazari, M.T. et al. (2023). Biochar-immobilized *Bacillus* spp. for heavy metals bioremediation: a review on immobilization techniques, bioremediation mechanisms and effects on soil. *Science of the Total Environment* 881: 163385.

Song, H. and Nam, K. (2023). Development of a potassium-based soil washing solution using response surface methodology for efficient removal of cesium contamination in soil. *Chemosphere* 332: 138854. doi: 10.1016/j.chemosphere.2023.138854.

Song, Z., Wang, C., Ding, L. et al. (2021). Soil mercury pollution caused by typical anthropogenic sources in China: evidence from stable mercury isotope measurement and receptor model analysis. *Journal of Cleaner Production* 288: 125687.

Srivastav, A.L., Pham, T.D., Izah, S.C. et al. (2022). Biochar adsorbents for arsenic removal from water environment: a review. *Bulletin of Environmental Contamination and Toxicology* 108: 616–628.

Suhani, I., Sahab, S., Srivastava, V. et al. (2021). Impact of cadmium pollution on food safety and human health. *Current Opinion in Toxicology* 27: 1–7.

Sun, X., Sun, M., Chao, Y. et al. (2023). Effects of lead pollution on soil microbial community diversity and biomass and on invertase activity. *Soil Ecology Letters* 5: 118–127.

Sun, W., Zhu, B., Yang, F. et al. (2021). Optimization of biosurfactant production from *Pseudomonas* sp. CQ2 and its application for remediation of heavy metal contaminated soil. *Chemosphere* 265: 129090.

Szabó, A., Perri, C., Csató, A. et al. (2010) Synthesis methods of carbon nanotubes and related materials. *Materials* 3: 3092.

Taghavi, M., Bakhshi, K., Zarei, A. et al. (2024). Soil pollution indices and health risk assessment of metal(loid)s in the agricultural soil of pistachio orchards. *Scientific Reports* 14: 1–14.

Teng, D., Mao, K., Ali, W. et al. (2020). Describing the toxicity and sources and the remediation technologies for mercury-contaminated soil. *RSC Advances* 10 (39): 23221–23232.

Thakur, S., Singh, L., Wahid, Z.A. et al. (2016). Plant-driven removal of heavy metals from soil: uptake, translocation, tolerance mechanism, challenges, and future perspectives. *Environmental Monitoring and Assessment* 188: 1–11.

Truskewycz, A., Shukla, R., and Ball, A.S. (2018). Phytofabrication of iron nanoparticles for hexavalent chromium remediation. *ACS Omega* 3: 10781–10790.
Tumolo, M., Ancona, V., De Paola, D. et al. (2020). Chromium pollution in European water, sources, health risk, and remediation strategies: an overview. *International Journal of Environmental Research and Public Health* 17: 5438.
Tyagi, S. and Annachhatre, A.P. (2023). A review on recent trends in solidification and stabilization techniques for heavy metal immobilization. *Journal of Material Cycles and Waste Management* 25 (2): 733–757.
Verma, A. and Bharadvaja, N. (2022). Plant-mediated synthesis and characterization of silver and copper oxide nanoparticles: antibacterial and heavy metal removal activity. *Journal of Cluster Science* 33: 1697–1712.
Wan, Y., Liu, J., Zhuang, Z. et al. (2024). Heavy metals in agricultural soils: sources, influencing factors, and remediation strategies. *Toxics* 12: 63.
Wang, X. (2012). Nanomaterials as sorbents to remove heavy metal ions in wastewater treatment. *Journal of Environmental and Analytical Toxicology* 2. doi: 10.4172/2161-0525.1000154.
WHO (2022). Arsenic. https://www.who.int/news-room/fact-sheets/detail/arsenic (accessed 13 June 2024).
Wu, W., He, Q., and Jiang, C. (2008). Magnetic iron oxide nanoparticles: synthesis and surface functionalization strategies. *Nanoscale Research Letters* 3: 397–415.
Xu, J., Li, Y., Wang, S. et al. (2023). Sources, transfers and the fate of heavy metals in soil-wheat systems: the case of lead (Pb)/zinc (Zn) smelting region. *Journal of Hazardous Materials* 441: 129863.
Yang, J., Hou, B., Wang, J. et al. (2019). Nanomaterials for the removal of heavy metals from wastewater. *Nanomaterials* 9: 424.
Ying, S., Guan, Z., Ofoegbu, P.C. et al. (2022). Green synthesis of nanoparticles: current developments and limitations. *Environmental Technology and Innovation* 26: 102336.
Yu, F., Du, Y., Guo, M. et al. (2023). Application of biosurfactant surfactin for the removal of heavy metals from contaminated water and soil via a micellar-enhanced ultrafiltration process. *Separation and Purification Technology* 327: 124947.
Zhang, Q. and Wang, C. (2020). Natural and human factors affect the distribution of soil heavy metal pollution: a review. *Water Air and Soil Pollution* 231: 1–13.
Zhang, C., Xia, T., Zhang, L. et al. (2024). Mercury pollution risks of agricultural soils and crops in mercury mining areas in Guizhou Province, China: effects of large mercury slag piles. *Environmental Geochemistry and Health* 46: 1–19.
Zhang, C., Yang, X., Yang, S. et al. (2023). Valorization of heavy metal enriched phytoremediation biomass using a deep eutectic solvent (DES). *Green Chemistry* 25: 771–778.
Zhang, Y., Zhang, Y., Akakuru, O.U. et al. (2021). Research progress and mechanism of nanomaterials-mediated in-situ remediation of cadmium-contaminated soil: a critical review. *Journal of Environmental Sciences* 104: 351–364.
Zhu, Y., Song, K., Cheng, G. et al. (2023). Changes in the bacterial communities in chromium-contaminated soils. *Frontiers in Veterinary Science* 9: 1066048.

10

Carbon Quantum Dots for the Efficient Degradation of Organic Contaminants

Vikky Kumar Mahto[1], Moumita Pal[2], Ved Prakash[3,4], Ankit Kumar Singh[1], Abhishek Rai[5], Vipendra Kumar Singh[6,#], and Vikas Kumar Singh[7]

[1] *University Department of Botany, Lalit Narayan Mithila University, Darbhanga, Bihar, India*
[2] *Department of Biosciences and Biomedical Engineering, Indian Institute of Technology Indore, India*
[3] *Ecotoxicology Laboratory, Regulatory Toxicology Group, CSIR-Indian Institute of Toxicology Research, Lucknow, Uttar Pradesh, India*
[4] *Academy of Scientific and Innovative Research (AcSIR), Ghaziabad, India*
[5] *University Department of Chemistry, Lalit Narayan Mithila University, Darbhanga, Bihar, India*
[6] *School of Biosciences and Bioengineering, Indian Institute of Technology Mandi, VPO Kamand, Mandi, Himachal Pradesh, India*
[7] *Department of Science and Humanities Supaul College of Engineering, Supaul, Bihar, India*
[#] *Present address: Department of Biochemistry and Molecular Medicine, The George Washington University, Washington, DC, USA*

10.1 Introduction

Recently, luminescent emitted nanomaterials, carbon quantum dots (CQDs), organic fluorophores, and aggregation-induced emission active materials have emerged and recognized by scientists due to having extraordinary applications in different areas of science and technology (Chen and Cao 2023). The accidental discovery of CQDs brought revolutionary change in the field of the nanomaterial world (Xu et al. 2004). The nontoxic nature of carbon nanoparticles (CNPs) of different sizes <10 nm was discovered through a laser ablation approach and named such kinds of nanoparticles (NPs) as CQDs (Sun et al. 2006). The fluorescent CQDs have been recognized globally due to their excellent structural dimension, and physicochemical properties (Wang et al. 2019). Additionally, the unique size distribution, low-cost, fast and facile synthesis method, tunable optical property, strong fluorescence, chemical inertness, high water solubility, strong photostability, and good biocompatibility made them valuable assets of nanotechnology (Figure 10.1) (Kumar et al. 2022).

Alternatively, the limited application of d-block metal quantum dots e.g. zinc selenide (ZnSe), cadmium sulfide (CdS), cadmium telluride (CdTe), cadmium selenide (CdSe), and lead selenide (PbSe) as compared to CQDs is due to their high-cost synthesis, toxic nature, and environmental hazard effect (Rastogi et al. 2021). However, these limitations also restricted their various applications in biomedicine and bioimaging; thus, CQDs are considered rising stars among multiple nanomaterials. The CQDs having various functional

Nanotechnology-based Sustainable Agriculture, First Edition. Edited by Pardeep Singh et al.

groups on their surface provide a suitable platform for the interaction of different organic and inorganic molecules and widen their applications in different field, including photovoltaic, optoelectronics, catalysis, drug delivery, bioimaging, biosensors, wastewater treatment, contrasting agents, light-emitting diodes, etc. (Figure 10.1) (Yuan et al. 2019; Singh et al. 2018a, 2018b; Boruah et al. 2023; Jia et al. 2019). However, surface passivation of CQDs is an excellent approach to enhance their electrical properties. However, CQDs do not contain only electrical properties compared to other NPs but also possess a distinct optical character, i.e. sharp fluorescence property (Kolanowska et al. 2022). Since fluorescence is a size-dependent property due to having a quantum confinement effect, smaller-sized CQDs have a high bandgap and lower wavelength to exhibit blue fluorescence, whereas bigger-sized quantum dots show fluorescence even in the infrared region of electromagnetic radiation (Qian et al. 2013). This chapter mainly focuses on two aspects: (i) the detail about CQDs structure, synthesis, and characterization and (ii) the application of CQDs in the organic contaminants degradation.

10.1.1 Structure of CQDs

To gain deep inside the structural arrangement of CQDs, it comprises a core-shell layout. The layout of CQDs is created through the nucleation methods, which generally requires slow and gradual escalation of the core while the shell surface is decorated with different

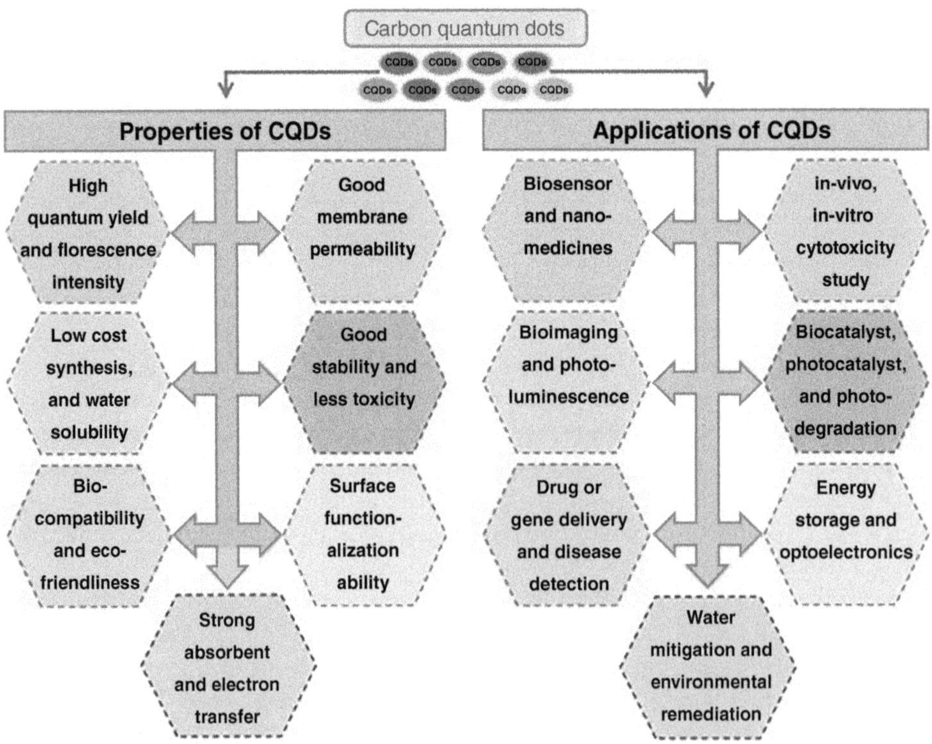

Figure 10.1 Properties and potential application of carbon quantum dots.

functional groups through the self-passivated process (Tang et al. 2012; Kumar et al. 2022). The intrinsic core of CQDs may be crystalline or amorphous, which depends upon the level of graphitization and the existence of sp^2/sp^3 carbon. The crystalline core of CQDs is mainly due to the presence of sp^2 carbon, whereas amorphous is due to mixed (sp^2/sp^3) carbon (Hola et al. 2014). In addition to this, the type of core formed also depends upon the behavior of the precursor used, the synthetic route, and the condition of the reaction temperature, pressure, reaction time, and pH (Emanuele et al. 2019). It has been found that a high temperature above 300 °C leads to excellent graphitization (sp^2 carbon), which results in the formation of a crystalline core, whereas at a lower temperature, the core consists of sp^2/sp^3 carbon leads to an amorphous core (Yu et al. 2018; Dhariwal, Rao, and Vaya 2024). Typically, the size of the core is very small in the range of 2–3 nm with a lattice distance of 0.2 nm (Zhang et al. 2015).

The shell of CQDs is typically amorphous and decorated with varieties of functional groups such as hydroxyl, amino, carboxyl, etc. (Dhariwal, Rao, and Vaya 2024). The presence of different functional groups on the shell surface of CQDs depends on the use of precursors or dopant species. Additionally, the composition ratio of different functional groups on the shell surface often varies with the type of synthetic route employed for their synthesis (Semeniuk et al. 2019). The optical behavior and electron donor–acceptor properties of CQDs are defined by the shell of the CQDs. Although the shell of CQDs has different functional groups that could be modulated and altered through different functionalities, the passivating agent leads to changes in their optical property. Therefore, the structure of CQDs is hole and sole for their intrinsic property (Mintz, Zhou, and Leblanc 2019; Gayen, Palchoudhury, and Chowdhury 2019).

10.2 Synthesis Method

To get the desired size, functionalities, and specific application of CQDs, many active research groups have developed economics and eco-friendly procedures for the synthesis of CQDs. However, the synthesis method of CQDs is generally categorized into two categories: "Top-down and Bottom-up" (Saputra et al. 2024). The synthesis method of CQDs is no doubt very facile, but some major challenges are also associated, such as controlling the uniformity and size during carbonization and tuning surface characteristics (Wang and Hu 2014). The synthesis of CQDs via the "Top-down" approach involves breaking or cutting down macroscopic precursors, such as graphene, graphene oxide, carbon nanotube, and polymeric materials into desirable nanoranges by physical, chemical, or electrochemical methods (Ikram, Azmat, and Perviaz 2024). Top-down synthesis approaches are achieved through several methods, which include chemical oxidation, arc discharge, laser ablation, electrochemical oxidations, ultrasonic synthesis, etc. (Tian et al. 2009; Xu et al. 2004; Deng et al. 2014; Li et al. 2011). In the Bottom-up approach, small precursor molecules, such as citric acid, carbohydrates, amino acids, phosphoric acids, and many more, are used for the synthesis of CQDs through chemical reactions. These strategies can be achieved by various methods, including hydrothermal, solvothermal, microwave, pyrolysis, thermal treatment, and plasma treatment (Li et al. 2011; Eskalen et al. 2020; Ma et al. 2015, 2019).

Although both outlooks are utilized for the synthesis of CQDs by many active research groups, there are some limitations. In the Top-down approach, there is a requirement for a sophisticated instrumentation setup and a costly precursor, whereas, in the Bottom-up approach, a cheaper precursor and a simple experimental setup are required for the synthesis of CQDs (Chae et al. 2017; Paulo, Palomares, and Martinez-Ferrero 2016). Furthermore, in both approaches, surface passivation of CQDs can be done through post-treatment to modulate the surface functional groups, which results in the enhancement of the quantum yield (QY) of CQDs. However, the Bottom-up approach has some edge over the Top-down in terms of cost-effectiveness and environmental friendliness, and thus this method is commonly used for the synthesis of CQDs. In this section, the most widely used techniques are explored in detail.

10.2.1 Hydrothermal Method

The hydrothermal method is the widely utilized technique for the synthesis of CQDs. In this method, precursor molecules, such as citric acid, carbohydrate molecules, amino acids, vegetable juice/pulp, fruit juice/pulp, and waste peel, have been carbonized in a customized stainless steel autoclave at high temperature and pressure (Li et al. 2011). In addition to this, no harmful gases were released during the chemical reaction, and the whole experiment was performed in an aqueous medium; therefore, this method has great advantages, such as being simple, nontoxic, cost-effective, one-step, and environment-friendly (Bian et al. 2014). In this method, varying the precursors and reaction conditions, such as temperature, pressure, and duration of reaction, can improve the optical properties of CQDs (Wang and Hu 2014).

10.2.2 Microwave Irradiation Method

In recent years, microwave irradiation methods have received more attention in research groups due to their rapid, eco-friendly, low-cost, and energy efficiency (Wang et al. 2012). Furthermore, this method does not require any sophisticated instrument, and the reaction temperature can be adjusted manually to avoid overheating. In this method, at high temperatures, small precursor molecules such as citric acid, urea, glucose, sucrose, etc. break and recombine to create a new molecule in the nanorange (Wang et al. 2023). This method is considered one of the best methods of Bottom-up approaches because the chemical bond of polar precursor molecules is easily broken in high-energy microwave radiation and converted into nanorange with distinct properties. The synthesis of stable luminescent CQDs using microwave radiation was first reported by Zhu et al. (2009).

10.2.3 Pyrolysis Method

The synthesis of CQDs through the pyrolysis way was reported by Liu et al. 2009. It is a novel method for the synthesis of CQDs, in which organic precursor molecules are carbonized above 430 °C under pressure in an oxygen-free environment. Furthermore, in this method, alkali and strong acids are conveniently used as the requirements of the reaction condition and serve as catalysts to assist the reaction at a faster rate. During pyrolysis, irreversible modifications in the chemical characteristics and physical phase of the precursor molecule occur, which leads to the formation of CQDs. Many of the active research groups, such as Martindale et al. (2015), Wang et al. (2016), and Rong et al. (2017), reported the synthesis of CQDs through the pyrolysis method using different precursor molecules

10.2.4 Chemical Ablation Method

In this method, organic precursor molecules are treated with strong oxidizing acids, such as sulfuric acid (H_2SO_4), nitric acid (HNO_3), hydrogen peroxide (H_2O_2), and perchloric acid ($HClO_4$), which undergo carbonization and are converted into CQDs (Kumar et al. 2022). For the synthesis of CQDs, it is one of the most accessible methods, but because of the use of strong oxidative conditions, harsh environments, and multiple steps; there is indigent control over the size of CQDs, which affects the optical properties of CQDs and therefore often requires post-passivation to enhance the fluorescence QY of CQDs (Wang and Hu 2014). For instance, CQDs were synthesized by Peng and Travas-Sejdic (2009) using carbohydrates as a precursor and strong oxidizing acid H_2SO_4/HNO_3. Firstly, H_2SO_4 carbonized the carbohydrates into carbonaceous material, then after HNO_3 broken down into individual CQDs. To improve their QY, they post-passivated the synthesized CQDs with amine-terminated compounds (4,7,10-trioxa-1,13-tridecanediaminePeng and Travas-Sejdic 2009).

10.2.5 Electrochemical Carbonization

It is the cost-effective, simple, nontoxic, and easily operated method employed for the synthesis of CQDs (Alaghmandfard et al. 2021). In this technique, a bulk precursor molecule, such as carbon fiber, graphite, or carbon nanotubes, is cut down into smaller particles under the influence of an electrochemical cell. The electrochemical cell consists of three components: a working electrode made up of carbon precursor, a reference electrode made up of platinum, and an electrolyte solution. When redox potential is generated due to the electrolysis of electrolyte solution, it results in the judicious cutting of carbon materials into smaller particles producing CQDs (Kumar et al. 2022). The completion of the synthesis of CQDs could be identified by an alteration in the color of the electrolyte solution. The great advantages of this method are that it can be operated at normal temperature and pressure, and the size of CQDs can be tuned by the controlled supply of voltage and current across aqueous or non-aqueous solutions (Paulo, Palomares, and Martinez-Ferrero 2016).

10.2.6 Arc Discharge Method

Xu and colleagues were the first to discover the arc discharge method for synthesizing fluorescent CQDs. They accidentally identified this method while using this approach to separate and purify a single-wall carbon nanotube. Nitric acid was employed as an oxidizing agent in this procedure to oxidize arch ash, forming various functional groups on

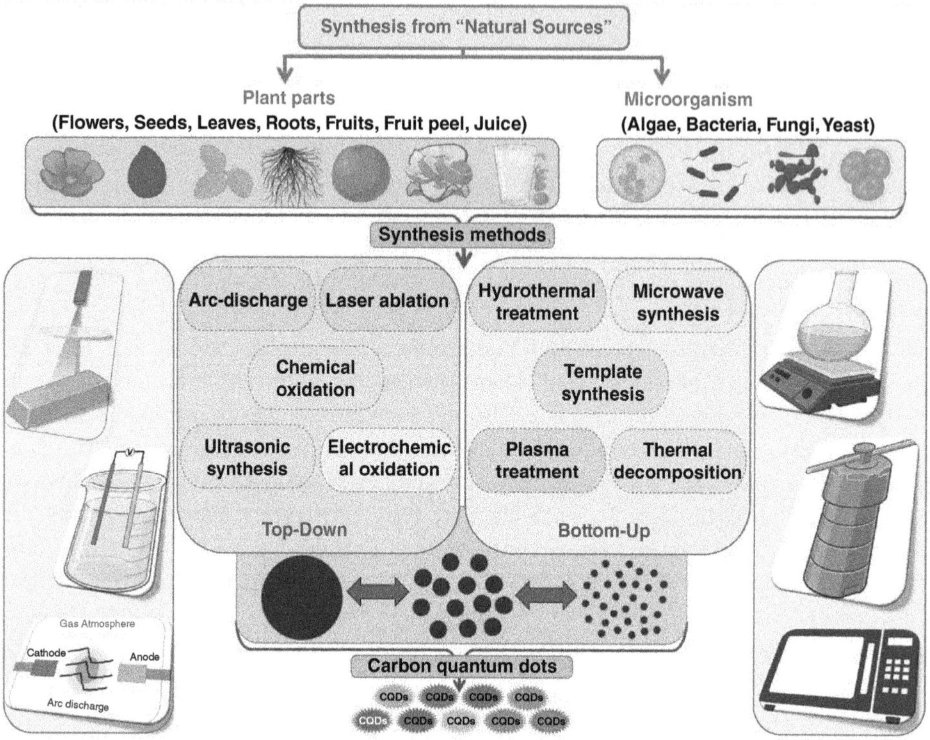

Figure 10.2 Different methods for synthesis of carbon quantum dots.

its surface and increasing its aqueous solubility. When excited at a wavelength of 366 nm, the QYs obtained were 1.66%. Carbon particles of various sizes are created during the discharge process (Figure 10.2). This approach produces CQDs with a wide range of particle sizes that are highly soluble in water. However, due to their intricate composition, CQDs produced using this method contain some contaminants that are challenging to remove (Yadav et al. 2023).

10.3 Organic Contaminants and Their Impacts on Plants and the Environment

Recently, organic contaminants acquired immense attraction owing to their detrimental effects on plants and the environment. These are toxic species that originate from natural and artificial processes, such as the production of petrochemicals, volcanoes, bushfires, the mining and processing of coal, the administration of pesticides, the dyeing of textile and leather goods, and the production of pharmaceuticals (Akeremale et al. 2023). These harmful substances are resistant to the natural remediation process and accumulate within the environment for several decades (Devendrapandi et al. 2024). There are several types of organic contaminants such as medications and personal hygiene items, dyes, polycyclic aromatic hydrocarbons (PAHs), perfluoroalkyl carboxylates, and sulfonates (Akeremale et al. 2023), polychlorinated biphenyls, some organochlorine pesticides and polychlorodibenzodioxins/furans (Choo et al. 2020), dioxins, and insecticides (Devendrapandi et al. 2024). Since these organic contaminants accumulate in nature, they are easily overtaken by plants and cause toxic effects. The ultrastructure, biosynthesis, membrane stability, and DNA of plant cells can all be adversely affected by organic pollutants (Zhang et al. 2017). The impact of some common organic pollutants is described as follows.

10.3.1 Dyes

Every year, the production of synthetic dyes amounts to 70 million tons worldwide, whereas over 10 000 tons are used in the textile sector. These dyes are only used to color textiles and other materials, including hair, fur, leather, waxes, and plastics (Dutta et al. 2024). Textile dyes are categorized into two types: hydrophilic and hydrophobic. Water-soluble hydrophilic dyes, also known as anionic dyes, are used to color textiles at an acidic pH in an aqueous solution. Certain dyes need particular solvents to be used because they are hydrophobic or water-insoluble. There are several problems associated with the use of dyes that are harmful to the environment. They are resistant to degradation due to high polarity and recalcitrance. Some are not biodegradable because the structure of anthraquinone-based dyes contains fused aromatic rings (Dutta et al. 2024). These dyes also have some unique properties, such as elevated thermal and photostability, which pose a severe threat to the environment. Certain metal constituents of the dyes are risks to soil fertility and plant development (Ramamurthy et al. 2024). Furthermore, oxidative stress, which is exacerbated by the chromium in textile dyes, is another issue that is resistant to resolution. The growth and development of a plant can be severely harmed by oxidative stress, especially in the areas of photosynthesis and CO_2 uptake (Lellis et al. 2019).

10.3.2 Polycyclic Aromatic Hydrocarbons

PAHs are a group of toxic pollutants that have recently become a great environmental issue due to their ubiquitous and highly hydrophobic nature. It consists of two or more benzene rings arranged in angular, linear, or cluster arrangements (Jajoo 2017; Onwujekwe et al. 2022). The characteristics of PAHs include their carcinogenicity, electrochemical stability, persistency, structural angularity, hydrophobicity, volatility, and resistance to biodegradation (Onwujekwe et al. 2022). It has been found that PAHs can cause tissue necrosis, oxidative stress, decreased growth, and leaf deformation (Jajoo 2017). The PAHs reduce the chlorophyll content of both C_3 and C_4 plants along with an increase in chlorophyll *a*/*b* ratio, which causes harm to the plants. It decreases the photosynthetic rate by inhibiting photosystem II (PSII). It inhibits the RuBisCO activity (Molina and Segura 2021). Phenanthrene and pyrene exposure reduce the growth, photosynthetic pigment levels, conductance of stomatal pores, effective QY of PSII, maximal QY, and photochemical quenching coefficient of higher plants (Jajoo et al. 2014).

10.3.3 Pesticides and Insecticides

Pesticides are synthetic chemicals used to get rid of insects, rodents, fungus, and weeds. Nematicides, insecticides, fungicides, rodenticides, molluscicides, and other substances are among them. The four groups of organic compounds that made up the majority of synthetic pesticides were organophosphates, carbamates, pyrethroids, and organochlorines (Pathak et al. 2022). Pesticides can persist in the environment for longer periods due to their long half-lives. They accumulate in nature due to excessive usage and lead to soil fertility disruption and decreased photosynthetic efficiency (Ahmad et al. 2024). Pesticide usage causes an array of enzymatic and non-enzymatic changes in antioxidants, both physiological and biochemical, that can affect plant yield (fruits, vegetables, and seeds) in the long run and initially impact plant growth from germination. Due to the production of reactive oxygen species, the usage of pesticides results in oxidative stress, which can ultimately cause a lack of development and decreased photosynthetic efficiency in plants (Pathak et al. 2022). The mechanisms by which insecticides and herbicides can exhibit phytotoxicity include preventing cell division, altering the permeability of membranes, obstructing respiration and photosynthesis, reducing the synthesis of proteins or hormones, and decreasing enzyme performance (Guedes et al. 2023). Therefore, it can be inferred that these organic contaminants have detrimental effects on plants and the environment; hence, they have to be efficiently removed from the environment (Figure 10.3).

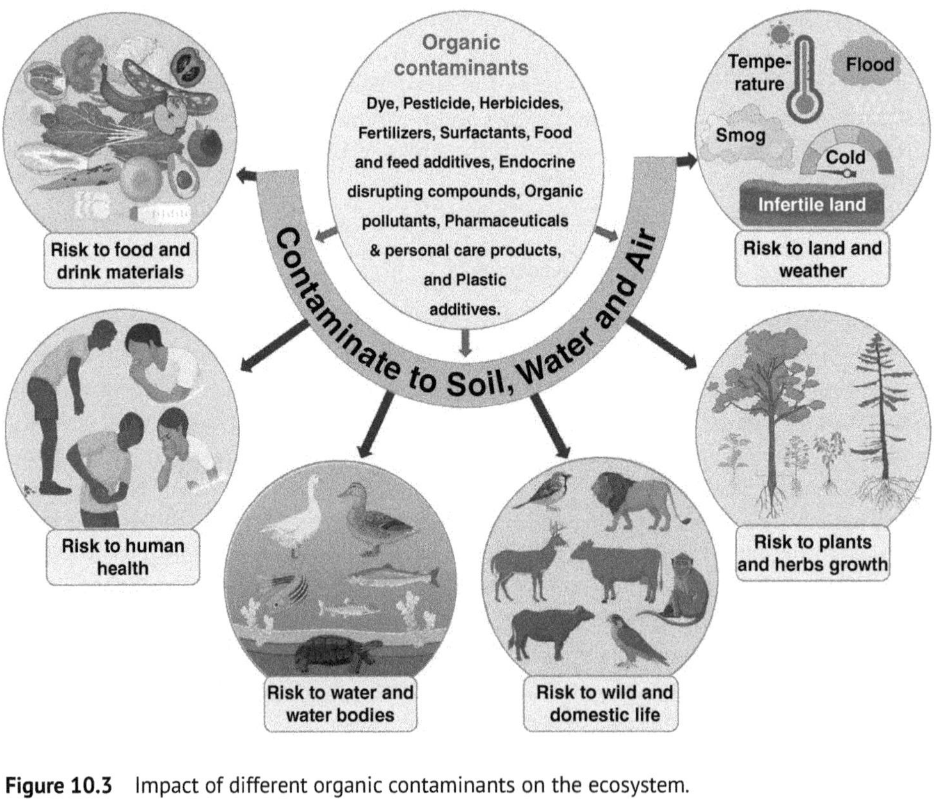

Figure 10.3 Impact of different organic contaminants on the ecosystem.

10.4 CQDs Application in Detection of Agrochemical Residues

CQDs are upcoming tools that promise the detection of agrochemical remnants using their unique optical and electrical properties. These fluorescent probes can be made in use for the identification of residual pesticides, hazardous to human health as well as ecosystem balance. These display advantages over traditional methods like chromatography and mass spectrometry for being rapid, sensitive, and cost-effective. In agricultural practices, the usage of pesticides and herbicides has a negative impact on health and the environment, CQDs have the potential to detect and provide remedies for these contaminants.

10.4.1 Pesticides and Herbicides

CQDs are emerging tools used in pesticides and herbicides detection by having specific characteristics, such as being highly water soluble with low toxicity and higher fluorescence. Detection of pesticides comes with a challenge, as it is hard to find naturally occurring receptors, synthetic receptors, and recognition elements. Rapid pesticide detection is done by keeping an eye on the inhibition of enzyme activity; however, poor pesticide selectivity towards enzymes fails to determine the nature of the interaction. This problem can be resolved with the recently developed 0-D nanomaterials. When pesticides interact with the fluorescent CQDs, it affects the fluorescence intensity, which depends on the nature of the interaction, that could be used as a mechanism of transduction (Mandal et al. 2019; Deka et al. 2019; Lin et al. 2018; Wang et al. 2015; Zhao et al. 2019). CQDs while interacting with pesticides perturb the electron–hole pair recombination; various kinds of transfer of charge, energy, and inner filter effect (IFE) could change the CQDs fluorescence (by either enhancing or quenching of fluorescence). For example, imidacloprid tightly binds with CQDs using the functional groups -NO_2 and -NH; this breaks the CQDs self-complexation and leads to switching the CQDs to a greater isolated state, thus enhancing their intensity of fluorescence (Mandal et al. 2019). Thiabendazole enhances the CQDs fluorescence in silica ingrained polymers (Kazemifard, Ensafi, and Rezaei 2020). However, pesticides, such as chloropyrifos, atrazine, lindane, and tetradifon, quench the CQDs fluorescence via covalent bond, hydrogen bond, or electrostatic bonds with the utilization of substitution reaction (Mandal et al. 2019).

It has been found that the probe selectivity depends on the differential pesticide binding affinity with CQDs. Most of the pesticides quench the green CQDs fluorescence through different mechanisms, viz., amicarbazone, glyphosate, and diazinon. These are involved in the quenching of CQDs green fluorescence, which is synthesized using juice of cauliflower through a hydrothermal approach, by fluorescent resonance energy transfer (FRET), in ng/mL limits of detection (LOD; Ashrafi et al. 2020). A drawback is that the CQDs were unable to differentiate with three pesticides together in a solution showing poor selectivity, which needs to be enhanced by specific ligand modulation on CQDs. One way to enhance the photoluminescent properties and selectivity of CQDs to modify the sensing mechanism toward various pesticides is their doping with numerous hetero atoms and flavonoid moiety, especially in dinoseb, dithianon, fenitrothion, and their related structural analogs (Li et al. 2020). It has been found that when heteroatoms are more diverse, the emission of CQDs wavelength varies to equalize the pesticide absorption wavelength to achieve greater IFE.

When magnetized silica beads are transformed along with graphene quantum dots (GQDs) and molecularly engraved polypyrrole, they become highly selective towards tributyltin (Zor et al. 2015). The nanocomposite magnetic properties are enriched with the analyte, one that further leads to the fluorescence quenching of GQDs in the composite using a mechanism of electron transfer with the 12.78 ppb LOD. However, pretilachlor was detected by increasement of CQD fluorescence acquired from aquatic hyacinth; an electron transfer mechanism was also found to have occurred to enhance the fluorescence (Deka et al. 2019).

For some pesticides, a linking agent to quench the CQDs fluorescence can be utilized for the pesticides detection when there is poor selectivity or direct alteration of the fluorescence of CQDs. One such linking agent used by Peng et al. (2019) was pralidoxime to the interaction of fluorescence-based nitrogen and magnesium-codoped CQDs (Mg, N-CQDs) paraoxon. Here, pralidoxime initiates redox reactions between paraxon and CQDs resulting in fluorescence quenching (Peng et al. 2019). With the help of Mg, and N-CQDs, paraoxon can be selectively detected in the nanomolar regime. Methyl-paraoxon can be decomposed into p-nitrophenol and phosphates using a catalysis reaction by CeO_2 NPs (Luo et al. 2020; Wei et al. 2019). The as-formed p-nitrophenol quenches CQDs fluorescence by IFE. In catalysis reaction, when combined with the CQDs-based IFE, it can particularly distinguish methyl-paraoxon up to 24.70 ng/mL. To detect carbendazim (CBM), a fluorescence-based CQDs-Au nanoclusters (Au-NCs) system has been designed and used as a probe (Yang et al. 2020). An Au-NCs quenches N-doped CQDs fluorescence using a mechanism of FRET, but it is different from CBM. When CBM interacts with Au-NCs, it leads to CQDs being free, and thus the fluorescence recovery takes place. Such kind of pesticide-mediated assemblage of Au-NPs separates CQDs and Au-NPs pair and detects cartap (Yang et al. 2019). Fe^{3+} ions generated from the oxidation of Fe^{2+} by the action of H_2O_2 can switch off CQDs fluorescence made from waste paper ashes (Lin et al. 2018). It is well known that the acetylcholinesterase enzyme activity employed for sensing of pesticides, is inhibited by organophosphorus pesticides obstructing the production of H_2O_2. Acetylcholinesterase/choline system and Fe^{2+} along with fluorescent CQDs can be utilized to recognize chlorpyrifos up to 3 ng/mL low level (Lin et al. 2018). Due to the presence of chlorpyrifos, the transition of Fe^{2+} to Fe^{3+} is hindered, enhancing the intensity (flourescence) of the probe system in contrast to the chlorpyrifos absence, where fluorescence is quenched via Fe^{3+} by interaction with oxygen (O_2) atoms on CQDs. It has been reported that an organophosphate pesticide using GQDs, choline oxidase-acetylcholinesterase enzyme systems was observed using a similar "turn-off"/"turn-on" perspective by Sahub et al (2018). The acetylcholinesterase/choline oxidase system generates H_2O_2, which in turn quenches the GQDs fluorescence; however, when organophosphate pesticides are present, this enzymatic reaction inhibits the displaying fluorescence of GQDs. The concentration of Dichlorvos up to 0.172 μg/mL at low levels can be detected using a probe. In a similar approach, N- and S-doped CQDs can sensitively detect carbaryl with a 5 μg/mL LOD, utilizing the choline oxidase and acetylcholinesterase enzymes that produce H_2O_2 (Li et al. 2016). The H_2O_2 produced from acetylcholine chloride prevents N- and S-doped CQDs from fluorescence. The presence of

N- and S-doped CQDs have been found to increase the reactivity of H_2O_2, thus resulting in high sensitivity.

An alachlor herbicide detection in samples of soil was done with high sensitivity and fluorescence by red-emissive nitrogen-doped CQDs (N-CQDs). The N-CQDs have been reported to detect traces of alachlor herbicide and demonstrated the detection range from 0.005 to 150 μM with 0.2 μM LOD. This probe has been used to detect alachlor from soil samples with significant recovery from 86.6 to 114.3%. This kind of probe shows promising fluorescence of alachlor detection with significant sensitivity and selectivity (Du et al. 2019). The herbicide atrazine was detected by a luminescent probe, N-CQDs, which has been designed with high selectivity. On the turning-on probe, fluorescence on atrazine addition is linked to the aggregation-mediated by increasement in fluorescence of the amine enacted CQDs. The network of atrazine and H bonding of surface amine groups of CQDs probes render the probe having the high sensitivity and selectivity for the atrazine even also with other metal ions and pesticides. The LOD for atrazine recognizing was noticed at 3 ppm, which is significantly at the minimal level of detection limit from other reported detection methods for atrazine. Due to this lowest detection limit property; this sensor was also being used in real on-site agricultural field samples. Due to the great luminescent potential and photostability, such a fluorescent probe will be explored to detect atrazine in microbial/bacterial cells by image analysis (Mohapatra, Bera, and Das 2018).

10.4.2 Fungicides and Insecticides

It has been reported that using luminescent CQDs in a lanthanide-sensitized luminescence (LSL) system can involve boosting selectivity and sensitivity. In this process, CQDs in the LSL system interact with the lanthanide ions by carboxylic groups, leading to the more efficient time-resolved luminescent probe. This method has been used to detect the propineb (PPN) fungicide, exhibiting PPN quenching the europium CQDs luminescence and showing the LOD of 0.03 μg/mL PPN and 3 mg/kg in capers, achieving the limit of the maximum residue of 25 mg/kg. This demonstrates that CQDs in the LSL system gives novel and simple approach for contamination screening in the agricultural and food sectors (Ruiz-Medina, Jiménez-López, and Llorent-Martínez 2022). CBM insecticide is widely used in agricultural practices to prevent fungal infection in plants. However, residual CBM in environmental matrices like water and food imposes significant risks to the ecosystem and human health. This CBM is detected by the N-CQDs and is developed by encapsulation in situ techniques using a zirconium metal-organic framework (UiO-66-NH_2) electrocatalyst. This electrocatalyst was based on selective electrochemical CBM detection. The N-CQDs@UiO-66-NH_2 coated screen-printed carbon electrode showed high selectiveness for the CBM detection, with a range of detection 0.02–126 μM and a LOD of 5.8 nM. The CQDs sensor showed more than 95% selectivity, with a real-time CBM retrieval rate of 96% in vegetables and 97% in environmental aqueous samples (Velmurugan et al. 2024).

CQDs have been shown to provide the promising detection of insecticides due to having better optical features, high QYs with biocompatibility. It has been reported that fluorescent CQDs have been synthesized from Jatropha fruit (J-CQDs) through a one-pot hydrothermal approach and shown emission with bright blue fluorescence. These J-CQDs have been employed as fluorometric sensors for pesticide detection. The property of the sensing for the insecticides is associated with irreversible suppression of AChE activity by regulated quenching events. With a simple design and short reaction time, these CQDs have provided excellent detection properties for the Chloprifos insecticide, with a LOD of 2.7 ng/mL with an acceptable recovery in environmental and agricultural samples (Chandra et al. 2022). Specific insecticides including; diazinon, glyphosate, and amicarbazone have been detected by fluorescence-based CQDs with LOD 0.25 ng/mL, and this enhancing/quenching the CQDs fluorescence in the presence of certain insecticides (Ashrafi et al. 2020). Another method was employed to detect the insecticides based on the concentration called ratiometric fluorescence sensors, where the intensity of the fluorescence ratio at two different wavelengths changes in response to concentrations. The acephate insecticides have been detected by using this method with a LOD 0.052 ppb (Li et al. 2022). Additionally, to enhance the detection potential of CQDs, combine these sensors with other materials, such as ionic liquids and covalent organic frameworks. These combined multidimensional CQDs can provide the property to detect various insecticides by producing fluorescence, phosphorescence, and light scattering signals based on different optical footprints (Zhang et al. 2021).

10.5 Photocatalytic Degradation of Organic Contaminants Using CQDs

The involvements of CQDs in photocatalytic degradation of organic contaminants is of utmost importance owing to their exclusive optical and the electronic properties (Yuan et al. 2016). In this context, Qu and his co-workers synthesized CNPs via a hydrothermal route. The synthesized CNPs were employed in multicolor bioimaging and as effective fluorescent sensors for detection of dopamine with a detection limit as low as nM (Qu et al. 2013). Miao and co-workers prepared CQDs decorated with TiO_2 (CQDs/TiO_2) via ultrasonic-hydrothermal methods. It degraded methylene blue (MB) 98% in one hour and *N*-benzylideneaniline 30% in two hours under the illumination of visible spectrum (Miao et al. 2016). Tian et al., synthesized hybrid photocatalyst CQDs/H-TiO_2 by hydrothermal and bath reflux strategies. It degrades the methyl orange (MO) 86% within 225 minutes under near-infrared radiation and UV-visible light (Tian et al. 2015). Ali and his associates synthesized a P25/CQD photocatalyst that removes the dyes like MO 40%, RhB 80%, and 4CP 49% under the visible light. The removal activity of P25/CQD photocatalyst was more than the bare P25 (Ali, Nair, and Sandhya 2019). Shen and co-workers synthesized CQDs modified P25 (commercial TiO_2) to form the CQDs/P25 heterojunction photocatalysts by facile and green hydrothermal methods using glucose and citric acid as precursor. CQDs/P25 degraded the phenol (~99%) under the illumination of UV light (Shen et al. 2018). Further, Ji and his co-workers synthesized CQD-decorated BiOBr (CQDs/BiOBr) photocatalysts using hydrothermal methods. It degrades antibiotic tetracycline (TC) 60% after 120 minutes, endocrine disrupter bisphenol A (BPA) 73% after 150 minutes, and dye RhB 91.8% after 20 minutes irradiation under the visible light. The activity was higher than the pristine BiOBr (Ji et al. 2018). Qu and his colleagues synthesized CQDs modified $KNbO_3$ (CQDs/$KNbO_3$) photocatalysts using hydrothermal and mixed-calcination methods that removes crystal violet dye (70%) after five hours under visible light illumination. On the other hand, the bare $KNbO_3$ degrades only 41.5% of the dye (Qu et al. 2018). Li and his associates fabricated the CQDs modified ZnO photocatalysts by sol-gel method. It degrades Rhodamine B (Rh B) and showed three times better photocatalytic activity in comparison to the bare ZnO (Li et al. 2013). Muthulingam and his associates prepared N-doped ZnO photocatalysts decorated with CQDs that degrades malachite green 100% after 30 minutes, MB, and fluorescein dyes 100% after 30 minutes under daylight irradiation (Muthulingam, Lee, and Uthirakumar 2015). Ding and his co-workers synthesized CQDs modified ZnO by facile hydrothermal process. It degrades MO, MB, and rRh B with k values of 0.0031, 0.0121, and 0.0092 minute^{-1}, respectively, under the irradiation of UV and visible light (Ding et al. 2016). Kaur and co-workers prepared CQDs modified ZnS (CQDs/ZnS) photocatalysts via precipitation methods. It degrades the Alizarin Red S dye 89% after 250 minutes under the illumination of visible spectrum. The activity was greater than the bare ZnS only (Kaur, Sharma, and Kansal 2016). Liu and co-workers prepared CQDs altered CdS (CQDs/CdS) photocatalysts via a facile hydrothermal activity using glutathione as a template. It degraded Rh B dye 90% after one hour under visible light irradiation, while bare CdS was 50% after one hour (Liu, Yu, and Zhang 2013). Sun et al. synthesized a CQD-decorated Bi_2MoO_6 (CQDs/Bi_2MoO_6) photocatalyst via a hydrothermal process. By using 0.1 g concentration, it degraded RhB 100% after 120 minutes and MB 91% after 120 minutes under visible light irradiation (Sun et al. 2017). Wang and colleagues fabricated CQDs modified Fe_3O_4 (CQDs/Fe_3O_4) photocatalyst composite via solvothermal method which degraded MB 100% after 15 minutes under the illumination of visible spectrum (Wang et al. 2014). Di and co-workers developed CQDs modified Bi_2WO_6 (CQDs/Bi_2WO_6) composites that degraded the Rh B 1.8 times more than the bare Bi_2WO_6 after 30 minutes, TC hydrochloride, BPA, and ciprofloxacin 87% after 120 minutes under the visible light (Di et al. 2015). The field of photocatalysis is vast and still much is yet to explore employing CQDs.

10.6 Conclusion and Future Outlook

Organic pollutants potentially affect the flora and fauna in the ecosystem at different trophic levels and together affect human health due to their mutagenic, carcinogenic, teratogenic, and endocrine-disrupting potential. For better management in terms of the detection and degradation of organic contaminants is a key topic. In brief, the major concerns associated with the upgrade in catalytic potentials and synthesis of carbon-based quantum dots. The second challenge related to economic feasibility of inventive and novel synthetic techniques is required. It is viable to initiate quantitative methods for rapid detection of the numerous organic pollutants, which are based on alteration in intensity as well as qualitative effects of the fluorescence enhancement and quenching. The oxidizing procedure and utilizing electrochemical support for alteration in carbon dots (CDs) could remarkably induce its absorption potential and quantity to eliminate organic chemicals or contaminants from agricultural residues and the environment. This could be measured from a novel viewpoint to bioremediation, but a detailed analysis to the toxicity-related concern is also warranted. The scalable techniques of purification, functionalization, and synthesis of carbon-based quantum dots must be introduced since characteristics of carbon-based quantum dots are greatly linked with the experimental setup, precursors, and dopants utilized. The photocatalytic activity has the benefits of good reproducibility, simple operation, excellent efficiency, environmental friendliness, and safety. They are fruitful and economically feasible to break down organic contaminants into least toxic forms at ambient pressure and temperature without generating toxic intermediates. In spite of lots of research work in the last few years related to the photocatalytic technologies has been done, there are still numerous hurdles for the development of potential photocatalytic techniques and investigation of fundamental mechanisms to enhance their potential that require further investigation in research.

By developing promising in situ characterization methods, it will enhance our understanding, how carbon-based quantum dots are established and what affects their photochemical properties and shape, and how they behave with their microenvironment. This would help researchers in developing methods for modifying the properties of nanomaterials to the detection and degradation of organic contaminants; various agrochemical residues. As per the majority of the published findings, CDs can break down or detect particular developing contaminants. So, it is critical to concentrate on purification steps and efficient synthesis, generating a comprehensive idea for the mechanisms establishing CDs detection, generation, and degradation, estimating the feasibility of CDs in wide applications, and predicting long-term consequences.

Abbreviations

4CP, 4-chlorophenol; AChE, acetylcholine esterase covalent; CBM, Carbendazim; COFs, Covalent organic frameworks; CQDs, Carbon Quantum Dots; GQDs, Graphene Quantum Dots; IFE, inner filter effect; LEDs, light-emitting diodes LEDs, LOD, Limits of detection; LSL, lanthanide-sensitized luminescence; MO, methyl orange; OCPs, Organochlorine pesticides; PAHs, Polycyclic aromatic hydrocarbons; PCBs, Polychlorinated biphenyls; PCDD/Fs, Polychlorodibenzodioxins/furans; PPCPs, Personal care products; PPN, Propineb; PSII, photosystem II; RhB, rhodamine B; ROS, reactive oxygen species; SPCE, screen-printed carbon electrode.

References

Ahmad, M.F., Ahmad, F.A., Alsayegh, A.A. et al. (2024). Pesticides impacts on human health and the environment with their mechanisms of action and possible countermeasures. *Heliyon* 10: e29128. doi: 10.1016/j.heliyon.2024.e29128.

Akeremale, O.K., Ore, O.T., Bayode, A.A. et al. (2023). Synthesis, characterization, and activation of metal organic frameworks (MOFs) for the removal of emerging organic contaminants through the adsorption-oriented process: a review. *Results in Chemistry* 5: 100866. doi: 10.1016/j.rechem.2023.100866.

Alaghmandfard, A., Sedighi, O., Rezaei, N.T. et al. (2021). Recent advances in the modification of carbon-based quantum dots for biomedical applications. *Materials Science and Engineering: C* 120: 111756. doi: 10.1016/j.msec.2020.111756.

Ali, M.M., Nair, J.A., and Sandhya, K.Y. (2019). Role of reactive oxygen species in the visible light photocatalytic mineralization of rhodamine B dye by P25–carbon dot photocatalyst. *Dyes and Pigments* 163: 274–284. doi: 10.1016/j.dyepig.2018.11.057.

Ashrafi Tafreshi, F., Fatahi, Z., Ghasemi, S.F. et al. (2020). Ultrasensitive fluorescent detection of pesticides in real sample by using green carbon dots. *PLoS One* 15 (3): e0230646. doi: 10.1371/journal.pone.0230646.

Bian, J., Huang, C., Wang, L. et al. (2014). Carbon dot loading and TiO_2 nanorod length dependence of photoelectrochemical properties in carbon dot/TiO_2 nanorod array nanocomposites. *ACS Applied Materials & Interfaces* 6 (7): 4883–4890. doi: 10.1021/am4059183.

Boruah, A., Bora, S., Thakur, A. et al. (2023). Solid-state phosphors from coal-derived carbon quantum dots. *ACS Omega* 8 (28): 25410–25423. doi: 10.1021/acsomega.3c02884.

Chae, A., Choi, Y., Jo, S. et al. (2017). Microwave-assisted synthesis of fluorescent carbon quantum dots from an A2/B3 monomer set. *RSC Advances* 7 (21): 12663–12669. doi: 10.1039/C6RA28176A.

Chandra, S., Bano, D., Sahoo, K. et al. (2022). Synthesis of fluorescent carbon quantum dots from Jatropha fruits and their application in fluorometric sensor for the detection of chlorpyrifos. *Microchemical Journal* 172: 106953. doi: 10.1016/j.microc.2021.106953.

Chen, W. and Cao, D. (2023). Luminescence nanomaterials and applications. *Nanomaterials* 13 (6): 1047. doi: 10.3390/nano13061047.

Choo, G., Wang, W., Cho, H.S. et al. (2020). Legacy and emerging persistent organic pollutants in the freshwater system: relative distribution, contamination trends, and bioaccumulation. *Environment International* 135: 105377. doi: 10.1016/j.envint.2019.105377.

Deka, M.J., Dutta, P., Sarma, S. et al. (2019). Carbon dots derived from water hyacinth and their application as a sensor for pretilachlor. *Heliyon* 5 (6): e01985. doi: 10.1016/j.heliyon.2019.e01985.

Deng, J., Lu, Q., Mi, N. et al. (2014). Electrochemical synthesis of carbon nanodots directly from alcohols. *Chemistry–A European Journal* 20 (17): 4993–4999. doi: 10.1002/chem.201304869.

Devendrapandi, G., Liu, X., Balu, R. et al. (2024). Innovative remediation strategies for persistent organic pollutants in soil and water: a comprehensive review. *Environmental Research* 249: 118404. doi: 10.1016/j.envres.2024.118404.

Dhariwal, J., Rao, G.K., and Vaya, D. (2024). Recent advancements towards the green synthesis of carbon quantum dots as an innovative and eco-friendly solution for metal ion sensing and monitoring. *RSC Sustainability* 2 (1): 11–36. doi: 10.1039/D3SU00375B.

Di, J., Xia, J., Ge, Y. et al. (2015). Novel visible-light-driven CQDs/Bi2WO6 hybrid materials with enhanced photocatalytic activity toward organic pollutants degradation and mechanism insight. *Applied Catalysis B: Environmental* 168: 51–61. doi: 10.1016/j.apcatb.2014.11.057.

Ding, D., Lan, W., Yang, Z. et al. (2016). A simple method for preparing ZnO foam/carbon quantum dots nanocomposite and their photocatalytic applications. *Materials Science in Semiconductor Processing* 47: 25–31. doi: 10.1016/j.mssp.2016.02.004.

Du, F., Cheng, Z., Lai, Z. et al. (2019). Red-emissive nitrogen doped carbon quantum dots for highly selective and sensitive fluorescence detection of the alachlor herbicide in soil samples. *New Journal of Chemistry* 43 (47): 18695–18701. doi: 10.1039/C9NJ04232F.

Dutta, S., Adhikary, S., Bhattacharya, S. et al. (2024). Contamination of textile dyes in aquatic environment: adverse impacts on aquatic ecosystem and human health, and its management using bioremediation. *Journal of Environmental Management* 353: 120103. doi: 10.1016/j.jenvman.2024.120103.

Emanuele, A., Cailotto, S., Campalani, C. et al. (2019). Precursor-dependent photocatalytic activity of carbon dots. *Molecules* 25 (1): 101. doi: 10.3390/molecules25010101.

Eskalen, H., Urus, S., Cömertpay, S. et al. (2020). Microwave-assisted ultra-fast synthesis of carbon quantum dots from linter: fluorescence cancer imaging and human cell growth inhibition properties. *Industrial Crops and Products* 147: 112209. doi: 10.1016/j.indcrop.2020.112209.

Gayen, B., Palchoudhury, S., and Chowdhury, J. (2019). Carbon dots: a mystic star in the world of nanoscience. *Journal of Nanomaterials* 2019 (1): 3451307. doi: 10.1155/2019/3451307.

Guedes, R.N.C., Biondi, A., Agathokleous, E. et al. (2023). (Systemic) insecticides in plants: phytotoxicity, bioactivation, or hormesis? *Agriculture Communications* 1 (1): 100002. doi: 10.1016/j.agrcom.2023.100002.

Hola, K., Bourlinos, A.B., Kozak, O. et al. (2014). Photoluminescence effects of graphitic core size and surface functional groups in carbon dots: COO^- induced red-shift emission. *Carbon* 70: 279–286. doi: 10.1016/j.carbon.2014.01.008.

Ikram, Z., Azmat, E., and Perviaz, M. (2024). Degradation efficiency of organic dyes on CQDs as photocatalysts: a review. *ACS Omega* 9 (9): 10017–10029. doi: 10.1021/acsomega.3c09547.

Jajoo, A. (2017). Effects of environmental pollutants polycyclic aromatic hydrocarbons (PAH) on photosynthetic processes. In: *Photosynthesis: Structures, Mechanisms, and Applications* (ed. H. Hou, M. Najafpour, G. Moore et al.), 249–259. Cham: Springer. doi: 10.1007/978-3-319-48873-8_11.

Jajoo, A., Mekala, N.R., Tomar, R.S. et al. (2014). Inhibitory effects of polycyclic aromatic hydrocarbons (PAHs) on photosynthetic performance are not related to their aromaticity. *Journal of Photochemistry and Photobiology B: Biology* 137: 151–155. doi: 10.1016/j.jphotobiol.2014.03.011.

Ji, M., Zhang, Z., Xia, J. et al. (2018). Enhanced photocatalytic performance of carbon quantum dots/BiOBr composite and mechanism investigation. *Chinese Chemical Letters* 29 (6): 805–810. doi: 10.1016/j.cclet.2018.05.002.

Jia, H., Wang, Z., Yuan, T. et al. (2019). Electroluminescent warm white light-emitting diodes based on passivation enabled bright red bandgap emission carbon quantum dots. *Advanced Science* 6 (13): 1900397. doi: 10.1002/advs.201900397.

Kaur, S., Sharma, S., and Kansal, S.K. (2016). Synthesis of ZnS/CQDs nanocomposite and its application as a photocatalyst for the degradation of an anionic dye, ARS. *Superlattices and Microstructures* 98: 86–95. doi: 10.1016/j.spmi.2016.08.011.

Kazemifard, N., Ensafi, A.A., and Rezaei, B. (2020). Green synthesized carbon dots embedded in silica molecularly imprinted polymers, characterization and application as a rapid and selective fluorimetric sensor for determination of thiabendazole in juices. *Food Chemistry* 310: 125812. doi: 10.1016/j.foodchem.2019.125812.

Kolanowska, A., Dzido, G., Krzywiecki, M. et al. (2022). Carbon quantum dots from amino acids revisited: survey of renewable precursors toward high quantum-yield blue and green fluorescence. *ACS Omega* 7 (45): 41165–41176. doi: 10.1021/acsomega.2c04751.

Kumar, P., Dua, S., Kaur, R. et al. (2022). A review on advancements in carbon quantum dots and their application in photovoltaics. *RSC Advances* 12 (8): 4714–4759. doi: 10.1039/D1RA08452F.

Lellis, B., Fávaro-Polonio, C.Z., Pamphile, J.A. et al. (2019). Effects of textile dyes on health and the environment and bioremediation potential of living organisms. *Biotechnology Research and Innovation* 3 (2): 275–290. doi: 10.1016/j.biori.2019.09.001.

Li, H., Deng, R., Tavakoli, H. et al. (2022). Ultrasensitive detection of acephate based on carbon quantum dot-mediated fluorescence inner filter effects. *Analyst* 147 (23): 5462–5469. doi: 10.1039/D2AN01552H.

Li, H., He, X., Liu, Y. et al. (2011). One-step ultrasonic synthesis of water-soluble carbon nanoparticles with excellent photoluminescent properties. *Carbon* 49 (2): 605–609. doi: 10.1016/j.carbon.2010.10.004.

Li, H., Sun, C., Vijayaraghavan, R. et al. (2016). Long lifetime photoluminescence in N, S co-doped carbon quantum dots from an ionic liquid and their applications in ultrasensitive detection of pesticides. *Carbon* 104: 33–39. doi: 10.1016/j.carbon.2016.03.040.

Li, Y., Zhang, B.P., Zhao, J.X. et al. (2013). ZnO/carbon quantum dots heterostructure with enhanced photocatalytic properties. *Applied Surface Science* 279: 367–373. doi: 10.1016/j.apsusc.2013.04.114.

Lin, B., Yan, Y., Guo, M. et al. (2018). Modification-free carbon dots as turn-on fluorescence probe for detection of organophosphorus pesticides. *Food Chemistry* 245: 1176–1182. doi: 10.1016/j.foodchem.2017.11.038.

Liu, R., Wu, D., Liu, S. et al. (2009). An aqueous route to multicolor photoluminescent carbon dots using silica spheres as carriers. *Angewandte Chemie International Edition* 48 (25): 4598–4601. doi: 10.1002/anie.200900652.

Liu, Y., Yu, Y.X., and Zhang, W.D. (2013). Carbon quantum dots-doped CdS microspheres with enhanced photocatalytic performance. *Journal of Alloys and Compounds* 569: 102–110. doi: 10.1016/j.jallcom.2013.03.202.

Luo, M., Wei, J., Zhao, Y. et al. (2020). Fluorescent and visual detection of methyl-paraoxon by using boron-and nitrogen-doped carbon dots. *Microchemical Journal* 154: 104547. doi: 10.1016/j.microc.2019.104547.

Ma, C.B., Zhu, Z.T., Wang, H.X. et al. (2015). A general solid-state synthesis of chemically-doped fluorescent graphene quantum dots for bioimaging and optoelectronic applications. *Nanoscale* 7 (22): 10162–10169. doi: 10.1039/C5NR01757B.

Ma, X., Li, S., Hessel, V. et al. (2019). Synthesis of luminescent carbon quantum dots by microplasma process. *Chemical Engineering and Processing-Process Intensification* 140: 29–35. doi: 10.1016/j.cep.2019.04.017.

Mandal, P., Sahoo, D., Sarkar, P. et al. (2019). Fluorescence turn-on and turn-off sensing of pesticides by carbon dot-based sensor. *New Journal of Chemistry* 43 (30): 12137–12151. doi: 10.1039/C9NJ03192H.

Martindale, B.C., Hutton, G.A., Caputo, C.A. et al. (2015). Solar hydrogen production using carbon quantum dots and a molecular nickel catalyst. *Journal of the American Chemical Society* 137 (18): 6018–6025. doi: 10.1021/jacs.5b01650.

Miao, R., Luo, Z., Zhong, W. et al. (2016). Mesoporous TiO_2 modified with carbon quantum dots as a high-performance visible light photocatalyst. *Applied Catalysis B: Environmental* 189: 26–38. doi: 10.1016/j.apcatb.2016.01.070.

Mintz, K.J., Zhou, Y., and Leblanc, R.M. (2019). Recent development of carbon quantum dots regarding their optical properties, photoluminescence mechanism, and core structure. *Nanoscale* 11 (11): 4634–4652. doi: 10.1039/C8NR10059D.

Mohapatra, S., Bera, M.K., and Das, R.K. (2018). Rapid "turn-on" detection of atrazine using highly luminescent N-doped carbon quantum dot. *Sensors and Actuators B: Chemical* 263: 459–468. doi: 10.1016/j.snb.2018.02.155.

Molina, L. and Segura, A. (2021). Biochemical and metabolic plant responses toward polycyclic aromatic hydrocarbons and heavy metals present in atmospheric pollution. *Plants* 10 (11): 2305. doi: 10.3390/plants10112305.

Muthulingam, S., Lee, I.H., and Uthirakumar, P. (2015). Highly efficient degradation of dyes by carbon quantum dots/N-doped zinc oxide (CQD/N-ZnO) photocatalyst and its compatibility on three different commercial dyes under daylight. *Journal of Colloid and Interface Science* 455: 101–109. doi: 10.1016/j.jcis.2015.05.046.

Onwujekwe, E.C., Ogbonna, J.C., Eze, C.C. et al. (2022). Biodegradation and impact of polycyclic aromatic hydrocarbons in the environment: a review. *Journal of Advances in Biology & Biotechnology* 25 (9): 35–47. doi: 10.9734/JABB/2022/v25i9597.

Pathak, V.M., Verma, V.K., Rawat, B.S. et al. (2022). Current status of pesticide effects on environment, human health and it's eco-friendly management as bioremediation: a comprehensive review. *Frontiers in Microbiology* 13: 962619. doi: 10.3389/fmicb.2022.962619.

Paulo, S., Palomares, E., and Martinez-Ferrero, E. (2016). Graphene and carbon quantum dot-based materials in photovoltaic devices: from synthesis to applications. *Nanomaterials* 6 (9): 157. doi: 10.3390/nano6090157.

Peng, H. and Travas-Sejdic, J. (2009). Simple aqueous solution route to luminescent carbogenic dots from carbohydrates. *Chemistry of Materials* 21 (23): 5563–5565. doi: 10.1021/cm901593y.

Peng, J., Yin, W., Shi, J. et al. (2019). Magnesium and nitrogen co-doped carbon dots as fluorescent probes for quenchometric determination of paraoxon using pralidoxime as a linker. *Microchimica Acta* 186: 1–9. doi: 10.1007/s00604-018-3147-1.

Qian, Z., Ma, J., Shan, X. et al. (2013). Surface functionalization of graphene quantum dots with small organic molecules from photoluminescence modulation to bioimaging applications: an experimental and theoretical investigation. *RSC Advances* 3 (34): 14571–14579. doi: 10.1039/C3RA42066C.

Qu, K., Wang, J., Ren, J. et al. (2013). Carbon dots prepared by hydrothermal treatment of dopamine as an effective fluorescent sensing platform for the label-free detection of iron (III) ions and dopamine. *Chemistry–A European Journal* 19 (22): 7243–7249. doi: 10.1002/chem.201300042.

Qu, Z., Wang, J., Tang, J. et al. (2018). Carbon quantum dots/KNbO3 hybrid composites with enhanced visible-light driven photocatalytic activity toward dye waste-water degradation and hydrogen production. *Molecular Catalysis* 445: 1–11. doi: 10.1016/j.mcat.2017.11.002.

Ramamurthy, K., Priya, P.S., Murugan, R. et al. (2024). Hues of risk: investigating genotoxicity and environmental impacts of azo textile dyes. *Environmental Science and Pollution Research* 31: 33190–33211. doi: 10.1007/s11356-024-33444-1.

Rastogi, A., Pandey, F.P., Parmar, A.S. et al. (2021). Effect of carbonaceous oil palm leaf quantum dot dispersion in nematic liquid crystal on zeta potential, optical texture and dielectric properties. *Journal of Nanostructure in Chemistry* 11: 527–548. doi: 10.1007/s40097-020-00382-6.

Rong, M., Feng, Y., Wang, Y. et al. (2017). One-pot solid phase pyrolysis synthesis of nitrogen-doped carbon dots for Fe^{3+} sensing and bioimaging. *Sensors and Actuators B: Chemical* 245: 868–874. doi: 10.1016/j.snb.2017.02.014.

Ruiz-Medina, A., Jiménez-López, J., and Llorent-Martínez, E.J. (2022). Luminescent determination of propineb fungicide by using a carbon quantum dots-europium ions system. *Talanta* 240: 123205. doi: 10.1016/j.talanta.2022.123205.

Sahub, C., Tuntulani, T., Nhujak, T. et al. (2018). Effective biosensor based on graphene quantum dots via enzymatic reaction for directly photoluminescence detection of organophosphate pesticide. *Sensors and Actuators B: Chemical* 258: 88–97. doi: 10.1016/j.snb.2017.11.072.

Saputra, A.M.A., Piliang, A.F.R., Goei, R. et al. (2024). Synthesis, properties, and utilization of carbon quantum dots as photocatalysts on degradation of organic dyes: a mini review. *Catalysis Communications* 187: 106914. doi: 10.1016/j.catcom.2024.106914.

Semeniuk, M., Yi, Z., Poursorkhabi, V. et al. (2019). Future perspectives and review on organic carbon dots in electronic applications. *ACS Nano* 13 (6): 6224–6255. doi: 10.1021/acsnano.9b00688.

Shen, T., Wang, Q., Guo, Z. et al. (2018). Hydrothermal synthesis of carbon quantum dots using different precursors and their combination with TiO_2 for enhanced photocatalytic activity. *Ceramics International* 44 (10): 11828–11834. doi: 10.1016/j.ceramint.2018.03.271.

Singh, V.K., Singh, V., Yadav, P.K. et al. (2018a). Bright-blue-emission nitrogen and phosphorus-doped carbon quantum dots as a promising nanoprobe for detection of Cr (VI) and ascorbic acid in pure aqueous solution and in living cells. *New Journal of Chemistry* 42 (15): 12990–12997. doi: 10.1039/C8NJ02126K.

Singh, V.K., Yadav, P.K., Chandra, S. et al. (2018b). Peroxidase mimetic activity of fluorescent NS-carbon quantum dots and their application in colorimetric detection of H_2O_2 and glutathione in human blood serum. *Journal of Materials Chemistry B* 6 (32): 5256–5268. doi: 10.1039/C8TB01286E.

Sun, C., Xu, Q., Xie, Y. et al. (2017). High-efficient one-pot synthesis of carbon quantum dots decorating Bi_2MoO_6 nanosheets heterostructure with enhanced visible-light photocatalytic properties. *Journal of Alloys and Compounds* 723: 333–344. doi: 10.1016/j.jallcom.2017.06.130.

Sun, Y.P., Zhou, B., Lin, Y. et al. (2006). Quantum-sized carbon dots for bright and colorful photoluminescence. *Journal of the American Chemical Society* 128 (24): 7756–7757. doi: 10.1021/ja062677d.

Tang, L., Ji, R., Cao, X. et al. (2012). Deep ultraviolet photoluminescence of water-soluble self-passivated graphene quantum dots. *ACS Nano* 6 (6): 5102–5110. doi: 10.1021/nn300760g.

Tian, J., Leng, Y., Zhao, Z. et al. (2015). Carbon quantum dots/hydrogenated TiO_2 nanobelt heterostructures and their broad spectrum photocatalytic properties under UV, visible, and near-infrared irradiation. *Nano Energy* 11: 419–427. doi: 10.1016/j.nanoen.2014.10.025.

Tian, L., Ghosh, D., Chen, W. et al. (2009). Nanosized carbon particles from natural gas soot. *Chemistry of Materials* 21 (13): 2803–2809. doi: 10.1021/cm900709w.

Velmurugan, S., Traiwatcharanon, P., Jiwanti, P.K. et al. (2024). Highly selective carbendazim fungicide sensing performance of nitrogen-doped carbon quantum dots encapsulated aminated-UiO-66 zirconium metal-organic framework electrocatalyst. *Electrochimica Acta* 479: 143911. doi: 10.1016/j.electacta.2024.143911.

Wang, H., Sun, P., Cong, S. et al. (2016). Nitrogen-doped carbon dots for "green" quantum dot solar cells. *Nanoscale Research Letters* 11: 1–6. doi: 10.1186/s11671-016-1231-1.

Wang, H., Wei, Z., Matsui, H. et al. (2014). Fe_3O_4/carbon quantum dots hybrid nanoflowers for highly active and recyclable visible-light driven photocatalyst. *Journal of Materials Chemistry A* 2 (38): 15740–15745. doi: 10.1039/C4TA03130J.

Wang, J., Jiang, J., Li, F. et al. (2023). Emerging carbon-based quantum dots for sustainable photocatalysis. *Green Chemistry* 25 (1): 32–58. doi: 10.1039/D2GC03160D.

Wang, Q., Liu, X., Zhang, L. et al. (2012). Microwave-assisted synthesis of carbon nanodots through an eggshell membrane and their fluorescent application. *Analyst* 137 (22): 5392–5397. doi: 10.1039/C2AN36059D.

Wang, W., Lu, Y.C., Huang, H. et al. (2015). Facile synthesis of N, S-codoped fluorescent carbon nanodots for fluorescent resonance energy transfer recognition of methotrexate with high sensitivity and selectivity. *Biosensors and Bioelectronics* 64: 517–522. doi: 10.1016/j.bios.2014.09.066.

Wang, X., Feng, Y., Dong, P. et al. (2019). A mini review on carbon quantum dots: preparation, properties, and electrocatalytic application. *Frontiers in Chemistry* 7: 671. doi: 10.3389/fchem.2019.00671.

Wang, Y. and Hu, A. (2014). Carbon quantum dots: synthesis, properties and applications. *Journal of Materials Chemistry C* 2 (34): 6921–6939. doi: 10.1039/C4TC00988F.

Wei, J., Yang, Y., Dong, J. et al. (2019). Fluorometric determination of pesticides and organophosphates using nanoceria as a phosphatase mimic and an inner filter effect on carbon nanodots. *Microchimica Acta* 186: 1–9. doi: 10.1007/s00604-018-3175-x.

Xu, X., Ray, R., Gu, Y. et al. (2004). Electrophoretic analysis and purification of fluorescent single-walled carbon nanotube fragments. *Journal of the American Chemical Society* 126 (40): 12736–12737. doi: 10.1021/ja040082h.

Yadav, P.K., Chandra, S., Kumar, V. et al. (2023). Carbon quantum dots: synthesis, structure, properties, and catalytic applications for organic synthesis. *Catalysts* 13 (2): 422. doi: 10.3390/catal13020422.

Yang, Y., Hou, J., Huo, D. et al. (2019). Green emitting carbon dots for sensitive fluorometric determination of cartap based on its aggregation effect on gold nanoparticles. *Microchimica Acta* 186: 1–8. doi: 10.1007/s00604-019-3361-5.

Yang, Y., Xing, X., Zou, T. et al. (2020). A novel and sensitive ratiometric fluorescence assay for carbendazim based on N-doped carbon quantum dots and gold nanocluster nanohybrid. *Journal of Hazardous Materials* 386: 121958. doi: 10.1016/j.jhazmat.2019.121958.

Yu, T., Wang, H., Guo, C. et al. (2018). A rapid microwave synthesis of green-emissive carbon dots with solid-state fluorescence and pH-sensitive properties. *Royal Society Open Science* 5 (7): 180245. doi: 10.1098/rsos.180245.

Yuan, F., Li, S., Fan, Z. et al. (2016). Shining carbon dots: synthesis and biomedical and optoelectronic applications. *Nano Today* 11 (5): 565–586. doi: 10.1016/J.NANTOD.2016.08.006.

Yuan, T., Meng, T., He, P. et al. (2019). Carbon quantum dots: an emerging material for optoelectronic applications. *Journal of Materials Chemistry C* 7 (23): 6820–6835. doi: 10.1039/C9TC01730E.

Zhang, C., Yao, F.E.N.G., Liu, Y.W. et al. (2017). Uptake and translocation of organic pollutants in plants: a review. *Journal of Integrative Agriculture* 16 (8): 1659–1668. doi: 10.1016/S2095-3119(16)61590-3.

Zhang, W., Yu, S.F., Fei, L. et al. (2015). Large-area color controllable remote carbon white-light light-emitting diodes. *Carbon* 85: 344–350. doi: 10.1016/j.carbon.2014.12.107.

Zhang, Y., Zhang, D., Zhao, Y. et al. (2021). An ionic liquid-assisted quantum dot-grafted covalent organic framework-based multi-dimensional sensing array for discrimination of insecticides using principal component analysis and clustered heat map. *Microchimica Acta* 188 (9): 298. doi: 10.1007/s00604-021-04936-5.

Zhao, Y., Zou, S., Huo, D. et al. (2019). Simple and sensitive fluorescence sensor for methotrexate detection based on the inner filter effect of N, S co-doped carbon quantum dots. *Analytica Chimica Acta* 1047: 179–187. doi: 10.1016/j.aca.2018.10.005.

Zhu, H., Wang, X., Li, Y. et al. (2009). Microwave synthesis of fluorescent carbon nanoparticles with electrochemiluminescence properties. *Chemical Communications* (34): 5118–5120. doi: 10.1039/B907612C.

Zor, E., Morales-Narváez, E., Zamora-Gálvez, A. et al. (2015). Graphene quantum dots-based photoluminescent sensor: a multifunctional composite for pesticide detection. *ACS Applied Materials & Interfaces* 7 (36): 20272–20279. doi: 10.1021/acsami.5b05838.

11

Carbon-based Nanomaterials: A Promising Tool for Sensing Toxic Metal Ions from Degraded Soil

Poorna Sneha M[1], *Mohit Biju*[2], *Aishwarya Thomas*[3], *and Parvathi Balachandran*[3]

[1] *Kusuma School of Biological Sciences, Indian Institute of Technology Delhi, New Delhi, India*
[2] *School of Biological and Marine Sciences, University of Plymouth, Plymouth, UK*
[3] *Department of Biotechnology, CMS College Kottayam (Autonomous), Kottayam, India*

11.1 Introduction

Environmental pollution is one of the primary issues affecting the ecosystem and human health. The primary contributors to environmental pollution are chemicals used in industrial processes, such as toxic gases (sulfur dioxide, hydrochloric acid), organophosphorus compounds, chemicals found in household and commercial wastewater (phenol, hydrogen peroxide), and heavy metal ions (Hg^{2+}, As^{3+}, Pb^{2+}, Cd^{2+}). These substances can potentially have an immediate or long-term effect on the environment by interfering with the regular functioning of ecological processes and through biomagnification due to their further accumulation in the food chain (Onyido, Norris, and Buncel 2004; Figure 11.1). Green fluorescent carbon nanomaterials and carbon dots are considered as emerging tools that hold immense promise for sensing toxic metal ions in soil due to their extraordinary characteristics. The carbon-based nanomaterials stand out as they offer high sensitivity and efficient surface area for detection; they also offer rapid in-situ analysis for environmental monitoring (Torrinha et al. 2023; Pereira Lopes Moreira and Ul Islam 2023; Thakur and Kumar 2023).

11.1.1 Contamination of Soil by Toxic Metal Ions

Due to natural as well as man-made conditions, the soil gets contaminated by toxic metal ions. Human-made processes like mining, use of fertilizers, smelting, spillage of petrochemicals, and industrial activities with naturally occurring processes like ash from volcanic eruption, and weathering contribute substantially to toxic heavy metal contamination in the soil (Xin et al. 2022; Singh 2021). This eventually leads to the accumulation of metals in the soil, impacting the health of plants, animals, and the environment (Azhar et al. 2022). The mineral phases, oxidation states, and the organic materials in the soil are

Nanotechnology-based Sustainable Agriculture, First Edition. Edited by Pardeep Singh et al.

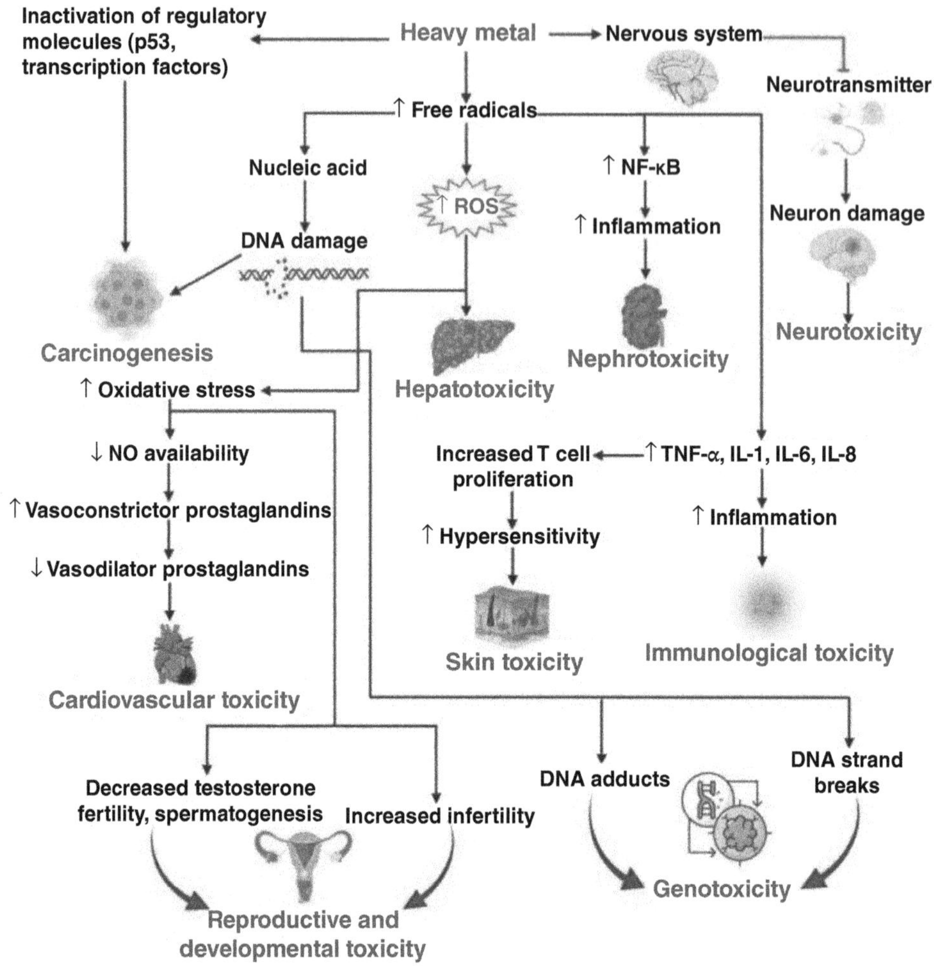

Figure 11.1 When heavy metals are ingested by the human body, they cause serious health issues, including neurotoxicity, nephrotoxicity, hepatotoxicity, skin toxicity, immunological toxicity, genotoxicity, cardiovascular toxicity, reproductive and developmental toxicity, and carcinogenesis (Mitra et al. 2022). CC BY 4.

critical elements influencing the reactions of heavy metals in the soil. The techniques such as X-ray diffraction (XRD), Fourier transform infrared spectroscopy (FTIR), and X-ray photoelectron spectroscopy (XPS) can be helpful in finding out the source of contamination and in analyzing the reaction between heavy metals and soil minerals (Irshad et al. 2021). Different factors available in the soil interact with the heavy metal ions differently. It was proven that zinc can retain at a low pH, while iron oxides can retain at a higher pH, a shift from zinc to iron oxide was noted in the case of soil enriched with aluminosilicate minerals (Matheus 2019). The adsorption of heavy metals like copper, cadmium, lead, zinc, and mercury by various factors of soil such as oxides, clay minerals, organic matter, and others affect the bioavailability and mobility of metal ions in soil (Caporale and Violante 2015).

11.1.2 Conventional Methods for the Detection of Toxic Metal Ions in Soil

The conventional methods available for the detection of toxic heavy metals in the soil may include X-ray fluorescence (XRF) spectroscopy, inductively coupled plasma-mass spectrometry (ICP-MS), electrochemical analysis, and monochromatic energy-dispersive X-ray fluorescence (MED-XRF), and among these techniques, XRF and ICP-MS are considered the most accurate ones for measuring heavy metal ions in the soil. The availability of silicon drift reactors and double-curved crystals in the MED-XRF system has increased the limits of detection of trace metal ions including Cd, Pb, and As. The electrochemical systems with screen-printed electrodes and potentiostat are some of the other techniques that are cheap, reliable, and effective in sensing metal ions from soil (Rani et al. 2020; Briffa et al. 2022; Xing et al. 2022).

The spectroscopic techniques applied for detecting toxic heavy metals from soil can be functionalized by different processes including the use of dithizone-anchored mesoporous TiO_2 for the trace detection and removal of heavy metals such as copper, mercury, lead, and cadmium, utilizing click chemistry for sensing copper ions by laser-induced breakdown, utilizing electrochemical spectral methods like electrochemical atomic absorption and electrochemical laser-induced breakdown for sensing trace heavy metals at low detection limits with high sensitivity, using gold particles at the nanoscale with surface plasmon resonance sensors can detect Zn and Cu at low concentrations and utilizing spark emission and machine learning techniques for sensing metal ions like Cr, Cu, Ni, and Pb in real time (Cao et al. 2022; Davari and Wexler 2020; Chen et al. 2018; Shahat, Ali, and El Shahat 2015).

11.1.3 Carbon-based Nanomaterials

Carbon-based nanomaterials such as graphene, carbon nanotubes (CNTs), and others have unique properties and applications, and based on these, they are synthesized by techniques like laser ablation, arc discharge, and chemical vapor decomposition (Ameta 2023; Torrinha et al. 2023). These carbon-based nanomaterials provide enhanced electrical, physical, mechanical, chemical, and sensing characteristics. Graphene and nanotubes have different forms and structures, such as nanotubes, nanofibers, helices, and nanospheres, and due to this structural variability, they have a crucial role in several applications such as gas sensors and water purification (Yaakob and Kamis 2022; Torrinha et al. 2023). Due to their specialized physiochemical characteristics, including superior mechanical strength, electrical conductivity, and flexibility, they have enormous sensing applications related to detection of toxic metal ions and environmental monitoring. For example, in the case of graphene with a two-dimensional (2D) nanostructure, they have a huge surface area with several applications in sensing, energy storage, and bioengineering (Yaakob and Kamis 2022). Guo and his colleagues with their experiment proved that carbon-based nanomaterials can immobilize heavy metals from soil successfully by the preparation of carbonaceous materials embedded with micro-nano and zero-valent ions (Guo et al. 2023). Ullah et al. (2022) pointed out that when nanomaterials are used in conjunction with aptamer-based sensors, they help in detecting toxic metal ions from the soil in real time.

In this chapter, we focused on providing an insight into carbon-based nanomaterials as a sensing tool, their different forms, and the sensing mechanisms. It also primarily covers the case studies or research works involving carbon-based nanomaterials associated with various processes and techniques in understanding its ability to sense toxic metal ions from soil. Apart from the abovementioned discoveries, this chapter also shares an account of the emerging trends and future prospects associated with the topic.

11.2 Carbon-based Nanomaterials for Sensing the Purpose of a Sensing Tool

11.2.1 Allotropy of Carbon

In the cosmos, carbon is considered as a plentiful and complex element. Carbon's allotrope qualities allow it to combine, depending on the arrangement of its carbon atoms, to generate compounds with entirely different properties. Due to their versatility, carbon nanomaterials are considered highly promising in the field of nanotechnology (Siqueira and Oliveira 2017). The carbon can be easily transformed into various forms and structures by different sp, sp^2, and sp^3 kinds of hybridization. Graphite became the subject of extensive research after the discovery of its structure by John D. Bernal at the start of twentieth century. During the 1950s and 1960s, significant theoretical and experimental work on the physical chemistry of graphite was performed. The discovery of fullerenes in 1985 declared the start of the age of nanocarbons, and the structural identification of nanotubes occurred next. The creation of carbon microfibers and graphite intercalation compounds became popular shortly after (Díez-Pascual 2021). The research based on carbon-based nanomaterials, including fullerenes, nanodiamonds, derivatives of CNTs and graphene, and others, has been rapidly expanded in the last several years (Dresselhaus and Terrones 2013).

11.2.2 Carbon-based Nanomaterials as an Alternative Strategy in Heavy Metal Sensing

The conventional methods for the detection of heavy metals in soil are usually in the range of femtomolar, which is lower than usual. The need for multiple samples requires challenging analytical processes, and dealing with sophisticated equipment requires skilled workers and expensive spectroscopy methods. These methods must be conjoined with other methods like chromatography for achieving metal ion speciation, and they are mostly suitable for quantitative analysis. These conventional techniques have merits, including high specificity and sensitivity, but they do have drawbacks, including the need for costly apparatus, high cost, and unsuitability for quick determination and other issues. Therefore, here arises a need for a less costly, easy, and reliable method that can be used for in situ as well as real-time sensing of heavy metals from soil. When compared with other techniques, electrochemical methods are found to be better for field use due to their intuitiveness, reliability, and affordability. With these electrochemical techniques, we could easily construct compact circuits as portable devices for the on-site surveillance of degraded soil samples. The electrodes modified with carbon-based nanomaterials are one of the foremost promising choices in case of electrochemical techniques due to their sensitive and selective detection of harmful heavy metals. CNTs, graphene, and others are extremely promising due to their unique forms and outstanding electrochemical characteristics (Bansod et al. 2017; Ma et al. 2015; Xie et al. 2019).

11.2.3 Unique Properties of Carbon-based Nanomaterials

The carbon-based nanomaterials can be used for the electrochemical detection of heavy metals with a low detection limit, a broad linear range, and high selectivity and sensitivity. There are several studies that proved that these materials can enhance electrode conductivity, stability, and sensitivity while increasing their effective active surface area (Li et al. 2024). The alteration of electrode surface through nanomaterials deposition enables nanoscale control over the desired material's size and form. The carbon nanomaterials, including CNTs and their types, graphene, carbon nanofibers, and nanoporous carbon, are extensively employed in the electrochemical analysis of a broad spectrum of targets, including heavy metals, primarily due to their significant electrical, thermal, mechanical, and chemical properties. The carbon nanostructures can form strong covalent bonds with other materials by chemical bonding, leading to their superior strength, hardness, and density. The sensitivity for the detection of heavy metals can be improved by modifying the functional groups on the surface of these nanomaterials, thus changing the material's density and structure. The carbon-based nanomaterials are crucial for developing sensors based on electrochemical processes as they significantly dampen the negative impacts of interference and selective recognition while enhancing the sensor's overall sensitivity and electrochemical performance of oxidation and reduction processes (Venkateswara Raju et al. 2023). The researchers are trying to examine these carbon-based nanomaterials more due to their extensive availability, excellent chemical and thermal stability, small size, significant reactivity, large surface area to volume ratio, and catalysis potential at the nanoscale (Madhura et al. 2019).

11.2.4 Types of Carbon-based Nanomaterials

CNTs, graphene, fullerene, carbon nanowires, carbon nano-onions, carbon nanofibers, carbon nanoribbons, carbon nanohorns, carbon-based quantum dots, carbon nanocones, nanodiamonds, and mesoporous carbon belong to the family of nanoscale carbon materials. The carbon-based nanomaterials can be primarily categorized into three groups: zero-dimensional (0D), one-dimensional (1D), and 2D depending up on their shapes, and graphene (2D), CNTs (1D), and fullerene (0D) are their representative members (Fritea et al. 2021). The CNTs and graphene are the extensively researched, produced, and used nanostructured carbon materials for numerous applications. In terms of its structure, graphite is an amorphous and disordered substance that remains stable at standard temperature and atmospheric pressure levels, and it serves as the foundation for the synthesis of other nanomaterials. During the controlled heating process, the massive thermal energy of the disordered carbon atoms leads them to form into a stable thermodynamic state until they reach the temperature of 3915 K (sublimation point). Graphene and CNTs are obtained based on the operating temperature (Proctor, Melendrez Armada, and Vijayaraghavan 2017). Here, we have discussed some of the relevant types of carbon-based nanomaterials.

11.2.4.1 Graphene

The graphene was discovered by Andre Geim and Novoselov's in 2004, which took place at the University of Manchester; the substance has gained attention and sparked the scientific curiosity of researchers worldwide. Graphene is the main structural component of many carbon-based compounds, and hence, it is appropriate that graphene is one of the most studied nanomaterials in the last decade especially due to its excellent qualities for its use in a wide variety of applications (Nasir et al. 2018).

Graphene oxide, reduced graphene oxide, and pristine graphene belong to the family of graphene. It has a 2D structure with sp^2 hybridization having hexagonal layer of carbon atom networks free from defects with delocalized π electrons that are just one atom thick. Its unique characteristics include a superior mechanical strength of 200 times higher than steel, a huge surface area, optical transparency, elasticity, high thermal conductivity, and excellent electrical conductivity of 6 times greater than Cu and 60 times greater than CNTs. Its electrical properties are affected by the layers of graphene and the number of edge flaws. In the case of graphene oxides, they are made of layered graphene sheets with basal planes and edges containing functional groups such as -OH, -ROH, -COOH, and others. The stacking of graphene produces three-dimensional (3D) graphite, its rolling produces 1D CNTs, and the wrapping produces 0D fullerenes (Chang et al. 2014; Fritea et al. 2021).

11.2.4.2 Nanodiamonds

The world mainly remained ignorant about nanoscale diamond particles until the late 1980s. They were first produced in the 1960s by an explosion in the USSR. Subsequently, these particles – now called nanodiamonds (NDs) – became more popular after a series of significant discoveries starting in the late 1990s. The variety of nanodiamonds accessible for study today is astounding. In many ways, diamond is an exceptional material. Most of the great qualities of bulk diamonds are preserved in nanodiamonds at the nanoscale. The distinct features of nanodiamonds include strong thermal conductivity and electrical resistivity, resistance to chemical decomposition, exceptional hardness and Young's modulus, optical quality, biocompatibility, fluorescence, and tolerance to harsh environmental conditions (Mochalin et al. 2012).

Nanodiamond's surface layer often consists of various functional groups; several atoms, including oxygen, hydrogen, and nitrogen, are present in addition to carbon. Low-resolution TEM studies of detonation ND aggregates have revealed that liquid-like carbon drops are recrystallized to form spherical NDs. NDs are sp^3 carbon nanoparticles with diameters of around 10 nm and crystal domains with a structure resembling a diamondoid. Three layers make up the intricate architecture of NDs: (i) a surface layer, (ii) a nonhomogeneous intervening carbon shell, and (iii) a carbon diamond core bound by sp^3. In an ND, most of the carbon atoms (approx. 70–90%) are in sp^3-hybridized carbon lattice, which ranges in size from 2 to 3 nm. It is believed that the intermediate layer of NDs, having a thickness between 0.4 and 1.0 nm, is a nonhomogeneous translational carbon shell. A continuous layer of onion-like carbon (OLC) occupying the interior surface and the graphite–carbon monolayers occupying the exterior surface together form the shell (Kumar et al. 2019).

11.2.4.3 Carbon Nanotubes

The CNTs were discovered by Lijima in the year 1991. There are two types of CNTs typically: multi-walled carbon nanotubes (MWCNTs) and single-walled carbon nanotubes (SWCNTs), which resemble rolled-up graphene in multilayered form and layered form, respectively. CNTs are 1D versions of 0D fullerene molecules. A graphene sheet that is few micrometers thick, rolled into a nanoscale-diameter cylinder, and topped with a spherical fullerene makes CNTs. Like graphene under tension, CNTs weigh just 1/66th of steel, yet their strength exceeds that of steel by two orders of magnitude. The melting temperature of nanotubes, estimated by computer simulations to be around 3700 °C, is greater than that of any metal yet relatively near graphite. SWCNTs can function as semiconductors or excellent electron conductors, largely depending upon their atomic structure and diameter. According to expert predictions, the extraordinarily high heat conductivity of isotopically pure diamonds is expected to be exceeded by CNTs (Mauter and Elimelech 2008; Vidu et al. 2014).

Since their discovery, the peculiar physicochemical features of CNTs have drawn attention in nanotechnology research. They are used in field emission, catalysis, biomedical applications, electronics, semiconductors, and air and water filters. The diameter of CNTs typically ranges from 1 nm to a few nanometers. Their mesopores and huge specific surface area help in the adsorption-based removal of heavy metals from soil (Baby, Saifullah, and Hussein 2019).

11.2.4.4 Fullerenes

The fullerenes were discovered by Richard Smalley, Harold Kroto, and Robert Curl in 1985, and later, they were honored with the Nobel Prize in Chemistry in 1996. The members belonging to the family of fullerene are carbon atoms in a range of Cn clusters ($n > 20$) on a spherical surface, where the carbon atoms are located close to the vertices of hexagons and pentagons. The fullerenes have a close-caged icosahedral symmetry with 20 hexagonal and 12 pentagonal rings. In a fullerene molecule, every carbon atom is sp^2 hybridized and forms covalent bonds with three other carbon atoms. Among the members of the fullerene family, C60 is the most studied one by researchers worldwide (Fatima and Mushtaq 2023; Zaytseva and Neumann 2016). C60 is considered the most prevalent and affordable among the all-carbon hollow-cluster materials of fullerenes (Tagmatarchis and Prato 2005). In a C60 molecule, there are two kinds of bond lengths: 6:6 ring bonds (short bonds) and 6:5 bonds (long bonds). The C60 molecule is hydrophobic, but can be made functionalized into either amphiphilic or hydrophilic. The C60 molecule is photoactive with numerous defects on its surface, high electron affinity, low bio-toxicity, and a large surface area to volume ratio (Fatima and Mushtaq 2023).

11.2.4.5 Carbon Dots

Carbon dots (CDs) are generally defined as quasi-zero-dimensional carbon-based nanomaterial and have fluorescence as one of their remarkable characteristics. These fluorescent carbon nanoparticles were accidentally produced during the purification of

SWNTs in 2004. Sun and his colleagues in 2006 used laser ablation of carbon targets to find surface-passivated CDs with a quantum yield of 10% as carbon nanomaterials for the first time in history. The development of carbon dots was hampered due to its complex preparation techniques and low quantum yield, but in the year 2013, Yang and his colleagues synthesized polymer-like carbon dots with 80% quantum yield by a single-step hydrothermal procedure using ethylenediamine and citric acid as precursors (Liu, Li, and Yang 2020).

Depending on the core carbon structure, functional groups, or other properties, CDs can be categorized into carbon quantum dots (CQDs), carbon nanodots (CNDs), carbonized polymer dots (CPDs), and graphene quantum dots (GQDs). GQDs are small pieces of graphene, showing anisotropic properties, and are comprised of one or more layers of graphene arranged in a network. The quantum confinement and edge effects occur due to various chemical modifications along their edges or within their interlayer defects. The carbon quantum dots are crystalline nanospheres with many chemical groups contributing to the quantum confinement effect and fluorescence of CDs (Mansuriya and Altintas 2021). The synthesis procedure may be tuned to adjust the functional groups on the surface and size of CQDs. The optical and fluorescent characteristics of CQDs are demonstrated by their physical dimension, which is lower than the Bohr radius of an exciton. In the case of CQDs, the majority of research works focused on sensing metal ions (Li et al. 2021).

CPDs have a spherical core-shell structure with a size of less than 20 nm. It has carbon at its center, and chains, or polymers, are attached to its carbon surface. In comparison to earlier CDs, these carbon arrangements show improved modifications and higher stability (Razavifar et al. 2024). CNDs occupy a position between graphene quantum dots and CPDs. This unique combination arises from their composition of a few layers of sp^2-hybridized graphene quantum dots arranged in a structured manner and the randomly interconnected carbon chains of CPDs (Mocci et al. 2022).

The ultraviolet region is generally associated with the strong absorption of CDs, which extends to visible and near-infrared wavelengths. The surface states, fluorophores, and element doping impact the fluorescence and electrical properties of carbon dots, which depend on the electronic transitions across the band gap of sp^2 domains. The electrochemical properties of CDs are regulated by their size, functions, and heteroatom doping (Ozyurt et al. 2023).

Comparison table showing differences between types of carbon-based nanomaterials.

Carbon-based nanomaterial	Year of discovery	Discoverers	Types and sizes	Dimension	Special features
Nanodiamonds (ND)	1963	K.V. Volkov V.V. Danilenko V.I. Elin	Diamondoids: 1–2 nm Ultrananocrystalline particles: 2–10 nm Nanocrystalline particles: 1–150 nm	3D	Superior hardness Fluorescence resistance chemically inert High surface area

Carbon-based nanomaterial	Year of discovery	Discoverers	Types and sizes	Dimension	Special features
Fullerenes	1985	H.W. Kroto R.F. Curl R.E. Smalley	Fullerene C60 ~0.7 nm Fullerene C70 ~0.8 nm Fullerenol ~0.7 or 0.8 nm	OD	Highly stable and resilient photostability Chemical reactivity and semiconductive behavior
Carbon nanotubes (CNTs)	1991	Lijima	SWCNT: 0.7–2 nm DWCNT: 2–5 nm MWCNT: 5–100 nm	1D	Excellent thermal conductivity High tensile strength and chemical inertness
Graphene	2004	Andre Geim Novoselov	Monolayer graphene: 0.34 nm Few-layer graphene: 0.68–3.4 nm Single-layer GO sheet: 0.8–1.2 nm	2D	Excellent electrical and thermal conductivity Biocompatibility flexibility
Carbon dots (CD)	2004	Xu et al.	CNDs: <10 nm CQDs: <10 nm CPDs: <10 nm GQDs: 10–100 nm	quasi-0D	Fluorescence and optical properties Water solubility Low toxicity

11.3 Sensing Mechanisms of Toxic Metal Ions by Nanomaterials

11.3.1 Nanocarbon in Electrochemical Sensing

In the case of electrochemical sensing of heavy metal ions, graphene and CNTs are the most studied nanocarbon derivatives. CNTs are widely used as pre-concentrator agents and transducer platforms because of their numerous benefits, which include a large efficient detection surface, reduced electrode fouling, superior electrocatalytic activity, and a fast transfer rate of electrons compared to bulk carbon electrodes (Aragay, Pons, and Merkoçi 2011; Cui, Wu, and Ju 2015). When CNTs are functionalized into polymeric films and deposited over the electrode surface, they exhibit a limited current, high potential, and chemical inertness (Profumo et al. 2006).

11.3.2 Existing Sensing Mechanisms Employed for Metal Ion Detection

Heavy metal detection is essential for environmental and public health monitoring. The methods developed for heavy metal detection have unique advantages and limitations. The following methods outline the mechanism of action employed for metal ion detection.

11.3.2.1 Graphene-based Sensor for Metal Ion Detection

The modification of functional groups in graphene could improve its ability to sense heavy metal ions with sensitivity and selectivity. Superparamagnetic Fe_3O_4 nanoparticles were used on RGO by Prakash et al. to detect chromium (Prakash, Chandra, and Bahadur 2012). Traditional mercury-based electrodes are notoriously controversial because of their toxicity and volatile nature, which negatively influence the environment. However, Zhao and his team have developed the mercury-graphene oxide (GO) modified glassy carbon electrode, which has better stripping capabilities and is promising for practical use in Pb detection (Zhou, Jiang, and Via 2014). Li and his colleagues developed a highly sensitive ion sensor using a graphene/silicon Schottky junction, which demonstrated the ability to detect ion concentration variations (Li, Hardaway, and Koley 2018). Alves et al. (2020) found a reusable ion-sensitive field-effect transistor based on functionalized graphene with high sensitivity to various metal ions at low concentrations. Sudibya et al. (2011) presented a field-effect transistor sensor using micropatterned, functionalized-protein-reduced graphene oxide film, which showed high sensitivity to metal ions. Almawgani et al. (2023) introduced an optical sensor inspired by a graphene metasurface that showed high sensitivity and low detection limits for heavy metal ions. These studies highlight the potential of graphene-based sensors as toxic metal ion detectors, with each approach offering unique advantages.

11.3.2.2 Nano-diamond-based Sensor for Metal Ion Detection

Diamond has unique electrochemical properties that make it different from other sp^2-hybridized carbon structures. The typical characteristics of these materials include good electrical resistivity, long-term stability in response, superior optical properties in the infrared and ultraviolet regions, stability in morphology and structure even at elevated temperature conditions, and stable background current (Brocenschi et al. 2014). The boron-doped nano-diamond sheets have demonstrated outstanding response precision, minimal overpotential for specific redox processes, electrocatalytic activity, and a low limit of detection. The properties that favor using diamond doped with boron include its wide low capacitance, the electrochemical potential window, and strong stability in electrochemistry (Wang et al. 2009). Su et al. (2020) prepared a nanodiamond electrode using spark plasma sintering technology, which proved effective in sensing heavy metals. Shellaiah and Sun (2021) provided a comprehensive review of diamond-based electrodes for the detection of metal ions, highlighting the potential of nano-diamond-based electrodes in the field of metal ion sensing. These studies demonstrate the potential of nano-diamond-based sensors for the detection of metal ions (Figure 11.2).

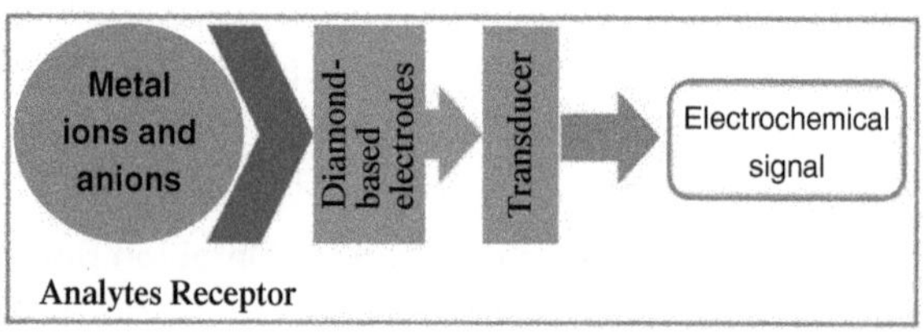

Figure 11.2 Illustration showing the detection of metal ions and anions by diamond-based sensors (Shellaiah and Sun 2021). CC BY 4.0.

11.3.2.3 CNT-based Sensor for Metal Ion Detection

Numerous investigations have been done to explore the detection capabilities of CNTs for sensing heavy metal ions. It was discovered that MWCNT modified with a gold electrode was able to detect heavy metals in salty water without the desalination of the sample. Heavy metal trace detections like arsenic and bismuth were also done using it (Profumo et al. 2006). In situ plating was used to deposit bismuth onto a screen-printed CNT electrode, resulting in a bismuth film (Hwang et al. 2008; Lee et al. 2008). The electrodes simultaneously provide lead, cadmium, and zinc readings that were distinct, precise, and repeatable. Glassy carbon electrodes modified with CNT/Bi, bismuth film plated with screen-printed CNTs, and a CNT/Bi nano-electrode array have all demonstrated notably improved performance for lead detection when compared to bare electrodes (Hwang et al. 2008; Liu et al. 2005).

MWCNTs containing ionic liquid and triphenylphosphine were employed to analyze cadmium, lead, and mercury ions rapidly and sensitively (Bagheri et al. 2013). Shin and Hong (2010) employed MWCNTs-Nafion nanocomposites film with electrodeposited platinum-iron to determine As (III) sensitively. The electrode built from MWCNT modified with carbon paste and Sb nanoparticles (NPs) was more sensitive and had a stronger peak signal than the carbon paste electrode with an Sb film (Ashrafi et al. 2014).

In addition, MWCNT electrodes containing Sb demonstrated an increased analytical signal (more than twice) for the detection of Cd (II) and Pb (II) when compared to MWCNT isolated. The authors postulated that the oxidation current and analytical signal were affected by the size of Sb particles on the coated layer, and these factors decreased in the following order: Sb-NP>Sb-film>Sb-microsphere. An increase in electrode surface area can create more active sites. As a result, it enables the collection and deposition of a more significant amount of metal on the electrode surface. Hg (II) had a strong adsorption capability on thiol-functionalized chitosan films with MWCNTs, and the stripping response was clearly defined (Deng et al. 2010).

A novel highly aligned MWCNT tower electrode was used in conjunction with Osteryoung square wave stripping voltammetry (OSWSV) to analyze lead, copper, zinc, and cadmium traces (Guo et al. 2011). Ashkenani et al. successfully detected copper using polymer-CNT film imprinted by Cu (II) ions, created by the synthesis of 1-(2-pyridyl azo)-2-naphthol complex. By comparing the ion accumulation in open- and closed-circuit scenarios, the stripping voltammetry method reveals that the open-circuit condition has a higher selectivity than the closed one (Ashkenani and Taher 2012).

Using an SWCNT film-gold electrode, Wei and coworkers testified to the strong chelation ability of thiol groups toward mercury ions and obtained a detection limit of 3.0 nM (Wei et al. 2014). A newly synthesized Schiff base, 4-methoxybenzylideneamino-2-thioxothiazolodin-4-one, was constructed with MWCNTs and performed the accurate sensing of mercury and lead ions in samples up to nano-molar levels (Afkhami et al. 2012).

Apart from the direct methods of detecting metal ions, indirect methods have also been in research that employ enzymes for detection. Cadmium detection was reported using a hydrogen peroxide biosensor prepared from a maize tassel (MT)-MWCNT composite by using an inhibitory measurement method by Moyo et al. The reduction of H_2O_2 through electrocatalytic horseradish peroxidase (HRP) enzyme activity was used to determine the concentration of Cd (II). Enzyme activity was found to be inhibited by a decrease in reduction current that was proportionate to the concentration of the trace metal ion (measurement in linear range: 5 to 30 mg/L; LOD: 0.51 mg/L). Similar MT-MWCNT electrodes were also used in a different study to determine lead and copper with LODs of 2.5 and 4.2 mg/L, respectively. It was determined that for both metal cations, this inhibition was reversible through a mix of competitive and noncompetitive mechanisms (Moyo, Okonkwo, and Agyei 2014).

11.3.2.4 Other Nanocarbons for Metal Ion Detection

In electrochemical sensing of heavy metal ions, there were limited recent uses of nanocarbon-based electrodes, such as carbon nanodots, fullerene, and graphene quantum dots. The application of gas chromatography–mass spectrometry and C60 fullerenes appeared to be helpful for the quantification of tin and mercury compounds present in water and sediment samples (Muñoz, Gallego, and Valcárcel 2005).

A fluorescence-based copper ion sensor has been created using novel donor-bridge acceptor dyads based on fullerene and ferrocene (Kumar and Menon 2009). This is an alternative fluorescence-based sensing method of mercury ions with CNDs, which are synthesized through an eco-friendly microwave-assisted process using flour as the source of carbon. It is now possible to effectively determine the presence of Hg (II) with accurate, selective identification and great sensitivity, which makes the detection limit as low as 0.5 nM, which is exceptionally low (Qin et al. 2013). To detect copper ions, an electrochemical method using (N-(2-550 aminoethyl)-N, N′, N′-Tris (pyridine-2-yl-methyl) ethane-1,2-diamine) TPEA-functionalized CDs (C-Dot-TPEA) has been developed (Shao et al. 2013).

GQDs have been utilized as an electrochemical sensor in recent times to detect metal ions with high sensitivity (Zhu et al. 2015). GQDs are preferable because of their many good qualities, including low toxicity, biocompatibility, and great photostability against photobleaching and blinking (Zhu et al. 2012).

Electrochemical sensing of copper (II) and lead (II) has been done with graphene nanodots embedded with porous gold electrodes, synthesized through ion beam sputtering deposition. The proposed platform demonstrated strong reusability and repeatability with an improved detection range of 0.009–4 mM for copper ions and 0.006–2.5 mM for lead ions due to the encaged graphene nanodots in a 3-D structure (Zhu et al. 2015).

11.3.3 Factors Influencing the Sensitivity of Sensing Mechanisms of Carbon-based Nanomaterials

Various factors, including material properties, environmental interactions, and functionalization techniques, influence carbon-based nanomaterials' sensitivity to sensing mechanisms.

11.3.3.1 Materialistic Properties

The carbon-based nanomaterials possess a large surface area, facilitating greater interaction with target materials for analysis (de Barros et al. 2023). The electronic properties of materials like graphene nanoribbons significantly change upon factors such as gas adsorption, transforming to a semi-metallic state from a metallic state, thereby enhancing sensitivity towards certain gases like CO and CO_2 (Husain, Ansari, and Almohammedi 2024).

11.3.3.2 Environmental Interactions

The presence of moisture can alter the charge carrier dynamics in carbon nanohorns, impacting their detection capabilities (SERBAN and Octavian BUIU 2024). The interaction between CNTs and their environments affects their temperature sensitivity; fewer nanotubes lead to higher sensitivity and vice versa due to changes in resistance (Ghomian et al. 2023).

11.3.3.3 Functionalization Techniques

The synthesis of the surface chemistry of carbon nanomaterials enhances the selectivity and sensitivity toward specific gases, which improves the performance of sensors (de Barros et al. 2023). The synthesis of nanostructures such as Ni-graphite composites has critical optimizing gas sensing properties, which allows for improved sensor performance (Sung et al. 2024).

While the above-mentioned factors improve the sensitivity, excessive functional group modifications or environmental interference can sometimes reduce the sensor performance.

11.4 Applications Related to Metal Ion Detection by Carbon-based Nanomaterials

Researchers have noted that CNTs are superior fluorescence quenchers when it comes to detecting metal ions. For example, a metal ion detection system based on multiplexed aptamers was developed using the fluorescence quenching capabilities of MWCNTs. ATTO-647N/aptamer-functionalized CNTs were developed for detecting Pb^{2+} ions in a related research (Singh et al. 2021).

For the first time, Wang and Si reported a multicolor fluorescent nanosensor based on aptamer-MWCNT that can detect Ag^{+}, Pb^{2+}, and Hg^{2+} ions in a homogenous solution. The ultrahigh quenching ability of MWCNTs and the precise binding ability of aptamers are combined in this nanosensor. Using this sensing platform, metal ion multiplex detection may be done simultaneously by observing the fluorescence signals of many fluorophores. This method shows high sensitivity and specificity. Also, the MWCNT-based test is less expensive than conventional assays (Wang and Si 2013).

Baghayeri et al. reported a new electrochemical sensing methodology for the simultaneous detection of Cd (II) and Pb (II) based on modifying the glassy carbon electrode (GCE) with poly (amidoamine) dendrimer-functionalized magnetic graphene oxide (GO-Fe_3O_4-PAMAM) nanocomposite. The GO-Fe_3O_4-PAMAM nanocomposite's excellent synergistic integration of nanomaterials allowed the current electrochemical device to sense analyte ions with great sensitivity and strong anti-interference (Baghayeri et al. 2019).

The thermal reduction of raw materials oil sludge and iron ore concentrate successfully produced carbonaceous material built-in micro-nano zero-valent iron (ZVI\NCM). The obtained ZVINCM exhibited remarkable performance of As, Cd, Cu, Pb, and Zn immobilization in the soil at the ideal conditions of limited oxygen and a temperature of 900 °C. Because of its high specific surface area and abundance of micro-nano FeO particles, ZVI-NCM is well-suited to offer a large number of sites for metal ions interaction (Guo et al. 2023).

It has been demonstrated that sustainable cysteamine covalently functionalized graphene (GSH) can increase the screen-printed carbon electrode's (SPCE's) reduced sensitivity to Cd^{2+} from Pb^{2+} and Cu^{2+}. While the sensitivity to Pb^{2+} and Cu^{2+} was unaffected, the Cd^{2+} sensitivity to GSH-SPCEs showed a difference from those of SPCEs alone. The selective rise in Cd^{2+} resulting from the free thiol groups in the GSH was based on the best Cd^{2+} sensitivity in the comparative testing of graphene acid (GA) and reduced graphene oxide (rGO). The HSAB hypothesis, which states that the thiol (soft base) exhibited a strong affinity for Cd^{2+} (soft acid). The GSH-SPCE could detect Cd^{2+}, Pb^{2+}, and Cu^{2+} at concentrations of up to 15, 11, and 6 parts per billion (ppb; Yang et al. 2023).

The laser-induced graphene fiber (LIGF)-based sensor detects lead and cadmium ions simultaneously. On the LIGF's working electrode surface, bismuth was placed in situ to enable the electrochemical detection of heavy metal ions. A heavy metal sensor was developed and utilized to detect Cd and Pb ions simultaneously in a pH 4.5, 0.1 M acetate buffer solution using the produced LIGF electrode. The resultant heavy-metal sensor, based on LIGF, showed increased sensitivity (0.19 and 0.20 μA μg/L for Cd and Pb ions, respectively; Jeong et al. 2022).

A laser-induced graphene (LIG)-based electrochemical sensor that detects Al^{3+} in soil has a linear detection range of 1.07–300 ppm for single-use measurements and 1.76–100 ppm for incremental measurements using the same electrode. This detection range is suitable for determining soil concentrations of Al^{3+} in agricultural fields and identifying potentially toxic environments for plants. The LIG-based bismuth-functionalized sensor platform can also detect Cd and Pb. The modifications in pH and bismuth concentration are done during the manufacturing process of these sensors to improve the sensitivity of Cd and Pb. This sensor is low-cost, simple to use, and has easy scalability, showing its potential to improve agricultural practices and monitor phytoremediation strategies (Reyes-Loaiza et al. 2024).

Au NCs-CDs are multi-colored fluorescent nanoprobes made of carbon dots (CDs) and Au nanoclusters (BSA-Au NCs). The Au NCs-CDs exhibit two peaks at 625 and 444 nm, respectively, under the single excitation condition ($\lambda ex = 365$ nm), which produced pink emission. The Au NCs-CDs may be employed for simultaneous and independent colorimetric and fluorescence detection of the heavy metal ions Hg^{2+} and Cr^{6+}. This probe makes quantitative detection of Hg^{2+} and Cr^{6+} easy and quick, with detection limits as low as 1.85 and 5.34 nM, respectively (Yu et al. 2018).

Using o-phosphorylethanolamine and citric acid as sources, highly photoluminescent nitrogen and phosphorus-co-doped carbon dots (N, P-CDs) have been prepared hydrothermally. With a detection limit of 0.16 μM, the N and P-CDs were utilized as a turn-on fluorescence probe to detect Cd (II) due to their optical qualities. The development of the Cd (II)-N, P-CDs complex, which results from the chelation of Cd (II) with surface functional groups, is considered the cause of the fluorescence enhancement with the addition of Cd (II). The Cd (II) detection findings in real samples indicated that N, P-CDs-based turn-on fluorescence probes would be highly beneficial to environmental protection (Lin et al. 2019).

CNTs are used to create an electrochemical sensor that serves as a base for film metal electrodes in stripping voltammetry. The sensor is made with a combination of multiwall CNTs, epoxy resin, and spherical glassy carbon (SGC) powder. This sensor was used to detect trace amounts of Ti (IV) on the lead film electrode in the study, and by employing cupferron as a complexing agent and conducting measurements in situ with adsorption

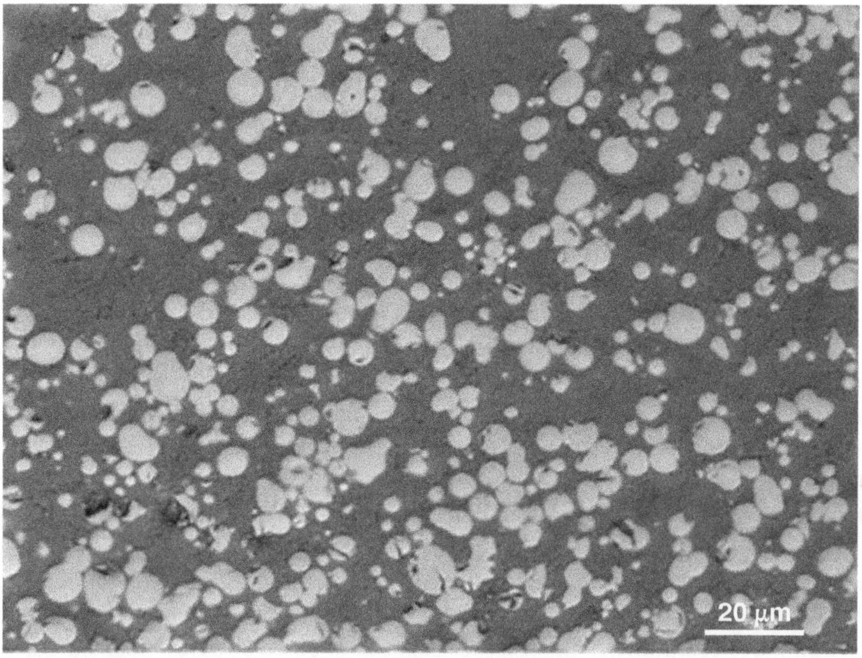

Figure 11.3 An inverted metallurgical microscope real image (Nikon) of the solid CNTs/SGC electrode. Grabarczyk et al. (2023) / Springer Nature / CC BY 4.0.

stripping voltammetry, it achieved a concentration range of $7 \times 10^{-10} - 7 \times 10^{-8}$ mol/L and a limit of detection (LOD) of 2.4×10^{-10}. These results show a robust linear response to Ti (IV) ions. The groundbreaking lead film CNTs/SGC electrode technique produces fast and reliable Ti (IV) detection with remarkable stability. Moreover, the repeatability, reproducibility, and relative standard deviations (RSDs) were consistently less than 4% (Grabarczyk, Wlazłowska, and Adamczyk 2023; Figure 11.3).

Zhu et al. developed a simple process for post-modifying graphene quantum dots with glycinin an alkaline environment to create glycine-functionalized graphene quantum dots (Gly- GQDs). The synthesized Gly-GQDs had a higher specific chelating capacity towards Hg^{2+}, a high QY of 35.7%, and a blue emission at 444 nm, independent of excitation. Moreover, Hg^{2+} was detected using Gly-GQDs, with a comparatively low detection limit of 8.3 nM (Zhu et al. 2021).

Using a hydrothermal process, Wang et al. produced gold nanoclusters (AuNCs) and N-doped carbon dots (NCDs) in a single step. Additionally, the fluorescence resonance energy transfer (FRET) effect was used to create AuNCs-NCDs complexes concurrently. A fluorescent assay sensing approach with three channels was developed to precisely identify Cd^{2+}, Pb^{2+}, and Hg^{2+} in soil and other complex environments. The precision of this method was demonstrated by the LOD values of 0.15, 0.20, and 0.09 μmol/L, which met with the national standard criteria (GB2762–2017). The partial least squares discriminant analysis (PLSDA) and partial least squares regression (PLSR) were used with the three-channel fluorescent channel array sensor to detect metal ions visually in complex samples. It has been discovered that AuNCs combine with Cd^{2+} and Pb^{2+} to generate complexes that show the Tyndall effect. These complexes can be precipitated and eliminated using centrifugation or sedimentation. The innovative approach that has been suggested can satisfy the need visually and helps in instrument-free, fast identification of Cd^{2+}, Pb^{2+}, and Hg^{2+}. This offers a fresh approach to detecting and eliminating various heavy metals in soil and other complex environments (Wang et al. 2022).

A multi-emission fluorescence sensor array based on carbon dots (QR-CDs) and lanthanide complexes (EDTA-Tb^{3+}) was used to detect the seven heavy metals, including Cr^{6+}, Fe^{2+}, Cu^{2+}, Fe^{3+}, Mn^{2+}, Co^{2+}, and Ni^{2+}. The most accurate technique was determined to be the tree-based pipeline optimization technique (TPOT) framework, which was able to identify the seven heavy metals at concentrations as low as 0.05 μM. When the sensor and SX model (a novel unified model using machine learning techniques in combination with the "stepwise prediction" strategy to produce optimal methods for screening) are coupled, the precise detection of various metal ion concentrations in actual samples could be done. Such a multi-emission fluorescence array shows good anti-interference, high sensitivity, and the ability to identify heavy metals in soil and lake water media. More significantly, the SX model demonstrated 100% accuracy in the concentration detection of unknown samples when it was originally used and assessed in the field of array sensing detection. This work demonstrated that a multi-emission fluorescence sensor array may be used quickly, efficiently, and with a low detection limit to sense heavy metals in complex environments (Xu et al. 2022).

An innovative dual-function fluorescence nanosensor based on silver/graphene quantum dots (Ag-GQDs) was created to research unconventional sensing techniques for the sensitive and selective detection of heavy metals. The fluorescence quenching of the nanocomposite serves as the basis for the detection. The nanoprobe can identify dangerous heavy metals, including Cd^{2+}, Pb^{2+}, and Hg^{2+} ions and their concentrations. The results clearly show that the fluorescence of the nanocomposite diminishes with increasing concentration, indicating the great sensitivity of the current sensing device, and when compared to conventional chromatographic detection techniques, the approach has been demonstrated to provide sensitivity, selectivity, efficiency, and cost-effectiveness (Moustafa et al. 2024).

Banna et al. developed a nontoxic, eco-friendly, and economically viable sensing strategy that uses microfabrication-free carbon fiber electrodes (CFEs) to detect microgram-per-liter levels of inorganic Hg, Cd, and Cd in soil solutions. The sensitivity of this sensor is suitable across a broad range of liquids and metal concentrations (10–1600 μg/L). Banna and co-workers also looked into the impact of solution pH and discovered that all metal ions are detectable and stable in a single run in the pH range of 4.0–5.0. One CFE was utilized for 100 trials and showed good repeatability over extended measurements. This CFE sensor will help in the non-toxic and economical in-situ prolonged measurement of toxic metal ions in the environment (Banna et al. 2024).

11.5 Challenges Associated with the Usage of Carbon-based Nanomaterials

The nanocarbon's distinct electrical and structural characteristics play a significant role in its electrochemical sensing capabilities. While graphene and CNTs have demonstrated their unique capacity to detect heavy metals in various matrixes, one of the primary barriers to developing a cost-effective sensor is keeping the electrode film stable over an extended period. Using nanocarbon electrodes in everyday applications requires an affordable material and a suitable sensing platform. Selectivity and sensitivity for metal analysis may be achieved with different functionalized layers of the same nanocarbon materials, but the operating factors are not consistent. For the best outcomes, it is therefore necessary to carefully consider factors including pH, deposition potential, duration, and electrode functionalized material concentration (Pandey et al. 2016). Restricted specificity, high contamination, and low recognition limits are some challenges and limitations associated with electrochemical sensing strategies. Because of the restricted penetration of sensing materials during the electrochemical process, they also restrict mass transmission. These challenges and limitations affect the selectivity and sensitivity of the real-time sensing application. By thoroughly comprehending the unique interactions and mechanistic insights between the carbon-based materials and the metal ions, the overall sensing performance has to be enhanced. Some carbon-based composite materials modified with screen-printed electrodes (SPE) are not as conductivity-enhancing as SPE. It is essential to create carbon-based nanomaterials free of any sensing mistakes, and a simple procedure for the preparation of electrodes (Venkateswara Raju et al. 2023). The loss of active sites and decreased conductivity in developed electrochemical sensor platforms for heavy metal detection may result from the separation of carbon nanomaterial leakage from the surface layer of the sensor. This highlights the necessity of creating new synthetic methods and materials to produce consistent and stable carbon nanomaterials (Song et al. 2022).

The widespread use of CNTs is restricted due to their high production cost. Recent research, however, suggests that it may be possible to produce CNTs of superior quality at a lower cost. As of now, catalytic chemical vapor deposition is the most widely used technique for creating affordable MWCNTs; however, it is not as scalable for large-scale SWCNT production. While other techniques like arc discharge and laser ablation may produce SWCNTs of excellent quality, their high operating costs and expensive components make them commercially unfeasible. To enable the broader use of CNTs in environmental protection, a novel synthesis technique that can efficiently and economically generate regulated SWCNTs is needed. The ability of CNTs to produce reactive oxygen species (ROS) in the presence of sunlight is well known. ROS can alter the charge and redox state, potentially making it dangerous. CNTs may also interact with proteins, RNA, enzymes, and other biomolecules to possibly have harmful effects, especially on plants and animals. As a result, both the creation of biocompatible CNTs and a thorough examination of safe disposal techniques are required (Ogunsola et al. 2024).

The current state of GQD-based heavy metal sensing research is based on its remarkable characteristic of fluorescence. Doping produced promising results, but not all of the doped GQDs' characteristics have yet been fully understood. Despite certain research focusing on the use of waste biomass for the environmentally friendly synthesis of GQDs, the need to meet green chemistry standards and promote sustainability is also essential (Bressi et al. 2021).

There are still critical barriers that need to be addressed to fully realize the potential of graphene-based sensors in real-world applications. These problems include concerns relating to scalability, repeatability, and cost-effectiveness. To overcome these obstacles and get graphene-based sensor technologies closer to long-term solutions that protect human health and the environment, cooperation and continuous innovation are crucial (Sengupta and Hussain 2024).

11.6 Future Prospects of Carbon-based Nanomaterials

Once the toxic metal ions reach a certain allowable limit, they have negative impacts on both the environment and individuals. Excessive amounts of them can potentially modify or disrupt some biological processes vital to living things. Various parameters, including exposure duration, route, dosage, and chemical species, influence the toxicity of these ions. Numerous man-made and natural processes allow these heavy metal ions to infiltrate our food chain. They can alter the conformation and render enzymes inactive by interacting with different functional sites in amino acids and other compounds. They can replenish vital metal ions from their active sites, which may result in dangerous illnesses (Lochab et al. 2021). The exceptional mechanical strength, strong sorption capabilities, large apparent surface area, electrocatalytic properties, and relative chemical inertness of carbon- based nanomaterials make them ideal for the detection of heavy metal ions. Improved and consistent ways of fabricating nanoscale carbon-based materials are being developed, and greater control of these materials' chemical and physical properties might result in more efficient sorption materials and sensing devices. There are still plenty of options to modify the surface of these materials to target specific contaminants and prevent them from aggregating in aqueous systems. Even though CNTs have shown significant advancements, more research with novel materials like graphene may be possible (Wanekaya 2021). Despite having such good qualities, carbon nanomaterials are difficult to apply at the industrial level to tackle environmental contaminants. Currently, the infrastructure put in place to address environmental contamination, particularly that caused by hazardous metals, is not built to fully utilize the advantages of nanostructured materials. Another problem is that nanoparticles are gradually regarded to pose a risk to human health. Therefore, ongoing efforts are needed to address the mentioned issues and improve the biocompatibility of carbon nanomaterials through simple modifications to enable their application in industry (Lochab et al. 2021).

Future research and developments with nanomaterials will require developing highly efficient materials characterized by excellent electrocatalytic properties that are strong enough to serve as perfect interfaces for simultaneously detecting multiple heavy metals. Heavy metal ions tend to form complexes and aggregates, making it difficult to quantify them accurately in soil. Therefore, future studies must address chemical fouling and interference (Venkateswara Raju et al. 2023). Disposable electrodes modified with nanomaterials should be used to develop a durable and portable potentiostat reader in the future. To provide on-site and real-time detection analysis of the samples for heavy metals, an Internet of Things (IoT) connectivity with a portable potentiostat reader will allow communication through a smartphone (Mohamad Nor et al. 2023). The anti-interference properties and the stability of the nanomaterials may be increased by improved synthesis methodology, geometric configuration, and surface properties. Combining several nanomaterial forms through hybridization is a smart strategy. Since nanocomposites, single and multicomponent nanostructures, including dual- or triple-composite nanostructures, generally possess better stability and sensitivity than single nanoparticles. One area of focus for heavy metal analysis in the future based on nanomaterials might be the integration and downsizing of in situ detection. A significant trend is the development of nanomaterial-based microfluidic systems for metal detection or integration with mass and spectral spectrometry. Simplification and miniaturization of the analysis system may substantially conserve samples and avoid lengthy processes, providing in-situ quick detection (Li et al. 2021). Researchers must grasp the fundamentals and possible uses of carbon nanostructures since this will undoubtedly drive the development of electrochemical sensors for environmental applications. It is also crucial to maximize the stability, selectivity, and sensitivity of existing sensor techniques. Moreover, an attempt must be made to explain the creation of transportable electrochemical sensors for metal ion detection in actual samples (Jose et al. 2023).

Economic costs must be carefully considered when implementing carbon-based materials. Because of their high preparation costs, CNTs and activated carbon are not appropriate for remediating heavy metals in soil. Therefore, the secret to developing high- cost carbon nanomaterials in the future is to achieve high-efficiency remediation at low doses. Furthermore, consideration should be given to creating production processes that are simpler and more cost-effective in terms of raw materials and preparation techniques. CBMs can harm humans and plants due to their oxidative stress toxicity and ability to influence the activity of soil bacterial populations. In the future, we must look at the possible health and environmental risks of carbon nanomaterials regarding soil heavy metal remediation (Fu et al. 2023). Scientists can push innovation in the design and production of graphene-based sensors so that sensors become more sensitive and selective and, therefore, have broader applications in protecting public health, managing resources, and controlling pollution. By offering proactive responses to new ecological issues and real-time data, integrating these sensors into extensive monitoring can completely transform methods for environmental management. The sensors based on graphene can guarantee environmental resilience and advanced human welfare worldwide with coordinated efforts by developing sustainable futures (Sengupta and Hussain 2024). For sensing technologies, precision is a critical need that must be met. Most fluorescent CD-based pollution sensors were used up until recently for real-world sample analysis on relatively basic matrixes like water. The majority of environmental and biological samples have significantly more intricate matrices that are made up of different organic substances and ions that might hinder the performance of fluorescence sensors. The accuracy of sensing can be affected by multiple interfering circumstances seen in real-world environmental scenarios, including temperature, chemical substances, pH, and others. When deploying such technologies in an environmental setting, one problem is avoiding inferences from these conditions. The creation of durable and trustworthy CD-based sensors to identify contaminants in intricate matrices is still a significant challenge (Ajith, Pardhiya, and Rajamani 2022).

11.7 Conclusion

Soil contamination by heavy metals like Sb, Pb, Zn, Cu, Hg, and others is highly hazardous and can significantly affect the ecosystem and, therefore, have implications on the health of humans, animals, and plants. The primary sources of heavy metals are gas emissions, fertilizers, sewage, and industrial processes among others. These pollutants can seep into the soil where they can persist for a very long period. Consequently, early HMI detection in the soil might be the key to contributing to the safety of the environment and public health. During the past decade, there has been dramatic progress in the utilization of nanotechnology and nanomaterials in a number of fields. This made the researchers to come up with nanotechnology-based platforms for sensing metal ions. The most significant part of this research is the development of carbon-based nanomaterial sensors for the detection of heavy metals. Carbon-based nanomaterials represent manifold advantages concerning sensitivity and selectivity, stability, high surface area, and distinctive physicochemical properties that allows them to be suitable for heavy metal ion detection applications. Hence, concluding, we covered the characteristics, applications, and recent advancements in carbon-based nanomaterials; apart from it, discussions were also made on various sensing methods mentioned in different case studies and research papers focusing on the use of carbon nanomaterials for the detection of heavy metal ions. Based on these studies, the challenges and future prospects on the topic were shared.

Abbreviations

FTIR, Fourier transform infrared spectroscopy; XPS, X-ray photoelectron spectroscopy; ICP-MS, inductively coupled plasma-mass spectroscopy; MED-XRF, monochromatic energy dispersive X-ray fluorescence; SPR, surface plasmon resonance; ICP-OES, inductively coupled plasma optical emission spectroscopy; XRD, X-ray diffraction; AAS, atomic absorption spectrometry; AES: atomic emission spectrometry; CNTs, carbon nanotubes; GO, graphene oxide; MWCNTs, multi-walled carbon nanotubes; CNF, carbon nanofibers; ND, nanodiamonds; CNDs, carbon nanodots; GDQ, generalized differential quadrature method; ZVI, zero-valent iron; SPCE, screen-printed carbon electrode; LOD, limit of detection.

References

Afkhami, A., Bagheri, H., Khoshsafar, H. et al. (2012). Simultaneous trace-levels determination of Hg(II) and Pb(II) ions in various samples using a modified carbon paste electrode based on multi-walled carbon nanotubes and a new synthesized Schiff base. *Analytica Chimica Acta* 746: 98–106. doi: 10.1016/j.aca.2012.08.024.

Ajith, M.P., Pardhiya, S., and Rajamani, P. (2022). Carbon dots: an excellent fluorescent probe for contaminant sensing and remediation. In: *Small*, 18 (15). John Wiley and Sons Inc. doi: 10.1002/smll.202105579.

Almawgani, A.H., Surve, J., Parmar, T. et al. (2023). A Graphene-metasurface-inspired optical sensor for the heavy metals detection for efficient and rapid water treatment. *Photonics* doi: 10.3390/photonics10010056.

Alves, A.P., Meireles, L.M., Ferrari, G.A. et al. (2020). Highly sensitive and reusable ion-sensor based on functionalized graphene. *Applied Physics Letters* 117: 033105. doi: 10.1063/5.0009555.

Ameta, R.K. (2023). Carbon-based nanomaterials for sensing applications. In: *Recent Advances in Biosensor Technology*, 30–44. Bentham Science Publishers EBooks. doi: 10.2174/97898151 23739123010005.

Aragay, G., Pons, J., and Merkoçi, A. (2011). Recent trends in macro-, micro-, and nanomaterial-based tools and strategies for heavy-metal detection. *Chemical Reviews* 111 (5): 3433–3458. doi: 10.1021/cr100383r.

Ashkenani, H. and Taher, M.A. (2012). Determination of cadmium(II) using carbon paste electrode modified with a Cd-ion imprinted polymer. *Mikrochimica Acta* 178 (1–2): 53–60. doi: 10.1007/s00604-012-0803-8.

Ashrafi, A.M., Cerovac, S., Mudrić, S. et al. (2014). Antimony nanoparticle-multiwalled carbon nanotubes composite immobilized at carbon paste electrode for determination of trace heavy metals. *Sensors and Actuators. B, Chemical* 191: 320–325. doi: 10.1016/j.snb.2013.08.087.

Azhar, U., Ahmad, H., Shafqat, H. et al. (2022). Remediation techniques for elimination of heavy metal pollutants from soil: a review. *Environmental Research* 214: 113918. doi: 10.1016/j.envres. 2022.113918.

Baby, R., Saifullah, B., and Hussein, M.Z. (2019). Carbon nanomaterials for the treatment of heavy metal-contaminated water and environmental remediation. *Nanoscale Research Letters* 14 (1): 341. doi: 10.1186/s11671-019-3167-8.

Baghayeri, M., Alinezhad, H., Fayazi, M. et al. (2019). A novel electrochemical sensor based on a glassy carbon electrode modified with dendrimer functionalized magnetic graphene oxide for simultaneous determination of trace Pb(II) and Cd(II). *Electrochimica Acta* 312: 80–88. doi: 10.1016/j.electacta.2019.04.180.

Bagheri, H., Afkhami, A., Khoshsafar, H. et al. (2013). Simultaneous electrochemical determination of heavy metals using a triphenylphosphine/MWCNTs composite carbon ionic liquid electrode. *Sensors and Actuators B: Chemical* 186: 451–460. doi: 10.1016/j.snb. 2013.06.051.

Banna, G.M.H.U., Siegenthaler, J., Benedict, A. et al. (2024). Heavy metal sensing in plant and soil solutions using carbon fiber electrode. *Sensors and Actuators A: Physical* 370: 115232. doi: 10.1016/j.sna.2024.115232.

Bansod, B., Kumar, T., Thakur, R. et al. (2017). A review on various electrochemical techniques for heavy metal ions detection with different sensing platforms. *Biosensors and Bioelectronics* 94: 443–455. doi: 10.1016/j.bios.2017.03.031.

Bressi, V., Ferlazzo, A., Iannazzo, D. et al. (2021). Graphene quantum dots by eco-friendly green synthesis for electrochemical sensing: recent advances and future perspectives. *Nanomaterials* 11 (5): 1120. doi: 10.3390/nano11051120.

Briffa, J., Blundell, R., Sinagra, E. et al. (2022). Validation of X-ray fluorescence spectrometer technique to determine heavy metal concentrations in soil samples. *Global Journal of Science Frontier Research* 43–48. doi: 10.34257/gjsfrhvol22is5pg43.

Brocenschi, R.F., Rocha-Filho, R.C., Li, L. et al. (2014). Comparative electrochemical response of estrone at glassy-carbon, nitrogen-containing tetrahedral amorphous carbon and boron-doped diamond thin-film electrodes. *Journal of Electroanalytical Chemistry* 712: 207–214. doi: 10.1016/j.jelechem.2013.11.014.

Cao, F., Jiao, F., Ma, S. et al. (2022). Laser-induced breakdown spectroscopy mediated amplification sensor for copper (II) ions detection using click chemistry. *Sensors and actuators B: Chemical* 371: 132594. doi: 10.1016/j.snb.2022.132594.

Caporale, A.G. and Violante, A. (2015). Chemical processes affecting the mobility of heavy metals and metalloids in soil environments. *Current Pollution Reports* 2 (1): 15–27. doi: 10.1007/s40726-015-0024-y.

Chang, J., Zhou, G., Christensen, E.R. et al. (2014). Graphene-based sensors for detection of heavy metals in water: a review. *Analytical and Bioanalytical Chemistry* 406 (16): 3957–3975. doi: 10.1007/s00216-014-7804-x.

Chen, S.-H., Li, Y.-X., Li, P.-H. et al. (2018). Electrochemical spectral methods for trace detection of heavy metals: a review. *TrAC Trends in Analytical Chemistry* 106: 139–150. doi: 10.1016/j.trac.2018.07.005.

Cui, L., Wu, J., and Ju, H. (2015). Electrochemical sensing of heavy metal ions with inorganic, organic and bio-materials. *Biosensors & Bioelectronics* 63: 276–286. doi: 10.1016/j.bios.2014.07.052.

Davari, S.A. and Wexler, A.S. (2020). Quantification of toxic metals using machine learning techniques and spark emission spectroscopy. *Atmospheric Measurement Techniques* 13 (10): 5369–5377. doi: 10.5194/amt-13-5369-2020.

de Barros, A., Braunger, M.L., de Oliveira, R.F. et al. (2023). Sensing materials: functionalized advanced carbon-based nanomaterials. In: *Encyclopedia of Sensors and Biosensors*, 254–268. Elsevier EBooks. doi: 10.1016/b978-0-12-822548-6.00014-5.

Deng, W., Tan, Y., Li, Y. et al. (2010). Square wave voltammetric determination of Hg(II) using thiol functionalized chitosan-multiwalled carbon nanotubes nanocomposite film electrode. *Microchim Acta* 169: 367–373. doi: 10.1007/s00604-010-0366-5.

Díez-Pascual, A.M. (2021). Carbon-based nanomaterials. *International Journal of Molecular Sciences* 22 (14): 7726. doi: 10.3390/ijms22147726.

Dresselhaus, M.S. and Terrones, M. (2013). Carbon-based nanomaterials from a historical perspective. *Proceedings of the IEEE* 101 (7): 1522–1535. doi: 10.1109/JPROC.2013.2261271.

Fatima, T. and Mushtaq, A. (2023). Efficacy and challenges of carbon-based nanomaterials in water treatment: a review. *International Journal of Chemical and Biochemical Sciences* 23: 232–248.

Fritea, L., Banica, F., Costea, T. et al. (2021). Metal nanoparticles and carbon-based nanomaterials for improved performances of electrochemical (Bio)sensors with biomedical applications. *Materials* 14 (21): 6319. doi: 10.3390/ma14216319.

Fu, T., Zhang, B., Gao, X. et al. (2023). Recent progresses, challenges, and opportunities of carbon-based materials applied in heavy metal polluted soil remediation. In: *Science of the Total Environment*, 856. Beijing, China: Elsevier B.V. doi: 10.1016/j.scitotenv.2022.158810.

Ghomian, T., Burdette, K., Kizilkaya, O. et al. (2023). A study on the temperature sensitivity of nanosensors based on single-walled carbon nanotubes in ambient conditions. *IEEE Transactions on Electron Devices* 70 (5): 2489–2494. doi: 10.1109/ted.2023.3259930.

Grabarczyk, M., Wlazłowska, E., and Adamczyk, M. (2023). Carbon nanotubes as a suitable material for electrochemical sensor used in voltammetric determinations of titanium. *Applied Nanoscience* 13 (10): 6841–6848. doi: 10.1007/s13204-023-02791-9.

Guo, S., Yang, H., Sun, Q. et al. (2023). Evaluation of a novel carbon-based micro-nano zero-valent iron composite for immobilization of heavy metals in soil. *Journal of Environmental Chemical Engineering* 11 (3): 109740. doi: 10.1016/j.jece.2023.109740.

Guo, X., Yun, Y., Shanov, V.N. et al. (2011). Determination of trace metals by anodic stripping voltammetry using a carbon nanotube tower electrode. *Electroanalysis* 23 (5): 1252–1259. doi: 10.1002/elan.201000674.

Husain, M.M., Ansari, M.T., and Almohammedi, A. (2024). Influence of varying carbon oxides concentrations on the selectivity of an electrical sensor utilizing graphene nanoribbons. *Micro and Nanostructures* 189 (2): 207813. doi: 10.1016/j.micrna.2024.207813.

Hwang, G.H., Han, W.K., Park, J.S. et al. (2008). Determination of trace metals by anodic stripping voltammetry using a bismuth-modified carbon nanotube electrode. *Talanta* 76 (2): 301–308. doi: 10.1016/j.talanta.2008.02.039.

Irshad, S., Liu, G., Yousaf, B. et al. (2021). Geochemical fractionation and spectroscopic fingerprinting for evaluation of the environmental transformation of potentially toxic metal(oid)s in surface–subsurface soils. *Environmental Geochemistry and Health* 43 (10): 4329–4343. doi: 10.1007/s10653-021-00932-z.

Jeong, S.-E., Kim, S., Han, J.-H. et al. (2022). Simple laser-induced graphene fiber electrode fabrication for high-performance heavy-metal sensing. *Microchemical Journal* 172: 106950. doi: 10.1016/j.microc.2021.106950.

Jose, J., Prakash, P., Jeyaprabha, B. et al. (2023). Principle, design, strategies, and future perspectives of heavy metal ion detection using carbon nanomaterial-based electrochemical sensors: a review. In: *Journal of the Iranian Chemical Society*, 20 (4): 775–791. Springer Science and Business Media Deutschland GmbH. doi: 10.1007/s13738-022-02730-5.

Kumar, A. and Menon, S.K. (2009). Fullerene–ferrocene dyad linked by rigid bilinkage: synthesis, photophysical properties and application as copper ion sensor. *Journal of Physical Organic Chemistry* 22 (7): 661–669. doi: 10.1002/poc.1497.

Kumar, S., Nehra, M., Kedia, D. et al. (2019). Nanodiamonds: emerging face of future nanotechnology. *Carbon* 143: 678–699. doi: 10.1016/j.carbon.2018.11.060.

Lee, C., Wei, X., Kysar, J.W. et al. (2008). Measurement of the elastic properties and intrinsic strength of monolayer graphene. *Science* 321 (5887): 385–388. doi: 10.1126/science.1157996.

Li, B., Xie, X., Meng, T. et al. (2024). Recent advance of nanomaterials modified electrochemical sensors in the detection of heavy metal ions in food and water. In: *Food Chemistry*, Volume 440. Elsevier Ltd. doi: 10.1016/j.foodchem.2023.138213.

Li, H., Hardaway, J.B., and Koley, G. (2018). Highly sensitive ion detection with Graphene/Si Schottky junction sensors. In: *2018 IEEE 13th Nanotechnology Materials and Devices Conference* (NMDC), 1–4. doi: 10.1109/NMDC.2018.8605920.

Li, Y.-K., Yang, T., Chen, M.-L. et al. (2021). Recent advances in nanomaterials for analysis of trace heavy metals. *Critical Reviews in Analytical Chemistry* 51 (4): 353–372. doi: 10.1080/10408347.2020.1736505.

Lin, L., Wang, Y., Xiao, Y. et al. (2019). Hydrothermal synthesis of carbon dots codoped with nitrogen and phosphorus as a turn-on fluorescent probe for cadmium(II). *Microchimica Acta* 186 (3): 147. doi: 10.1007/s00604-019-3264-5.

Liu, J., Li, R., and Yang, B. (2020). Carbon dots: a new type of carbon-based nanomaterial with wide applications. *ACS Central Science* 6 (12): 2179–2195. doi: 10.1021/acscentsci.0c01306.

Liu, W., Zhao, J., Ouyang, Z. et al. (2005). Impacts of sewage irrigation on heavy metal distribution and contamination in Beijing, China. *Environment International* 31 (6): 805–812. doi: 10.1016/j.envint.2005.05.042.

Lochab, A., Sharma, R., Kumar, S. et al. (2021). Recent advances in carbon based nanomaterials as electrochemical sensor for toxic metal ions in environmental applications. *Materials Today Proceedings* 45: 3741–3753. doi: 10.1016/j.matpr.2021.01.271.

Ma, K., Qu, F., Chen, Z. et al. (2015). A method for rapid detection of heavy metals in soil and its application. *Proceedings of the World Congress on Intelligent Control and Automation (WCICA)*: 2015-March (March): 109–113. doi: 10.1109/WCICA.2014.7052696.

Madhura, L., Singh, S., Kanchi, S. et al. (2019).Nanotechnology-based water quality management for wastewater treatment. *Environmental* Chemistry *Letters* 17 (1): 65–121. doi: 10.1007/s10311-018-0778-8.

Mansuriya, B.D. and Altintas, Z. (2021). Carbon dots: classification, properties, synthesis, characterization, and applications in health care – an updated review (2018–2021). *Nanomaterials* 11 (10): 2525. doi: 10.3390/nano11102525.

Matheus. (2019). Chemical interaction mechanisms of metals, inorganics anions, organic acids and protein on purified material, artificial soil and cropland soil. *LA Referencia* (Red Federada de Repositorios Institucionales de Publicaciones Científicas). doi: 10.11606/t.11.2019.tde-19112019-100331.

Mauter, M.S. and Elimelech, M. (2008). Environmental applications of carbon-based nanomaterials. *Environmental Science & Technology* 42 (16): 5843–5859. doi: 10.1021/es8006904.

Mitra, S., Chakraborty, A.J., Tareq, A.M. et al. (2022). Impact of heavy metals on the environment and human health: novel therapeutic insights to counter the toxicity. *Journal of King Saud University-Science* 34 (3): 101865. doi: 10.1016/j.jksus.2022.101865.

Mocci, F., de Villiers Engelbrecht, L., Olla, C. et al. (2022). Carbon nanodots from an in silico perspective. *Chemical Reviews* 122 (16): 13709–13799. doi: 10.1021/acs.chemrev.1c00864.

Mochalin, V.N., Shenderova, O., Ho, D. et al. (2012). The properties and applications of nanodiamonds. In: *Nature Nanotechnology*, 7 (1): 11–23. Nature Publishing Group. doi: 10.1038/nnano.2011.209.

Mohamad Nor, N., Ramli, N.H., Poobalan, H. et al. (2023). Recent advancement in disposable electrode modified with nanomaterials for electrochemical heavy metal sensors. In: *Critical Reviews in Analytical Chemistry*, 53 (2): 253–288. Taylor and Francis Ltd. doi: 10.1080/10408347.2021.1950521.

Moustafa, A.H.E., Mousa, M.A., Abdelrahman, H.H. et al. (2024). A novel successful strategy for the detection of antibiotics and toxic heavy metals based on fluorescence silver/graphene quantum dots nanocomposites. *Applied Nanoscience* 14 (1): 1–20. doi: 10.1007/s13204-023-02921-3.

Moyo, M., Okonkwo, J.O., and Agyei, N.M. (2014). An amperometric biosensor based on horseradish peroxidase immobilized onto maize tassel-multi-walled carbon nanotubes modified glassy carbon electrode for determination of heavy metal ions in aqueous solution. *Enzyme and Microbial Technology* 56: 28–34. doi: 10.1016/j.enzmictec.2013.12.014.

Muñoz, J., Gallego, M., and Valcárcel, M. (2005). Speciation analysis of mercury and tin compounds in water and sediments by gas chromatography–mass spectrometry following preconcentration on C60 fullerene. *Analytica Chimica Acta* 548 (1–2): 66–72. doi: 10.1016/j.aca.2005.05.062.

Nasir, S., Hussein, M., Zainal, Z. et al. (2018). Carbon-based nanomaterials/allotropes: a glimpse of their synthesis, properties and some applications. *Materials* 11 (2): 295. doi: 10.3390/ma11020295.

Ogunsola, S.S., Oladipo, M.E., Oladoye, P.O. et al. (2024). Carbon nanotubes for sustainable environmental remediation: a critical and comprehensive review. *Nano-Structures & Nano-Objects* 37: 101099. doi: 10.1016/j.nanoso.2024.10109.

Onyido, I., Norris, A. R., and Buncel, E. (2004). Biomolecule—mercury interactions: modalities of DNA base—mercury binding mechanisms. Remediation strategies. *Chemical Reviews* 104 (12): 5911–5930. doi: 10.1021/cr030443w.

Ozyurt, D., Al Kobaisi, M., Hocking, R.K. et al. (2023). Properties, synthesis, and applications of carbon dots: a review. *Carbon Trends* 12: 100276. doi: 10.1016/j.cartre.2023.100276.

Pandey, S.K., Singh, P., Singh, J. et al. (2016). Nanocarbon-based electrochemical detection of heavy metals. *Electroanalysis* 28 (10): 2472–2488. doi: 10.1002/elan.201600173.

Pereira Lopes Moreira, R. and Ul Islam, S. (2023). Carbon-dots based sensors for detection of pollutants from soil. *Acs Symposium Series* 139–162. doi: 10.1021/bk-2023-1441.ch006.

Prakash, A., Chandra, S., and Bahadur, D. (2012). Structural, magnetic, and textural properties of iron oxide-reduced graphene oxide hybrids and their use for the electrochemical detection of chromium. *Carbon* 50 (11): 4209–4219. doi: 10.1016/j.carbon.2012.05.002.

Proctor, J.E., Melendrez Armada, D., and Vijayaraghavan, A. (2017). An introduction to graphene and carbon nanotubes. Boca Raton, FL: CRC Press. doi: 10.1201/9781315368191.

Profumo, A., Fagnoni, M., Merli, D. et al. (2006). Multiwalled carbon nanotube chemically modified gold electrode for inorganic As speciation and Bi(III) determination. *Analytical Chemistry* 78 (12): 4194–4199. doi: 10.1021/ac060455s.

Qin, X., Lu, W., Asiri, A.M. et al. (2013). Microwave- assisted rapid green synthesis of photoluminescent carbon nanodots from flour and their applications for sensitive and selective detection of mercury(II) ions. *Sensors and Actuators. B, Chemical* 184: 156–162. doi: 10.1016/j.snb.2013.04.079.

Rani, K.U., Katiyar, R., Manjunatha, C. et al. (2020). Heavy metal-ion detection in soil using anodic stripping voltammetry. In: *2020 International Conference for Emerging Technology (INCET)*. doi: 10.1109/incet49848.2020.9154169.

Razavifar, M., Khoshsima, A., Riazi, M. et al. (2024). Recent developments, challenges, and prospects of carbon dots (CDs) for fluid flow investigation in porous media. *Petroleum Research* 9 (4): 553–564. doi: 10.1016/j.ptlrs.2024.04.004.

Reyes-Loaiza, V., De La Roche, J., Hernandez-Renjifo, E. et al. (2024). Laser-induced graphene electrochemical sensor for quantitative detection of phytotoxic aluminum ions (Al3+) in soils extracts. *Scientific Reports* 14 (1): 5772. doi: 10.1038/s41598-024-56212-0.

Sengupta, J. and Hussain, C.M. (2024). Sensitive and selective detection of heavy metal ions and organic pollutants with graphene-integrated sensing platforms. *Nanoscale* 16 (30): 14195–14212. doi: 10.1039/D4NR00956H.

Serban, B.-C. and Octavian, BUIU. (2024). Some considerations about the sensing mechanisms and electrical response of carbon nanohorns – based gas sensors. *Romanian Journal of Information Science and Technology* 27 (2): 137–150. doi: 10.59277/romjist.2024.2.02.

Shahat, A., Ali, E.A., and El Shahat, M.F. (2015). Colorimetric determination of some toxic metal ions in post-mortem biological samples. *Sensors and Actuators B: Chemical* 221: 1027–1034. doi: 10.1016/j.snb.2015.07.032.

Shao, X., Gu, H., Wang, Z. et al. (2013). Highly selective electrochemical strategy for monitoring of cerebral Cu^{2+} based on a carbon Dot- TPEA hybridized surface. *Analytical Chemistry* 85 (1): 418–425. doi: 10.1021/ac303113n.

Shellaiah, M. and Sun, K.W. (2021). Diamond-based electrodes for detection of metal ions and anions. *Nanomaterials* 12: 64. doi: 10.3390/nano12010064.

Shin, S.H. and Hong, H.G. (2010). Anodic stripping voltammetric detection of Arsenic(III) at Platinum-Iron(III) nanoparticle modified carbon nanotube on glassy carbon electrode. *Bulletin of the Korean Chemical Society* 31 (11): 3077–3083. doi: 10.5012/BKCS.2010.31.11.3077.

Singh, H., Bamrah, A., Bhardwaj, S.K. et al. (2021). Nanomaterial-based fluorescent sensors for the detection of lead ions. *Journal of Hazardous Materials* 407: 124379. doi: 10.1016/j.jhazmat.2020.124379.

Singh, N.B. (2021). Remediation of metal ions contaminated soil by nanomaterials. In: *Nanomaterials for Soil Remediation*, 409–426. Elsevier EBooks. doi: 10.1016/B978-0-12-822891-3.00019-0.

Siqueira Jr, J.R. and Oliveira, O.N. (2017). Carbon-based nanomaterials. In: *Nanostructures*, 233–249. William Andrew Publishing. doi: 10.1016/B978-0-323-49782-4/00009-7.

Song, H., Huo, M., Zhou, M. et al. (2022). Carbon nanomaterials-based electrochemical sensors for heavy metal detection. *Critical Reviews in Analytical Chemistry* 1–20. doi: 10.1080/10408347.2022.2151832.

Su, S., Han, L., Gao, L. et al. (2020). Spark plasma sintering compaction of hybrid nanodiamond/carbon nanotubes/metal electrode and its application. *Diamond and Related Materials* 104: 107746. doi: 10.1016/j.diamond.2020.107746.

Sudibya, H.G., He, Q., Zhang, H. et al. (2011). Electrical detection of metal ions using field-effect transistors based on micropatterned reduced graphene oxide films. *ACS Nano* 5(3): 1990–1994. doi: 10.1021/nn103043v.

Sung, J., Gong, B.Y., Yu, H. et al. (2024). Mechanistic insights into the synthesis of nickel-graphene nanostructures for gas sensors. *Small Methods* 19: e2400245. doi: 10.1002/smtd.202400245.

Tagmatarchis, N. and Prato, M. (2005). Carbon-based materials: from fullerene nanostructures to functionalized carbon nanotubes. *Pure and Applied Chemistry* 77 (10): 1675–1684. doi: 10.1351/pac200577101675.

Thakur, A. and Kumar, A. (2023). Recent trends in nanostructured carbon-based electrochemical sensors for the detection and remediation of persistent toxic substances in real-time analysis. *Materials Research Express* 10 (3): 034001–034001. doi: 10.1088/2053-1591/acbd1a.

Torrinha, Á., Oliveira, F., Shahid, N. et al. (2023). Green carbon (Nano) materials-based sensors for analysis of hazardous metal ions. *ACS Symposium Series* 91–138. doi: 10.1021/bk-2023-1441.ch005.

Ullah, S., Zahra, Q., Mansoorianfar, M. et al. (2022). Heavy metal ions detection using nanomaterials-based aptasensors. *Critical Reviews in Analytical Chemistry* 1–17. doi: 10.1080/10408347.2022.2115287.

Venkateswara Raju, C., Hwan Cho, C., Mohana Rani, G. et al. (2023). Emerging insights into the use of carbon-based nanomaterials for the electrochemical detection of heavy metal ions. *Coordination Chemistry Reviews* 476: 214920. doi: 10.1016/j.ccr.2022.214920.

Vidu, R., Rahman, M., Mahmoudi, M. et al. (2014). Nanostructures: a platform for brain repair and augmentation. *Frontiers in Systems Neuroscience* 8 (June). doi: 10.3389/fnsys.2014.00091.

Wanekaya, A.K. (2011). Applications of nanoscale carbon-based materials in heavy metal sensing and detection. *The Analyst* 136 (21): 4383. doi: 10.1039/c1an15574a.

Wang, S. and Si, S. (2013). Aptamer biosensing platform based on carbon nanotube long-range energy transfer for sensitive, selective and multicolor fluorescent heavy metal ion analysis. *Analytical Methods* 5 (12): 2947. doi: 10.1039/c3ay40360b.

Wang, S., Deng, G., Yang, J. et al. (2022). Carbon dot- and gold nanocluster-based three-channel fluorescence array sensor: visual detection of multiple metal ions in complex samples. *Sensors and Actuators B: Chemical* 369: 132194. doi: 10.1016/j.snb.2022.132194.

Wang, S., Swope, V.M., Butler, J.E. et al. (2009). The structural and electrochemical properties of boron-doped nanocrystalline diamond thin-film electrodes grown from Ar-rich and H2-rich source gases. *Diamond and Related Materials* 18 (4): 669–677. doi: 10.1016/j.diamond.2008.11.033.

Wei, J., Yang, D., Chen, H. et al. (2014). Stripping voltammetric determination of mercury(II) based on SWCNT-PhSH modified gold electrode. *Sensors and Actuators. B, Chemical* 190: 968–974. doi: 10.1016/j.snb.2013.09.083.

Xie, F., Yang, M., Jiang, M. et al. (2019). Carbon- based nanomaterials – A promising electrochemical sensor toward persistent toxic substance. In: *TrAC – Trends in Analytical Chemistry*, Volume 119. Elsevier B.V. doi: 10.1016/j.trac.2019.115624.

Xin, X., Shentu, J., Zhang, T., Yang, X. et al. (2022). Sources, indicators, and assessment of soil contamination by potentially toxic metals. *Sustainability* 14 (23): 15878. doi: 10.3390/su142315878.

Xing, Y., Zhang, H., Yang, Z. et al. (2022). Evaluation of 20 elements in soils and sediments by ED-XRF of monochromatic excitation. *Metals* 12 (11): 1798. doi: 10.3390/met12111798.

Xu, Z., Chen, J., Liu, Y. et al. (2022). Multi-emission fluorescent sensor array based on carbon dots and lanthanide for detection of heavy metal ions under stepwise prediction strategy. *Chemical Engineering Journal* 441: 135690. doi: 10.1016/j.cej.2022.135690.

Yaakob, Y. and Kamis, S.L. (2022). Carbon-based nanomaterials: synthesis and characterizations. In: *Enhanced Carbon-Based Materials and Their Applications*, 9–36. Hoboken, NJ: Wiley. doi: 10.1002/9783527829682.ch2.

Yang, Q., Nguyen, E.P., Panáček, D. et al. (2023). Metal-free cysteamine-functionalized graphene alleviates mutual interferences in heavy metal electrochemical detection. *Green Chemistry* 25 (4): 1647–1657. doi: 10.1039%2Fd2gc02978b.

Yu, L., Zhang, L., Ren, G. et al. (2018). Multicolorful fluorescent-nanoprobe composed of Au nanocluster and carbon dots for colorimetric and fluorescent sensing Hg^{2+} and Cr^{6+}. *Sensors and Actuators B: Chemical* 262: 678–686. doi: 10.1016/j.snb.2018.01.192.

Zaytseva, O. and Neumann, G. (2016). Carbon nanomaterials: production, impact on plant development, agricultural and environmental applications. *Chemical and Biological Technologies in Agriculture* 3 (1): 17. doi: 10.1186/s40538-016-0070-8.

Zhou, C., Jiang, W., and Via, B.K. (2014). Facile synthesis of soluble graphene quantum dots and its improved property in detecting heavy metal ions. *Colloids and surfaces. B, Biointerfaces* 118: 72–76. doi: 10.1016/j.colsurfb.2014.03.038.

Zhu, H., Xu, Y., Liu, A. et al. (2015). Graphene nanodots-encaged porous gold electrode fabricated via ion beam sputtering deposition for electrochemical analysis of heavy metal ions. *Sensors and Actuators. B, Chemical* 206: 592–600. doi: 10.1016/j.snb.2014.10.009.

Zhu, Q., Mao, H., Li, J. et al. (2021). A glycine-functionalized graphene quantum dots synthesized by a facile post-modification strategy for a sensitive and selective fluorescence sensor of mercury ions. *Spectrochimica Acta Part A: Molecular and Biomolecular Spectroscopy* 247: 119090. doi: 10.1016/j.saa.2020.119090.

Zhu, S., Tang, S., Zhang, J. et al. (2012). Control the size and surface chemistry of graphene for the rising fluorescent materials. *Chemical Communications (Cambridge, England)* 48 (38): 4527–4539. doi: 10.1039/c2cc31201h.

12

Breaking Barriers of Conventional Disease Protection: Impact of Nanopathology

Puja Kumari[1,2], Sawant Shraddha Bhaskar[1], Jeetu Narware[3], and Abhijeet Ghatak[1]

[1] Department of Plant Pathology, Bihar Agricultural University, Sabour, Bhagalpur, Bihar, India
[2] ICAR-National Research Centre on Litchi, Mushahari, Muzaffarpur, Bihar, India
[3] Department of Mycology and Plant Pathology, IAS, Banaras Hindu University, Varanasi, Uttar Pradesh, India

12.1 Introduction

Ensuring food security is essential to meet the growing needs of both human and animal populations (Valoppi et al. 2021). Food security is achieved through four key components: the availability of food, access to it, proper consumption, and its stability. The primary pillars of food security include food access, availability, utilization, and stability (Savary et al. 2019). Existing, emerging, and reemerging plant pathogens have a detrimental effect on these pillars. Emerging and reemerging plant diseases are characterized by their increased virulence, expanded host range, and geographical spread, often leading to epidemics. In crop production systems, plant diseases are managed through conventional field practices, such as crop rotation and plowing, along with the use of disease-resistant varieties and pesticides (Engelhardt, Stam, and Huckelhoven 2018). Nanotechnology is the science, engineering, and application of materials and devices at the nanoscale, which is typically between 1 and 100 nm. To put that in perspective, a nanometer is one-billionth of a meter, or about 100 000 times smaller than the diameter of a human hair. The chemical and physical properties of nanoparticles (NPs) are influenced by shape, size, surface charge, and crystallinity. During a science conference, for the first time, the term "nanotechnology" was coined by Taniguchi, a professor at Tokyo University of Science (Taniguchi 1974). In 1959, American scientist and Nobel Prize winner Richard Feynman popularized the term "nanotechnology." Feynman gave a talk titled "There's Plenty of Room at the Bottom" at the California Institute of Technology (Caltech) during the annual conference of the American Physical Society. Feynman proposed a hypothesis, "Why can't we write the entire 24 volumes of the Encyclopedia Britannica on the head of a pin?" at this lecture. He also outlined his vision for the use of machines to build smaller machines and even down to the molecular level (Feynman 1960). He is considered as the father of nanotechnology.

Nanotechnology-based Sustainable Agriculture, First Edition. Edited by Pardeep Singh et al.

Nanotechnology is a multidisciplinary field that involves manipulating and utilizing materials at the nanoscale to create innovative solutions and technologies. Its diverse applications and unique properties offer significant potential across various sectors, but careful consideration of safety, ethics, and costs is essential for its responsible advancement. This field involves manipulating matter at the atomic and molecular levels to create new materials and devices with unique properties and functions (Elmer and White 2018). Nanotechnology is currently being explored in agriculture for applications, such as the delivery of plant chemicals, water management, seed treatment, nanobarcoding, gene transfer, controlled agrochemical release, and the use of nanosensors (Maluin et al. 2020). Nanopathology is not just a game-changer in human health, and it is also making waves in plant disease management. By applying nanotechnology to agriculture, researchers and farmers are discovering innovative ways to enhance plant health, increase crop yields, and reduce losses due to diseases (Ghatak 2021; Mandal, Beura, and Ghatak 2021). Agricultural pest and pathogen control often relies on agrochemicals, which can contaminate water, soil, and food. The rise of nanomaterial engineering has introduced innovative nanotools for sustainable agriculture, enabling precise pathogen detection and controlled delivery of nutrients and agrochemicals to enhance food production (Lang et al. 2021, Figure 12.1). The application of nanotechnology in the fields

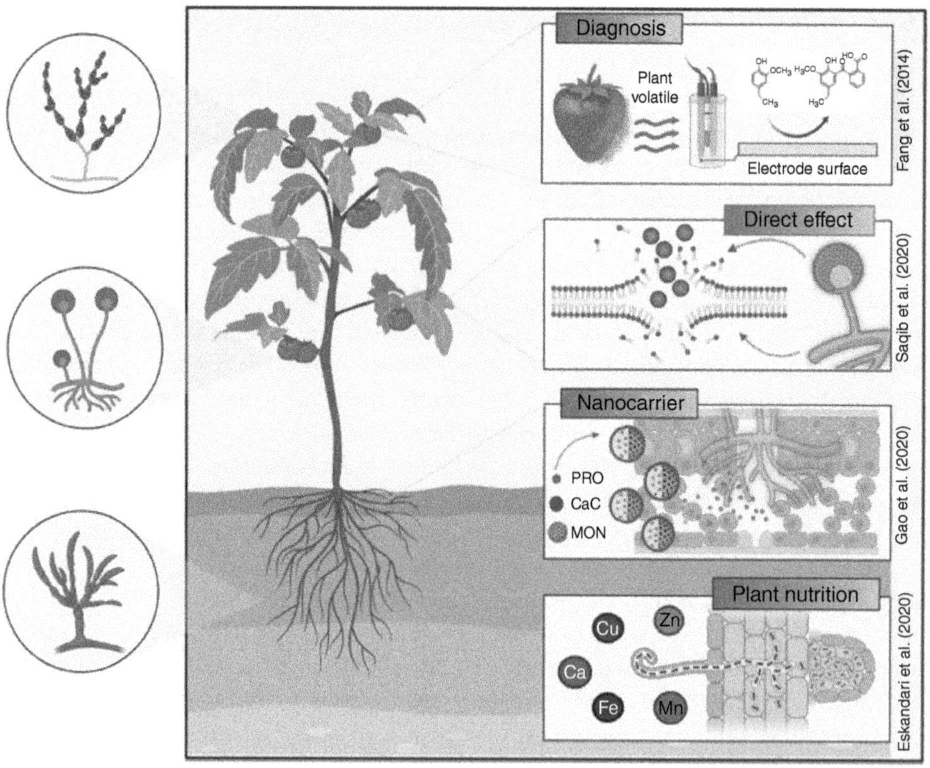

Figure 12.1 Nanopathology help plants to fight with the pathogens and help in growth and development of plants by the use of NPs in a different form by different mode of action (Muthuramalingam et al. 2024).

of genetic alterations, crop protection, and diagnostics is still in its infancy, but it has recently drawn attention in research on plant pathology (Dutta et al. 2022). Plant disease management is the primary obstacle in attaining food security in agriculture. Because of their distinct chemical and physical characteristics, NPs are used in crop protection, particularly to fight plant diseases. Outstanding outcomes were achieved with the use of NPs in the control of plant diseases, particularly in the treatment of bacterial and fungal plant pathogens. Fungal nanotechnology excels in its precision for targeting and delivering therapeutic agents. This approach effectively differentiates between harmful and beneficial microorganisms, improving treatment effectiveness while minimizing unintended effects on nontarget organisms (Lang et al. 2021). Nanoencapsulation stands out for its capacity to regulate fungicide release, providing long-lasting protection and minimizing the frequency of applications (Kutawa et al. 2021). The power of nanotechnology to create enormous structures with novel molecular architecture and to function at the molecular level is what makes it so crucial. The primary goal of nanotechnology is to take advantage of these characteristics and effectively apply them in various fields. Nanotechnology in agriculture could revolutionize fungicides and active ingredients for controlling fungal diseases in plants.

12.2 Evolution of Nanotechnology in the Agriculture Field

About 50 years ago, the idea of using nanotechnology in the field of agriculture first emerged. It has several advantages, including nanofertilizers to minimize the use of synthetic fertilizers and their loss by runoff, nanopesticides to kill infective plant pathogens and destructive insect pests and to minimize rampant use of agrochemicals by smart delivery systems, and nanosensors as an intelligent detection tool for different plant pathogens as their initial stage of infection and as bio-nanocomposites.

12.3 Key Characteristics and Aspects of Nanotechnology

12.3.1 Key Characteristics

At the nanoscale, materials can exhibit properties that differ significantly from those observed at larger scales, including the following:

- *Quantum effects*: NPs may exhibit quantum mechanical behaviors, such as discrete energy levels and quantum tunneling.
- *Increased surface area*: Nanomaterials have a much higher surface area to volume ratio, which can enhance their reactivity and interaction with other substances.
- *Size-dependent properties*: Properties such as color, strength, and conductivity can change as materials are reduced to nanoscale dimensions.
- *NPs can protect plants through two main mechanisms*: (i) as individual particles providing direct protection or (ii) as carriers for fungicides and fertilizers applied via soaking, drenching, or spraying on foliage, roots, or seeds. They enhance several aspects, including (i) solubility of low water-soluble fungicides, (ii) shelf life, (iii) targeted uptake in microbes, and (iv) reduced toxicity (Kutawa et al. 2021).

12.3.2 Aspects of Nanotechnology

12.3.2.1 Nanoscale Materials

Nanotechnology involves working with materials at the nanoscale, which often exhibit unique physical and chemical properties compared to their bulk counterparts. These properties can include enhanced strength, increased reactivity, or unique optical effects. Examples of nanoscale materials include the following:

- *NPs*: Tiny particles with sizes ranging from 1 to 100 nm.
- *Nanotubes*: Cylindrical nanostructures, such as carbon nanotubes, with extraordinary strength and electrical conductivity.
- *Nanosheets*: Thin layers of materials, such as graphene, which have unique electrical and mechanical properties.

12.3.2.2 Nanofabrication Techniques or NP Synthesis Techniques

Nanofabrication is the assembly of structures at the nanometer scale, and its techniques are vital for advancements in molecular adsorption, catalysis, magnetism, luminescence, nonlinear optics, and molecular sensing, enabling reproducible mass production (Ebad-Sichani et al. 2023), which includes the following:

- *Top-down approaches*: Methods that involve etching or milling larger materials down to nanoscale sizes.
- *Bottom-up approaches*: Techniques that build up nanoscale structures from smaller units, such as atoms or molecules, often through chemical reactions or self-assembly.
- *Green synthesis of NPs*: Conventional techniques for synthesizing NPs are costly and produce potentially harmful byproducts. Since high temperatures and pressures are used in chemical and physical processes, it is imperative to reduce the danger of chemical toxicity in the environment. Microbial NP synthesis, which combines nanotechnology and microbiology, is one of the green production techniques. Numerous studies have reported on an eco-friendly and quick way of NP biosynthesis using different species (Figure 12.2) of biocontrol agents *Trichoderma* and *Hypocrea* (Teleomorphs). This method has the potential to overcome the limitations of the chemical and physical methods of NP biosynthesis. Mycogenic NPs, which are generated from fungal cells or cell extracts, can be used as nanofertilizers,

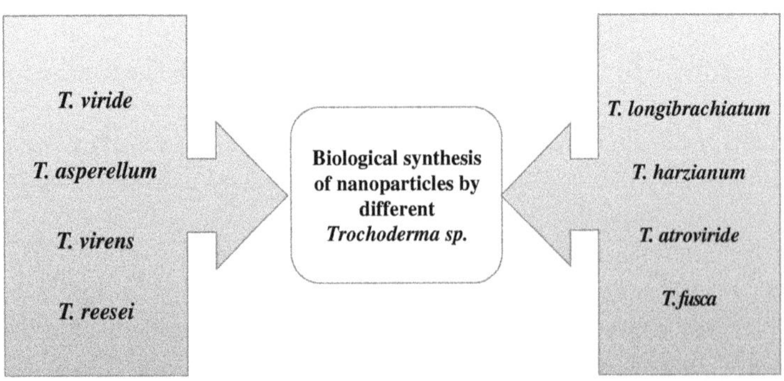

Figure 12.2 Different species of *Trichoderma* used for the biological synthesis of nanoparticles.

nanofungicides, plant growth regulators, nanocoatings, and other applications. Moreover, *Trichoderma*-mediated NPs have been applied to environmental remediation techniques, such as heavy metal contamination detection and pollutant removal. The *Trichoderma* sp. are said to release different secondary metabolites, which act as a reducing and capping agents (Alghuthaymi et al. 2022). Apart from fungi, different bacteria, such as *Pseudomonas* sp., *Serratia* sp. (Mishra et al. 2014), *Cyanobacterium, Anabaena* strain (Singh et al. 2014), and different plant parts are also used for the synthesis of NPs.
- *Chemical synthesis*: Different acidic and basic chemicals are used as reducing and capping agents for the synthesis of NPs.

12.3.2.3 Applications of Nanotechnology

Nanotechnology has a wide range of applications across various fields, including the following:

- *Medicine*: For drug delivery systems, diagnostic tools, and imaging agents. NPs can target specific cells or tissues, providing precise treatment and reducing side effects.
- *Electronics*: In the development of smaller, faster, and more efficient electronic devices. Nanomaterials can enhance the performance of semiconductors and sensors.
- *Energy*: To improve energy efficiency and storage. Nanotechnology is used in developing advanced batteries, fuel cells, and solar panels with better performance.
- *Materials science*: Creating stronger, lighter, and more durable materials. Nanocomposites can offer improved mechanical, thermal, and chemical properties.
- *Environmental protection*: In applications such as water purification, pollution control, and soil remediation. Nanomaterials can help remove contaminants from water and air.

12.4 Nanopathology

12.4.1 Early Detection and Diagnosis

Nanotechnology offers advanced tools for the early detection and diagnosis of plant diseases, which is crucial for effective management (Figure 12.3):

- *Nanosensors*: It can be used to detect specific pathogens or disease markers at very low concentrations. These sensors can be deployed in the field to provide real-time monitoring of plant health. Timely on-site diagnosis of plant pathogens using portable nanobiosensors can also result from research on disease epidemiology and the creation of disease-prevention strategies. These sensors can be positioned throughout the field and linked to a GPS unit to enable real-time tracking of crop health, disease, and soil conditions (Nezhad 2014). Combining biotechnology and nanotechnology could lead to the development of biosensors with increased sensitivity, allowing for an earlier response to changes in the environment and the rising incidence of disease. It will be simpler to treat diseases in plants if we can identify them before any outward signs show up. Biosensors will advance precision farming by giving precise data and

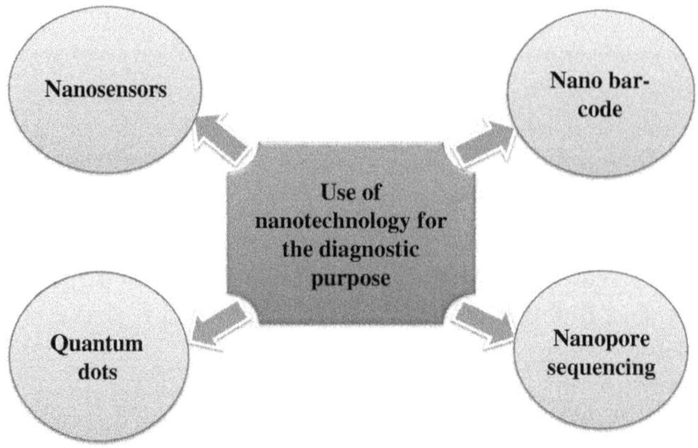

Figure 12.3 Potential use of nanotechnology for plant disease diagnosis.

helping farmers make better decisions, which will boost agricultural productivity and output (Rai and Ingle 2012). A fascinating type of optical biosensor is the colorimetric biosensor, which requires no additional chemicals or equipment to operate. It uses a color shift to quickly and reliably confirm the presence of pathogenic microorganisms in a sample. Functioning similarly to the lateral flow test is the solution-based colorimetric sensor. The color changes from red to purple due to NP aggregation when a colloidal gold NP-mounted receptor interacts with the target infections. For a range of crop diseases, such as *Fusarium* species in maize (Xu et al. 2019), Potato Virus X in potatoes (Drygin et al. 2012), and *Pantoea stewartii* bacteria in corn (Feng et al. 2015), horizontal flow immunoassays utilizing colloidal gold-plated NPs have been developed. One major drawback of all lateral flow assay–based biosensors is their lack of high sensitivity (Pohlmann, Dieser, and Sprinzl 2014).

- *Nano bar-code*: "Bio-barcoding" is a novel technique for identifying ultra-sensitive, enzyme-free proteins and DNAs. Early plant disease detection is facilitated by the pathogen identification sensitivity of the NP-based bio-barcode test.

  Two different probes are used for the nano bar-code method:
  1. Magnetic micro-beads: possess DNA or an antibody as a bio probe that is intended to recognize target pathogens.
  2. Gold NPs containing polyclonal antibodies or an oligonucleotide bio-barcode: enzyme and polymerase chain reaction (PCR) free, and it is a sensitive technique to identify proteins and DNA.

Nanotechnology advancements are driving changes in bio-barcode technology. It is an extremely effective method for identifying protein complexes and nucleic acids without the use of enzymes or PCR. Improvements have been done in a number of areas to boost the usefulness of bio-barcodes in many fields. It is advised to use DNA barcoding to identify fungi (Xu 2016).

- *Nanopore sequencing*: The plant diseases caused by pathogenic bacteria, fungi, and viruses can be quickly diagnosed with the use of the nanopore sequencing platform. It expedited the diagnosing process by indicating potential disease agents that were subsequently confirmed by additional diagnostic techniques. The benefits of this approach include real-time data, field or laboratory sequencing capabilities, and the ability to analyze results for diagnostic purposes without requiring an extensive understanding of bioinformatics tools (Chalupowicz et al. 2019).

Two major problems that nanopore sequencing needs to overcome are (i) recognizing the nucleotides as the strand moves through the nanopore and (ii) controlling the DNA strand's speed as it moves through the nanopore.

Quantum dots (QDs): QDs are spherical and a few nanometer in size. The QDs have been widely used in various applications, such as cell labeling, cell tracking, in vivo imaging, and DNA detection (Sharon, Choudhary, and Kumar 2010). Cadmium selenide (CdSe) QDs were used as signal amplifiers for more accurate and improved detection of banana bunchy top virus (BBTV) in a QD-fluorescence resonance energy transfer (FRET)-based sensor that was shown to be able to identify *Polymyxa betae*, the only known vector of beet necrotic yellow vein virus (BNYVV) for virus infection to beet plants. The primary antibody utilized in this study was generated from the recombinant coat protein of BBTV. CdSe QDs may greatly enhance the electrical signals in this assay, which qualifies the process for use in a laboratory. While QD-based biosensors have emerged as a new class of sensors and are expected to open up new avenues for plant pathogen identification (Majumder et al. 2020), NP-based diagnostic kits can identify pathogens quickly and accurately, allowing for faster response and treatment.

12.4.2 Targeted Delivery of Agrochemicals

Nanocarriers can deliver fungicides, pesticides, and herbicides directly to infected or susceptible plant tissues, reducing the quantity of chemicals needed and minimizing environmental impact. Organic solvents are needed to aid solubilize insecticides with limited water solubility, which raises the insecticide's toxicity and expense. Alternatively, the solubility can be increased by using NPs, which lowers the toxicity. Thus far, it has been effective to incorporate low-water-soluble pesticides into modified chitosan and porous silica (Campos et al. 2018). Nanocapsules can protect active ingredients in agrochemicals from degradation, ensuring more effective treatment over time. Nano-formulations of pesticides and fungicides can enhance their penetration into plant tissues and increase their efficacy against pathogens. NPs can be engineered to release active ingredients in a controlled manner, providing prolonged protection against diseases.

12.4.3 Advanced Plant Disease Management

12.4.3.1 Enhanced Resistance and Tolerance

Nanotechnology can help plants develop resistance or tolerance to diseases.

NP-based treatments can stimulate plant immune responses or introduce resistance genes more efficiently than traditional methods. Nanomaterials can be used to deliver gene-editing tools like Clustered Regularly Interspaced Short Palindromic Repeats (CRISPR) to enhance plant resistance to specific pathogens. CRISPER is a family of DNA sequences found in the genomes of prokaryotic organisms such as bacteria and archaea. The direct application of different types of NPs on plants to fight with various plant pathogens is a new and innovative topic of research. Many research works have already been conducted in this regard against different pathogens, and positive results were obtained in this regard. Different metallic NPs, such as silver, zinc, zinc oxide, copper, titanium, and titanium dioxide NPs, are effective against a wide range of plant pathogens (Kriti, Ghatak, and Mandal 2020a, 2020b; Kumar et al. 2022; Table 12.1). NPs have demonstrated significant potential in managing various biotic diseases by disrupting pathogen activity or enhancing plant defense mechanisms. For instance, silver NPs have been shown to be effective against *Alternaria alternata*, a fungal pathogen causing leaf spot diseases in crops. These NPs disrupt the fungal cell membrane, leading to cell death. Similarly, chitosan NPs have been used to control *Fusarium oxysporum*, a soil-borne fungus responsible for wilt diseases in plants. Chitosan NPs inhibit fungal growth by interfering with the cell wall synthesis of the pathogen.

In another example, zinc oxide NPs have shown efficacy against *Xanthomonas oryzae*, the causative agent of bacterial leaf blight in rice. These NPs generate reactive oxygen species that damage the bacterial cells (Elmer and White 2018). Additionally, silica NPs have been used to manage powdery mildew in crops by creating a physical barrier on the leaf surface, preventing fungal spores from penetrating the plant tissues. These examples highlight the diverse mechanisms through which NPs can be employed in disease management, offering a sustainable and targeted approach to controlling biotic diseases in plants.

12.4.4 Soil and Water Treatment

NPs can be used to decontaminate soil and water from pathogens or pollutants that might harm plants. Nano-enhanced fertilizers can improve soil health and nutrient availability, reducing the likelihood of disease outbreaks. NPs offer innovative solutions for soil and water treatment in agriculture by enhancing plant resilience and mitigating biotic stresses. For instance, chitosan NPs releasing nitric oxide have been shown to protect soybean plants from copper-induced stress, promoting root health and growth (Tortella et al. 2023). Metal and metal oxide NPs, such as copper and zinc oxide, act as effective nanofertilizers and antimicrobial agents, controlling soil-borne pathogens like *F. oxysporum* and reducing plant wilt diseases (Table 12.1). Silica NPs enhance

Table 12.1 Nanoparticle (NP) types and their roles in disease resistance.

S. No.	Nanoparticles	Mechanism	Properties	Example of disease resistance	References
1	Silver NPs	Release of silver ions that disrupt microbial cell walls and inhibit growth	Antimicrobial, cytotoxic to pathogens	Fungal infections in plants (*Fusarium*, *Botrytis*)	Rai et al. (2009)
2	Zinc oxide NPs	Generation of reactive oxygen species (ROS) that damage microbial cells	Ultraviolet (UV)-blocking, antimicrobial	Bacterial blight in rice (*Xanthomonas oryzae*)	Sadiq, Abdullah, and Khan (2019)

S. No.	Nanoparticles	Mechanism	Properties	Example of disease resistance	References
3	Gold NPs	Interaction with cellular components leading to cellular damage	Biocompatible, customizable size	Powdery mildew in grapevines (*Erysiphe necator*)	Bafakeeh, Alshammari, and Khan (2020)
4	Iron oxide NPs	Enhanced magnetic properties for targeted delivery and destruction of pathogens	Magnetic, biocompatible	Rust disease in wheat (*Puccinia triticina*)	Khorram, Sadeghi, and Baghery (2021)
5	Titanium dioxide NPs	Photocatalytic activity under UV light leading to pathogen inactivation	UV-responsive, high surface area	Bacterial wilt in tomatoes (*Ralstonia solanacearum*)	Wu, Zhang, and Li (2018)
6	Copper NPs	Release of copper ions that disrupt microbial metabolism	Antimicrobial, fungicidal	Downy mildew in cucumbers (*Pseudoperonospora cubensis*)	Khan, Khan, and Sultana (2013)
7	Silica NPs	Enhanced delivery of antimicrobial agents to pathogen sites	High surface area, biocompatible	Leaf spot disease in soybeans (*Septoria glycines*)	Garcia, Martinez, and Gomez (2022)
8	Cerium oxide NPs	Generation of ROS and scavenging of free radicals to inhibit pathogens	Antioxidant, UV-resistant	Fusarium wilt in tomatoes (*Fusarium oxysporum*)	Xu, Wang, and Zhang (2020)
9	Graphene oxide NPs	Disruption of microbial cell membranes and inhibition of cellular functions	High surface area, antimicrobial	Leaf blight in potatoes (*Phytophthora infestans*)	Ahmad, Khan, and Shah (2021)
10	Platinum NPs	Interaction with cellular DNA leading to pathogen cell death	Antimicrobial, catalytic properties	Root rot in beans (*Fusarium solani*)	Singh, Kumar, and Singh (2019)

soil health by promoting plant growth and inducing resistance against biotic stresses, thereby improving crop productivity. Additionally, NPs interact with root exudates and modify rhizospheric conditions, facilitating better nutrient uptake and enhancing soil structure. In water treatment, silver NPs exhibit strong antimicrobial properties, helping to reduce microbial contamination in irrigation water and thereby decreasing the incidence of waterborne plant diseases. Polymeric NPs serve as carriers for biocontrol agents, enabling targeted delivery and sustained release of these agents to combat soil and water pathogens effectively (Tortella et al. 2023). These applications demonstrate the versatility of NPs in improving soil and water conditions, leading to healthier plants and increased agricultural productivity.

12.4.5 Biofortification and Plant Health

Nano-nutrients can be used to enhance the nutritional content of crops, making plants more resilient to diseases and improving their overall health. Since nanofertilizers are more readily adsorbed on clay surfaces and are thought to migrate more slowly in the soil than conventional slow-release fertilizers, they do not leach into groundwater (Fraser 2019). Additionally, only in particular soil moisture, temperature, and pH conditions can nitrogen (N), phosphorus (P), potassium (K), copper (Cu), iron (Fe), manganese (Mn), molybdenum (Mo), and zinc (Zn) nanofertilizers be developed to provide nutrients more efficiently (Nongbet et al. 2022).

12.5 Challenges and Considerations

While the potential benefits of nanopathology in plant disease management are significant, there are challenges to address, which are as follows:

Environmental impact: The long-term effects of NPs on ecosystems need to be carefully studied to avoid unintended consequences.

Regulation and safety: Ensuring that nano-based products are safe for use in agriculture and comply with regulations is essential for widespread adoption.

Cost and accessibility: Developing and implementing nanotechnology solutions can be costly, which might limit their accessibility for small-scale farmers.

12.6 Conclusion

Nanopathology is breaking new ground in plant disease management by providing advanced tools for detection, treatment, and prevention. Their role in different aspects of plant pathology can be well explored by doing some intensive research work. With the increase in population, there is immense pressure on researchers to find some alternatives for increasing production with keeping in mind about the safety of human, animal, plant health, and the environment. The flexibility of nanotechnology can be used for both disease control and plant science research. The study of plant genomics and gene functions can be advanced through the application of nanotechnology and NPs in agricultural science. NPs can be utilized to improve crop species, generate disease-resistant plants, and transfer genes into plants. When it comes to delivering nucleotides via virus-induced gene silencing, nucleotide-engineered NPs will offer numerous benefits. In the future, crops may be improved by using NP-mediated gene transfer to increase their resistance to pests and diseases. These innovations hold the promise of more effective, sustainable, and precise approaches to managing plant diseases, ultimately leading to healthier crops and more resilient agricultural systems. This chapter gives a brief information about the different aspects of nanotechnology and its broad application in the field of phytopathology and agriculture. Nanotechnology opens the door of new innovative technology towards the field of precision agriculture, which not only help in managing different stresses affecting the crop but also provide new options for boosting the agriculture production (Figure 12.3).

References

Ahmad, I., Khan, M.A., and Shah, A.A. (2021). Graphene oxide nanoparticles for plant disease control: insights and applications. *Nanotechnology Reviews* 10 (1): 157–172.

Alghuthaymi, M.A., Abd-Elsalam, K.A., AboDalam, H.M. et al. (2022). *Trichoderma*: an eco-friendly source of nanomaterials for sustainable agroecosystems. *Journal of Fungi* 8 (4): 367.

Bafakeeh, O.T., Alshammari, A.S., and Khan, M.I. (2020). Gold nanoparticles as a tool for plant disease management. *Materials Science and Engineering: C* 114: 111068.

Campos, E.V., Proença, P.L., Oliveira, J.L. et al. (2018). Chitosan nanoparticles functionalized with β-cyclodextrin: a promising carrier for botanical pesticides. *Scientific Reports* 8 (1): 2067.

Chalupowicz, L., Dombrovsky, A., Gaba, V. et al. (2019). Diagnosis of plant diseases using the Nanopore sequencing platform. *Plant Pathology* 68 (2): 229–238.

Drygin, Y.F., Blintsov, A.N., Grigorenko, V.G. et al. (2012). Highly sensitive field test lateral flow immunodiagnostics of PVX infection. *Applied Microbiology* and *Biotechnology* 93: 179–189.

Dutta, P., Kumari, A., Mahanta, M. et al. (2022). Advances in nanotechnology as a potential alternative for plant viral disease management. *Frontiers in Microbiology* 13: 935193.

Ebad-Sichani, S., Sharafi-Badr, P., Hayati, P., Soltanian-Fard, M.J. (2023). Nano-Fabrication Methods. In: Nanofabrication Techniques - Principles, Processes and Applications. Ed. Sahu, D.R. IntechOpen. doi:10.5772/intechopen.112429.

Elmer, W. and White, J.C. (2018). The future of nanotechnology in plant pathology. *Annual Review of Phytopathology* 56: 111–133.

Engelhardt, S., Stam, R., and Huckelhoven, R. (2018). Good riddance? Breaking disease susceptibility in the era of new breeding technologies. *Agronomy* 8 (7): 114.

Eskandari, S., Höfte, H., and Zhang, T. (2020). Foliar manganese spray induces the resistance of cucumber to *Colletotrichum lagenarium*. *Journal of Plant Physiology* 246: 153129.

Fang, Y., Umasankar, Y., and Ramasamy, R.P. (2014). Electrochemical detection of p-ethylguaiacol, a fungi infected fruit volatile using metal oxide nanoparticles. *Analyst* 139 (15): 3804–3810.

Feng, M., Kong, D., Wang, W. et al. (2015). Development of an immunochromatographic strip for rapid detection of *Pantoea stewartii* subsp. *stewartii*. *Sensors* 15 (2): 4291–4301.

Feynman, R.P. (1960). There's plenty of room at the bottom. *Engineering and Science* 23: 22–36.

Fraser, A. (2019). Land grab/data grab: precision agriculture and its new horizons. *The Journal of Peasant Studies* 46 (5): 893–912.

Gao, Y., Liang, Y., Dong, H. et al. (2020). A bioresponsive system based on mesoporous organosilica nanoparticles for smart delivery of fungicide in response to pathogen presence. *ACS Sustainable Chemistry & Engineering* 8 (14): 5716–5723.

Garcia, L., Martinez, A., and Gomez, J. (2022). Silica nanoparticles for enhanced delivery of antimicrobial agents in agriculture. *Agricultural Sciences* 13 (5): 456–467.

Ghatak, A. (2021). Plant disease management through application of nanoscience. In: *Sustainable Agriculture: Advances in Technological Interventions* (ed. A.K. Singh and V.B. Patel), 367–377. Palm Bay, FL: Apple Academic Press.

Khan, Y., Khan, M.A., and Sultana, N. (2013). Copper nanoparticles in plant disease management. *Journal of Nanoparticle Research* 15(6): 1686.

Khorram, S., Sadeghi, M., and Baghery, A. (2021). Iron oxide nanoparticles: a review of their applications in agriculture and disease management. *Journal of Agricultural and Food Chemistry* 69 (12): 3476–3492.

Kriti, A., Ghatak, A., and Mandal, N. (2020a). Antimycotic efficacy of zinc nanoparticle on dark-spore forming phytopathogenic fungi. *Journal of Pharmacognosy and Phytochemistry* 9 (2): 750–754.

Kriti, A., Ghatak, A., and Mandal, N. (2020b). Inhibitory potential assessment of silver nanoparticle on phytopathogenic spores and mycelial growth of *Bipolaris sorokiniana* and *Alternaria brassicicola*. *International Journal of Current Microbiology and Applied Sciences* 9 (3): 692–699.

Kumar, A., Ghatak, A., Nazneen, H. et al. (2022). Testing inhibitory efficacy of copper and silver nanoparticles against *Alternaria brassicicola* and *Bipolaris sorokiniana*. *The Pharma Innovation Journal* 11 (2): 2690–2693.

Kutawa, A.B., Ahmad, K., Ali, A. et al. (2021). Trends in nanotechnology and its potentialities to control plant pathogenic fungi: a review. *Biology* 10 (9): 881.

Lang, C., Mission, E.G., Ahmad Fuaad, A.A.H. et al. (2021). Nanoparticle tools to improve and advance precision practices in the agrifoods sector towards Sustainability – a review. *Journal of Cleaner Production* 293: 126063.

Majumder, S., Bhattacharya, B., Singh, P.K. et al. (2020). Impedimetric detection of *Banana bunchy top virus* using CdSe quantum dots for signal amplification. *SN Applied Sciences* 2 (4): 638.

Maluin, F.N., Hussein, M.Z., Azah Yusof, N. et al. (2020). Chitosan-based agronanofungicides as a sustainable alternative in the basal stem rot disease management. *Journal of Agricultural and Food Chemistry* 68 (15): 4305–4314.

Mandal, N., Beura, K., and Ghatak, A. (2021). Nanotechnology in agricultural science. In: *Sustainable Agriculture: Advances in Technological Interventions* (ed. A.K. Singh and V.B. Patel), 189–202. Palm Bay, FL: Apple Academic Press.

Mishra, S., Singh, B.R., Singh, A. et al. (2014). Biofabricated silver nanoparticles act as a strong fungicide against *Bipolaris sorokiniana* causing spot blotch disease in wheat. *PLoS One* 9 (5): e97881.

Muthuramalingam, R., Barroso, K., Milagres, J. et al. (2024). Tiny but mighty: nanoscale materials in plant disease management. *Plant Disease* 108 (2): 241–255.

Nezhad, A.S. (2014). Future of portable devices for plant pathogen diagnosis. *Lab a Chip* 14 (16): 2887–2904.

Nongbet, A., Mishra, A.K., Mohanta, Y.K. et al. (2022). Nanofertilizers: a smart and sustainable attribute to modern agriculture. *Plants* 11 (19): 2587.

Pohlmann, C., Dieser, I., and Sprinzl, M. (2014). A lateral flow assay for identification of *Escherichia coli* by ribosomal RNA hybridisation. *Analyst* 139: 1063–1071.

Rai, M., Yadav, A., and Gade, A. (2009). Silver nanoparticles as a new generation of antimicrobials. *Biotechnology Advances* 27 (1): 76–83.

Rai, M., and Ingle, A. (2012). Role of nanotechnology in agriculture with special reference to management of insect pests. *Applied Microbiology and Biotechnology* 94 (2): 287–293.

Sadiq, M.B., Abdullah, A., and Khan, M.A. (2019). Zinc oxide nanoparticles for plant disease management: a review. *Journal of Nanomaterials* 6: 66–70.

Saqib, S., Zaman, W., Ayaz, A. et al. (2020). Postharvest disease inhibition in fruit by synthesis and characterization of chitosan iron oxide nanoparticles. *Biocatalysis and Agricultural Biotechnology* 28: 101729.

Savary, S., Willocquet, L., Pethybridge, S.J. et al. (2019). The global burden of pathogens and pests on major food crops. *Nature Ecology & Evolution* 3 (3): 430–439.

Sharon, M., Choudhary, A.K., and Kumar, R. (2010). Nanotechnology in agricultural diseases and food safety. *Journal of Phytology* 2 (4): 83–92.

Singh, G., Babele, P.K., Kumar, A. et al. (2014). Synthesis of ZnO nanoparticles using the cell extract of the *Cyanobacterium, Anabaena* strain L31 and its conjugation with UV-B absorbing compound shinorine. *Journal of Photochemistry and Photobiology B: Biology* 138: 55–62.

Singh, S., Kumar, V., and Singh, P. (2019). Platinum nanoparticles: their role in plant disease management and bioactivity. *Materials Today Chemistry* 13: 242–254.

Taniguchi, N. (1974). On the basic concept of nanotechnology. In: *Proceedings of the International Conference on Production Engineering*, August 26–29, Tokyo 18–23.

Tortella, G., Rubilar, O., Pieretti, J.C. et al. (2023). Nanoparticles as a promising strategy to mitigate biotic stress in agriculture. *Antibiotics* 12: 338. doi: 10.3390/antibiotics12020338.

Valoppi, F., Agustin, M., Abik, F. et al. (2021). Insight on current advances in food science and technology for feeding the world population. *Frontiers in Sustainable Food Systems* 5. doi: 10.3389/fsufs.2021.626227.

Wu, X., Zhang, Y., and Li, J. (2018). Titanium dioxide nanoparticles: their role in plant disease management. *Journal of Environmental Management* 223: 310–317.

Xu, J. (2016). Fungal DNA barcoding. *Genome* 59 (11): 913–932.

Xu, L., Wang, M., and Zhang, X. (2020). Cerium oxide nanoparticles in plant disease management: mechanisms and applications. *Frontiers in Plant Science* 11: 1092.

Xu, S., Zhang, G., Fang, B. et al. (2019). Lateral flow immunoassay based on polydopamine-coated gold nanoparticles for the sensitive detection of zearalenone in maize. *ACS Applied Materials & Interfaces* 11: 31283–31290.

13

Role of Nanoparticles in Plant Disease Management

Umesh Kumar[1], *Prince Kumar Singh*[1], *Parvati Madheshiya*[1], *and Indrajeet Kumar*[2]

[1] *Laboratory of Ecotoxicology, Department of Botany, Institute of Science, Banaras Hindu University, Varanasi, India*
[2] *Department of Botany, Government Degree College, Pawanikala, Sonbhadra, India*

13.1 Introduction

Nanoparticles (NPs) are tiny particles with dimensions ranging from 1 to 100 nm. By maintaining certain physical, chemical, and biological properties, NPs can be developed to stand distinguished from their bulk and molecular counterparts (Yang, Peters, and Williams III 2008). NPs can be developed from a variety of different substances, such as proteins, polysaccharides, and synthetic polymers. A wide range of variables, including drug solubility, stability, permeability, biodegradability, and toxicity, determine the selection of matrix composition. Drugs that are soluble, identified, provided, or linked to a matrix of NPs belong within this group of compounds (Reverchon and Adami 2006). The potential for the NPs to influence poorly soluble substances, labile and poorly absorbed physiological substances, more effective solubility, decreased toxicity, extended shelf life, and potentially beneficial methods of delivery is considerable (Mohanraj and Chen 2006; Hayles et al. 2017). NPs can encapsulate and transfer fungicides, insecticides, or other active ingredients directly to the plant's targeted region. This targeted delivery can decrease the quantities of chemicals required, minimizing surrounding impact while maximizing efficacy. Plant cells can be entered by NPs through either ion channels, cytoplasmic membrane proteins, or endocytosis, a cellular process that forms vesicles surrounding the item being carried. Solute uptake in plant roots occurs mainly at the nerve ends of the roots through root hairs because suberin makes the rest of the root surface impermeable. These permeable areas provide for the effective absorption and transmission of NPs (Chichiriccò and Poma 2015; Yadav and Yadav 2018). NPs can be employed to protect plants through two distinct mechanisms: (a) They can be applied directly to plant parts, such as seeds, leaves, or roots to defend against insects and pathogens, including microbes like fungi, bacteria, and viruses, or (b) as carriers for existing pesticides or other bioactive agents, such as double-stranded RNA (dsRNA). Research on the antibacterial, antifungal, and recognized antiviral characteristics of metal NPs such as silver, copper, zinc oxide, and

Nanotechnology-based Sustainable Agriculture, First Edition. Edited by Pardeep Singh et al.

titanium dioxide has been extensive (Kim et al. 2018; Worrall et al. 2018). The use of nanocomposites, nanotools, and nano-enabled biosensors could be more effectively used across a range of fields, including developing crop plants' ability to absorb minerals and nutrients, managing plant diseases by enabling the targeted and precisely controlled delivery of active ingredients to particular sites, and enabling effective and immediate disease diagnostic tools as well as highly regulated and accurate plant active ingredient transportation (Sathiyabama 2019). Nanotechnology, with its multidisciplinary applications across various fields such as agricultural biotechnology, the food industry, and disease diagnosis, has prompted the design of many tools with nanotechnology features. The development of nutrient-enriched meals is facilitated by these technologies, which also improve disease management and encourage plant growth. However, the real-world implementation of these tools, specifically in the food supply and agriculture sectors, is related to significant problems that need to be resolved prior to their commercial exploitation and use (Zehra et al. 2021). Traditionally, plant diseases are initially identified visually by human experts and then evaluated under a microscope. Those procedures are frequently carried out in companies and research labs, require a laboratory, are costly, time-consuming, and require expert workers. Approaches that have been automated, affordable, accessible, accurate, dependable, and delicate are needed for the future advancement of phytopathological disease diagnostics (Sankaran et al. 2010). Innovative sensors based on NPs that could quickly, economically, and efficiently diagnose plant diseases in the modern era have recently been produced (Omanović-Mikličanina and Maksimović 2016). It creates protective coatings on plant surfaces, acting as barriers against pathogens and pests also by improving the plant's defenses towards disease, these coatings may boost its capacity to resist disease. Farmers are preparing to recognize diseases, volatiles, crop residues from chemicals, and developments in the environment. The integration of nanosensing methods in conjunction with advanced new approaches is certain to transform agriculture. It offers a promising avenue for improving plant disease management and increasing crop yields. The need to provide environmentally friendly, high-performing insecticides at a reasonable price has grown (Worrall et al. 2018). The application of nanotechnology in agriculture for target gene transfer, water management, seed germination, plant hormone management, nanobarcoding, nanosensors, as well as controlled release of agrichemicals is currently being studied (Hayles et al. 2017) that may help with plant health, problems with agriculture, and in particular, lowering financial losses, raising food production, and enhancing nutritional security. An overview of possible nanotechnological instruments and methods for managing plant diseases is given in this chapter.

13.2 Types of NPs used in Plant Disease Management

NPs are frequently utilized as antibacterial agents and have demonstrated efficacy against a wide range of diseases caused by soil pathogens. They are commonly used on soil, seeds, roots, or leaves to protect plants from pathogenic organisms such as viruses, bacteria, and fungi (Figure 13.1). These NPs function as elicitor molecules that play defensive mechanisms on a regional and systemic level in plant systems against pathogens. Several studies have been reported as antibacterial, antifungal, and antiviral properties of metallic NPs

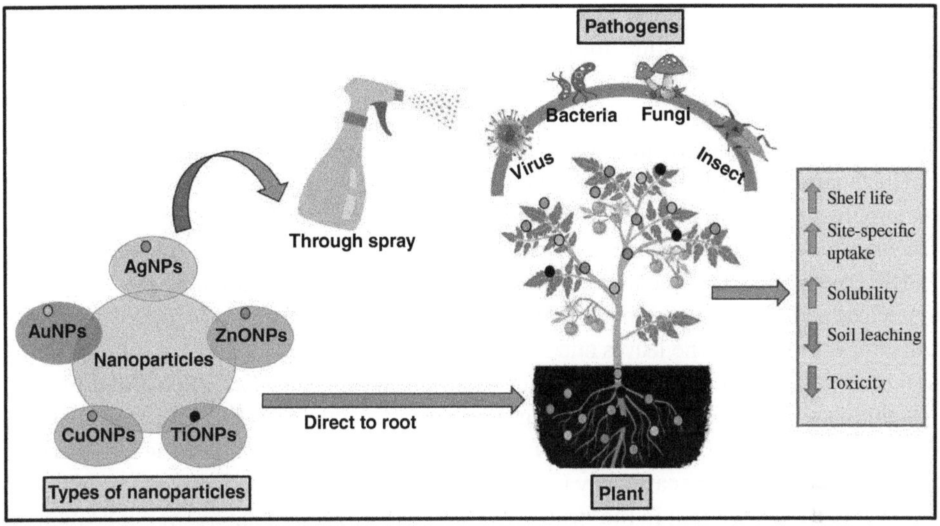

Figure 13.1 Various types of nanoparticles offer numerous advantages, including extended shelf life, precise delivery to target sites, enhanced solubility, and reduced soil leaching and toxicity (Modified from Ray et al., 2023).

such as silver, gold, titanium oxide, zinc oxide, and copper oxide (Kim et al. 2018). Much previous research has examined the effective use of NPs in the management of soil phytopathogen. Kaman and Dutta (2019) found that the AgNPs at a concentration of 100 ppm showed antifungal activity against soil-borne pathogens, including *Sclerotium rolfsii, Rhizoctonia solani, Fusarium oxysporum*, and *Sclerotinia sclerotiorum*. Similarly, Desai et al. (2021) reported the antifungal action of AgNPs against *S. rolfsii* in wheat by complete reduction of sclerotial germination and mycelial growth at 100 ppm in both field and laboratory conditions. The dose of AgNPs at 50 ppm showed more beneficial for plant growth and disease control, but wheat plant roots showed phytotoxic effects at this concentration. The ZnONPs showed antifungal activity when synthesized with *Trichoderma* species (Zaki et al. 2022). The study also reported the antifungal activity in vitro on cotton seedlings against soil-borne pathogens, such as *Macrophomina phaseolina, R. solani*, and *F. fujikuroi.* Min et al. (2009) reported that when the AgNPs crossed threshold dose for the mycelial growth of fungus, *R. solani, S. sclerotiorum*, and *Sarracenia minor* showed detrimental effects such as disintegrated fungal cells, separated the layers of the hyphal wall, and ultimately caused the hyphal to collapse and die. Tomah et al. (2020) found that the AgNPs showed antifungal activity generated with *Trichoderma* spp. against *Sclerotinia sclerotiorum*. The AgNPs at the dose of 200 μg/mL in vitro showed 93.8, 100, and 100% reduction of sclerotial formation, myceliogenic germination, and hyphal development, respectively. The scanning electron microscope (SEM) and energy-dispersive X-ray spectroscopy (EDS) analysis showed the AgNPs' interaction with fungal cells, causing lamellar fragmentation, the formation of microspores or fissures on cell walls, and their accumulation. Guilger-Casagrande et al. (2021) also reported the antifungal activity of AgNPs against *Sclerotinia sclerotiorum*. Chen et al. (2020) reported that magnesium oxide nanoparticle (MgONPs) showed the in vitro suppression of antifungal activity such as fungal growth, spore germination, and sporangium development of *Phytophthora nicotianae* and *Thielaviopsis basicola*.

The antifungal mechanism was detected by SEM, transmission electron microscopy (TEM), and EDS examination through direct contact and adsorption of NPs to fungal hyphae, which leads to alterations in cell structure. MgONPs were found to be effective in the suppression of black root rot by 42.35% and black shank disease by 36.58% caused by fungus at 500 μg/mL in greenhouse environments. Moreover, Dutta et al. (2023) found that CuONPs improved efficiency by 33.69% at 100 mg/L against *Phytophthora nicotianae* by the suppression of black shanks and causing no phytotoxicity. They also reported that the accumulation of intracellular reactive oxygen species (ROS) causes damage to plant tissues and SOD enzyme activity led to the reduction of these radicles. The application of CuONPs to tobacco plants leads to an increment in the defense activity, resistance gene expression, and copper buildup in the leaves and roots significantly. The silver-chitosan NPs' inoculation to tomato seedlings showed a reduction in disease severity after 14 days and up to a 70% inhibition of *F. oxysporum* mycelial growth (Alvarez-Carvajal et al. 2020). The CuONPs were found to be superior to ZnO and FeO NPs for their ability to reduce the incidence of bacterial wilt in tomatoes caused by *R. solanacearum* (Jiang et al. 2021). Alkubaisi and Aref (2017) found the alteration in the structure of the virus capsid protein by the application of AuNPs on the Barley yellow dwarf virus (BaYDV) after 24 and 48 hours through the detection of TEM. The reduction in the virus particle by 75.3 and 24.7% depended on the size of AuNPs measuring 3.151 and 31.67 nm, respectively. Major virus particle degradation was seen in the crop *Hordeum vulgare*; after 48 hours, totally lysed virus-like particles appeared. According to Gopalakrishnan Nair, Kim, and Chung (2014), mung bean plants treated with CuONPs showed a considerable rise in lipid peroxidation and hydrogen peroxide levels. Likewise, Souza et al. (2019) discovered that FeONPs exhibited phytotoxicity in *Lemna minor*, resulting in a dose-dependent increase in ROS and malondialdehyde production. After conducting an analysis, Czarnocka and Karpiński (2018) discovered that plants have ROS scavenging mechanisms that are mediated by several antioxidant enzymes, such as GR, CAT, APX, and SOD. ROS production in plants and its elimination via antioxidant defense mechanisms are determined by the kind of NPs, amount, period of exposure, and plant type. The NPs made up of cellulose, lignin, and chitosan are recognized as polymeric nanoparticles (PNPs) for their antimicrobial, antifungal, and antibacterial action in addition to stimulating plant development. Due to their natural abundance and biodegradability, these materials are regarded as environmentally beneficial. Dawwam, Al-Shemy, and El-Demerdash (2022) reported the use of coprecipitation technique to make ZnO NPs and cellulose nanocrystals (CNCs) from palm sheath fiber agricultural waste.

The CNS-ZnO bionanocomposite was found as an antibacterial property against both Gram-positive (*Staphylococcus aureus, Listeria monocytogenes*) and Gram-negative (*Escherichia coli*) soil-borne bacteria. The CNS-ZnO composite examined the effectiveness at low concentrations, with minimum inhibitory concentration (MIC) values ranging from 0.5 to 1.0 μg/mL against *E. coli* and *L. monocytogenes* and 0.25 to 1 μg/mL against *Salmonella* and *S. aureus*. Although, these bacteria's downregulation of genes linked to virulence and toxicity showed that the nanocomposites may have antioxygenic activity. Schiavi et al. (2022) make CNCs from olive pruning waste in a different study through the use of a chemical bleaching method. The olive knot disease caused by bacterial pathogen *Pseudomonas savastanoi* pv. *savastanoi* in vitro were reduced by nanocrystals. The study found that CNCs suppressed bacterial biofilm formation and bacterial epiphytic survival, they had no negative effects on leaf development or root uptake. The lignin aromatic polymer occurs naturally has antibacterial qualities and functions as a nanocarrier. Lignin nanocarriers were made by combining solvent displacement and sonochemistry methods, which effectively suppressed Gram-positive *Bacillus megaterium* and Gram-negative *E. coli* (Paul et al. 2021). Chitosan (ChNPs) plays an important role against soil-borne bacteria as an antibacterial and plant growth stimulant. The biopolymer-based ChNPs were found more efficient against infection caused by many pathogens including *Sclerotium rolfsii* (Boruah and Dutta 2021), *Fusarium oxysporum* (Sathiyabama 2019), and *R. solani* (Boruah and Dutta 2021). Kheiri et al. (2016) makes ChNPs with different molecular weights and examined their antifungal efficiency against *Fusarium* head blight-causing *F. graminearum*. They reported that five thousand ChNPs reduced 77% the growth of mycelial cells. ChNPs were found to be a dose-dependent decrease in pathogen radial development, demonstrating their effectiveness against *Macrophomina phaseolina* and *R. solani* (Saharan et al. 2013). The application of chitosan to tomato plants produced root exudates that inhibited the nematode *Meloidogyne javanica*, which feeds on knotted roots, and the soil-borne fungal disease *F. oxysporum* f. sp. *radicis-lycopersici* (Suarez-Fernandez et al. 2020). The mycelial growth of fungal pathogens was found to be two times slower than that of the control, and after 72 hours, the hatching rate of *M. javanica* eggs was 1.5 times lower. The ChNPs were found more efficient in reducing the bacterial wilt in potatoes and tomatoes caused by *R. solanacearum* (Khairy et al. 2022). The in vivo test revealed that ChNP treatments were more successful in reducing the frequency and severity of wilt-infected plant disease. It was discovered that ChNPs were more successful in inhibiting bacteria by directly interacting with the bacterial cell wall to alter the cell's structure, eliminate the flagella, and induce cell lysis. Gene expression differences between the *R. solanacearum* strains treated with and without ChNPs were investigated using RAPD-PCR. According to Kim et al. (2013), the chitosan-lignosulfonate (CS-LS) nanohybrid was found to be more effective than either substance alone at inhibiting the development of *Bacillus subtilis, Staphylococcus aureus*, and *E. coli*.

13.3 Plant Disease Management Through NPs

There is a vast diversity of microbes such as bacteria, fungi, viruses, and nematodes which are known to provide serious health concerns to plants and lead to a decrease in yield. Incorporating nanotechnology provides an innovative mitigation strategy to control various plant diseases caused by pests. The use of NPs provides novel mitigating strategies by direct interaction with pathogens and boosting plant defense systems. Due to its small, NPs easily penetrate the host tissues and disrupt the cellular processes of pathogens. NPs are known to generate free radicals, leading to pest cell death. Furthermore, NPs can also accelerate plant immune response, resulting plant's resistance to infections. NPs are a useful strategy for long-term disease management because of their simultaneous effects of immune boosting and pathogen inhibition. Numerous in vitro investigations have demonstrated that NPs can inhibit fungal mycelium development and spore germination in addition to bacterial growth, motility, and proliferation, hence regulating the survival, growth, and multiplication of plant diseases (Masum et al. 2019; Malandrakis et al. 2021). Initially, Park et al. (2006) studied the antimicrobial efficacy of Ag-NPs by making nanosized silica-silver particles effective against plant pests such as *Pythium* sp., *Colletotrichum* sp., *P. syringae*, and *Xanthomonas campestris*. Furthermore, studies have also confirmed the antifungal and nematocidal efficacy of silver NPs (Lamsal et al. 2011; Ul Haq and Ijaz 2019). Chitosan, a polysaccharide is another widely studied NPs, known for its biocompatibility, biodegradability, antimicrobial potential, and low toxicity (Cota-Arriola et al. 2013). Curcumin-copper oxide (Cur-Cu) NPs were shown to enhance the disease resistance in *Cicer arietinum* infected with *F oxysporum* f. sp. *ciceri* (Sathiyabama 2019. Similarly, antibacterial assays demonstrated that magnesium oxide (MgO) NPs effectively check the growth of *R. solanacearum*, a bacterium responsible for bacterial wilt in tomato plants (Imada et al. 2016). A comparative study conducted by Chen et al. (2020) found that MgO-NPs are more effective against soil-borne fungal pathogens than their macroscale counterparts. The antimicrobial action of NPs is a growing area of research, with some researchers focusing on understanding the mechanism of action of NPs to control pathogens invasion in plant diseases. For example, Hajipour et al. (2012) overviewed that NPs can disrupt proteins and membranes, generate ROS, damage cell walls, and cause leakage of nuclear and cytoplasm (Masum et al. 2019). These mechanisms work together to resist various plant pathogens. The ROS generated by NPs can affect plant biomolecules, while the systemic resistance genes linked to salicylic acid, ethylene, jasmonic acid, and phenylalanine ammonia are switched on as part of a plant's defense response (Derbalah and Elsharkawy 2019). The following sections discuss some of the mechanisms by which metal-based nanomaterials inhibit the cell structures of bacteria and fungi.

13.3.1 Inhibition of Biofilm Formation

Biofilms are complex structures made by microbes such as bacteria or fungal cells stacked in their own extracellular polymers. These biofilms, which are composed of biomolecules i.e. proteins, polysaccharides, lipids, and nucleic acids, can be single- or multi-layered (Gupta et al. 2016). Most of the bacteria form biofilms to protect themselves from extreme environments and immune responses, making them highly resistant to several antibiotics. Two main strategies for biofilm remediation are biofilm inhibitors, which prevent early-stage formation, and biofilm dispersants, which break up established biofilm communities (Karygianni et al. 2020; Verderosa, Totsika, and Fairfull-Smith 2019). When subjected to a rotating magnetic field, magneto-responsive gallium-based liquid metal (LM) NPs can physically disturb biofilms. Research has demonstrated that plant- or microbe-produced NPs such as Ag, ZnO, and Fe_2O_3 greatly prevent the production of biofilms in plant diseases such as tomato wilt, bacterial leaf blight, and rice brown stripe (Elbourne et al. 2020).

13.3.2 Cell Wall and Cell Membrane Destruction

Cell wall provide the shape to bacteria and fungi save from hostile surroundings, it is essential to their survival, development, and replication. It may also function as a target for many drugs including antibiotics (Mueller and Levin 2020). Bacterial cell walls are chiefly made up of peptidoglycan, a polymer that forms a protective mesh around the cell. Penicillin, an antibiotic kills bacterium by disrupting cell wall formation (Romaniuk and Cegelski 2015). Because of their tiny size, NPs can enter and harm the internal structures and cell walls of bacteria. TEM and other microscopic techniques have demonstrated that

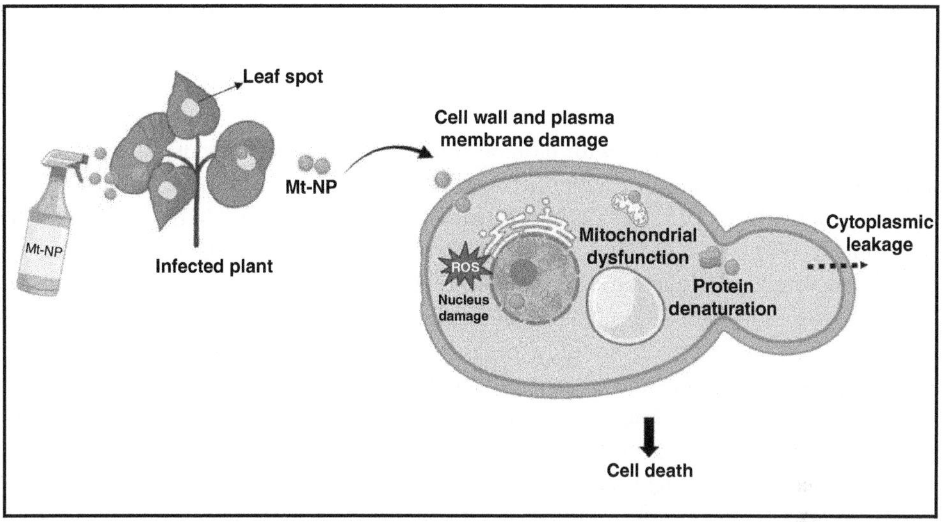

Figure 13.2 Mechanism of metallic nanoparticles' (Mt-NPs) action as nanofungicides.

metal-based NPs such as Ag, Fe_2O_3, and ZnO produced through biosynthetic processes cause plant pathogenic bacteria's cell walls to rupture, wrinkle, leak nutrients, and ultimately die, whereas the walls of untreated bacteria remain intact (Abdallah et al. 2020).

The cell membrane, a complex structure that forms the outer layer of the cytoplasm, primarily constitutes of lipids and proteins. Moreover, fungal cell membranes are rich in lipids such as glycerophospholipids, sphingolipids, and sterols, whereas bacterial membranes mainly contain phospholipids (Strahl and Errington 2017). Cell membranes play vital roles as barriers and in cellular processes i.e. substance transport, cell recognition, electron transfer, etc. Besides these, they are also common targets for antibacterial drug design (Shi et al. 2018). NPs have been shown to disrupt the cell membranes of plant pathogens, causing damage, vacuolization, and leakage of cytoplasmic contents. This disruption may result in electrostatic interactions that alter membrane surface potential, enhancing NPs adhesion (Figure 13.2). Number of studies have demonstrated that MgO-NPs cause significant changes in the zeta potential of fungal cell membranes, resulting in deformation and finally cell death (Chen et al. 2020; Linklater et al. 2020).

13.3.3 Interaction with Biomolecules

Silver (Ag) has long been recognized for its antimicrobial defense and its potent antibacterial properties. Ag-NPs' antibacterial properties may be due to the release of silver ions in biological systems. When silver zeolite, a ceramic material based on silver, comes into contact with bacterial cells, it releases silver ions. This release causes the creation of ROS inside the cell, which interferes with biological processes and stops the growth of bacteria. Furthermore, respiratory enzymes are inhibited by silver ions, which increases the formation of ROS and harms cells (Matsumura et al. 2003). Additionally, by specifically targeting two important locations in the respiratory chain – between cytochrome b and cytochrome a2 – these ions disrupt cellular respiration. At a dosage of 13 μM, inhibition at this location can lower cellular respiration by as much as 80%. By stopping the oxidation of cellular substrates, silver ions, such as those present in silver nitrate, also impede respiration; the degree of inhibition varies with substrate concentration (Dunford et al. 1997). Furthermore, NPs can interact with biomolecules like DNA and proteins, affecting their structure and function. Studies, both in vivo and in vitro, have shown that TiO_2 NPs, when exposed to sunlight, can damage DNA by causing unwinding or breaks in the cellular DNA of pathogens (Tang et al. 2012). Besides impacting DNA, these nanomaterials may also alter the structure and function of proteins. For example, when lysozyme interacts with TiO_2-NPs, part of its structure shifts from an α-helix to a β-fold, reducing its solubilization potential (Xu et al. 2010). Furthermore, prolonged exposure to TiO_2-NPs has been found to downregulate the expression of genes encoding key enzymes involved in glycolysis in artificial wetland pests, leading to lower NADH and ATP production and inhibiting microbial denitrification processes (Yang et al. 2020).

Besides bacterial and fungal diseases, NPs also play a major role in boosting plant defense mechanisms against viral infections by modulating antioxidant defense systems and phytohormone pathways. ROS are created during times of stress to prevent the spread of pathogens, but too many ROS can lead to oxidative stress and upset the equilibrium between ROS and antioxidant enzymes such as ascorbate peroxidase, catalase, and superoxide dismutase (Singh et al. 2022). Depending on their kind and concentration, metal NPs may change this equilibrium by either increasing or decreasing oxidative stress (Cai et al. 2019). For instance, SiO_2 and NiO NPs enhance the defense response genes in cucumber plants against papaya ringspot virus (PRSV) and cucumber mosaic virus (CMV), respectively, while Fe_3O_4 NPs raise ROS levels in tobacco plants, increasing viral resistance. In a similar vein, tomato plants treated with AgNP showed increased antioxidant enzymatic activity after being inoculated with viruses like potato virus Y (PVY) and tobacco mosaic virus (TMV). Furthermore, NPs may also affect phytohormones such as salicylic acid (SA), jasmonic acid (JA), ethylene, and others like auxins and cytokinins, which regulate plant defense responses. AgNPs were shown to increase cytokinin levels in *Capsicum annum* L., and Fe_2O_3 and TiO_2 NPs influenced phytohormone levels in tobacco infected with TMV, enhancing brassinosteroid (BR) and zeatin ribose (ZR) while reducing abscisic acid (ABA; Hao et al. 2019). Moreover, ZnO, SiO_2, and Fe_3O_4 NPs have been reported to upregulate SA-induced pathogenesis-related proteins, further enhancing plant defense against plant pathogens (Derbalah and Elsharkawy 2019; Farooq et al. 2021).

13.4 Emerging Strategies for Mitigating Plant Diseases via NPs

Nowadays, NPs are increasingly employed in agriculture as efficient methods of managing plant diseases (Humbal and Pathak 2023). In addition to improving crop productivity, food processing, and packing, nanobiofertilizers are utilized to encourage plant growth, strain resistance, and seed germination. Moreover, NPs work as pesticide carriers, control bacterial, viral, and fungal infections, promote the development of transgenic crops, and enable precision farming by utilizing nanosensors to transform contemporary agriculture.

13.4.1 Techniques to Better Seed Germination and Plant Development

The large impact of seed quality on crop production and plant growth, germination rate and ultimately total yield is well known. To improve the efficiency of rapeseed breeding, alternative germination methods are needed since in vitro germination may not always reflect results obtained in the field. Improved development of plants, higher agricultural yields, and rapid seed germination have all been revealed benefits of nanotechnology (Acharya and Pal 2020). Decades of research and development demonstrate the effectiveness of NPs in improving resistance, absorption of nutrients as well as plant growth (A. Singh et al. 2024). This increase in proline contents could be one of the factors responsible for improving seed germination, dry mass, and chlorophyll content performance of tomato and squash plants under saline conditions irrigated with some salt of SiO_2-NPs treatment by Siddiqui and Al-Whabi (2014). Nano-zeolite ($Na_2Al_2Si_3O_{10}$-$2H_2O$-NPs) also improves long-term nutrient availability, supporting better plant development (Rambabu et al. 2020). Nevertheless, the detailed mechanisms underlying how NPs assist in water absorption and nutrient uptake in plants continue to be elusive. Nanozymes, a supremely competitive nanomaterial in the field of science, have evolved as an enzyme mimicry such as cerium dioxide NPs (CD-NPs), thus mimicking enzyme behaviors signaling improved germination phenotypes in pine seed, which had 1.3-time improvement by CD nanozyme applications. However, the effect of NPs on seeds varies with nano-size, surface properties, and preparation procedures, so all these factors should be cautiously evaluated for their ecotoxicity and toxic potentialities (Altammar 2023).

13.4.2 Advancements in Food Processing and Packaging Technologies

Mycotoxins are fungal contaminants that commonly affect agricultural plants, especially cereals and fruits, during pre- and post-harvest stages. Fungal-derived mycotoxins, which contain secondary metabolites, pose significant health risks to both humans and animals (Gurikar et al. 2023). Studies have revealed that currently 25% of produced grains are contaminated with deoxynivalenol (DON), and 40% of crops are lost annually due to the number of infections and pests (Worrall et al. 2018). Nanomaterials (NMs) and NPs generate a sophisticated pathway to manage a high level of mycotoxin pollution. At the post-harvest stage, the application of NPs to crop's fruits or NM's leaves inhibited mycotoxin generation. Magnetic iron NPs are more efficient than silver or copper NPs for the absorption of Aflatoxin B1 (AFB1); however, further toxicity research is necessary to ensure safe and healthy utilization (Kelidari et al. 2018). Moreover, outperforming conventional zeolites, magnetic zeolite nanocomposites (MZNC) removed over 99% of aflatoxins from barley flour (Karami-Osboo, Maham, and Nasrollahzadeh 2020). If food is not consumed immediately after generation or harvest, proper food packaging is necessary for avoidance of contamination. Food packaging must be lightweight, cost-effective, and safe ensuring protection from bacteria, chemicals, and airborne pollutants. Due to the nonbiodegradable and nonrecyclable nature of plastic, food packaging is a serious concern for environmental safety. By embedding in active and intelligent packaging systems, nanotechnology gives a practical solution (Nile et al. 2020). These approaches minimize plastic consumption and provide enhanced protection, an extended lifespan, and the ability to inhibit the growth of pathogens. Additionally, by improving the sustainability of its packaging, NPs can encourage environmental friendliness in the food business (de Sousa et al. 2023). As the need for safe, long-lasting, and sustainable packaging grows, nanotechnology is emerging as a potent instrument to transform sectors and solve ecological and economic issues.

13.4.3 Stress Tolerance as a Key to Optimizing Crop Productivity

Under unfavorable environmental conditions, nano-agriculture has the potential to boost crop yields. About 23% of the world's agricultural yield is impacted by salinity stress, which restricts agricultural production. (Onaga and Wydra 2016). Furthermore, the application of iron sulfate NPs ($FeSO_4$-NPs) to sunflower leaves increased shoot dry weight, salt resistance, and expanded leaf surface area (Torabian, Zahedi, and Khoshgoftar 2017). Application of $FeSO_4$-NPs led to improvements in substomatal CO_2 level, chlorophyll content, and CO_2 absorption rates. Furthermore, these NPs increased the iron richness of the leaves and reduced salt content. They may improve the stress tolerance of crops by increasing enzyme activity (Shojaei et al. 2019). By limiting oxidative damage and increasing biomass preservation and plant growth, emerging nanozyme techniques – particularly Mn_4O_4 NPs with antioxidant enzyme-like characteristics – have helped plants like cucumbers cope with salt stress (Lu et al. 2013).

13.4.4 Nano-biofertilizers: Revolutionizing Soil Enhancement

Inorganic fertilizers are the major source of mineral content for conventional agriculture, and they may cause a negative impact on the environmental integrity (Srivastav et al. 2024). In the case of the agriculture sector, overconsumption of fertilizers, especially phosphate, ammonium salts, nitrate, and urea may cause runoff and leaching, especially of nitrogen, which deteriorates soil quality and pollutes water quality, triggering climate change (Lin et al. 2021). Slow-release fertilizers, also known as nanofertilizers, represent a sustainable option that supports food and energy demands while increasing crop development (Raliya et al. 2017). In addition to their controlled-release mechanism, which lowers costs, improves nutrient absorption efficiency, and lessens adverse environmental effects, they are a good substitute for sustainable agriculture (Subramanian et al. 2015). Nanocoated fertilizers utilize a thin layer of nanoporous material to encapsulate and stabilize nutrients. By enhancing surface tension, this layer ensures increased nutrient retention and the gradual breakdown of the nano-layer allows for a controlled nutrient release that surpasses standard fertilizers (Subramanian et al. 2015). The effectiveness of nano-biofertilizers (NBFs) is substantially higher than that of conventional fertilizers (Rezaei and Khanmirzaei 2024). These fertilizers mix biofertilizers, such as rhizobacteria, with organic compounds and nanomaterials to encourage plant development and provide some advantageous effects. NBFs support plants at every stage of development because of their regulated and gradual nutrient supply. Additionally, it enhances nutrient uptake in crops, shortens the growth time, and improves the impacts of biofertilizers (Gouda et al. 2018), providing a sustainable option for modern agriculture.

13.4.5 Next-generation Delivery Mechanisms for Fertilizers and Nutrients

To address plant nutrient deficits, a smart delivery system that can precisely release nutrients in a controlled manner while navigating biological barriers may be designed. Nanofertilizers greatly boost crop productivity by improving nutrient efficiency, enhancing the amount of organic matter in the soil, and addressing issues with uneven fertilization and multi-nutrient deficiencies. In dry environments, foliar application of nano phosphorus has significantly increased the crop yields like pearl millet and cluster beans (Santra and Banerjee 2024). Furthermore, Corradini, De Moura, and Mattoso (2010) optimized the interaction and stability of chitosan NPs by formulating solutions with inorganic fertilizers like N, P, and K. Kottegoda et al. (2011) synthesized hydroxyapatite NPs modified with urea for sustained nitrogen release over a period of 60 days. Furthermore, zeolites treated with hexadecyl trimethyl ammonium and nanocapsules containing nano and sub-nano-composites have demonstrated efficient nutrient release control (Acharya and Pal 2020).

13.4.6 NP-based Approaches to Bacterial Disease Management

The antimicrobial capabilities of NPs are significant against a diverse array of bacteria (Table 13.1; Yilmaz et al. 2023). NPs in the size range of 10–100 nm exhibit strong antibacterial activity against both Gram-positive and Gram-negative bacteria (Braithwaite et al. 2014). They are particularly useful against multidrug-resistant (MDR) pathogens, such as methicillin-resistant *Staphylococcus aureus*, erythromycin-resistant *Streptococcus pyogenes*, ampicillin-resistant *E. coli*, and vancomycin-resistant *S. aureus* (Bose and Chatterjee 2016). The antimicrobial action of NPs involves a Trojan horse mechanism, where they attach to bacterial cell surfaces, disrupt membranes, and release toxic metal ions (Burduşel et al. 2018). The extensive surface area of NPs allows for efficient microbial contact, achieving significant effects even at low concentrations (Li et al. 2024). Gram-negative bacteria, such as *E. coli*, exhibit reduced susceptibility due to their external lipopolysaccharide layer, necessitating higher NP concentrations for effective treatment compared to Gram-positive bacteria (da Rosa et al. 2020).

13.4.7 Role of NPs in Managing Viral Infections

Viruses contribute to approximately 50% of all emerging plant diseases, which can be difficult to control and sometimes lack effective cures. Effective management of viral infections requires an understanding of their complex replication cycle, including cell entry, genome replication, assembly, and release of new virus particles (Table 13.1; Mazzon and Marsh 2019). NPs have shown potential in combating viruses through mechanisms such as binding to viral proteins, preventing attachment to host cells, degrading virus particles before they enter cells, and interacting with viral DNA (Bhatti and DeLong 2023). Research by Elbeshehy, Elazzazy, and Aggelis (2015) demonstrated that NPs from Bacillus species effectively reduced yellow bean mosaic virus symptoms. Additionally, AgNPs significantly mitigated viral replication and disease severity in plants like *Cymopsis tetragonoloba* and tomatoes, particularly when combined with salicylic acid or applied prior to infection (El-Shazly et al. 2017; Noha, Bondok, and El-Dougdoug 2018).

Table 13.1 Mechanism of NPs' action in plants conferring tolerance against bacterial, viral, fungal, and insect-borne infections.

Pathogen	Types of NPs	Host	Mechanism of action	References
Bacterial infection				
Xanthomonas euvesicatoria	Cu composites	Pepper	Total suppression of bacterial growth and significant advancement in plant bacterial spot disease	Fan et al. (2021)
Xanthomonas oryzae pv. oryzae	ZnO (biosynthesis)	Rice	Hampered the cell wall, inhibited bacterial growth and biofilm structure, and increased the generation of ROS	Ogunyemi et al. (2020)
Erwinia cacticida	Ag (biosynthesis)	Watermelon	Bacterial cell wall disruption and bacterial cell content leaks	Chikte et al. (2019)
Acidovorax oryzae	Ag (biosynthesis)	–	Suppression of bacterial survival, proliferation, biofilm formation, and swarming behavior	Masum et al. (2019)
Ralstonia solanacearum	Fe_2O_3 (biosynthesis)	Tomato	Damage to the bacterial cell wall lowered disease severity, promoted plant growth, and increased fresh biomass	Alam et al. (2019)
Dickeya dadantii	ZnO and TiO_2	Sweet potato	Suppression of bacterial growth	Hossain et al. (2019)
Acidovorax oryzae	MgO and MnO_2	Rice	Constraining bacterial development	Ogunyemi et al. (2019)
Viral infection				
Broad bean stain virus	TiO_2 NPs	Broad bean	Induce the expression of defense genes responsive to salicylic acid	Lavanya and Arun (2021)

Pathogen	Types of NPs	Host	Mechanism of action	References
Alfalfa mosaic virus	Chitosan/dextran NPs	Peruvian tobacco	Elevate total carbohydrates and phenolic compounds and induce the expression of peroxidase, PAL, and *PR-1* genes	Abdelkhalek et al. (2021)
Tobacco mosaic virus	Pomegranate peel (PP) extract and PP NPs	Datura	Decreased the number of local lesions and the amount of virus-infected plants, exceeding the use of PP extract alone	Dawi et al. (2021)
Tobacco mosaic virus	Fe NPs	Benthi	Triggering plant oxidants and SA-related *PR* gene expression	Cai et al. (2020)
Bean yellow mosaic virus	Carboxymethyl chitosan-Ti nanocomposites	Broad bean	Increased amounts of total soluble proteins and nonenzymatic and enzymatic antioxidants	Sofy et al. (2020)
Tomato yellow leaf curl virus	SiO_2 NPs	Tomato	Elevate the concentration of polyphenols and peroxidase	Cai et al. (2019)
Tomato bushy stunt virus	Ag NPs	Potato	Decline in the prevalence of disease and viral concentration	El-Shazly et al. (2017)
Fungal infection				
Fusarium solani	ZnO	Tomato	ZnO NPs induced spore development and mycelial growth	Malandrakis et al. (2021)
Candida albicans	AgNPs	–	Suppressed pathogen proliferation	Shafea and Mahmoud (2021)
Aspergillus terreus	AgNPs	–	Retarded fungal development and lower biomass	Shafea and Mahmoud (2021)

(Continued)

Table 13.1 (Continued)

Pathogen	Types of NPs	Host	Mechanism of action	References
Fusarium oxysporum f. sp. chrysanthemi	CuO	Chrysanthemum	Inhibition of fungal growth	Ibrahim et al. (2020)
Fusarium oxysporum	AgNPs	–	Antifungal activity	Ghojavand, Madani, and Karimi (2020)
Cladosporium fulvum	Ag (biosynthesis)	–	The effectiveness of fungal growth inhibition increases with concentration	Elgorban et al. (2017)
Alternaria solani	Ag (biosynthesis)	Tomato	Reduced both fungal spore numbers and dry weight, with a marked increase in fresh weight and chlorophyll content	Kumari et al. (2017)
Insecticidal disease				
Heterodera sacchari	AgNPs	Paddy	Decreased nematode infection in the soil and roots and enhanced rice plant vegetative growth	Fabiyi et al. (2020)
Meloidogyne javanica	AgNPs	Tomato	Upregulated the plant's defense response by increasing chitinase gene expression	Ghareeb et al. (2020)
Cnaphalocrocis medinalis	Hollow mesoporous silica NPs	Rice	Enhanced toxicity and regulated release, with no detrimental impact on the development of plants	Gao et al. (2019)
Plutella xylostella larva	Hollow mesoporous silica NPs	Cabbage	Expressing no impacts on plant development, higher toxicity, and regulated release	Kaziem et al. (2018)
Meloidogyne incognita	AgNPs	Brinjal	Inhibition of both *M. incognita* eggs and the J2 stage brought about varying increases in vegetative growth and fruit weight	Abbassy et al. (2017)

Pathogen	Types of NPs	Host	Mechanism of action	References
Meloidogyne incognita	AgNPs	Bermuda grass	An increase in turfgrass productivity appeared during the initial year, and the development of root galls was reduced over the course of two years; no phytotoxicity was detected	Cromwell et al. (2014)
Tuta absoluta	SiNPs	Tomato	Increased toxicity	El-Samahy, El-Ghobary, and Khafagy (2014)

13.4.8 Utilization of NPs for Managing Fungal Infections

Plant disease management can be significantly increased by the innovative strategy of using NPs as a substitute for synthetic fungicides. The demand for novel antifungal drugs is driven by rising resistance in pathogenic fungi, despite the lack of studies on the effectiveness of NP against phytopathogenic fungi (Jo et al. 2015; Table 13.1; Lockhart, Chowdhary, and Gold 2023). Conventional pesticide usage has caused fungal habitats to be disturbed, resulting in resistant fungal mutants (Guilger et al. 2017). The use of *A. flavus* to produce AgNPs revealed inhibitory effects on *Candida albicans, C. tropicalis*, and *Trichoderma* spp. (Mallmann et al. 2015; Fatima et al. 2016). The effectiveness of NPs in antifungal applications is increased by their improved permeability and retention qualities. While addressing the drawbacks of synthetic fungicides, ongoing research attempts to provide better antifungal techniques that reduce threats to humans and animals (Hahn 2014). ROS generation contributes to NP-induced toxicity and destruction of fungal cellular structures, which is why NPs have also demonstrated efficacy against fungi such as *Alternaria alternata* and *Trichosporonsahii* (Xia et al. 2016).

13.4.9 NP-based Platforms for Effective Insecticide Application

The integration of NPs into insecticide formulations in the early 2000s provided a solution for low-solubility insecticides, traditionally reliant on toxic and costly organic solvents, by enhancing solubility and reducing toxicity (Table 13.1; Nasif, Nuruzzaman, and Naidu 2024). Studies have successfully loaded such insecticides into modified chitosan (Campos et al. 2018) and porous silica though environmental impacts were not investigated. For example, Lu et al. (2013) found that azadirachtin-loaded chitosan NPs inhibited cell proliferation in *Spodoptera litura* and provided sustained release. Furthermore, Liu et al. (2015) reported increased mortality in *H. armigera* larvae with thiamethoxam-loaded dendrimers. Essential oils, widely used for their insecticidal properties, tend to evaporate quickly due to their volatile nature. This rapid evaporation reduces their effectiveness over time, limiting their duration of action. To enhance their stability and prolong efficacy, encapsulation techniques or combining them with slow-release carriers are often employed. Additionally, Lai et al. (2006) showed that SLN-encapsulated *Artemisia arborescens* oil had a 45.5% evaporation rate compared to 80% for nonencapsulated oil. Yang et al. (2009) observed 80% mortality in rice beetles with polyethylene glycol (PEG)-encapsulated garlic oil, versus 11% with the oil alone. Moreover, Song et al. (2012) demonstrated that silica-encapsulated chlorfenapyr was as effective as or more effective than conventional methods in controlling *Plutella xylostella* on *Brassica* over three days.

13.4.10 Enhancing Crop Performance Through Genetic Improvement

Biotechnological approaches for crop genetic modification involve transferring beneficial genes to improve yield, pest resistance, and climate resilience. This process consists of dynamic transformation, where desired DNA is inserted into plant cells, followed by stable transformation, where the DNA is integrated into the plant genome (Godbey 2022). The efficacy of this integration determines the success of creating transgenic plants with enhanced traits (Arencibia et al. 2019). Recent advances include Zhao et al. (2017), who utilized magnetic NPs for DNA delivery in cotton, and Kwak et al. (2019), who employed chitosan-coated single-walled carbon nanotubes for efficient chloroplast transformation, demonstrating a practical, cost-effective method for plant bioengineering (Kwak et al. 2019).

13.4.11 Harnessing Nanosensors for Smart Agricultural Practices

Food safety remains a crucial concern in agriculture and the food industry. Traditional methods for detecting hazardous substances and microbial pathogens are often expensive and time-consuming, highlighting the need for alternative, more efficient detection approaches. Nanosensors have emerged as a promising technology for enhancing quality monitoring in the agri-food sector (Jafarizadeh-Malmiri et al. 2019). These sensors facilitate precise evaluation of pesticides, pathogens, and toxins throughout the entire supply chain – from production and processing to transportation and marketing (Inbaraj and Chen 2016). Nano-based sensors also track critical parameters like temperature, pH, humidity, and product degradation (Hooshmand et al. 2023). Despite being in the early stages of development, nanosensors, including nanofibers, thin nanofilms, NPs, and nanorods, are showing promise in biosensor technology (Premnath et al. 2024). Biosensor operation relies on biological recognition and sensing principles, where nanomaterials such as gold and iron oxide NPs, quantum dots, graphene oxide, and carbon nanotubes serve as transducers (Srivastava, Dev, and Karmakar 2018). The nanoconfinement of these materials enhances their optical, mechanical, electrical, and thermal properties, improving their transducing capabilities. Optical nanosensors benefit from the enhanced optical properties of metallic NPs, up-converting NPs (UCNPs), and quantum dots, while physical and piezoelectric characteristics are used in mass nanosensors, which have limited application in the agri-food industry (Srivastava, Dev, and Karmakar 2018).

13.5 Mitigation Strategies for Addressing NP-related Risks

Due to their long accumulation capacity, persistence, and dispersal, NPs in the environment pose significant risks to species by contributing to ecological harm (El-Kady et al. 2023). The generation of ROS plays a major role in terms of NP toxicity, and NPs solubilize with some ions and interact with cell surfaces leading to detrimental consequences. To resolve the NP-posed environmental problem, the number of extensive studies is required to identify dangers and create effective preventive strategies. These sections explore a framework of innovative engineering technologies, different environmental approaches, and legislative attempts to reduce NPs' toxicity.

13.5.1 Interaction of NPs with Cellular Surfaces

The potential harmful impacts of NPs due to their interaction with organism's surroundings can be controlled by designing NPs to decrease their bonding capacity with biological surfaces. Several studies have shown that altering the NPs' surface charge may affect their toxicity. Cationic NPs, for instance, are more dangerous to living systems because they migrate to negatively charged cells (Lai et al. 2017). Depending on how they are used, applying negatively charged ligands can lessen the detrimental effects of NPs. Another tactic is to coat NPs with PEG, which can lower toxicity by limiting their interaction with organisms. However, this tactic may also improve bioavailability and environmental persistence for aquatic species (Xu et al. 2017). Proteins and other biomolecules can be used as coatings to reduce the toxicity of NPs, just like PEG (Ravindran et al. 2012). It may also be possible to directly modify their form by influencing how NPs interact with cell membranes. Li, Kröger, and Liu (2015) further revealed that with low ligand coverage, biological membranes mimic endocytosis around gold NPs in this sequence: nanospheres < nanocubes < nanorods < nanodisks. This pattern reflects the membrane bending energy specific to each NP's shape.

13.5.2 Innovative Solutions for Sustainable Practices

The transformation of organic components, particularly plant parts, into NPs has emerged as a well-established green technology, with NPs derived from natural plant materials like fruits, stems, foliage, and antioxidant-rich medicinal plants (Hait, Kashyap, and Giri 2024). This eco-friendly strategy provides a variety of benefits, including lower toxicity, economic savings, efficiency, and positive impacts on humanity. Furthermore, there is a plentiful availability of raw materials like nutrient-rich plant-derived waste. Different research works indicate that plant-derived NPs have the potential to successfully restore soils contaminated by hazardous waste, including heavy metals along with various organic and inorganic contaminants (Machado et al. 2013). Various plants, i.e. grapes, bananas, tea, and coffee, can synthesize NPs such as Ag, Fe, and Co, which is attributed to green chemistry principles. In addition, Shahwan et al. (2011) revealed that Fe-NPs from tea leaves can decolorize dyes including methyl orange and methylene blue. This shows how green NPs can be an affordable, one-step method that enables prolonged pollutant degradation with reduced toxicity risks. Furthermore, bacterial NPs, referred to as bio-NPs, are effective and safe agents for biodegradation (Johnson et al. 2012). In the green biosynthesis of NPs, there are generally three important phases: selecting a suitable reaction medium, identifying a natural reducing agent, and adding benign stabilizing components (Buchman et al. 2019).

Naturally occurring NPs can display a greater surface area, higher catalytic activity, and improved interactions with metal salts and enzymes, thanks to the microbial carrier matrix (Tauseef et al. 2023). Because of their greater growth rate, ease of cultivation, and capacity to flourish under typical conditions of pH, pressure, and temperature, microorganisms like bacteria, actinobacteria, fungi, and yeasts are primarily employed for the synthesis of NPs (Fariq, Khan, and Yasmin 2017). Bacteria with functional groups like -OH, -NH+, -COOH, and -SH are essential for the reduction and stability of NPs because they offer ion adhesion binding sites that facilitate reduction processes in the periplasm or on the surface of cell walls (Shahzad et al. 2019).

13.5.3 Regulatory Strategies for Controlling NPs' Risks

A variety of legal and regulatory frameworks that control the processing and dispersion of NPs in ecosystems must develop in tandem with environmentally friendly and cutting-edge engineering solutions (Khan et al. 2024). A major obstacle is the wide distribution of NPs and the ensuing need for regulation to guarantee the best results. NPs have been recognized as emerging pollutants in several countries and are currently being monitored by regulatory agencies (G. Singh et al. 2024). The substantial dangers associated with the production and disposal of NPs are not lessened by the current legal procedures. To effectively regulate NPs and reduce their toxicity, a comprehensive strategy that includes integrated legislation and public awareness campaigns is required (Barpujari 2011). To gather reliable data, researchers have also underlined the significance of including ecotoxicology and the environmental fate of NPs in their investigations. They have done this by supporting the creation of regulations, standardizing procedures, and conducting environmental monitoring (Chávez-Hernández et al. 2024). It is crucial to update these assessments utilizing contemporary techniques to guarantee accuracy and eradicate the environmental harm posed by NPs.

13.6 Conclusion and Future Prospects

Nematodes, bacteria, viruses, and fungi are examples of plant diseases that spread via the soil and can do significant harm to a plant's roots and collar, leading to significant economic harm. Chemical pesticides and other traditional techniques sometimes do not address these illnesses, causing environmental degradation. Because of its antibacterial properties, which include producing ROS and creating toxicity from NPs, nanotechnology has emerged as a viable substitute for managing soil-borne illnesses. NPs' tiny size and large surface area allow them to interact and infiltrate soil bacteria more efficiently, increasing control efficacy. Insecticides and fertilizers can be effectively transported by them, lowering the quantity of chemicals released into the environment. Nanomaterials help stimulate plant defenses and promote growth, while nanodiagnostics enables even earlier detection of diseases that are carried by the soil. However, there are still problems in understanding how NPs interact with beneficial organisms and soil systems, as well as their phytotoxicity and ecotoxicity. The effects on soil structure, fertility, and microbiological processes must be evaluated. The safety, long-term environmental impacts, and possible combination of NPs with biological control agents require further research. Although the global market for nanomaterials is expanding, their use in agriculture is restricted because of their high cost, complicated production process, and potential negative effects on human and environmental health. Future development requires a thorough investigation of the safety, application, and sustainability of agricultural products based on nanotechnology. Comprehensive risk assessments, regulatory frameworks, and long-term field experiments are essential. Interdisciplinary collaboration among biologists, engineers, and pathologists will be key to developing safe, effective, and environmentally friendly soil disease management strategies.

References

Abbassy, M.A., Abdel-Rasoul, M.A., Nassar, A.M. et al. (2017). Nematicidal activity of silver nanoparticles of botanical products against root-knot nematode. *Meloidogyne incognita. Archives of Phytopathology and Plant Protection* 50 (17–18): 909–926.

Abdallah, Y., Liu, M., Ogunyemi, S.O. et al. (2020). Bioinspired green synthesis of chitosan and zinc oxide nanoparticles with strong antibacterial activity against rice pathogen *Xanthomonas oryzae* pv. oryzae. *Molecules* 25 (20): 4795.

Abdelkhalek, A., Qari, S.H., Abu-Saied, M.A.A.R. et al. (2021). Chitosan nanoparticles inactivate alfalfa mosaic virus replication and boost innate immunity in *Nicotiana glutinosa* plants. *Plants* 10 (12): 2701.

Acharya, A. and Pal, P.K. (2020). Agriculture nanotechnology: translating research outcome to field applications by influencing environmental sustainability. *NanoImpact* 19: 100232.

Alam, T., Khan, R.A.A., Ali, A. et al. (2019). Biogenic synthesis of iron oxide nanoparticles via *Skimmialaureola* and their antibacterial efficacy against bacterial wilt pathogen *Ralstonia solanacearum. Materials Science and Engineering: C* 98: 101–108.

Alkubaisi, N.A. and Aref, N.M. (2017). Dispersed gold nanoparticles potentially ruin gold barley yellow dwarf virus and eliminate virus infectivity hazards. *Applied Nanoscience* 7: 31–40.

Altammar, K.A. (2023). A review on nanoparticles: characteristics, synthesis, applications, and challenges. *Frontiers in Microbiology* 14: 1155622.

Alvarez-Carvajal, F., Gonzalez-Soto, T., Armenta-Calderón, A.D. et al. (2020). Silver nanoparticles coated with chitosan against *Fusarium oxysporum* causing the tomato wilt. *Biotecnia* 22 (3): 73–80.

Arencibia, A.D., D'Afonseca, V., Chakravarthi, M. et al. (2019). Learning from transgenics: advanced gene editing technologies should also bridge the gap with traditional genetic selection. *Electronic Journal of Biotechnology* 41: 22–29.

Barpujari, I. (2011). Attenuating risks through regulation: issues for nanotechnology in India. *Journal of Biomedical Nanotechnology* 7 (1): 85–86.

Bhatti, A. and DeLong, R.K. (2023). Nanoscale interaction mechanisms of antiviral activity. *ACS Pharmacology & Translational Science* 6 (2): 220–228.

Boruah, S. and Dutta, P. (2021). Fungus mediated biogenic synthesis and characterization of chitosan nanoparticles and its combine effect with *Trichoderma asperellum* against *Fusarium oxysporum, Sclerotium rolfsii* and *Rhizoctonia solani*. *Indian Phytopathology* 74 (1): 81–93.

Bose, D. and Chatterjee, S. (2016). Biogenic synthesis of silver nanoparticles using guava (*Psidium guajava*) leaf extract and its antibacterial activity against *Pseudomonas aeruginosa*. *Applied Nanoscience* 6 (6): 895–901.

Braithwaite, M.C., Tyagi, C., Tomar, L.K. et al. (2014). Nutraceutical-based therapeutics and formulation strategies augmenting their efficiency to complement modern medicine: an overview. *Journal of Functional Foods* 6: 82–99.

Buchman, J.T., Hudson-Smith, N.V., Landy, K.M. et al. (2019). Understanding nanoparticle toxicity mechanisms to inform redesign strategies to reduce environmental impact. *Accounts of Chemical Research* 52 (6): 1632–1642.

Burduşel, A.C., Gherasim, O., Grumezescu, A.M. et al. (2018). Biomedical applications of silver nanoparticles: an up-to-date overview. *Nanomaterials* 8 (9): 681.

Cai, L., Cai, L., Jia, H. et al. (2020). Foliar exposure of Fe_3O_4 nanoparticles on *Nicotiana benthamiana*: evidence for nanoparticles uptake, plant growth promoter and defense response elicitor against plant virus. *Journal of Hazardous Materials* 393: 122415.

Cai, L., Liu, C., Fan, G. et al. (2019). Preventing viral disease by ZnONPs through directly deactivating TMV and activating plant immunity in *Nicotiana benthamiana*. *Environmental Science: Nano* 6 (12): 3653–3669.

Campos, E.V., Proença, P.L., Oliveira, J.L. et al. (2018). Chitosan nanoparticles functionalized with β-cyclodextrin: a promising carrier for botanical pesticides. *Scientific Reports* 8 (1): 2067.

Chávez-Hernández, J.A., Velarde-Salcedo, A.J., Navarro-Tovar, G. et al. (2024). Safe nanomaterials: from their use, application and disposal to regulations. *Nanoscale Advances*, 6; 1575–1792.

Chen, J., Wu, L., Lu, M. et al. (2020). Comparative study on the fungicidal activity of metallic MgO nanoparticles and macroscale MgO against soilborne fungal phytopathogens. *Frontiers in Microbiology* 11: 365.

Chichiriccò, G. and Poma, A. (2015). Penetration and toxicity of nanomaterials in higher plants. *Nanomaterials* 5 (2): 851–873.

Chikte, R.G., Paknikar, K.M., Rajwade, J.M. et al. (2019). Nanomaterials for the control of bacterial blight disease in pomegranate: Quo vadis? *Applied Microbiology and Biotechnology* 103: 4605–4621.

Corradini, E., De Moura, M.R., and Mattoso, L.H.C. (2010). A preliminary study of the incorparation of NPK fertilizer into chitosan nanoparticles. *Express Polymer Letters* 4 (8): 509–515.

Cota-Arriola, O., Onofre Cortez-Rocha, M., Burgos-Hernández, A. et al. (2013). Controlled release matrices and micro/nanoparticles of chitosan with antimicrobial potential: development of new strategies for microbial control in agriculture. *Journal of the Science of Food and Agriculture* 93 (7): 1525–1536.

Cromwell, W.A., Yang, J., Starr, J.L. et al. (2014). Nematicidal effects of silver nanoparticles on root-knot nematode in bermudagrass. *Journal of Nematology* 46 (3): 261.

Czarnocka, W. and Karpiński, S. (2018). Friend or foe? Reactive oxygen species production, scavenging and signalling in plant response to environmental stresses. *Free Radical Biology and Medicine* 122: 4–20.

da Rosa, C.G., de Melo, A.P.Z., Sganzerla, W.G. et al. (2020). Application in situ of zein nanocapsules loaded with *Origanum vulgare* Linneus and *Thymus vulgaris* as a preservative in bread. *Food Hydrocolloids* 99: 105339.

Dawi, F., El-Beltagi, H.S., Abdel-Mobdy, Y.E et al. (2021). Synergistic impact of the pomegranate peels and its nanoparticles against the infection of tobacco mosaic virus (TMV). *Fresenius Environmental Bulletin* 30 (1): 731–746.

Dawwam, G.E., Al-Shemy, M.T., and El-Demerdash, A.S. (2022). Green synthesis of cellulose nanocrystal/ZnO bio-nanocomposites exerting antibacterial activity and downregulating virulence toxigenic genes of food-poisoning bacteria. *Scientific Reports* 12 (1): 16848.

de Sousa, M.S., Schlogl, A.E., Estanislau, F.R. et al. (2023). Nanotechnology in packaging for food industry: past, present, and future. *Coatings* 13 (8): 1411.

Derbalah, A.S.H. and Elsharkawy, M.M. (2019). A new strategy to control Cucumber mosaic virus using fabricated NiO-nanostructures. *Journal of Biotechnology* 306: 134–141.

Desai, P., Jha, A., Markande, A. et al. (2021). Silver nanoparticles as a fungicide against soil-borne *sclerotium rolfsii*: a case study for wheat plants. In: *Biobased Nanotechnology for Green Applications*, 513–542. Springer.

Dunford, R., Salinaro, A., Cai, L. et al. (1997). Chemical oxidation and DNA damage catalysed by inorganic sunscreen ingredients. *FEBS Letters* 418 (1–2): 87–90.

Dutta, P., Kumari, A., Mahanta, M. et al. (2023). Nanotechnological approaches for management of soil-borne plant pathogens. *Frontiers in Plant Science* 14: 1136233.

Elbeshehy, E.K., Elazzazy, A.M., and Aggelis, G. (2015). Silver nanoparticles synthesis mediated by new isolates of *Bacillus* spp., nanoparticle characterization and their activity against Bean Yellow Mosaic Virus and human pathogens. *Frontiers in Microbiology* 6: 453.

Elbourne, A., Cheeseman, S., Atkin, P. et al. (2020). Antibacterial liquid metals: biofilm treatment via magnetic activation. *ACS Nano* 14 (1): 802–817.

Elgorban, A.M., El-Samawaty, A.E.R.M., Abd-Elkader, O.H. et al. (2017). Bioengineered silver nanoparticles using *Curvularia pallescens* and its fungicidal activity against *Cladosporium fulvum*. *Saudi Journal of Biological Sciences* 24 (7): 1522–1528.

El-Kady, M.M., Ansari, I., Arora, C. et al. (2023). Nanomaterials: a comprehensive review of applications, toxicity, impact, and fate to environment. *Journal of Molecular Liquids* 370: 121046.

El-Samahy, M.F.M., El-Ghobary, A.M., and Khafagy, I.F. (2014). Using silica nanoparticles and neem oil extract as new approaches to control *Tuta absoluta* (meyrick) in tomato under field conditions. *International Journal of Plant & Soil Science* 3 (10): 1355–1365.

El-Shazly, M.A., Attia, Y.A., Kabil, F.F. et al. (2017). Inhibitory effects of salicylic acid and silver nanoparticles on potato virus Y-infected potato plants in Egypt. *Middle East Journal of Agriculture Research* 6 (3): 835–848.

Fabiyi, O.A., Olatunji, G.A., Atolani, O. et al. (2020). Preparation of bio-nematicidal nanoparticles of *Eucalyptus officinalis* for the control of cyst nematode (*Heterodera sacchari*). *The Journal of Animal & Plant Sciences*, 30 (5): 1172–1177.

Fan, Q., Liao, Y.Y., Kunwar, S. et al. (2021). Antibacterial effect of copper composites against *Xanthomonas euvesicatoria*. *Crop Protection* 139: 105366.

Fariq, A., Khan, T., and Yasmin, A. (2017). Microbial synthesis of nanoparticles and their potential applications in biomedicine. *Journal of Applied Biomedicine* 15 (4): 241–248.

Farooq, T., Adeel, M., He, Z. et al. (2021). Nanotechnology and plant viruses: an emerging disease management approach for resistant pathogens. *ACS Nano* 15 (4): 6030–6037.

Fatima, F., Verma, S.R., Pathak, N. et al. (2016). Extracellular mycosynthesis of silver nanoparticles and their microbicidal activity. *Journal of Global Antimicrobial Resistance* 7: 88–92.

Gao, Y., Zhang, Y., He, S. et al. (2019). Fabrication of a hollow mesoporous silica hybrid to improve the targeting of a pesticide. *Chemical Engineering Journal* 364: 361–369.

Ghareeb, R.Y., EL-Saedy, M.A., Fathy, M.M. et al. (2020). Green biosynthesized silver nanoparticles using macroalgae and their usage as a bioagent on Meloidogyne incognita. *Plant Archives* 20: 9682–9692.

Ghojavand, S., Madani, M., and Karimi, J. (2020). Green synthesis, characterization and antifungal activity of silver nanoparticles using stems and flowers of felty germander. *Journal of Inorganic and Organometallic Polymers and Materials* 30: 2987–2997.

Godbey, W.T. (2022). Chapter 18-Transgenics and genetically modified organisms in agriculture. In: *Biotechnology and its Applications*, 2e, 411–428. Elsevier.

Gopalakrishnan Nair, P.M., Kim, S.H., and Chung, I.M. (2014). Copper oxide nanoparticle toxicity in mung bean (*Vigna radiata* L.) seedlings: physiological and molecular level responses of in vitro grown plants. *Acta Physiologiae Plantarum* 36: 2947–2958.

Gouda, S., Kerry, R.G., Das, G. et al. (2018). Revitalization of plant growth promoting rhizobacteria for sustainable development in agriculture. *Microbiological Research* 206: 131–140.

Guilger, M., Pasquoto-Stigliani, T., Bilesky-Jose, N. et al. (2017). Biogenic silver nanoparticles based on *Trichoderma harzianum*: synthesis, characterization, toxicity evaluation and biological activity. *Scientific Reports* 7 (1): 44421.

Guilger-Casagrande, M., Germano-Costa, T., Bilesky-José, N. et al. (2021). Influence of the capping of biogenic silver nanoparticles on their toxicity and mechanism of action towards *Sclerotinia sclerotiorum*. *Journal of Nanobiotechnology* 19: 1–18.

Gupta, P., Sarkar, S., Das, B. et al. (2016). Biofilm, pathogenesis and prevention—a journey to break the wall: a review. *Archives of Microbiology* 198: 1–15.

Gurikar, C., Shivaprasad, D.P., Sabillón, L. et al. (2023). Impact of mycotoxins and their metabolites associated with food grains. *Grain & Oil Science and Technology* 6 (1): 1–9.

Hahn, M. (2014). The rising threat of fungicide resistance in plant pathogenic fungi: botrytis as a case study. *Journal of Chemical Biology* 7: 133–141.

Hait, M., Kashyap, N.K., and Giri, A.K. (2024). Synthesis of metal NPs using plant flowers and fruits. In: *Biogenic Wastes-Enabled Nanomaterial Synthesis: Applications in Environmental Sustainability*, 195–237. Cham: Springer Nature Switzerland.

Hajipour, M.J., Fromm, K.M., Ashkarran, A.A. et al. (2012). Antibacterial properties of nanoparticles. *Trends in Biotechnology* 30 (10): 499–511.

Hao, Y., Fang, P., Ma, C. et al. (2019). Engineered nanomaterials inhibit *Podosphaera pannosa* infection on rose leaves by regulating phytohormones. *Environmental Research* 170: 1–6.

Hayles, J., Johnson, L., Worthley, C. et al. (2017). Nanopesticides: a review of current research and perspectives. In: *New Pesticides and Soil Sensors*, 193–225. Elsevier.

Hooshmand, S., Kassanos, P., Keshavarz, M. et al. (2023). Wearable nano-based gas sensors for environmental monitoring and encountered challenges in optimization. *Sensors* 23 (20): 8648.

Hossain, A., Hong, X., Ibrahim, E. et al. (2019). Green synthesis of silver nanoparticles with culture supernatant of a bacterium *Pseudomonas rhodesiae* and their antibacterial activity against soft rot pathogen *Dickeya dadantii*. *Molecules* 24 (12): 2303.

Humbal, A. and Pathak, B. (2023). Application of nanotechnology in plant growth and diseases management: tool for sustainable agriculture. In: *Agricultural and Environmental Nanotechnology: Novel Technologies and Their Ecological Impact*, 145–168. Singapore: Springer Nature Singapore.

Ibrahim, E., Luo, J., Ahmed, T. et al. (2020). Biosynthesis of silver nanoparticles using onion endophytic bacterium and its antifungal activity against rice pathogen *Magnaporthe oryzae*. *Journal of Fungi* 6 (4): 294.

Imada, K., Sakai, S., Kajihara, H. et al. (2016). Magnesium oxide nanoparticles induce systemic resistance in tomato against bacterial wilt disease. *Plant Pathology* 65 (4): 551–560.

Inbaraj, B.S. and Chen, B.H. (2016). Nanomaterial-based sensors for detection of foodborne bacterial pathogens and toxins as well as pork adulteration in meat products. *Journal of Food and Drug Analysis* 24 (1): 15–28.

Jafarizadeh-Malmiri, H., Sayyar, Z., Anarjan, N. et al. (2019). *Nanobiotechnology in Food: Concepts, Applications and Perspectives*, 81–94. Berlin, Germany: Springer.

Jiang, H., Lv, L., Ahmed, T. et al. (2021). Effect of the nanoparticle exposures on the tomato bacterial wilt disease control by modulating the rhizosphere bacterial community. *International Journal of Molecular Sciences* 23 (01): 414.

Jo, Y.K., Cromwell, W., Jeong, H.K. et al. (2015). Use of silver nanoparticles for managing *Gibberella fujikuroi* on rice seedlings. *Crop Protection* 74: 65–69.

Johnson, A., Merilis, G., Hastings, J. et al. (2012). Reductive degradation of organic compounds using microbial nanotechnology. *Journal of the Electrochemical Society* 160 (1): G27.

Kaman, P.K. and Dutta, P. (2019). Synthesis, characterization and antifungal activity of biosynthesized silver nanoparticle. *Indian Phytopathology* 72: 79–88.

Karami-Osboo, R., Maham, M., and Nasrollahzadeh, M. (2020). Synthesised magnetic nano-zeolite as a mycotoxins binder to reduce the toxicity of aflatoxins, zearalenone, ochratoxin A, and deoxynivalenol in barley. *IET Nanobiotechnology* 14 (7): 623–627.

Karygianni, L., Ren, Z., Koo, H. et al. (2020). Biofilm matrixome: extracellular components in structured microbial communities. *Trends in Microbiology* 28 (8): 668–681.

Kaziem, A.E., Gao, Y., Zhang, Y. et al. (2018). α-Amylase triggered carriers based on cyclodextrin anchored hollow mesoporous silica for enhancing insecticidal activity of avermectin against *Plutella xylostella*. *Journal of Hazardous Materials* 359: 213–221.

Kelidari, H.R., Babaei, R., Nabili, M. et al. (2018). Improved delivery of voriconazole to Aspergillus fumigatus through solid lipid nanoparticles as an effective carrier. *Colloids and Surfaces A: Physicochemical and Engineering Aspects* 558: 338–342.

Khairy, A.M., Tohamy, M.R., Zayed, M.A. et al. (2022). Eco-friendly application of nano-chitosan for controlling potato and tomato bacterial wilt. *Saudi Journal of Biological Sciences* 29 (4): 2199–2209.

Khan, J., Kaur, A., Alam, M.A. et al. (2024). Nanopesticides, nanoherbicides, and nanofertilizers: risks and environmental acceptability. In: *Nanopesticides, Nanoherbicides, and Nanofertilizers*, 94–113. CRC Press.

Kheiri, A., Jorf, S.M., Malihipour, A. et al. (2016). Application of chitosan and chitosan nanoparticles for the control of *Fusarium* head blight of wheat (*Fusarium graminearum*) *in vitro* and greenhouse. *International Journal of Biological Macromolecules* 93: 1261–1272.

Kim, D.Y., Kadam, A., Shinde, S. et al. (2018). Recent developments in nanotechnology transforming the agricultural sector: a transition replete with opportunities. *Journal of the Science of Food and Agriculture* 98 (3): 849–864.

Kim, S., Fernandes, M.M., Matamá, T. et al. (2013). *Chitosan–lignosulfonates* sono-chemically prepared nanoparticles: characterisation and potential applications. *Colloids and Surfaces B: Biointerfaces* 103: 1–8.

Kottegoda, N., Munaweera, I., Madusanka, N. et al. (2011). A green slow-release fertilizer composition based on urea-modified hydroxyapatite nanoparticles encapsulated wood. *Current Science* 101 (1): 73–78.

Kumari, M., Pandey, S., Bhattacharya, A. et al. (2017). Protective role of biosynthesized silver nanoparticles against early blight disease in *Solanum lycopersicum*. *Plant Physiology and Biochemistry* 121: 216–225.

Kwak, S.Y., Lew, T.T.S., Sweeney, C.J. et al. (2019). Chloroplast-selective gene delivery and expression in planta using chitosan-complexed single-walled carbon nanotube carriers. *Nature Nanotechnology* 14 (5): 447–455.

Lai, F., Wissing, S.A., Müller, R.H. et al. (2006). *Artemisia arborescens* L essential oil-loaded solid lipid nanoparticles for potential agricultural application: preparation and characterization. *Aaps Pharmscitech* 7: E10–E18.

Lai, L., Li, S.J., Feng, J. et al. (2017). Effects of surface charges on the bactericide activity of CdTe/ZnS quantum dots: a cell membrane disruption perspective. *Langmuir* 33 (9): 2378–2386.

Lamsal, K., Kim, S.W., Jung, J.H. et al. (2011). Application of silver nanoparticles for the control of *Colletotrichum* species in vitro and pepper anthracnose disease in field. *Mycobiology* 39 (3): 194–199.

Lavanya, R. and Arun, V. (2021). Detection of Begomovirus in chilli and tomato plants using functionalized gold nanoparticles. *Scientific Reports* 11 (1): 14203.

Li, H., Rehman, A., ur Rahman, S. et al. (2024). Biosynthesized zinc oxide nanoparticles modulate the phytoremediation potential of *Pennisetum giganteum* and its rhizocompartments associated microbial community structure. *Journal of Cleaner Production* 434: 140346.

Li, Y., Kröger, M., and Liu, W. K. (2015). Shape effect in cellular uptake of PEGylated nanoparticles: comparison between sphere, rod, cube and disk. *Nanoscale* 7 (40): 16631–16646.

Lin, S.S., Shen, S.L., Zhou, A. et al. (2021). Assessment and management of lake eutrophication: a case study in Lake Erhai, China. *Science of the Total Environment* 751: 141618.

Linklater, D.P., Baulin, V.A., Le Guével, X. et al. (2020). Antibacterial action of nanoparticles by lethal stretching of bacterial cell membranes. *Advanced Materials* 32 (52): 2005679.

Liu, X., He, B., Xu, Z. et al. (2015). A functionalized fluorescent dendrimer as a pesticide nanocarrier: application in pest control. *Nanoscale* 7 (2): 445–449.

Lockhart, S.R., Chowdhary, A., and Gold, J.A. (2023). The rapid emergence of antifungal-resistant human-pathogenic fungi. *Nature Reviews Microbiology* 21 (12): 818–832.

Lu, W., Lu, M.L., Zhang, Q.P. et al. (2013). Octa hydrogenated retinoic acid-conjugated glycol chitosan nanoparticles as a novel carrier of azadirachtin: synthesis, characterization, and in vitro evaluation. *Journal of Polymer Science Part A: Polymer Chemistry* 51 (18): 3932–3940.

Machado, S., Pinto, S.L., Grosso, J.P. et al. (2013). Green production of zero-valent iron nanoparticles using tree leaf extracts. *Science of the Total Environment* 445: 1–8.

Malandrakis, A.A., Kavroulakis, N., Avramidou, M. et al. (2021). Metal nanoparticles: phytotoxicity on tomato and effect on symbiosis with the *Fusarium solani* FsK strain. *Science of the Total Environment* 787: 147606.

Mallmann, E.J.J., Cunha, F.A., Castro, B.N. et al. (2015). Antifungal activity of silver nanoparticles obtained by green synthesis. *Revista do instituto de Medicina Tropical de sao Paulo* 57 (2): 165–167.

Masum, M.M.I., Siddiqa, M.M., Ali, K.A. et al. (2019). Biogenic synthesis of silver nanoparticles using *Phyllanthus emblica* fruit extract and its inhibitory action against the pathogen *Acidovorax oryzae* strain RS-2 of rice bacterial brown stripe. *Frontiers in Microbiology* 10: 820.

Matsumura, Y., Yoshikata, K., Kunisaki, S.I. et al. (2003). Mode of bactericidal action of silver zeolite and its comparison with that of silver nitrate. *Applied and Environmental Microbiology* 69 (7): 4278–4281.

Mazzon, M. and Marsh, M. (2019). Targeting viral entry as a strategy for broad-spectrum antivirals. *F1000Research* 1–10.

Min, J.S., Kim, K.S., Kim, S.W. et al. (2009). Effects of colloidal silver nanoparticles on sclerotium-forming phytopathogenic fungi. *The Plant Pathology Journal* 25 (4): 376–380.

Mohanraj, V.J. and Chen, Y.J.T.J.O.P.R. (2006). Nanoparticles-a review. *Tropical Journal of Pharmaceutical Research* 5 (1): 561–573.

Mueller, E.A. and Levin, P.A. (2020). Bacterial cell wall quality control during environmental stress. *MBio* 11 (5): 10–1128.

Nasif, S.O., Nuruzzaman, M., and Naidu, R. (2024). Porous silica nanocarriers: advances in structural orientation and modification to develop sustainable pesticide delivery systems. *ACS Agricultural Science & Technology* 4 (2): 144–172.

Nile, S.H., Baskar, V., Selvaraj, D. et al. (2020). Nanotechnologies in food science: applications, recent trends, and future perspectives. *Nano-micro Letters* 12: 1–34.

Noha, K., Bondok, A.M., and El-Dougdoug, K.A. (2018). Evaluation of silver nanoparticles as antiviral agent against ToMV and PVY in tomato plants. *Sciences* 8 (01): 100–111.

Ogunyemi, S.O., Zhang, F., Abdallah, Y. et al. (2019). Biosynthesis and characterization of magnesium oxide and manganese dioxide nanoparticles using *Matricaria chamomilla*

L. extract and its inhibitory effect on *Acidovorax oryzae* strain RS-2. *Artificial Cells, Nanomedicine, and Biotechnology* 47 (1): 2230–2239.
Ogunyemi, S.O., Zhang, M., Abdallah, Y. et al. (2020). The bio-synthesis of three metal oxide nanoparticles (ZnO, MnO_2, and MgO) and their antibacterial activity against the bacterial leaf blight pathogen. *Frontiers in Microbiology* 11: 588326.
Omanović-Mikličanina, E. and Maksimović, M. (2016). Nanosensors applications in agriculture and food industry. *Bulletin of the Chemists and Technologists of Bosnia and Herzegovina* 47: 59–70.
Onaga, G. and Wydra, K. (2016). Advances in plant tolerance to abiotic stresses. *Plant Genomics* 10 (9): 229–272.
Park, H.J., Kim, S.H., Kim, H.J. et al. (2006). A new composition of nanosized silica-silver for control of various plant diseases. *Plant Pathology Journal* 22 (3): 295–302.
Paul, S., Thakur, N.S., Chandna, S. et al. (2021). Development of a light activatable lignin nanosphere based spray coating for bioimaging and antimicrobial photodynamic therapy. *Journal of Materials Chemistry B* 9 (6): 1592–1603.
Premnath, B.J., Pratima, B.J., Ravichandiran, R. et al. (2024). Recent advancements and applications of nanosensors in various fields. In: *Metal Oxide–Based Carbon Nanocomposites for Environmental Remediation and Safety*, 207–229. London: CRC Press.
Raliya, R., Saharan, V., Dimkpa, C. et al. (2017). Nanofertilizer for precision and sustainable agriculture: current state and future perspectives. *Journal of Agricultural and Food Chemistry* 66 (26): 6487–6503.
Rambabu, K., Banat, F., Pham, Q.M. et al. (2020). Biological remediation of acid mine drainage: review of past trends and current outlook. *Environmental Science and Ecotechnology* 2: 100024.
Ravindran, A., Prathna, T.C., Verma, V.K. et al. (2012). Bovine serum albumin mediated decrease in silver nanoparticle phytotoxicity: root elongation and seed germination assay. *Toxicological & Environmental Chemistry* 94 (1): 91–98.
Ray, M.K., Mishra, A.K., Mohanta, Y.K. et al. (2023). Nanotechnology as a promising tool against phytopathogens: a futuristic approach to agriculture. *Agriculture* 13 (9): 1856.
Reverchon, E. and Adami, R. (2006). Nanomaterials and supercritical fluids. *The Journal of Supercritical Fluids* 37 (1): 1–22.
Rezaei, S. and Khanmirzaei, A. (2024). Nano-biofertilizers: a new area for enhancing plant nutrition. In: *Plant Holobiome Engineering for Climate-Smart Agriculture*, 617–632. Singapore: Springer Nature Singapore.
Romaniuk, J.A. and Cegelski, L. (2015). Bacterial cell wall composition and the influence of antibiotics by cell-wall and whole-cell NMR. *Philosophical Transactions of the Royal Society B: Biological Sciences* 370 (1679): 20150024.
Saharan, V., Mehrotra, A., Khatik, R. et al. (2013). Synthesis of chitosan-based nanoparticles and their in vitro evaluation against phytopathogenic fungi. *International Journal of Biological Macromolecules* 62: 677–683.
Sankaran, S., Mishra, A., Ehsani, R. et al. (2010). A review of advanced techniques for detecting plant diseases. *Computers and Electronics in Agriculture* 72 (1): 1–13.
Santra, H.K. and Banerjee, D. (2024). Microbial extracellular polymeric substance: function and role against environmental stress. In: *Bacterial Secondary Metabolites*, 83–106. Elsevier.

Sathiyabama, M. (2019). Biopolymeric nanoparticles as a nanocide for crop protection. In: *Nanoscience for Sustainable Agriculture*, 139–152. Springer.

Schiavi, D., Francesconi, S., Taddei, A.R. et al. (2022). Exploring cellulose nanocrystals obtained from olive tree wastes as sustainable crop protection tool against bacterial diseases. *Scientific Reports* 12 (1): 6149.

Shafea, T.M.M. and Mahmoud, K.M. (2021). Green synthesis, characterization and antimicrobial activity of silver nanoparticles using locally growing *Allium sativum* extract. *ZANCO Journal of Pure and Applied Sciences* 33 (1): 35–41.

Shahwan, T., Sirriah, S.A., Nairat, M. et al. (2011). Green synthesis of iron nanoparticles and their application as a Fenton-like catalyst for the degradation of aqueous cationic and anionic dyes. *Chemical Engineering Journal* 172 (1): 258–266.

Shahzad, A., Saeed, H., Iqtedar, M. et al. (2019). Size-controlled production of silver nanoparticles by *Aspergillus fumigatus* BTCB10: likely antibacterial and cytotoxic effects. *Journal of Nanomaterials* 2019 (1): 5168698.

Shi, Y., Cai, M., Zhou, L. et al. (2018, January). The structure and function of cell membranes studied by atomic force microscopy. *Seminars in Cell & Developmental Biology* 73: 31–44.

Shojaei, T.R., Salleh, M.A.M., Tabatabaei, M. et al. (2019). Applications of nanotechnology and carbon nanoparticles in agriculture. In: *Synthesis, Technology and Applications of Carbon Nanomaterials*, 247–277. Elsevier.

Siddiqui, M.H. and Al-Whaibi, M.H. (2014). Role of nano-SiO2 in germination of tomato (*Lycopersicum esculentum* seeds Mill.). *Saudi Journal of Biological Sciences* 21 (1): 13–17.

Singh, A., Sharma, A., Singh, O. et al. (2024). In-depth exploration of nanoparticles for enhanced nutrient use efficiency and abiotic stresses management: present insights and future horizons. *Plant Stress* 14: 100576.

Singh, G., Thakur, N., and Kumar, R. (2024). Nanoparticles in drinking water: assessing health risks and regulatory challenges. *Science of the Total Environment* 949: 174940.

Singh, R., Kuddus, M., Singh, P.K. et al. (2022). Nanotechnology for nano phytopathogens: from detection to the management of plant viruses. *BioMed Research International* 2022 (1): 8688584.

Sofy, A.R., Hmed, A.A., Abd EL-Aleem, M.A. et al. (2020). Mitigating effects of Bean yellow mosaic virus infection in faba bean using new carboxymethyl *chitosan-titania* nano bio composites. *International Journal of Biological Macromolecules* 163: 1261–1275.

Song, M.R., Cui, S.M., Gao, F. et al. (2012). *Dispersible silica* nanoparticles as carrier for enhanced bioactivity of chlorfenapyr. *Journal of Pesticide Science* 37 (3): 258–260.

Souza, L.R.R., Bernardes, L.E., Barbetta, M.F.S. et al. (2019). Iron oxide nanoparticle phytotoxicity to the aquatic plant Lemna minor: effect on reactive oxygen species (ROS) production and chlorophyll a/chlorophyll b ratio. *Environmental Science and Pollution Research* 26: 24121–24131.

Srivastav, A.L., Patel, N., Rani, L. et al. (2024). Sustainable options for fertilizer management in agriculture to prevent water contamination: a review. *Environment, Development and Sustainability* 26 (4): 8303–8327.

Srivastava, A.K., Dev, A., and Karmakar, S. (2018). Nanosensors and nanobiosensors in food and agriculture. *Environmental Chemistry Letters* 16: 161–182.

Strahl, H. and Errington, J. (2017). Bacterial membranes: structure, domains, and function. *Annual Review of Microbiology* 71 (1): 519–538.

Suarez-Fernandez, M., Marhuenda-Egea, F.C., Lopez-Moya, F. et al. (2020). Chitosan induces plant hormones and defenses in tomato root exudates. *Frontiers in Plant Science* 11: 572087.

Subramanian, K.S., Manikandan, A., Thirunavukkarasu, M. et al. (2015). Nano-fertilizers for balanced crop nutrition. In: *Nanotechnologies in Food and Agriculture*, 69–80. Cham: Springer.

Tang, W., Fan, J., He, Y. et al. (2012). The cadmium–mercaptoacetic acid complex contributes to the genotoxicity of mercaptoacetic acid-coated CdSe-core quantum dots. *International Journal of Nanomedicine* 7; 2631–2640.

Tauseef, A., Hisam, F., Hussain, T. et al. (2023). Nanomicrobiology: emerging trends in microbial synthesis of nanomaterials and their applications. *Journal of Cluster Science* 34 (2): 639–664.

Tomah, A.A., Alamer, I.S.A., Li, B. et al. (2020). Mycosynthesis of silver nanoparticles using screened *Trichoderma* isolates and their antifungal activity against *Sclerotinia sclerotiorum*. *Nanomaterials* 10 (10): 1955.

Torabian, S., Zahedi, M., and Khoshgoftar, A.H. (2017). Effects of foliar spray of nano-particles of FeSO4 on the growth and ion content of sunflower under saline condition. *Journal of Plant Nutrition* 40 (5): 615–623.

Ul Haq, I. and Ijaz, S. (2019). Use of metallic nanoparticles and nanoformulations as nanofungicides for sustainable disease management in plants. *Nanobiotechnology in Bioformulations* 289–316.

Verderosa, A.D., Totsika, M., and Fairfull-Smith, K.E. (2019). Bacterial biofilm eradication agents: a current review. *Frontiers in Chemistry* 7: 495483.

Worrall, E.A., Hamid, A., Mody, K.T. et al. (2018). Nanotechnology for plant disease management. *Agronomy* 8 (12): 285.

Xia, Z.K., Ma, Q.H., Li, S.Y. et al. (2016). The antifungal effect of silver nanoparticles on *Trichosporon asahii*. *Journal of Microbiology, Immunology and Infection* 49 (2): 182–188.

Xu, L., Yang, J., Xue, B. et al. (2017). Molecular insights for the biological interactions between polyethylene glycol and cells. *Biomaterials* 147: 1–13.

Xu, Z., Liu, X.W., Ma, Y.S. et al. (2010). Interaction of nano-TiO_2 with lysozyme: insights into the enzyme toxicity of nanosized particles. *Environmental Science and Pollution Research* 17: 798–806.

Yadav, A. and Yadav, K. (2018). Nanoparticle-based plant disease management: tools for sustainable agriculture. In: *Nanobiotechnology Applications in Plant Protection*, 29–61. Springer.

Yang, F.L., Li, X.G., Zhu, F. et al. (2009). Structural characterization of nanoparticles loaded with garlic essential oil and their insecticidal activity against *Tribolium castaneum* (Herbst) (Coleoptera: Tenebrionidae). *Journal of Agricultural and Food Chemistry* 57 (21): 10156–10162.

Yang, W., Peters, J.I., and Williams III, R.O. (2008). Inhaled nanoparticles – a current review. *International Journal of Pharmaceutics* 356 (1–2): 239–247.

Yang, X., Chen, Y., Guo, F. et al. (2020). Metagenomic analysis of the biotoxicity of titanium dioxide nanoparticles to microbial nitrogen transformation in constructed wetlands. *Journal of Hazardous Materials* 384: 121376.

Yılmaz, G.E., Göktürk, I., Ovezova, M. et al. (2023). Antimicrobial nanomaterials: a review. *Hygiene* 3 (3): 269–290.

Zaki, S.A., Ouf, S.A., Abd-Elsalam, K.A. et al. (2022). Trichogenic silver-based nanoparticles for suppression of fungi involved in damping-off of cotton seedlings. *Microorganisms* 10 (2): 344.

Zehra, A., Rai, A., Singh, S.K. et al. (2021). An overview of nanotechnology in plant disease management, food safety, and sustainable agriculture. In: *Food Security and Plant Disease Management*, 193–219. London: Elsevier.

Zhao, X., Meng, Z., Wang, Y. et al. (2017). Pollen magnetofection for genetic modification with magnetic nanoparticles as gene carriers. *Nature Plants* 3 (12): 956–964.

14

Challenges and Risk Assessment of Nanomaterial-based Chemicals Used for Sustainable Agriculture

Ranjani Ravikumar[1], *Jayavardhini M*[1], *Sai Sidharth A*[1], *Vikky Rajulapati*[2], *and Philip Bernstein Saynik*[1]

[1] *Department of Chemical Engineering, Sri Venkateswara College of Engineering, Sriperumbudur, India*
[2] *School of Science and Technology, Örebro University, Örebro, Sweden*

14.1 Introduction

Agriculture, though not fully formed, has been practiced since the dawn of humanity. In ancient times, humans depended entirely on hunting and gathering for food and nourishment. Over time, they learned and mastered the skill of harnessing the natural elements to make tools to aid in hunting and resource gathering easier. The basis of agriculture was formed in the Neolithic Age, which succeeded the Golden Age (the Palaeolithic or Stone Age) and the Silver Age (the Ice Age; Bello 2014). During this period, humans started farming agricultural products and forming civilizations. As agriculture evolved, people started cultivating crops in a more systematic and organized manner. However, with the growth of civilizations, crops are cultivated repeatedly, resulting in the soil losing fertility, thereby affecting the growth of the next batch of crops. To combat this requirement, creating more farmlands by clearing forests was essential. Additionally, techniques such as crop rotation or planting leguminous plants like alfalfa were practiced to induce nitrogen fixation in the soil (to improve its fertility). Farmers started using ground-up bones as it improved productivity (Whitehill 1888). Furthermore, farmers in European countries realized the potential of bird dropping, also known as guano, as a nutrient for plant growth to such an extent that it was called white gold (Schnug, Jacobs, and Stöven 2018).

Agricultural practices underwent significant changes with agricultural industrialization since the early 1900s. The development of equipment such as tractors and other advancements helped reduce the cost of production. Additionally, they developed complex methods for planting, irrigation, and crop rotation, which increased productivity to a certain extent. It was, however, only after agricultural scientists understood that potassium, nitrogen, and phosphorus played a significant and critical role in crop

Nanotechnology-based Sustainable Agriculture, First Edition. Edited by Pardeep Singh et al.

growth. As a result, chemical fertilizers containing nitrogen, potassium, and phosphorus were used. With the fourfold increase in population in the twentieth century, the use of fertilizers becomes necessary.

Justus Liebig formed the basis for current fertilizers in 1840 (Russel and Williams 1977). Chemists like Fritz Haber and Carl Bosch invented the Haber–Bosch process for ammonia production, which accelerated fertilizer production, thus saving humankind from a crisis of famine due to a lack of nitrate resources (Tamaru 1991). Phosphate chemicals entered the fertilizer domain through the work of John Lawes, who converted insoluble phosphate rock into water-soluble calcium phosphate (Roberts 2019). Fertilizer production saw a significant increase after World War II when industries initially set up to produce ammonia for ammunition shifted their focus to fertilizer manufacturing. Even though the use of fertilizers was prevalent, it was only during the Green Revolution that the use of chemical fertilizers to increase plant growth became popular (Malik, Rahman, and Rais 2022). In recent decades, crop productivity has significantly risen due to the increasing use of fertilizers. With these benefits aside, they also have their fair share of downsides (Moulick et al. 2020). In the recent decade, the focus on sustainable agriculture has resulted in innovations to minimize the adverse effects of using conventional chemical fertilizers; nanomaterial-based chemicals have emerged as a promising solution. These chemicals, characterized by their nanoscale size and unique properties, provide an environmentally friendly approach to enhancing soil fertility and productivity. Nanomaterial-based chemicals significantly advance sustainable agriculture by minimizing dependence on traditional chemical fertilizers, which pose environmental and health risks. Considering the brief overview of nanomaterials used in agriculture discussed in earlier chapters, we can categorize the types of nanomaterials used, which include nanopesticides, nanoherbicides, and fertilizers. Additionally, nanosensor-based farming is a newly developed technology.

Due to their extremely small size, the characteristics of nanofertilizers would be completely different from those of macroscopic particles. Combined with conventional methods, this unique characteristic of nanofertilizers has led to many breakthroughs in the agricultural sector. In nanofertilizers, the nutrients are coated using nanomaterials. This is more beneficial than traditional fertilizers in many ways. Some benefits are controlled release of fertilizers, high surface area, minimized nutrient losses, increased fertility of soil, and reduced hazard levels (Nongbet et al. 2022).

Similar to nanofertilizers, nanopesticides incorporate nanoparticles to enhance their effectiveness and make them more beneficial. Pests pose a threat to the entire agricultural sector. Many products (pesticides) have been developed in the past to eradicate pests. However, these pesticides focused primarily on eradicating pests and not their effect on the environment and the health of living beings. Nanopesticides can improve water solubility and availability and prevent the degradation of agrochemicals (Yadav et al. 2020). Even though they are beneficial, their effects have yet to be fully known and must be studied thoroughly to know their downsides (Chaud et al. 2021).

The application of nanoherbicides has demonstrated superior control over the release of biomolecules, resulting in enhanced efficacy. This improvement is attributed to their ability to adhere to and maintain prolonged contact with the leaf surfaces. Nanoherbicides, when synthesized using polymers, exhibit increased compliance due to the biocompatibility and biodegradability of these organic nanomaterials. The polymers facilitate the production of nanoherbicides with improved performance and environmental compatibility (Forini et al. 2022).

Despite all the benefits, nanofertilizers are hard to commercialize due to limitations like higher production costs and safety and environmental concerns. Further studies must be done on this subject to further understand its effects on the environment, plants, and the living beings that consume it. This chapter focuses on different types of nanofertilizers, the risks, challenges, and uncertainties associated with them (Yadav, Yadav, and Abd-Elsalam 2023).

14.2 Nanofertilizers – Types

Nanofertilizers can be classified into several types based on different definitions. It must be noted that there is an inconsistency in classifying nanofertilizers due to the various definitions available that are ambiguous. However, some broad classifications are made based on their actions, nutrients provided, and consistency (Yadav, Yadav, and Abd-Elsalam 2023) as shown in Figure 14.1. The three broad classifications of nanofertilizers are as follows:

- Action-based
- Nutrient-based
- Consistency-based

The application technique and benefits of nanofertilizers are depicted in Figure 14.2.

14.2.1 Action-based Nanofertilizers

As the name suggests, action-based fertilizers are classified based on the actions that they perform. Nanofertilizers can be grouped into four types: controlled-release targeted delivery, plant growth stimulating, and water and nutrient loss controlling. Some types of such fertilizers are listed in the following and their key advantages and disadvantages are given in Table 14.1:

1. Carbon-based
2. Chitosan-based
3. Clay-based
4. Layer Double Hydroxides
5. Nanocapsule-based
6. Nanogel-based
7. Polyurethane-based
8. Starch-based
9. Zeolite-based

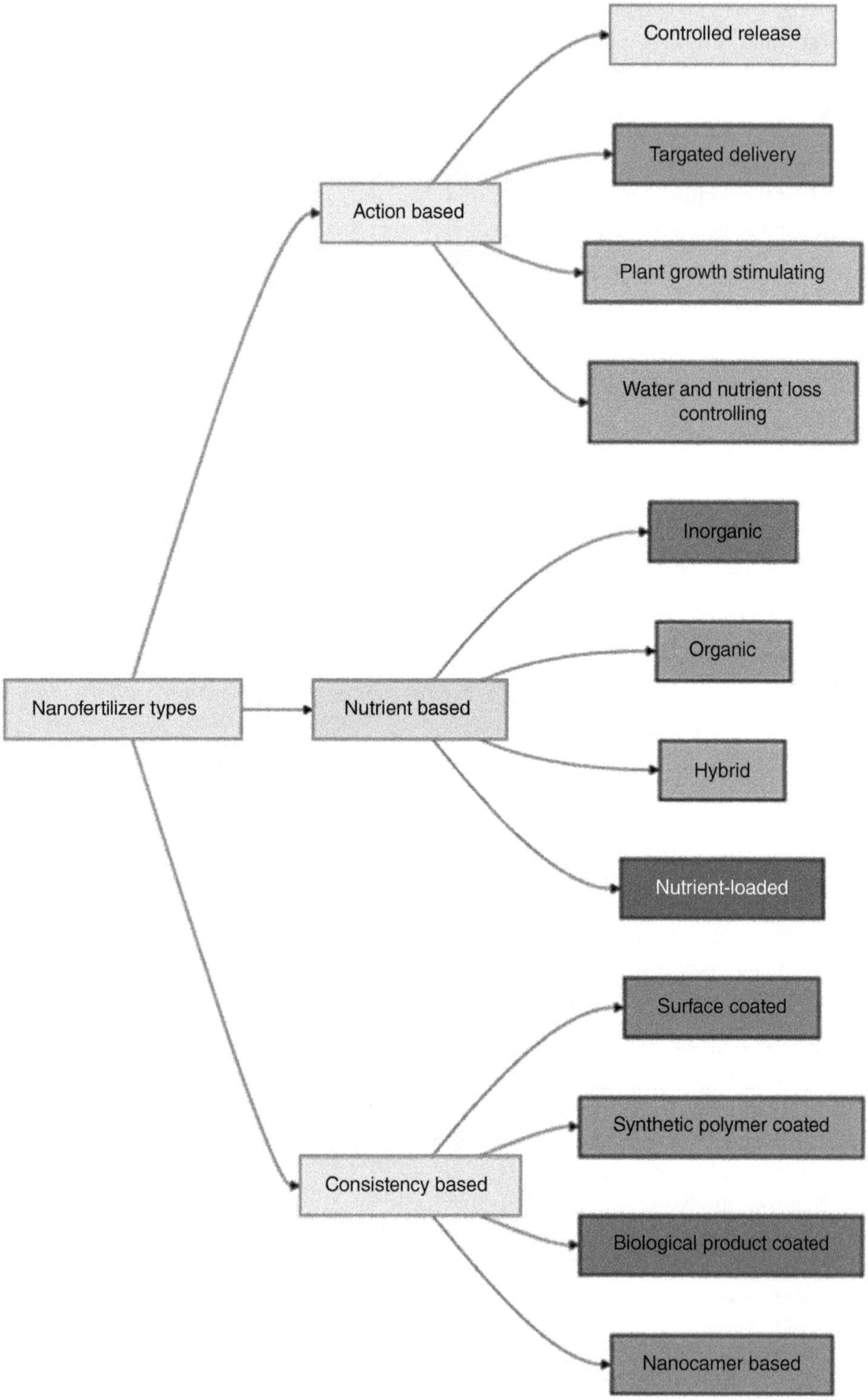

Figure 14.1 Classification of nanofertilizers. *Source:* Yadev et al., 2023 / MDPI / CC BY 4.0.

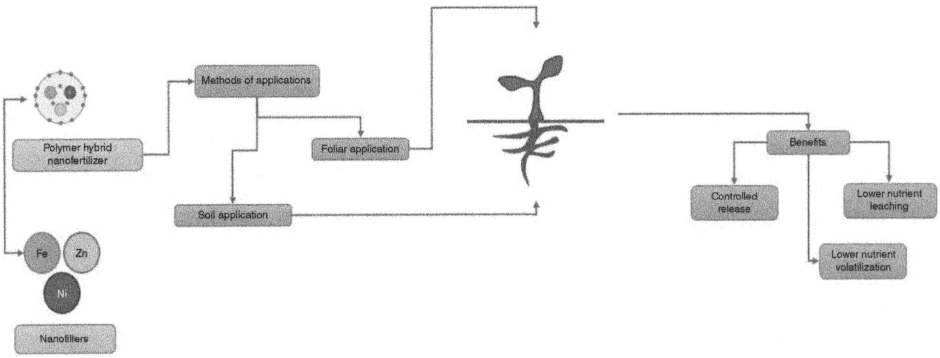

Figure 14.2 Application and benefits of nanofertilizers. *Source:* Referenced from Jakhar et al. (2022) and Yadav, Yadav, and Abd-Elsalam (2023).

Table 14.1 The advantages and disadvantages of the nanofertilizer.

Nanofertilizer type	Key advantages	Key disadvantages
Carbon-based	Enhances plant growth and improves water and nutrient retention and drought resistance	Slow and complex to synthesize
Chitosan-based	Biodegradable, adaptable, and protective against environmental stress	Low efficiency due to weak structure and permeability issues
Clay-based	Large surface area and good for nutrient regulation	Can interfere with plant processes like leaf growth and water absorption
Nanocapsule-based	Efficient in nutrient delivery and prevents leaching	Complicated to produce and material limitations
Nanogel-based	Soluble, biodegradable, and enhances water retention	Challenges with degradation and potential toxicity
Polyurethane-based	Controls nutrient release and improves water retention and soil health	Unstable over time and susceptible to rapid breakdown
Starch-based	Renewable, delivers nutrients efficiently, and eco-friendly	Expensive, unstable, and time-consuming to produce
Zeolite-based	Tailored nutrient delivery and cost-effective	Requires specific formulations and less effective for anionic nutrients

14.2.2 Nutrient-based Nanofertilizers

Nutrient-based fertilizers are nanofertilizers that are classified based on how much it enhances the nutrient availability in plants. Some nutrient-based nanofertilizers are as follows:

1. Inorganic Nanofertilizers
    - Macronutrient Nanofertilizers

  a. Nitrogen-based
  b. Phosphorous-based
  c. Potassium-based
  d. Calcium-based
  e. Magnesium-based
  f. Sulfur-based

- Micronutrient Nanofertilizers
  a. Boron-based
  b. Copper-based
  c. Iron-based
  d. Nickel-based
  e. Titanium-based
  f. Zinc-based

2. Organic Nanofertilizers
3. Hybrid Nanofertilizers
4. Nutrient-loaded Nanofertilizers

14.2.3 Consistency-based Nanofertilizers

Consistency-based nanofertilizers, including surface-coated and nanocarrier-based technologies, have several benefits like enhanced nutrient utilization, reduction of nutrient loss, and less impact on the environment. Some types are as follows:

1. Surface-coated Nanofertilizers
2. Synthetic Polymer-coated
3. Biological Product-coated
    - Organic compound-coated
    - Microbe-coated (Nanobiofertilizers)
4. Nanocarrier-based Nanofertilizers

14.3 Risk Assessment

The utilization of nanomaterials as fertilizers presents substantial challenges, primarily associated with risk assessment. A critical issue is that most research on nanomaterials is conducted at the laboratory scale, thereby neglecting the myriad potential risks linked to their deployment in open-field environments. The diverse stress factors present in open-field conditions, in contrast to the controlled and optimal settings of experimental procedures, pose significant concerns regarding the scalability and practical applicability of nanomaterials in agricultural practices (Iqbal, Umar, and Mahmooduzzafar 2019).

Nanofertilizers have been observed to potentially induce adverse effects, including inhibited plant growth and chromosomal abnormalities (Dimkpa et al. 2013).

The nonfertilizer application may hinder water transport and photosynthetic rates in plants, accompanied by a reduction in growth hormone concentrations (Iqbal, Umar, and Mahmooduzzafar 2019).

Several studies have documented the various effects of different dosages of nanofertilizer on plants. The studies emphasize that, while not universal, some applications result in reduced root growth and decreased chlorophyll molecule formation at specific dosages. These issues present significant challenges for upscaling nanofertilizers to agricultural operations (Seleiman et al. 2020).

Research on the toxic consequences of using nanofertilizers is still ongoing. Due to their increased penetration capabilities, nanofertilizers could pose significant risks if they produce undesired responses within biological systems. The study conducted on silver nanoparticles (AgNPs) when combined with urea and evaluated for their potential phytotoxic effects on plants demonstrated that AgNPs can negatively impact seed germination and root growth and highlighted a crucial need for comprehensive risk assessment to be conducted before their widespread use in agriculture (Jaskulski et al. 2022).

Given their unique properties, previously established risk assessment strategies need to be improved for the application of nanofertilizers. Consequently, developing new strategies altered for this range of fertilizers is necessary. Such assessments must be conducted on a balanced basis, considering the potential both the merits and demerits to ensure a comprehensive understanding of their impacts (Kah 2015).

14.4 Uncertainties

The agricultural sector is making significant strides in integrating nanomaterials in numerous ways. Some of the applications include monitoring of nutrition content in the soil, plant growth using nanosensors, focused weed control using nanoherbicides, pest management using nanopesticides, and improved nutrient absorption efficiency using nanofertilizers. Shahid et al. (2024) have explained the role of nanomaterials in a sustainable agriculture approach. The delivery of the target pest chemical has a low environmental impact. However, quite a few uncertainties and concerns are associated with their functions and use. These uncertainties can be in significant areas such as environmental impact, human health risks, agronomic performances, regulatory and policy considerations, economic considerations, and technical challenges (Singh et al. 2022).

Nanomaterial-based chemicals used in the agricultural sector vary widely based on their composition, surface coating, size, and other factors. Certain nanomaterials, especially those incorporating metals and special coatings, may not be easily water-soluble (Peralta-Videa et al. 2009). Environmental factors such as temperature, pH, and the presence of other chemicals also affect their solubility. For example, Fe_3O_4 nanofertilizers are more readily soluble in acidic conditions compared to neutral pH conditions. The resulting condition of prolonged persistence in neutral pH conditions leads to uncertainties about their long-term impact. The transport of nanomaterials through environmental compartments such as soil, water, and air has yet to be studied extensively. The changes in the properties of nanoparticles with times at the point of application are yet to be studied. Metal oxide nanoparticles may have unintended consequences, including decreased germination rates, reduced expression of photosynthesis-related genes, lower chlorophyll content, and diminished soil fertility (Pelegrino et al. 2020; Rajput et al. 2018; Yang et al. 2020).

Minerals in the form of nanoparticles get absorbed from the soil and travel to various parts of the plant. These nanoparticles, which the plants absorb, can bioaccumulate even to toxic concentrations and even more dangerously enter the animal and human food chain (Yadav et al. 2021). This could pose a significant risk to ecosystems and human health in the long term (Rabalao et al. 2022). The long-term health effects of chronic exposure to nanoparticles are unknown. There are growing concerns about the potential health impacts of nanoparticles, particularly their effects on respiratory, cardiovascular, and neurological systems. Ecotoxicity is an essential consideration for the impact of nanoparticles on beneficial soil organisms, including nontarget soil microbes, water species, and insects. The measurement of the ecotoxicity levels of various nanoparticles can be challenging (Mgadi et al. 2024).

The synthesis and production of nanoparticles and their characterizations would be costly, encompassing expenses for advanced manufacturing and detailed studies of their physicochemical and biological properties (Singh et al. 2022). Complying with existing and forthcoming regulations for the safe use of nanomaterials may incur additional costs and challenges, such as the need for extensive safety and environmental impact testing.

Most of the research on nanomaterials, especially nanofertilizers, has been on their short-term impacts. Long-term studies are necessary to fully understand their effect on soil health, crop productivity, and the environment. Further research is needed to address biosafety concerns and ensure the sustainability of agricultural practices.

14.5 Risk Management

Risk management in nanofertilizers shouldn't be overlooked, and countries should establish regulatory frameworks to monitor and control their use (Avila-Quezada et al. 2022). The European Food Safety Authority has set limits on the dosage of silicon dioxide nanomaterial content in consumables. While the Registration, Evaluation, Authorization and Restriction of Chemicals (REACH) legislation by the European Commission is well-suited for the risk management of nanomaterials, it lacks specific requirements for assessing the unique complexities of nanomaterials' sizes, focusing instead only on their chemical compositions. However, the latest provisions took effect on January 1, 2020, and more stringent measures have been implemented to ensure that all parties utilizing nanomaterials comply with the newly developed protocols. These protocols emphasize the impact on the environment and the inherent properties of nanomaterials. Additionally, in a move to improve safety, the Organization for Economic Cooperation and Development (OECD) has suggested further evaluation of the regulatory hazards associated with nanomaterials (Avila-Quezada et al. 2022; Malakar et al. 2021; Rodríguez 2018).

A study conducted on pea plant seedlings emphasized the significance of evaluating nanocarriers in agricultural applications before commercial use. The application of nano-chitosan-based NPK to pea plants adversely affected their growth and had potential genotoxic effects (Khalifa and Hasaneen 2018).

A thorough risk–benefit analysis and a complete life cycle assessment (LCA) of nanofertilizers are necessary despite the numerous benefits associated with their usage. Through these evaluations, farmers may be able to combat the lower efficacy of traditional fertilizers and create more effective fertilizers. The efficiency drop in nitrogen-based fertilizers can be related to a low nitrogen recovery efficiency (REN) in agriculture, which significantly contributes to environmental issues. REN is defined as the quantity of applied nitrogen absorbed by the crop. Studies show that most cropping systems, such as the corn-based systems in the Midwestern United States, have REN <50%. The lower yield and higher consumption of traditional fertilizers can be reduced through the application of fertilizers, leading to more efficient agricultural practices (Avila-Quezada et al. 2022; Singh, Singh, and Lichtfouse 2021).

Under the Classification, Labelling, and Packaging (CLP) Regulations, there is a recognized need to reduce the potential environmental hazards posed by nanomaterials. New requirements have been put in place to list out the nano-ingredients used in products, along with warnings to customers regarding their presence and potential hazards. Additionally, the Scientific Committee on Emerging and Newly Identified Health Risks (SCENIHR) emphasizes the importance of studying the effects of nanomaterials on human health to achieve a comprehensive assessment of their impacts. This thorough approach aims to increase consumer awareness and ensure the safe application of nanomaterials in various sectors (Baran 2016).

Legislation on the application of nanofertilizers is crucial, as it ensures that producers follow the regulations and maintain specific standards. Major hazards related to nanofertilizers can be avoided if these fertilizers are applied only after confirming their nontoxicity and safety for commercial usage (Singh, Singh, and Lichtfouse 2021).

14.6 Regulations and Safety Measures

Nano-based technology has brought in improvements in many sectors, including agriculture. However, regulation is essential to mitigate possible risks through new products introduced in the market. Through regulations, sustainable agricultural practices can be promoted while promoting the safety of the environment and public health. Each country has created dedicated organizations to oversee the usage of nanomaterial-based chemicals in the agricultural sector. These organizations play a pivotal role in developing the regulations, assessing the risk factors such as toxicity level and impact on human health, and supporting compliance with safety and environmental standards before approval of any nanotechnology-based agricultural products. The end users/consumers are informed about the product details, such as its regulatory compliance, through a specification label controlled by the regulatory body (Adholaya et al. 2017; Kumari et al. 2023).

Listed below are some prominent regulatory bodies established by countries to ensure the proper use of nanotechnology in developing sustainable agriculture.

14.6.1 United States

1. The Toxic Substances Control Act (TSCA, 2016) authorizes the Environmental Protection Agency (EPA) to monitor the usage of nanoscale materials in agricultural products. The EPA mandates documenting and maintaining existing data on

exposure, health, and safety related to nanoscale chemical substances (United States Environment Protection Agency 2016).

The industries that manufacture, import, or process such substances are obligated to inform the EPA of the following: precise chemical composition, manufacturing technique, information related to exposure, health, safety data, and environmental effects (Adholaya et al. 2017; Kumari et al. 2023).

2. The Federal Insecticide, Fungicide, and Rodenticide Act (FIFRA, 1947) governs the registration and usage of pesticides in the United States. The safe and adequate usage of nanopesticides is assessed by the EPA under this act (Kumari et al. 2023).

14.6.2 United Kingdom

The United Kingdom, like many other countries, has numerous regulatory bodies and voluntary organizations dedicated to overseeing the use of nanotechnology in sustainable agriculture. The Food Standard Agency (FSA); the UK Expert Committee on Pesticides (ECP); the Department for Environment, Food and Rural Affairs (DEFRA); and the Health and Safety Executive (HSE) are some of them. These entities establish regulations and guidelines to govern the use of nanotechnology in sustainable agriculture. They are responsible for ensuring that all organizations and individuals adhere to these rules.

1. Nanotechnology Safety Guidance by HSE (2011): This guidance assists in handling nano-based products used in industries and agriculture in compliance with the Control of Substances Hazardous to Health (COSHH, 2022) Regulations. They provide detailed norms highlighting the various procedures and challenges in the safe operation of the nanoscale substances.
2. Although the application of nanotechnology in the agricultural and food sectors is still emerging, it is poised for significant expansion shortly. However, the possible risks related to the environment and human health are to be considered. The FSA is tasked with the safety and health of edible goods and coving products developed with nanotechnology. So, in 2004, the FSA released a report containing the information and the protocols to be followed when using nanomaterials in agricultural products.
3. DEFRA is a regulatory body that ensures the safe handling and disposal of nanomaterial-based agricultural products by understanding their fundamental properties, toxicokinetics, and the fate of nanomaterials in the environment (DEFRA 2007; Kumari et al. 2023).
4. The task of evaluating and advising on the use of pesticides, including nanopesticides, is carried out by ECP. They independently provide scientific advice on their usage to the UK government.

14.6.3 Canada

Similar to many other countries, no specific regulations solely exist for nano-based products, but some regulations within the existing legislative and regulatory framework apply to nanomaterials. Two key regulatory bodies of Canada, namely, the Canadian Food Inspection Agency (CFIA) and the Public Health Agency of Canada (PHAC), share responsibility for regulating products for use in the food and agriculture sector, which include nanotechnology-based materials. These agencies set specific objectives for each sector to ensure the safety and the appropriate dosage of the nanomaterial.

1. Canadian Environmental Protection Act (CEPA, 1999) provides the legislative basis for programs whose activities relate to assessing and managing risks from chemicals, including nanotechnology-based products, used in the environment. This establishes a boundary for the users to ensure the safe usage of the materials.

   The New Substances Notification Regulations (NSNR) under the CEPA further allows for controlling chemicals, including nanomaterials, that exceed allowable thresholds and may pose environmental and public health risks (Justice Laws Website, Government of Canada, 2018).
2. Canada Agricultural Products Act (CAPA, 1985): The primary goal of this act is to provide consumers with high-quality agricultural products. It provides the regulatory framework for the industry to ensure that materials do not compromise the consumer in terms of safety and quality (Adholaya et al. 2017; Kumari et al. 2023).

14.6.4 Europe

The European Parliament and the Council of the European Union, in consultation with key bodies and agencies such as the European Food Safety Authority (EFSA), European Commission (EC), and European Chemical Agency (ECHA), set regulations related to the use of the material, including nanomaterials. A few of the key regulations are discussed below.

1. Regulation (EC) No. 396/2005: The maximum residual levels (MRLs) of pesticides in or on food are governed by this regulation. It is also applicable to pesticides containing nanomaterials (Kumari et al. 2023).
2. Regulation (EC) No 2019/1009: This regulation, also known as the regulation on fertilizer products, covers the production of nonstandardized chemically produced inorganic fertilizers, including nanofertilizers. The regulation guarantees soil protection and reduction of environmental and safety risks by limiting the impurities in the fertilizing products (Kumari et al. 2023).
3. Regulation (EC) No 1907/2006: Also known as the "REACH Regulation," this regulation provides the framework for the safe usage, assessment, and evaluation of hazardous chemicals. This is to ensure the safety of the end user and the environment (Adholaya et al. 2017).
4. Regulation (EC) No. 1107/2009: This regulation seeks to set rules for the authorization for the use of active components used in the manufacture of plant protection products, which include nanopesticides. It also lists the labelling requirements for providing essential information to the end user (Adholaya et al. 2017; Kumari et al. 2023).
5. Article 4(f), Treaty of the Functioning of the European Union (TFEU): This article focuses on customer protection and provides for sharing competencies among European Union (EU) member states. It allows EU institutes to formulate regulations to ensure the public safety and quality of nanomaterials (Baranowska 2018).

14.6.5 Australia

Multiple regulatory bodies play a role in managing the regulations related to nanotechnology-based agricultural products. Australian Pesticides and Veterinary Medicines Authority (APVMA), Food Standards Australia New Zealand (FSANZ), and Work Health and Safety (WHS) are a few among them.

1. Australian Pesticides and Veterinary Medicines Authority (APVMA, 2014): This organization is responsible for establishing regulations to ensure the safe use of nanotechnology-based pesticides and veterinary medicine. It also serves as the granting authority for permits related to the manufacture, import, and sale of agricultural and veterinary chemicals (Annual Report 2022-23, APVMA, Australian Government 2023; Kumari et al. 2023).

14.6.6 Switzerland

The Federal Office of Public Health (FOPH), Federal Office for Agriculture (FOAG), and Federal Office for the Environment (FOEN) together ensure that the use of nanomaterials in the agricultural sector is safe, effective, and environmentally sustainable.

1. FOPH of Switzerland: This regulatory body of Switzerland maintains a record of the nanomaterials used in the agricultural sector. It has designed a methodology for assessing synthetic nanomaterials used in sectors that have an impact on public health. The central nanotechnology information hub, InfoNano, provides information on the risks and opportunities associated with nanoscale materials. Information regarding the composition, shape, size, surface area, aggregation state, coating, and functionalization of the nanoproduct must be provided for approval by the FOPH of Switzerland (Adholaya et al. 2017).
2. FOAG: One of their primary objectives is to assess the safety of nanomaterials used in agricultural products to ensure they do not pose risks to human health or the environment. In addition, they develop and enforce regulations for using nanomaterials in agriculture, ensuring compliance with international and national standards.

14.6.7 Russia

Russian Corporation of Nanotechnologies (RUSNANO Group): Although they primarily focus on developing nanomaterials and nanotechnology through R&D, they actively help establish standards for nanomaterials. The agency is responsible for ensuring that the safe reference provision provided by the World Health Organization (WHO) is adhered to (Adholaya et al. 2017).

14.6.8 China

The Ministry of Agriculture and Rural Affairs (MARA), Ministry of Environmental Protection (MEP), Standardization Administration of China (SAC), State Food and Drug Administration (SFDA), the National Centre for Nanoscience and Technology (NCNST), and the Commission on Nanotechnology Standardization are some agencies and regulatory bodies of China that are responsible for the safe usage of nanotechnology-based agricultural products. Specific to nanomaterials and nanotechnology, National Technical Committee 279 on Nanotechnology (SAC/TC279) is set up to review and approve regulations. These organizations also take responsibility for building up the regulatory framework to assure the consumers of good quality and safe products (Adholaya et al. 2017; Kumari et al. 2023).

Regulation on Administration of Agricultural Genetically Modified Organisms Safety, enforced in 2001, specifies the safety criteria and approval processes for the application of genetically modified organisms in agriculture. It also includes the provision of safety assessments for nanomaterial-based agricultural products.

14.6.9 South Korea

In South Korea, regulations regarding nanotechnology are quite limited and still in the process of development.

1. National Nano-safety Strategic Plan (2012/2016): This regulation directs the safety of nanomaterial-based agricultural products (Adholaya et al. 2017).

14.6.10 India

In India, the field of nanotechnology is one of the most promising domains of research. Nanotechnology in the agricultural sector is growing and has a greater impact on sustainability. The Department of Biotechnology (DBT), the Ministry of Environment, Forest and Climate Change (MoEFCC), the Food Safety and Standards Authority of India (FSSAI), and the Indian Council of Agricultural Research (ICAR) are the regulatory bodies in India that construct regulations and ensure the safety.

1. The Environment (Protection) Act, 1986: Trade and commerce are crucial parts of any manufacturing company. This act provides a paved path for the manufacturers to produce and trade nanomaterial-based agricultural products under the regulated laws by the MoEFCC.
2. The Hazardous and Other Wastes (Management and Transboundary Movement) Rules, 2016: Disposal of the nanomaterial-based waste is always a challenge to users and manufacturers. This rule clearly provides guidelines for the safe disposal of hazardous waste. This also ensures that the production and storage are done by following proper standard operating procedure (SOP) and is safe.
3. The Insecticides Act, 1968: The sales and distribution of the nanomaterial-based insecticides are done by following certain protocols. This law provides guidelines for the proper packaging and storing techniques before sales or distribution. This also ensures that all nanomaterial-based insecticides are registered before sale or coming into the market.
4. MoEFCC Notification on Manufacture, Storage, and Import of Hazardous Chemical Rules (1989): This law ensures that the fine quality of the product reaches the customer safely. This also imparts some regulations on the manufacturers so that environmental safety is not compromised (Adholaya et al. 2017; Kumari et al. 2023).

5. Department of Science and Technology (DST) Nano mission: The Government of India launched a mission on Nanoscience in order to ensure the safe and best practices in the use of nanomaterials and their storage (DST, Government of India).
6. Ministry of Agriculture and Farmers Welfare, Government of India: This organization was initiated for the formulation of regulations pertaining to agriculture, which includes the safe usage and research related to nanopesticides and nanofertilizers (Annual Report 2023–24, Ministry of Agriculture & Farmers Welfare, Government of India 2024).

14.7 Ethical and Safety Concerns of Nanofertilizers and Nanopesticides

Although nanofertilizers and nanopesticides are deemed highly beneficial and even instrumental in changing the course of agriculture in the future, their effects on the environment and human and animal health are still uncertain. This uncertainty makes the general public wary of all nanofertilizers and nanopesticides. Despite having less information about nanofertilizers and nanopesticides, people in the United States believe they have much potential. The Europeans, however, do not share the same views (Coles and Frewer 2013; Mwaanga 2018; Siegrist et al. 2008). The risks associated with nanofertilizers and nanopesticides must be studied first to understand the risks they impose before making them commercial. "Nanotoxicology" is thus developed to assess the nanoproducts and determine their toxicity. It also helps in formulating products that are safe to use. Also, EU member states have started implementing initiatives to make people more aware of the developments in nanotechnology. In the United Kingdom, a project named Deepening Ethical Engagement and Participation with Emerging Nanotechnologies (DEEPEN) was created to create acceptance among the people. Similarly, the Nanopodium program was created in the Netherlands, and the Nanosoc program was created in Belgium as a forum for discussing nanotechnology. The Nanologue program was created in Germany to accentuate the benefits of nanotechnology and the ethical, safety, and legal issues associated with it (Baran 2016). The tiny size of the nanofertilizers and pesticides, which is their characteristic feature, is considered a safety risk as it is assumed that they can penetrate tissues and cells as most nanoparticles have sizes comparable to biological molecules. Also, their effect on the environment is not thoroughly understood yet. If nanofertilizers and pesticides are incorporated into agriculture, they affect an entrance into the food chain, affecting all the organisms associated with it. So, these aspects must be considered and studied to understand and be sure of their effects. Other ethical, social, and political concerns must also be considered (Nanofertilizers, Bio-fit.eu.). A leading risk with using nano-based products is the lack of authority over it. Also, there are no technical tools for monitoring the levels of nanoparticles in the environment (Baran 2016). Numerous government bodies have formed rules and regulations to reduce the risks associated with nanomaterials (Jeevanandam et al. 2018). "Nano mission" by the Government of India provides a structural framework regarding Environment, Health and Safety (EH&S). The guidelines provide comprehensive protocols for hazard identification, control strategies when exposed to nanomaterials, safety measures, and best practices for handling (Haydar, Ghosh, and Roy 2024). The fear and uncertainty instilled in the people must be eliminated first by providing awareness about nanofertilizers and pesticides and explaining how beneficial they are compared to conventional fertilizers and pesticides.

14.8 Conclusion

Nanotechnology is the promising frontier for addressing the most challenging problems faced by the agricultural sector. The integration of nanotechnology into sustainable agriculture holds the promise to overhaul the methods of food production and environmental conservation. The continuous research in the field of nanotechnology has provided real-time monitoring and targeted inventions and reduced the overuse of chemicals.

Nanofertilizers and nanopesticides are instrumental in developing the agricultural sector. This innovation has various advantages over the existing fertilizers and pesticides by enhancing productivity and stability. Despite various advantages, integrating nanotechnology into agricultural methods has risks. The risk assessments are conducted to ascertain the challenges and repercussions on agriculture and human health. Continuous monitoring is done to understand the involvement of the nanoparticles with the biological system and the ecosystem and the uncertainties in the long-term use of the nanomaterial. These uncertainties are addressed in the chapters of the book. Managing risk plays a vital role in developing agriculture towards sustainability. A regulatory framework is set by the government and some volunteer organizations to keep pace with the advancement in nanotechnology. The regulations are followed to guarantee the harmless usage of the nanofertilizers. Public perception and ethical debates are crucial in determining the future of nanotechnology in agricultural practices.

Despite various challenges in the field of nanotechnology, the future of nanotechnology-based sustainable agriculture is bright. With the development of nanotechnology, it's possible for us to develop a stronger and more resilient framework and sustainable agricultural system tailored to address the needs of a growing population. It holds a greater responsibility to transform agriculture into a more sustainable and productive endeavor. By using these developing strategies in the field of nanotechnology, the new era of agricultural innovation that supports both the environment and human health is improvised.

References

Adholaya, A., Das, R.K., Dubey, M.K. et al. (2017). Regulation of Nanoproducts in Agriculture. *The Energy and Resources Institute (TERI)*. https://www.teriin.org/sites/default/files/2018-03/zero-draft-policy.pdf

Annual Report 2022–23. (2023). *Australian Pesticides and Veterinary Medicines Authority, Australian Government*. https://www.apvma.gov.au/sites/default/files/2023-10/APVMA Annual Report 2022-23.pdf

Annual Report 2023–24. (2024). *Department of Agriculture & Farmers Welfare, Ministry of Agriculture & Farmers Welfare, Government of India*. https://agriwelfare.gov.in/Documents/AR_English_2023_24.pdf

Avila-Quezada, G.D., Ingle, A.P., Golińska, P. et al. (2022). Strategic applications of nano-fertilizers for sustainable agriculture: benefits and bottlenecks. *Nanotechnology Reviews* 11 (1): 2123–2140. doi: 10.1515/ntrev-2022-0126.

Baran, A. (2016). Nanotechnology: legal and ethical issues. *Ekonomia i Zarzadzanie* 8 (1): 47–54. doi: 10.1515/emj-2016-0005.

Baranowska, N. (2018). Public authority liability and the regulation of nanotechnology: a European perspective. *Canadian Journal of Law and Technology* 16. https://ojs.library.dal.ca/CJLT/article/view/9004/7780

Bello, R.S. (2014). *Timeline of Agricultural Mechanization*. CreateSpace Independent Publishing Platform.

Chaud, M., Souto, E.B., Zielinska, A. et al. (2021). Nanopesticides in agriculture: benefits and challenge in agricultural productivity, toxicological risks to human health and environment. *Toxics* 9 (6): 131. doi: 10.3390/toxics9060131.

Coles, D. and Frewer, L.J. (2013). Nanotechnology applied to European food production – a review of ethical and regulatory issues. *Trends in Food Science & Technology* 34 (1): 32–43. doi: 10.1016/j.tifs.2013.08.006.

DEFRA (2007). Characterizing the potential risks posed by engineered nanoparticles. Department of Environment, Food and Rural Affairs (DEFRA). https://assets.publishing.service.gov.uk/media/5a78c3ab40f0b63247699fd3/pb12901-nanoparticles-riskreport-071218.pdf

Dimkpa, C.O., McLean, J.E., Martineau, N. et al. (2013). Silver nanoparticles disrupt wheat (*Triticum aestivum* L.) growth in a sand matrix. *Environmental Science & Technology* 47 (2): 1082–1090. doi: 10.1021/es302973y.

Forini, M.M.L., Pontes, M.S., Antunes, D.R. et al. (2022). Nano-enabled weed management in agriculture: from strategic design to enhanced herbicidal activity. *Plant Nano Biology* 1 (100008): 100008. doi: 10.1016/j.plana.2022.100008.

Haydar, M.S., Ghosh, D., and Roy, S. (2024). Slow and controlled release nanofertilizers as an efficient tool for sustainable agriculture: recent understanding and concerns. *Plant Nano Biology* 7: 100058. doi: 10.1016/j.plana.2024.100058.

Iqbal, M., Umar, S., and Mahmooduzzafar, F. (2019). Nano-fertilization to enhance nutrient use efficiency and productivity of crop plants. In: *Nanomaterials and Plant Potential*, 473–505. Springer International Publishing.

Jaskulski, D., Jaskulska, I., Majewska, J. et al. (2022). Silver nanoparticles (AgNPs) in urea solution in laboratory tests and field experiments with crops and vegetables. *Materials* 15 (3): 870. doi: 10.3390/ma15030870.

Jeevanandam, J., Barhoum, A., Chan, Y.S. et al. (2018). Review on nanoparticles and nanostructured materials: history, sources, toxicity and regulations. *Beilstein Journal of Nanotechnology* 9: 1050–1074. doi: 10.3762/bjnano.9.98.

Kah, M. (2015). Nanopesticides and nanofertilizers: emerging contaminants or opportunities for risk mitigation? *Frontiers in Chemistry*, 3. doi: 10.3389/fchem.2015.00064.

Khalifa, N.S. and Hasaneen, M.N. (2018). The effect of chitosan–PMAA–NPK nanofertilizer on Pisum sativum plants. *3 Biotech* 8 (4). doi: 10.1007/s13205-018-1221-3.

Kumari, R., Suman, K., Karmakar, S. et al. (2023). Regulation and safety measures for nanotechnology-based agri-products. *Frontiers in Genome Editing* 5. doi: 10.3389/fgeed.2023.1200987.

Malakar, A., Kanel, S.R., Ray, C. et al. (2021). Nanomaterials in the environment, human exposure pathway, and health effects: a review. *The Science of the Total Environment* 759 (143470): 143470. doi: 10.1016/j.scitotenv.2020.143470.

Malik, J.A., Rahman, M.M., and Rais, N. (2022). Microbial biotechnology: an introduction. In: *Microbes and Microbial Biotechnology for Green Remediation*, 3–11. Elsevier.

Mgadi, K., Ndaba, B., Roopnarain, A. et al. (2024). Nanoparticle applications in agriculture: overview and response of plant-associated microorganisms. *Frontiers in Microbiology* 15. doi: 10.3389/fmicb.2024.1354440.

Moulick, R.G., Das, S., Debnath, N. et al. (2020). Potential use of nanotechnology in sustainable and 'smart' agriculture: advancements made in the last decade. *Plant Biotechnology Reports* 14 (5): 505–513. doi: 10.1007/s11816-020-00636-3.

Mwaanga, P. (2018). Risks, uncertainties, and ethics of nanotechnology in agriculture. In: *New Visions in Plant Science*. InTech.

National Programme on Nano Science and Technology (earlier Nano Mission). (n.d.). gov.in. Retrieved October 8, 2024, from https://dst.gov.in/scientific-programmes/mission-nano-science-and-technology-nano-mission

New substances notification regulations (organisms). (2018, February 2). *Justice.Gc.Ca*. https://laws-lois.justice.gc.ca/eng/regulations/SOR-2005-248

Nongbet, A., Mishra, A.K., Mohanta, Y.K. et al. (2022). Nanofertilizers: a smart and sustainable attribute to modern agriculture. *Plants* 11 (19). Available from: doi.org/10.3390/plants 1119258.

Pelegrino, M.T., Kohatsu, M.Y., Seabra, A.B. et al. (2020). Effects of copper oxide nanoparticles on growth of lettuce (*Lactuca sativa* L.) seedlings and possible implications of nitric oxide in their antioxidative defense. *Environmental Monitoring and Assessment* 192 (4). doi: 10.1007/s10661-020-8188-3.

Peralta-Videa, J.R., Lopez, M.L., Narayan, M. et al. (2009). The biochemistry of environmental heavy metal uptake by plants: implications for the food chain. *The International Journal of Biochemistry & Cell Biology* 41 (8–9): 1665–1677. doi: 10.1016/j.biocel.2009.03.005.

Rabalao, T.M., Ndaba, B., Roopnarain, A. et al. (2022). Towards a circular economy: the influence of extraction methods on phytosynthesis of metallic nanoparticles and their impact on crop growth and protection. *JSFA Reports* 2 (5): 208–221. doi: 10.1002/jsf2.44.

Rajput, V., Minkina, T., Fedorenko, A. et al. (2018). Toxicity of copper oxide nanoparticles on spring barley (*Hordeum sativum distichum*). *The Science of the Total Environment* 645: 1103–1113. doi: 10.1016/j.scitotenv.2018.07.211.

Roberts, T. (2019). Phosphorus: past history and contributions to the global food supply. *Better Crops with Plant Food* 103 (1): 6–8. doi: 10.24047/bc10316.

Rodríguez, H. (2018). Nanotechnology and risk governance in the European Union: the constitution of safety in highly promoted and contested innovation areas. *Nanoethics* 12 (1): 5–26. doi: 10.1007/s11569-017-0296-3.

Russel, D.A. and Williams, G.G. (1977). History of chemical fertilizer development. *Soil Science Society of America Journal* 41 (2): 260–265. doi: 10.2136/sssaj1977.03615995004100020020x.

Schnug, E., Jacobs, F., and Stöven, K. (2018). Guano: the white gold of the seabirds. In: *Seabirds*. InTech.

Seleiman, M.F., Almutairi, K.F., Alotaibi, M. et al. (2020). Nano-fertilization as an emerging fertilization technique: why can modern agriculture benefit from its use? *Plants* 10 (1): 2. doi: 10.3390/plants10010002.

Shahid, S., Khan, M.S., Kumar, A. et al. (2024). Role of nanomaterials in sustainable agriculture. In: *Sustainable Materials and Technology*, 227–248. Springer Nature Singapore.

Siegrist, M., Stampfli, N., Kastenholz, H. et al. (2008). Perceived risks and perceived benefits of different nanotechnology foods and nanotechnology food packaging. *Appetite* 51 (2): 283–290. doi: 10.1016/j.appet.2008.02.020.

Singh, P.M., Tiwari, A., Maity, D. et al. (2022). Recent progress of nanomaterials in sustainable agricultural applications. *Journal of Materials Science* 57 (24): 10836–10862. doi: 10.1007/s10853-022-07259-9.

Singh, V.K., Singh, R., and Lichtfouse, E. (eds.) (2021). *Sustainable Agriculture Reviews 50: Emerging Contaminants in Agriculture*, 1e. Springer Nature.

Tamaru, K. (1991). The history of the development of ammonia synthesis. In: *Fundamental and Applied Catalysis*, 1–18. US: Springer.

Whitehill, A.R. (1888). *The History, Properties, Source of the Ingredients, Mode of application, and Uses of Commercial Fertilizers*. Morgantown, WV: West Virginia Agricultural Experimental Station.

Yadav, A., Yadav, K., and Abd-Elsalam, K.A. (2023). Nanofertilizers: types, delivery and advantages in agricultural sustainability. *Agrochemicals* 2 (2): 296–336.

Yadav, J., Jasrotia, P., Kashyap, P.L. et al. (2021). Nanopesticides: current status and scope for their application in agriculture. *Ochrana Rostlin [Plant Protection Science]* 58 (1): 1–17. doi: 10.17221/102/2020-pps.

Yadav, R.K., Singh, N.B., Singh, A. et al. (2020). Expanding the horizons of nanotechnology in agriculture: recent advances, challenges and future perspectives. *Vegetos (Bareilly, India)* 33 (2): 203–221. doi: 10.1007/s42535-019-00090-9.

Yang, Z., Xiao, Y., Jiao, T. et al. (2020). Effects of copper oxide nanoparticles on the growth of rice (*Oryza Sativa L.*) seedlings and the relevant physiological responses. *International Journal of Environmental Research and Public Health* 17 (4): 1260. doi: 10.3390/ijerph17041260.

Index

Note: Page numbers in *italics* and **bold** refers to figures and tables respectively.

Nanotechnology-based Sustainable Agriculture, First Edition. Edited by Pardeep Singh et al.

d

n

t